Number Theory

Pure and Applied
UNDERGRADUATE // TEXTS · 48

Number Theory

Róbert Freud
Edit Gyarmati

AMERICAN
MATHEMATICAL
SOCIETY

Providence, Rhode Island USA

EDITORIAL COMMITTEE

2010 *Mathematics Subject Classification.* Primary 11-00,
11-01, 11A05, 11A07, 11A25, 11A41.

For additional information and updates on this book, visit
www.ams.org/bookpages/amstext-48

Library of Congress Cataloging-in-Publication Data

Names: Freud, Róbert, author.
Title: Number theory / Róbert Freud, Edit Gyarmati.
Description: Providence, Rhode Island: American Mathematical Society, [2020] | Series: Pure and
   applied undergraduate texts, 1943-9334; volume 48 | Includes bibliographical references and
   index.
Identifiers: LCCN 2020014015 | ISBN 9781470452759 (paperback) | ISBN 9781470456917 (ebook)
Subjects: LCSH: Number theory. | AMS: Number theory – General reference works (handbooks,
   dictionaries, bibliographies, etc.). | Number theory – Instructional exposition (textbooks,
   tutorial papers, etc.). | Number theory – Elementary number theory – Multiplicative structure;
   Euclidean algorithm; greatest common divisors. | Number theory – Elementary number theory
   – Congruences; primitive roots; residue systems. | Number theory – Elementary number theory
   – Arithmetic functions; related numbers; inversion formulas. | Number theory – Elementary
   number theory – Primes.
Classification: LCC QA241 .F74 2020 | DDC 512.7–dc23
LC record available at https://lccn.loc.gov/2020014015

Contents

Introduction

The book is intended to serve several purposes; being a

(A) Theoretical textbook for teaching number theory at universities and colleges, mostly for majors in mathematics, applied mathematics, mathematics education, and computer science.

(B) Collection of exercises and problems for the above audience.

(C) Handbook for those interested in more detail in some chapters of number theory beyond the compulsory and elective courses and/or writing a thesis in this subject.

(D) Manual summarizing the most important chapters of (elementary) number theory for mathematicians and mathematics teachers.

Structure of the book

To achieve the above goals, the discussion starts at an absolutely basic level and the first two chapters are based solely on high school mathematics. This part uses elementary and non-abstract tools, and instead of overly compact reasoning, detailed explanations facilitate better understanding for beginners. On the other hand, we lay stress on presenting theorems illustrating the deeper coherence of the material and on proofs containing nice and difficult ideas.

The subsequent chapters enter more and more deeply into the discussion of various topics in number theory. We strive to present a wide panorama of this extremely multi-colored world (including many old but still unsolved problems) and to discuss many methods elaborated through many centuries to treat these questions. Where possible, the newest results of number theory are inserted. Several parts apply some results and methods from other fields of mathematics too, mostly from (classical, linear, and abstract) algebra, analysis, and combinatorics.

The book is structured to systemize the material and to provide a close relation between the individual chapters as much as possible.

As a general guideline, the notions and statements are thoroughly illuminated from various aspects beyond the formal phrasing, they are illustrated by examples and connections to the previous material. Their essential features are strongly emphasized pointing out the complications and analyzing the motives for introducing a given notion. Careful attention is paid to start from the concrete where possible and to proceed towards the general only afterwards. We try to give a broad perspective about the strong and colorful relations of number theory to other branches of mathematics.

Exercises

Each section in every chapter is followed by exercises. They serve several purposes: some of them check the comprehension of the notions, theorems, and methods, and give a deeper understanding; others present new examples, relations, and applications; again others study further problems related to the topic. They often include also theorems disguised as exercises revealing some interesting aspects or more remote connections not treated in the text in detail.

Exercises vary in quantity and in difficulty within fairly large limits depending on the topic, size, and depth of the material. The hard and extra-hard exercises (in our judgement) are marked with one and two asterisks, resp. (The difficulty of an exercise is always relative, of course: besides the abilities, interests, and preliminary general knowledge of the solver, it depends strongly also on the exercises already solved.)

Answers and/or some hints to nearly all exercises can be found in the chapter Answers and Hints. To some (mostly harder) problems detailed solutions are presented in an online chapter available at www.ams.org/bookpages/amstext-48. These exercises are marked with a letter **S** in the text.

The reader is advised to consult a hint or solution only if an exercise turns out to be absolutely unmanageable, or to return to the same problem later, or to solve first some special case of it.

It is important to unravel the message and background of an exercise, its position and role in the mathematical environment. Also a generalization or raising new problems are very useful (even if it is not clear how to solve them).

Short overview of the individual chapters

The first two chapters are introductory, discussing the divisibility of integers, the greatest common divisor, unique prime factorization, and elementary facts about congruences. A firm mastery of this material is indispensable for understanding the later chapters.

In Chapters 3 and 4 we continue to develop the theory of congruences.

Chapter 5 deals with prime numbers. This simply defined set is one of the most mysterious objects in mathematics. We discuss Euclid's theorems (more than two thousand years old) and the sensational discovery of the last decades, the public key cryptosystems based on the contrast of quick primality testing and awfully slow prime factorization. In this chapter we rely both on previously acquired knowledge in number theory and the results and methods of elementary analysis.

In Chapter 6 we study arithmetic functions. Besides investigating some concrete important functions, we present several general constructions and applications.

Chapter 7 is about Diophantine equations. After discussing the simplest types (linear equations, Pythagorean triples), we look at Waring's problem and prove the special cases of Fermat's Last Theorem for exponents three and four. The methods require the theory of Gaussian and Eulerian integers that will be generalized in Chapters 10 and 11.

The topic of Chapter 8 is Diophantine approximation that is important for certain applications. We briefly consider also the connection with the geometry of numbers and continued fractions.

Chapters 9–11 are closely related to each other. The basic properties of algebraic numbers and algebraic integers from Chapter 9 are essential for understanding the next two chapters. Chapter 10 studies field extensions, focusing on the arithmetic properties of algebraic integers in a simple extension of the rational field by an algebraic number. Here, an intensive use is made of the notions and theorems of elementary linear algebra. Finally, in Chapter 11 the arithmetic aspects of ideals are investigated. On the one hand, ideals constitute a fine tool for exhibiting some necessary and sufficient, or useful sufficient, conditions for the validity of unique prime factorization in general rings, and on the other hand, the validity of unique prime factorization for ideals of algebraic integers (though in general not for the algebraic integers themselves) plays an important role in studying algebraic number fields.

In Chapter 12 several interesting problems from combinatorial number theory are presented. Some of these can be discussed even at a high school study circle, whereas others require deeper methods from various branches of mathematics. We hope that the selection gives an idea also about the fundamental role of Paul Erdős in the progress of this field with thrilling questions and ingenious proofs.

Throughout the text, we often refer to interesting aspects of the history of number theory and this purpose is served also by the short Historical Notes at the end of the book.

As is clear also from the above description, the different subfields of number theory are closely interrelated to each other and to other branches of mathematics. This causes a serious difficulty since, on the one hand, it is important to emphasize this tight connection during the discussion of the individual topics, but, on the other hand, it is desirable that every chapter be self-contained and complete. We tried to achieve a balance that makes it possible to get a gradually growing full picture of a mathematical field rich in problems and ideas for continuous readers, but allows those who just pick a few chapters to acquire interesting, substantial, and useful knowledge.

Technical details

The chapters are divided into sections. Definitions, theorems, and formulas are numbered as $k.m.n$ where k refers to the chapter, m to the section, and n is the serial number within the given section. Definitions and theorems have a common list, thus, for example, Definition 6.2.1 is followed by Theorem 6.2.2. Examples, exercises, etc. are numbered with a single number restarting in each section. The statement of a definition or theorem is closed by a ♣ sign and the end of a proof is denoted by □.

The search for notations, notions, and theorems can be facilitated by the very detailed Index at the end of the book.

We distinguish the floor and ceiling of (real) numbers, denoted by ⌊ ⌋ and ⌈ ⌉, resp., thus e.g. $\lfloor \pi \rfloor = 3$, $\lceil \pi \rceil = 4$ (we do not use the notation $[\pi]$). The fractional part is denoted by { }, i.e. $\{c\} = c - \lfloor c \rfloor$. Divisibility, greatest common divisor, and least common multiple are denoted as usual, so e.g. $7 \mid 42$, $(9, 15) = 3$, and $[9, 15] = 45$. Square brackets [ ] can mean a least common multiple, a closed interval, or just a replacement for (round) parentheses (this latter function occurs frequently in Chapter 11 where round parentheses ( ) stand for an ideal; to avoid confusion, the greatest common divisor is denoted here by $\gcd\{a, b\}$).

Polynomials and functions are denoted generally without indicating the argument: f, g, etc. but sometimes also $f(x)$, $g(x)$, etc. can occur. The degree of a polynomial is denoted by "deg," so e.g., $\deg(x^3 + x) = 3$. As usual, **Q**, **R**, and **C** stand for the rational, real, and complex numbers. **Z**, $\mathbf{Z}_m$, and $F[x]$ mean the integers, the modulo m residue classes, and the polynomials over F. At field extensions, $\mathbf{Q}(\vartheta)$ and $I(\vartheta)$ denote the simple extension of the rationals by ϑ and (in case ϑ is algebraic) the ring of algebraic integers in this extension. The letter p denotes nearly exclusively a (positive) prime and the log (without a lower index) stands for natural logarithm (of base e). For (finite and infinite) products and sums we often use the signs $\prod$ and $\sum$, e.g.

$$\prod_{i=1}^{r} p_i^{\alpha_i}, \qquad \prod_{p \le n} p, \qquad \sum_p \frac{1}{p^2}$$

mean the product $p_1^{\alpha_1} \dots p_r^{\alpha_r}$, the product of primes not greater than n, and the sum of reciprocals of squares of primes.

Commemoration

The book is dedicated to the memory of Paul Turán, Paul Erdős, and Tibor Gallai (who were close friends and collaborators).

Both authors enjoyed the privilege to be in touch with two giants of 20th century number theory, Paul Turán and Paul Erdős.

We were educated in Paul Turán's legendary seminars where we learned how to explore, elaborate, and explain to others the essential components of a mathematical problem. Turán taught us that connecting seemingly remote areas can often result in new, efficient methods.

Edit Gyarmati wrote a number theory textbook (in Hungarian) some fifty years ago using Turán's lectures among several other sources that can be considered as a predecessor of this book in a certain sense. The experiences of our lectures, the students' broadening preliminary knowledge (e.g. in linear algebra), and the new scientific achievements in this field during the past decades necessitated the creation of a new book instead of a long-due revision. The spirit and structure of the two books show several similar features, of course.

Both of us were largely influenced by the mathematical and human greatness of Paul Erdős sharing his enthusiastic devotion towards "nice" mathematical problems and proofs, talking about these (and many more things) equally naturally and openly with great scientists or just interested beginners. Róbert Freud owes many adventures in doing joint mathematics and a great deal of his professional progress to Erdős.

Edit Gyarmati's choosing mathematics as a profession is mostly due to her unforgettable high school teacher, Tibor Gallai, who was a world-famous expert in graph theory. Gallai was a brilliant personality whose wonderful classes both in high school and at universities helped to start mathematical research for the best students, and offered the joy of understanding and creation for all pupils.

Acknowledgements

We are very thankful for the great job the reviewers Imre Ruzsa (Chapter 12), András Sárközy (Chapters 1–12), and Mihály Szalay (Chapters 1–11) did. All three of them checked the manuscript with extreme thoroughness and suggested many general, concrete, and stylistic improvements nearly all of which were accepted by us. The conceptual remarks of András Sárközy helped us in unifying some notions, homogenizing the structure, and mentioning several further results. Mihály Szalay checked every tiny detail carefully, solved all the exercises without a solution given in the book, noted even the smallest inaccuracies, and his concretely worded suggestions made it possible to correct many lesser or greater errors and discrepancies. Imre Ruzsa added many valuable observations on Chapter 12.

In spite of all the efforts of the authors (and reviewers) there probably remain errors and imperfections in the book. Any comments or suggestions are gratefully accepted.

The book in its present form is an English translation and an improved and corrected version of the two Hungarian editions used by all universities of science in Hungary. Edit Gyarmati, who was not only my coauthor but also my wonderful wife for many decades, passed away in 2014, and could not participate in preparing this manuscript. I devote this work to her memory.

Budapest, February 2019
Róbert Freud
Institute of Mathematics, University Eötvös Loránd
1117 Budapest, Pázmány Péter sétány 1c, Hungary
freud@caesar.elte.hu

Chapter 1

Basic Notions

In this chapter, we survey some basic notions, theorems, and methods about the divisibility of integers. When introducing the concepts, we mostly rely on general divisibility properties only and keep the special features of the integers to a minimum. Using the even numbers and some other examples, we point out that certain well known facts, including the unique factorization into primes (the Fundamental Theorem of Arithmetic), are by no means obvious.

To prove the Fundamental Theorem, we start from the division algorithm, then describe the Euclidean algorithm yielding the special property of the greatest common divisor, which is the key to verify the equivalence of the irreducible and prime elements among the integers. We provide also a direct proof for the Fundamental Theorem using induction, that does not rely on the division algorithm. Finally, we discuss some important consequences.

1.1. Divisibility

If a and b are rational numbers, where $b \neq 0$, then dividing a by b, we get a rational number again. A similar statement does not hold for integers, hence the following definition makes sense:

Definition 1.1.1. An integer b is called a *divisor* of an integer a if there exists some integer q satisfying $a = bq$. ♣

Notation: $b \mid a$. This relation can be expressed also saying that a is *divisible* by b, or a is a *multiple* of b. If there is no integer q satisfying $a = bq$, then b is not a divisor of a, which is denoted by $b \nmid a$.

In the following, we shall use the words "integer" and "number" as synonyms unless stated otherwise.

The number 0 is divisible by every integer (including 0 itself!) as $0 = b \cdot 0$ for any integer b. The other extreme contains those numbers which divide every integer:

Definition 1.1.2. A number dividing every integer is called a *unit*. Multiplying an integer c by a unit, we get an *associate* of c.   ♣

Theorem 1.1.3. *There are two units among the integers: 1 and* -1.   ♣

Proof. 1 and -1 are units, since for any integer a, we have $a = (\pm 1)(\pm a)$. Hence $\pm 1 \mid a$.

Conversely, if ε is a unit, then ε divides 1, i.e. $1 = \varepsilon q$ for some q. Since $|\varepsilon| \geq 1$ and $|q| \geq 1$, therefore only

$$|\varepsilon| = 1, \quad \text{i.e.} \quad \varepsilon = \pm 1$$

is possible.   □

Remark: Divisibility can be introduced also in other sets of numbers (moreover, in any integral domain, see Exercise 1.1.23). Consider, for example, the even numbers. Here $b \mid a$ means that there exists an *even* number q satisfying $a = bq$. Hence, here $2 \mid 20$, but $2 \nmid 10$, and 10 has no divisors at all. This implies that there are no units among the even numbers. On the other hand, there are infinitely many units among the (special real) numbers $c + d\sqrt{2}$ where c and d are arbitrary integers (see Exercise 1.1.22). This means that the units may show very different forms and are related not (only) to the sign changes as Theorem 1.1.3 could suggest falsely.

Theorem 1.1.4. *If ε and δ are units and $b \mid a$, then also $\varepsilon b \mid \delta a$ holds.*   ♣

Proof. As ε divides also 1, therefore $1 = \varepsilon r$ with a suitable r. If $a = bq$, then $\delta a = (\varepsilon b)(\delta qr)$, hence $\varepsilon b \mid \delta a$, as claimed.   □

By Theorem 1.1.4, a number and its associates behave identically concerning divisibility, i.e. the units "do not count" in this respect. This makes possible to deal (later) only with non-negative or (after clarifying the special role of 0) with positive integers in divisibility investigations.

The next theorem summarizes some simple but important properties of divisibility of integers.

Theorem 1.1.5.   (i) *For every a, we have $a \mid a$.*

 (ii) *If $c \mid b$ and $b \mid a$, then $c \mid a$.*

(iii) *Both $a \mid b$ and $b \mid a$ hold simultaneously if and only if a is an associate of b.*

(iv) *If $c \mid a$ and $c \mid b$, then $c \mid a + b$, $c \mid a - b$, $c \mid ka$ for any (integer) k, and $c \mid ra + sb$ for any (integers) r and s.*   ♣

Properties (i)–(iii) express that divisibility of integers is a reflexive and transitive relation that is not symmetric (in fact, it is nearly antisymmetric). From (iv), we mostly use the first three implications, each of which is a special case of the last one ($r = s = 1$; $r = 1, s = -1$; and $r = k, s = 0$, respectively).

Proof. We verify only (iii). The others can be easily proven using just the definition of divisibility.

If $a = \varepsilon b$ where ε is a unit, then $b \mid a$ is straightforward. Also, $1 = \varepsilon r$ implies $ra = b$, hence $a \mid b$ is valid as well.

Conversely, if $a \mid b$ and $b \mid a$, i.e. $b = aq$ and $a = bs$ with suitable integers q and s, then $b = b(qs)$. If $b = 0$, then necessarily $a = 0$, thus $a = \varepsilon b$. If $b \neq 0$, then $qs = 1$, hence s is a unit (and so is q), yielding $a = \varepsilon b$. $\qquad\qquad\square$

Exercises 1.1

(Unless stated otherwise, all numbers are integers, the exponents are non-negative integers, and the digits are understood to be in decimal representation.)

1. Write a three-digit number twice as one string. Show that the resulting six-digit number is divisible by 91.

2. Verify that 8 always divides the difference of the squares of two odd numbers.

3. Assume that the three digit number $\overline{abc}$ (having digits a, b, and c in this order) is a multiple of 37. Prove that the number $\overline{bca}$ is also divisible by 37.

4. Show that if $5a + 9b$ is divisible by 23, then $3a + 10b$ is also divisible by 23.

5. True or false?

   (a) $c \mid a + b \Longrightarrow c \mid a, \ c \mid b$

   (b) $c \mid a + b, \ c \mid a \Longrightarrow c \mid b$

   (c) $c \mid a + b, \ c \mid a - b \Longrightarrow c \mid a, \ c \mid b$

   (d) $c \mid 2a + 5b, \ c \mid 3a + 7b \Longrightarrow c \mid a, \ c \mid b$

   (e) $c \mid ab \Longrightarrow c \mid a$ or $c \mid b$

   (f) $c \mid a, \ d \mid b \Longrightarrow cd \mid ab$

   (g) $c \mid a, \ d \mid a \Longrightarrow cd \mid a$.

6. Verify the following:

   (i) $a - b \mid a^n - b^n$

   (ii) $a + b \mid a^{2k+1} + b^{2k+1}$

   (iii) $a + b \mid a^{2k} - b^{2k}$.

7. Determine all integers c for which $(c^6 - 3)/(c^2 + 2)$ is an integer.

8. Prove that $133 \mid 11^{n+2} + 12^{2n+1}$ for every n.

9. Find infinitely many n satisfying $29 \mid 2^n + 5^n$.

10. Show that $(b - 1)^2 \mid b^k - 1$ holds if and only if $b - 1 \mid k$.

* 11. Assume $2^b - 1 \mid 2^a + 1$. Prove that $b = 1$ or 2.

12. Prove the following propositions.

   (a) If $b \mid a$ and $a \neq 0$, then $|b| \leq |a|$.

   (b) Every non-zero integer has only finitely many divisors.

13. Which numbers are equal to the sum of their (a) two; (b) three (not necessarily distinct) positive divisors?

14. Verify the following divisibility laws. A number is divisible by

   (a) 3 or 9 if and only if the sum of its digits is divisible by 3 or 9, respectively;

   (b) 4 or 25 if and only if the number formed of its last two digits is divisible by 4 or 25, respectively;

   (c) 8 or 125 if and only if the number formed of its last three digits is divisible by 8 or 125, respectively;

   (d) 11 if and only if the sum of its digits with alternating signs is divisible by 11.

15. Does there exist a power of 2 (with a positive integer exponent) containing all the ten digits with the same multiplicity?

* 16. Does there exist a multiple of 2^{1000} having only the digits 1 and 2?

17. Show that

   (a) the product of any three consecutive integers is divisible by 6

   * (b) the product of any k consecutive integers is divisible by $k!$.

S 18. Let $n > 1$ be an arbitrary integer. Romeo picks one of the positive divisors of n, let it be d_1. Then Juliet chooses a positive divisor d_2 that does not divide d_1. Again, Romeo takes d_3 that divides neither d_1, nor d_2, etc. Whoever must pick n itself loses the game. Who has a winning strategy if n is

   (a) 16

   (b) 3^{1111}

   (c) 10

   (d) 50

   ** (e) 123456789101112131415?

* 19. Prove that taking any $n + 1$ elements from 1, 2, ..., $2n$, one of the numbers will divide another one.

20. Though the *divisibility* 0 | 0 holds, why does the *division* 0/0 make no sense?

21. Restricting ourselves to the set of even numbers, characterize those elements that have

   (a) no divisors at all

   (b) exactly two (positive or negative) divisors?

22. We investigate divisibility relations among the (special real) numbers $c + d\sqrt{2}$ where c and d are arbitrary integers.

   (a) Determine whether or not $12 - 7\sqrt{2}$ is divisible by $3 + 4\sqrt{2}$.

    (b) Verify that $1 + \sqrt{2}$ is a unit.

    (c) Demonstrate that there are infinitely many units.

    (d) What is the number of divisors of any element?

    (e) Prove that $c + d\sqrt{2}$ is a unit if and only if $|c^2 - 2d^2| = 1$.

S* (f) Show that the units are exactly the elements $\pm(1+\sqrt{2})^k$ where k is an arbitrary integer.

    (g) How many times does it occur among the *integers* that the double of a square number is bigger or smaller by one, than another square?

23. An *integral domain* is a commutative ring without zero divisors (containing at least two elements), i.e. where addition and multiplication are commutative and associative, there exists a zero element, every element has a negative (an additive inverse), the distributive law is valid, and the product of two non-zero elements is never zero. (Roughly speaking, we have the usual "nice" properties seen in the integers.) We can define divisibility and unit according to Definitions 1.1.1 and 1.1.2. Prove the following propositions (a)-(c).

  S (a) There exists a unit if and only if multiplication has an identity element (i.e. an element e satisfying $ea = a$ for every a).

    (b) The units are exactly the divisors of the identity element, or, stated otherwise, the units are those elements that have a multiplicative inverse.

    (c) Any divisor of a unit and the product or quotient of two units are units.

    (d) Investigate the statements of Theorem 1.1.5.

1.2. Division Algorithm

Theorem 1.2.1. *To any integers a and $b \neq 0$, there exist some uniquely determined integers q and r satisfying*

$$a = bq + r \quad and \quad 0 \le r < |b|. \qquad \clubsuit$$

Proof. Assume first $b > 0$. The condition

$$0 \le r = a - bq < b$$

holds if and only if

$$bq \le a < b(q + 1),$$

i.e.

$$q \le a/b < q + 1.$$

Clearly, there exists exactly one such integer q namely the *floor* (or lower integer part) of a/b, i.e. the biggest integer that is not greater than a/b: $q = \lfloor a/b \rfloor$.

If $b < 0$, then the condition

$$0 \le r = a - bq < |b| = -b$$

is equivalent to

$$q \ge a/b > q - 1$$

which again holds for exactly one integer q (then q is the "ceiling" (or upper integer part) of a/b: $q = \lceil a/b \rceil$, i.e. the smallest integer that is still greater than or equal to a/b). □

The number q is called the *quotient* and r is called the (least non-negative) *remainder* (or *residue*) of the division algorithm. The divisibility $b \mid a$ holds (for $b \neq 0$) if and only if the remainder is 0.

It is often more convenient to allow also negative remainders. The following variant of Theorem 1.2.1 refers to this situation and can be proven similarly.

Theorem 1.2.1A. *To any integers a and $b \neq 0$, there exist some uniquely determined integers q and r satisfying*

$$a = bq + r \quad and \quad -\frac{|b|}{2} < r \leq \frac{|b|}{2}.$$ ♣

In this case r is called the *remainder of least absolute value*.

Example. Take $a = 30$, $b = -8$, then

$$30 = (-8)(-3) + 6 = (-8)(-4) - 2,$$

thus the least non-negative remainder is 6 and the remainder of least absolute value is -2.

The proof of the next theorem shows how the division algorithm provides the representation of positive integers in a *number system*.

Theorem 1.2.2. *Let $t > 1$ be a fixed integer. Then any positive integer A has a unique representation as*

$$A = a_n t^n + a_{n-1} t^{n-1} + \cdots + a_1 t + a_0, \quad where \quad 0 \leq a_i < t \quad and \quad a_n \neq 0.$$ ♣

Proof. From $0 \leq a_0 < t$ and $t \mid A - a_0$, we have that a_0 is the least non-negative remainder when A is divided by t in the division algorithm, hence there exists exactly one appropriate a_0. Denoting the quotient by q_0, we get

$$q_0 = \frac{A - a_0}{t} = a_n t^{n-1} + a_{n-1} t^{n-2} + \cdots + a_2 t + a_1.$$

As in the previous situation, we find a_1 as the least non-negative remainder when q_0 is divided by t. Continuing the process, we obtain the existence and uniqueness of every other a_i, as well. □

In this representation

$$A = a_n t^n + a_{n-1} t^{n-1} + \cdots + a_1 t + a_0,$$

the numbers a_i are the *digits* of A in the *number system of base t* (if $t > 10$, then we have to extend $0, 1, \ldots, 9$ with further digits). The above representation is denoted by

$$A = a_n a_{n-1} \ldots a_1 a_{0[t]} \quad or \quad A = \overline{a_n a_{n-1} \ldots a_1 a_0}_{[t]}$$

(the overline may be needed to avoid ambiguity, i.e. not to confuse the string of digits with a product). If $t = 10$, then we generally omit the notation of the base of the number system.

Example. $38 = 38_{[10]} = 123_{[5]}$ since $38 = 1 \cdot 5^2 + 2 \cdot 5 + 3 \cdot 1$.

In everyday life, we generally use the decimal system, but e.g. the binary system can often be more useful in computers, among others. In the binary system we have only two digits, 0 and 1, and to perform addition and multiplication we need only the following simple tables (however, the representation of a number requires many more digits than in the decimal case):

$\oplus$	0	1
0	0	1
1	1	10

$\odot$	0	1
0	0	0
1	0	1

Despite its simplicity, the division algorithm (independently of the least non-negative or least absolute value character of the remainder) has a great significance both from the practical and theoretical points of view. It can be efficiently used for divisibility problems since only "the remainder counts" in many cases. Its most important application is perhaps the Euclidean algorithm, which consists of a sequence of division algorithms and will be treated in the next section.

Exercises 1.2

(Unless stated otherwise, all numbers are in decimal representation.)

1. Dividing 10849 and 11873 by the same three digit positive integer, we obtain the same (non-negative) remainder. What is this remainder?

2. Show that to every m, there exist infinitely many powers of 2 such that the difference of any two of them is divisible by m.

3. Prove that given n integers, we can always select some of them (one, or more, or all) so that their sum is divisible by n.

4. Show that every positive integer has a non-zero multiple consisting of digits 0 and 1 only.

* 5. The sequence of *Fibonacci numbers* is defined by the recursion

$$\varphi_0 = 0, \qquad \varphi_1 = 1, \qquad \varphi_{j+1} = \varphi_j + \varphi_{j-1}, \qquad j = 1, 2, \dots.$$

The first few elements are

$$0, 1, 1, 2, 3, 5, 8, 13, 21, 34, \dots.$$

Prove that every m has infinitely many multiples among the Fibonacci numbers.

(*Remark*: Some books do not consider 0 as a Fibonacci number and define the sequence by the above recursion starting with $\varphi_1 = \varphi_2 = 1$. This causes no confusion if we agree that by the "nth Fibonacci number" we always mean φ_n.)

6. What are the possible remainders of a square when divided by (a) 3 (b) 4 (c) 5 and (d) 8?

7. Show that the sum of squares of 12 consecutive integers is never a square.

8. (a) Can all digits of a square (greater than 9) be the same?

   * (b) Find all squares greater than 81 having an even number of digits where all digits of the first half are the same and also all digits of the second half are the same.

9. Verify that the sum of three odd powers of an integer is always divisible by 3.

* 10. Take eight arbitrary distinct integers and form the product of their pairwise differences. What is the largest k for which this product is divisible by 2^k in any case?

11. How many positive integers with at most 10 digits are divisible by the floor of their square root? (E.g. 12 has this property since $\lfloor\sqrt{12}\rfloor = 3$ divides 12, but 22 does not because 22 is not a multiple of $\lfloor\sqrt{22}\rfloor = 4$.)

12. What is the connection between $\lfloor a + b\rfloor$ and $\lfloor a\rfloor + \lfloor b\rfloor$?

13. Can we perform the division algorithm among the even numbers (i.e. are the analogs of Theorems 1.2.1–1.2.1A valid)?

14. Show that by rephrasing the rules in Exercise 1.1.14 suitably, we can determine also the remainder (and not just check divisibility). How do these laws generalize for number systems of other bases?

15. We find that $23 + 46 + 12 + 18 = 99$ and 99 divides 23461218, obtained by joining the above terms into a string. Is this just a fortunate coincidence?

16. Form the sum of digits of 1223^{1001}, then the sum of digits of the number obtained, etc. till we arrive at a one digit number. What is this final integer?

17. How can we transform quickly the representations of an integer between number systems of base 3 and 9 into each other? Between which other pairs of number systems can we establish similar quick conversions?

18. A positive integer n has four digits in some number system and two digits in the number system of base one larger. Determine n.

19. Converting 740 into a number system of base t, we obtain a four digit integer whose last digit is 5. Determine t.

20. We want to devise ten weights by which a two-armed balance can measure all integer grams up to a limit as large as possible. How should we choose these weights if we can put them

    (a) only onto one of the pans of the balance

    * (b) onto both pans?

21. Examine roughly how many more digits are needed to represent a large integer in base 2 than in base 10. The precise formulation of the problem is: Let $B(n)$ and $D(n)$ be the number of digits of n in binary and decimal representations. Show that the sequence $B(n)/D(n)$ tends to a limit as $n \to \infty$ and determine its value.

22. *Number systems with varying base.* Let $t_1, t_2, \ldots$ be arbitrary integers greater than 1. Show that every positive integer A has a unique representation as

$$A = a_n t_n t_{n-1} \ldots t_1 + a_{n-1} t_{n-1} \ldots t_1 + \cdots + a_1 t_1 + a_0$$

where $0 \le a_i < t_{i+1}$ and $a_n \ne 0$.

23. Write a positive integer in base $b_1 = 2$. Then subtract 1 and consider the string as a number in a larger base b_2. Subtract 1 again (in base b_2) and read the string as a number in a base $b_3 > b_2$, etc. For example, we start with $23_{[10]} = 10111_{[2]}$, then subtracting 1 and switching to $b_2 = 5$, we obtain $10110_{[5]} = 655_{[10]}$. Subtracting 1 again (in base 5) and introducing $b_3 = 9$, we get $10104_{[9]} = 6646_{[10]}$, etc. What happens if we continue this process indefinitely?

1.3. Greatest Common Divisor

Definition 1.3.1. The *greatest common divisor* of a and b is d if

(i) $d \mid a$, $d \mid b$

(ii) if c satisfies $c \mid a$ and $c \mid b$, then $c \leq d$. ♣

We often abbreviate the expression greatest common divisor as gcd using its initials. The notation is: $d = (a, b)$, or $d = \gcd(a, b)$, or $d = \gcd\{a, b\}$.

There is no greatest common divisor of 0 and 0 since every integer is a common divisor and there is no maximal number among these.

In any other case, however, exactly one d satisfies Definition 1.3.1 (for given a and b), namely the maximal element of the set D of common divisors; D is not empty since 1 is always a common divisor and D is finite since a non-zero integer has only finitely many divisors (see Exercise 1.1.12b).

Definition 1.3.2. A *special common divisor* of a and b is δ, if

(i') $\delta \mid a, \delta \mid b$

(ii') if c satisfies $c \mid a$ and $c \mid b$, then $c \mid \delta$. ♣

Thus, a special common divisor is a common divisor which is a multiple of all common divisors.

The definition implies that if two integers possess a special common divisor, then it is unique apart from a unit factor. This means that on the one hand, any associate of a special common divisor is a special common divisor again, and on the other hand, two special common divisors must be associates. Exercise 1.3.10 requires the verification of this fact.

For $a = b = 0$, the special common divisor is 0 by definition.

In what follows, we disregard this case and assume that at least one of a and b differs from zero.

Now we show that if there exists a special common divisor δ, then it can only be an associate of the greatest common divisor d. By (ii) we have

$$|\delta| \leq d,$$

but (ii') implies $d \mid \delta$, hence

$$d \leq |\delta|.$$

Combining the two inequalities, we get $d = |\delta|$, so $\delta = \pm d$.

It is not at all straightforward, however, to show that the greatest common divisor satisfies also the special property (ii′), i.e. that any two integers possess a special common divisor.

Theorem 1.3.3. *Any two integers have a special common divisor.*   ♣

Proof. We prove the existence of a special common divisor via the *Euclidean algorithm*, which is one of the most ancient procedures in mathematics. We divide the first number by the second one, then we divide the second number by the remainder, etc., and continue to divide the actual divisor by the actual remainder till we obtain 0 as a remainder. We show that the procedure terminates and the last non-zero remainder is a special common divisor of the two numbers.

Let us see the details. Assume that (e.g.) $b \neq 0$. If $b \mid a$, then $\delta = b$.

If $b \nmid a$, then we obtain for suitable integers q_i and r_i

$$a = bq_1 + r_1 \quad \text{where} \quad 0 < r_1 < |b|$$
$$b = r_1q_2 + r_2 \quad \text{where} \quad 0 < r_2 < r_1$$
$$r_1 = r_2q_3 + r_3 \quad \text{where} \quad 0 < r_3 < r_2$$
$$\cdots$$
$$r_{n-2} = r_{n-1}q_n + r_n \quad \text{where} \quad 0 < r_n < r_{n-1}$$
$$r_{n-1} = r_nq_{n+1} \qquad (r_{n+1} = 0).$$

The procedure terminates in finitely many steps since the remainders form a strictly decreasing sequence of non-negative integers:

$$|b| > r_1 > r_2 > \ldots.$$

Now we verify that r_n is a special common divisor of a and b, indeed.

Proceeding through the equalities of the algorithm upwards, first we establish that r_n is a common divisor of a and b. The last equality implies $r_n \mid r_{n-1}$. Using the next to last equality, we get

$$r_n \mid r_{n-1}, \; r_n \mid r_n \Longrightarrow r_n \mid r_{n-1}q_n + r_n = r_{n-2}.$$

Continuing upwards similarly, finally we arrive at $r_n \mid b$ and (from the first equality) $r_n \mid a$.

To show the special property, we proceed now downwards. Let $c \mid a$ and $c \mid b$, then we have $c \mid a - bq = r_1$ from the first equality. Turning to the second equality, we obtain

$$c \mid b, \; c \mid r_1 \Longrightarrow c \mid b - r_1q_2 = r_2.$$

Continuing downwards similarly, the next to last equality implies $c \mid r_n$.   □

Remarks: (1) Instead of least non-negative remainders, we can perform the Euclidean
    algorithm also with remainders of least absolute value; then the *absolute values*
    of the remainders form a strictly decreasing sequence of non-negative integers,
    hence the procedure terminates in finitely many steps in this case, too.

(2) As an integer and its negative behave equivalently concerning divisibility, we can restrict ourselves to the positive value of the special common divisor which is (as we have seen) equal to the greatest common divisor. Hence the notations (a, b) and $\gcd(a, b)$ will mean this uniquely determined positive integer, and we shall (generally) use the greatest common divisor name also for the special common divisor.

(3) For a practical computation of the greatest common divisor, it is often more convenient to use the variant

$$(a, b) = (b, r_1) = (r_1, r_2) = \cdots = (r_{n-1}, r_n) = (r_n, 0) = r_n$$

of the Euclidean algorithm that is based on the simple relation $(a, b) = (b, a - kb)$.

(4) At first sight, Definition 1.3.2, including the special property (ii'), might seem artificial and unnecessary, but it is justified by the fact that it relies on divisibility relations only in contrast to Definition 1.3.1 which uses also ordering relations (greater-smaller). Therefore, it is not surprising that—as it will soon turn out—we can apply rather the special property (ii') instead, both for theoretical and practical purposes. A further advantage of building the notion purely on divisibility is that in certain sets of numbers (or more generally in most integral domains) Definition 1.3.1 does not even make sense. An obvious reason for this is if we cannot define an order (satisfying the usual "good" properties) in the set as, for example, in certain subsets of the complex numbers. But we can run into a problem with Definition 1.3.1 also in sets that can be ordered, e.g., among the numbers $c + d\sqrt{2}$ (where c and d are integers). Here we have infinitely many units (see Exercise 1.1.22) and there is no maximal one among them. (If we consider only common divisors where no two are associates, Definition 1.3.1 still makes no sense since taking any two common divisors we can multiply the first one by a unit so that the resulting associate will exceed the second one.) Therefore, in the further chapters of number theory we shall always define the greatest common divisor according to Definition 1.3.2.

Now we prove some important properties of the greatest common divisor (among the integers).

Theorem 1.3.4. *If* $c > 0$, *then* $(ca, cb) = c(a, b)$. ♣

Proof. Consider the Euclidean algorithm determining (a, b) and let $r_n = (a, b)$ be the last non-zero residue. Multiplying each equality by c, we obtain the Euclidean algorithm producing (ca, cb). Hence, here the last non-zero residue is $(ca, cb) = cr_n = c(a, b)$. □

For another proof of Theorem 1.3.4, see Exercise 1.3.11.

Theorem 1.3.5. *The greatest common divisor of integers* a *and* b *can be expressed as* $(a, b) = au + bv$ *with suitable integers* u *and* v. ♣

Proof. From the first equality of the Euclidean algorithm, we can express r_1 as

$$r_1 = a - bq_1.$$

This and the second equality imply

$$r_2 = b - r_1 q_2 = b - (a - bq_1)q_2 = a(-q_2) + b(1 + q_1 q_2),$$

i.e. r_2 can be written in the form $aU + bV$. Proceeding similarly, the next to last inequality guarantees that $(a, b) = r_n$ can be expressed as $au + bv$.   □

An important consequence of Theorem 1.3.5 is the following theorem about the solvability of a *linear Diophantine equation* $ax + by = c$ in two variables. In general, algebraic equations are called Diophantine when both the coefficients and the solutions are among the integers. We shall study some important types in detail in Chapter 7. Hence, in the equation $ax + by = c$, the coefficients a, b, and c are fixed integers and a solution means a pair of integers x, y.

Theorem 1.3.6. *Let a, b, and c be fixed integers, where a and b are not both zero. The Diophantine equation $ax + by = c$ is solvable if and only if $(a, b) \mid c$.*   ♣

Proof. Assume first that there exists a solution x_0, y_0. Then $(a, b) \mid a$ and $(a, b) \mid b$ imply

$$(a, b) \mid ax_0 + by_0 = c.$$

Conversely, assume $(a, b) \mid c$, i.e. $(a, b)t = c$ for some integer t. By Theorem 1.3.5, we have

$$(a, b) = au + bv$$

with suitable integers u and v. Multiplying this equality by t, we get

$$c = a(ut) + b(vt),$$

i.e. $x = ut, y = vt$ is a solution of the Diophantine equation $ax + by = c$.   □

Note that the Euclidean algorithm serves also as a procedure to find a solution of a linear Diophantine equation.

We deal with further questions (the number of solutions, a survey of all solutions, another method to find the solutions) concerning a linear Diophantine equation in Section 7.1 and discuss its relation to congruences in Section 2.5.

We define the greatest common divisor of more than two integers by the special property immediately as a common divisor that is a multiple of every common divisor. We denote the positive greatest common divisor of $a_1, a_2, \ldots, a_k$ (not all zero) by $(a_1, a_2, \ldots, a_k)$. Its existence can be proven simply, using that the set of all common divisors of two numbers is the same as the set of divisors of the greatest common divisor of the two numbers. Hence

$$(a_1, a_2, \ldots, a_k) = \left(\left(\ldots ((a_1, a_2), a_3), \ldots, a_{k-1} \right), a_k \right).$$

Definition 1.3.7. The integers $a_1, a_2, \ldots, a_k$ are *relatively prime* or *coprime* if they have no other common divisors than units, i.e. $(a_1, a_2, \ldots, a_k) = 1$.   ♣

Definition 1.3.8. The integers $a_1, a_2, \ldots, a_k$ are *pairwise relatively prime* or *pairwise coprime* if no two have other common divisors than units, i.e. $(a_i, a_j) = 1$ for every $1 \leq i \neq j \leq k$.   ♣

Evidently, pairwise coprime integers are coprime as well, but the converse is false (for $k > 2$); see Exercise 1.3.5.

We saw already in Exercise 1.1.5e that if an integer divides a product and does not divide one of the factors, then this does **not** imply that it divides the other factor. The correct condition is contained in the following theorem, that occurs already in Euclid's *Elements*, and, besides its usefulness in divisibility problems, plays a key role in the proof of the Fundamental Theorem of Arithmetic.

Theorem 1.3.9. *If $c \mid ab$ and $(c, a) = 1$, then $c \mid b$.* ♣

Proof. Clearly, we may assume that a, b, and c are positive. Using the special property of the greatest common divisor and Theorem 1.3.4, the divisibilities $c \mid ab$ and $c \mid cb$ imply

$$c \mid (ab, cb) = (a, c)b = b. \qquad \square$$

Exercises 1.3

(Using here the notation (c, d), we assume automatically that c and d cannot be both zero.)

1. Compute $(3794, 2226)$ and write it in the form $3794u + 2226v$.

2. Show that the following fractions are in reduced form for every positive integer n:

   (a) $\dfrac{3n + 5}{7n + 12}$

   (b) $\dfrac{3n^2 + 1}{4n^2 + 3}$

   (c) $\dfrac{n! - 1}{(n + 1)! - 1}$

   (d) $\dfrac{7^n - 2}{7^{n+1} - 5}$.

3. Find all possible values of $(n^2 + 2, n^4 + 4)$ if n assumes all positive integers.

4. What are the possible values of

   (a) $(a + b, a - b)$

   (b) $(a + 2b, 4a - b)$

   if $(a, b) = 5$?

5. Exhibit three coprime integers no two of which are coprime.

6. True or false?

   (a) If $(a, b) = d$, then $(\frac{a}{d}, \frac{b}{d}) = 1$.

   (b) If $(a, b) = d$, then at least one of $(\frac{a}{d}, b) = 1$ and $(a, \frac{b}{d}) = 1$ holds.

   (c) $c \mid ab$ if and only if $\frac{c}{(c,a)} \mid b$.

   (d) $c \mid ab, (a, b) = 1 \Longrightarrow c \mid a$ or $c \mid b$.

7. Let a and b be positive integers. How many numbers are divisible by b among the integers $a, 2a, 3a, \ldots, ba$?

8. Let a and b be distinct positive integers. True or false?

    (a) $(a + n, b + n) = 1$ holds for infinitely many integers n.

    (b) $(a + n, b + n) = (b + n, bn) = 1$ holds for infinitely many integers n.

    (c) $(a + n, bn) = (b + n, bn) = 1$ holds for infinitely many integers n.

9. Let a and b be given integers.

    (a) How many pairs of integers u, v satisfy $(a, b) = au + bv$?

    (b) What is the greatest common divisor of u and v in the representation $(a, b) = au + bv$?

    (c) Let H be the set of numbers $au + bv$ where u and v assume all integer values. What is the smallest positive element of H?

10. *Uniqueness of the special common divisor.* Let δ be a special common divisor of integers a and b. Using the definition of the special common divisor, prove the following propositions.

    (a) For any unit ε, $\varepsilon\delta$ is a special common divisor of a and b.

    (b) If δ_1 is another special common divisor of a and b, then $\delta_1 = \varepsilon\delta$ for some unit ε.

S 11. Give an alternative proof for Theorem 1.3.4 that uses only the notion (and existence) of the special common divisor and does not rely (directly) on the Euclidean algorithm.

12. We call *repunits* those positive integers where every digit is 1 (in decimal representation).

    (a) Which numbers have a repunit multiple?

    (b) Which is the smallest repunit multiple of 3^{1000}?

S* 13. Show that
$$(a^n - 1, a^k - 1) = a^{(n,k)} - 1$$
holds for any integers $n > 0$, $k > 0$, and $a > 1$.

14. Let a be a positive integer.

    (a) Verify that if n and k are distinct powers of two an a is an even number, then $(a^n + 1, a^k + 1) = 1$.

   * (b) Determine $(a^n + 1, a^k + 1)$ in general.

15. Prove that any two consecutive Fibonacci numbers (see Exercise 1.2.5) are coprime. What about the second neighbors? And the third neighbors?

** 16. Let φ_m be the mth Fibonacci number. Verify
$$k \mid n \iff \varphi_k \mid \varphi_n, \quad \text{moreover}, \quad \varphi_{(k,n)} = (\varphi_k, \varphi_n).$$

17. *Commensurability of segments.* In his *Elements*, Euclid investigates also *common measures* of segments besides the common divisors of integers. A common measure of two segments is a segment that can be measured an integer number of times onto both segments (without remainders). Two segments are *commensurable* if they have a common measure.

   (a) Prove that two segments are commensurable if and only if the ratio of their lengths is a rational number.

   (b) How many common measures do two commensurable segments possess?

   (c) Formulate the division algorithm for segments and show that the Euclidean algorithm terminates in finitely many steps if and only if the two original segments are commensurable.

   (d) Verify that commensurable segments have a greatest common measure and any common measure can be measured an integer number of times onto this greatest one (without remainder).

   (e) Show that the side and the diagonal of a square are not commensurable (thus giving a geometric proof for the irrationality of $\sqrt{2}$).

1.4. Irreducible and Prime Numbers

We have seen that 0 and the units play special roles in divisibility: every integer divides 0 and the units divide every integer. Consider now any integer a different from 0 and units. By the definition of units, $\varepsilon \mid a$ and $\varepsilon a \mid a$ for every unit ε. These are called the *trivial divisors* of a. The numbers having only trivial divisors are of distinguished importance:

Definition 1.4.1. An integer p different from units (and zero) is called *irreducible* if it can be factored into the product of two integers **only** so that one of the factors is a unit:

$$p = ab \Longrightarrow a \text{ or } b \text{ is a unit.} \qquad \clubsuit$$

We do not have to prescribe $p \neq 0$ because 0 has non-trivial factorizations too, e.g. $0 = 5 \cdot 0$. We note further that in the product $p = ab$, both factors cannot be units since then their product, i.e. p, would be a unit as well. (Hence, the word "or" occurs at the end of Definition 1.4.1 in an "exclusive" sense.)

Thus, the irreducible numbers are those integers distinct from units that can be factored into the product of two integers only trivially, or otherwise stated, are divisible only by their associates and units. Such numbers are e.g. 2, 3, -17, etc. If a non-zero integer has a non-trivial divisor, then it is called a *composite* number.

Before introducing the following notion, we recall that if an integer c divides a factor of a product, then c necessarily divides also the product, but the converse is false: e.g. for $c = 6$ we have $6 \mid 3 \cdot 4$, but $6 \nmid 3$ and $6 \nmid 4$. The numbers satisfying the converse are of special significance:

Definition 1.4.2. An integer p different from units and zero is called a *prime* number (or shortly, just a *prime*) if it can divide the product of two integers **only** if it divides at least one of the factors:

$$p \mid ab \Longrightarrow p \mid a \text{ or } p \mid b. \qquad \clubsuit$$

At the end of Definition 1.4.2, the word "or" occurs in an inclusive sense since it can happen that p divides both factors of the product. We also note that the restriction $p \neq 0$ was necessary here since 0 would otherwise satisfy the property required in Definition 1.4.2:

$$0 \mid ab \Longrightarrow ab = 0 \Longrightarrow a = 0 \text{ or } b = 0 \Longrightarrow 0 \mid a \text{ or } 0 \mid b.$$

Definition 1.4.2 implies that if a prime divides a product of more (than two) factors, then it must divide at least one of them.

Theorem 1.4.3. *Among the integers, p is a prime if and only if it is irreducible.* $\qquad \clubsuit$

Proof. We may clearly assume that p is not zero and not a unit.

I. First, we take a prime p and prove that it is irreducible. Given a product $p = ab$, we have to verify that a or b is a unit.

The equality $p = ab$ implies that $p \mid ab$. Since p is a prime, therefore we infer that $p \mid a$ or $p \mid b$. The first case means that $ab \mid a$ and hence $b \mid 1$ (since $a \neq 0$), i.e. b is a unit. The second case yields similarly that a is a unit.

II. We assume now that p is irreducible and prove that it is a prime. Given $p \mid ab$, we have to verify that at least one of $p \mid a$ and $p \mid b$ holds.

If $p \mid a$, then we are done. If $p \nmid a$, then the irreducibility of p and $(p, a) \mid p$ yield $(p, a) = 1$. The conditions $p \mid ab$ and $(p, a) = 1$ imply $p \mid b$ by Theorem 1.3.9. $\qquad \square$

Thus we have shown that the irreducible and prime numbers coincide among the integers. Therefore we can define the prime numbers as in high school by the irreducible property and to use either of the two adjectives irreducible and prime for these numbers. For brevity, we shall generally use the word prime except if we want to emphasize the irreducible property.

The two notions, however, are not equivalent in many other sets of numbers. E.g. among the even numbers, 6 is irreducible since it cannot be written as the product of two even numbers, but it is not a prime because it divides $18 \cdot 2$ without dividing either of the factors. We shall see further examples in Chapter 10.

Among the integers, the study of prime numbers is one of the most important areas in number theory. Euclid proved that there exist infinitely many primes (Theorem 5.1.1), but on the other hand, there are many easily formulated and yet unsolved problems concerning the prime numbers. We shall deal with these more in detail in Chapter 5.

Exercises 1.4

According to the conventions, we shall use the word prime or prime number also for the irreducible numbers among the integers. We note, however, that Exercises 1.4.1–1.4.7 refer to irreducible numbers.

1. Determine all positive integers n for which each of the following numbers is a prime:

    (a) n, $n + 2$, and $n + 4$

    (b) n and $n^2 + 8$

    (c) n, $n + 6$, $n + 12$, $n + 18$, and $n + 24$

    (d) n, $n^3 - 6$, and $n^3 + 6$.

2. Does there exist an infinite arithmetic progression with a non-zero difference consisting purely of primes?

3. Captain Immortal has three immortal grandchildren whose ages are three distinct primes and the sum of the squares of their ages is a prime. How old is the captain's youngest grandchild? (Do not forget about the immortality of the grandchildren, they can be several million years old!)

4. Let a and k be integers greater than one. Prove the following assertions.

    (a) If $a^k - 1$ is a prime, then $a = 2$ and k is a prime.

    (b) If $a^k + 1$ is a prime, then k is a power of two.

    Remark: The primes of the form $2^k - 1$ are called *Mersenne primes* and the primes of the form $2^k + 1$ are called *Fermat primes*. We shall study them in detail in Section 5.2.

S 5. Determine all integers $t > 1$ and *odd* numbers $k > 0$ for which $1^k + 2^k + 3^k + \cdots + t^k$ is a prime.

6. Find all positive integers n for which

    (a) $n^3 - n + 3$

    (b) $n^3 - 27$

    (c) $n^8 + n^7 + n^6 + n^5 + n^4 + n^3 + n^2 + n + 1$

    (d) $n^4 + 4$

    (e) $n^8 + n^6 + n^4 + n^2 + 1$

    is a prime.

7. Let $n > 1$. Prove the following assertions.

    (a) If n has no divisor t satisfying $1 < t \le \sqrt{n}$, then n is a prime.

    (b) The smallest divisor of n greater than 1 is a prime.

    (c) If n is composite but has no divisor t satisfying $1 < t \le \sqrt[3]{n}$, then n is the product of two primes.

8. Prove that $(n - 5)(n + 12) + 51$ is never divisible by 289 if n is an integer.

9. Which will be the irreducible and prime elements among the even numbers?

10. The notion of irreducible and prime elements can be defined in any integral domain I (see Exercise 1.1.23). Prove the following propositions.

    (a) If multiplication has no identity element in I, then there are no primes in I.

    (b) If multiplication has an identity element in I, then every prime is irreducible in I.

1.5. The Fundamental Theorem of Arithmetic

Theorem 1.5.1 (The Fundamental Theorem of Arithmetic). *Every integer different from 0 and units is the product of finitely many irreducible numbers and this decomposition is unique apart from the order of the factors and associates. (Uniqueness means that if*

$$a = p_1 p_2 \dots p_r = q_1 q_2 \dots q_s$$

where all p_i and q_j are irreducible, then $r = s$ and the numbers p_i and q_j can be coupled into associate pairs.) ♣

Remarks: (1) The units and 0 had to be excluded because these cannot be decomposed into the product of irreducible numbers: the units can be written only as a product of units, and writing 0 as a product at least one of the factors must be 0 (and then this factor is not irreducible).

(2) To interpret the theorem for an irreducible number, it should be considered as a product of a single factor.

(3) A few remarks concerning the uniqueness. Assume that the integer a is the product $a = p_1 p_2 \dots p_r$ of irreducible numbers. Then changing the order of the factors we obtain the same product. Also, if $\varepsilon_1, \dots, \varepsilon_r$ are arbitrary units whose product is 1, then $\varepsilon_1 p_1, \dots, \varepsilon_r p_r$ are irreducible as well and their product is a again. The uniqueness part of the theorem claims that apart from these trivial variants there is no other way to write a as the product of irreducible elements. Taking e.g. 12, a few such decompositions are

$$12 = 2 \cdot 2 \cdot 3 = 2 \cdot (-3) \cdot (-2) = 3 \cdot (-2) \cdot (-2).$$

(4) When stating the theorem, we should definitely use the notion of irreducible numbers since the theorem declares that (nearly) every integer can be assembled essentially in a unique way from these bricks. For clarity, we shall strictly distinguish the notions irreducible and prime during the proof. We shall see that their equivalence is crucial for the validity of the Fundamental Theorem.

(5) The Fundamental Theorem is false in many sets of numbers (and integral domains). Taking e.g. the even numbers, 100 has two essentially different decompositions into the product of irreducible elements: $100 = 2 \cdot 50 = 10 \cdot 10$. We shall see further examples in Chapter 10.

Now we turn to the proof of the Fundamental Theorem. We shall give two proofs for the uniqueness part.

Proof of decomposability. Consider an integer a different from 0 and units. If a is irreducible, then we are done.

If a is composite, then it has a non-trivial irreducible divisor since its smallest non-trivial positive divisor must be irreducible (see Exercise 1.4.7b). Then $a = p_1 a_1$ where p_1 is irreducible and a_1 is not a unit.

If a_1 is irreducible, then we are done; otherwise there exists an irreducible number p_2 satisfying $a_1 = p_2 a_2$ where a_2 is not a unit.

We proceed similarly with a_2, etc. Our algorithm must terminate in finitely many steps since the integers $|a_i|$ are positive and form a strictly decreasing sequence:

$$|a| > |a_1| > |a_2| > \dots,$$

hence some a_k must be irreducible: $a_k = p_{k+1}$.

Then we get the decomposition $a = p_1 p_2 \dots p_{k+1}$. $\qquad\square$

First proof of uniqueness. Our main tool is that every irreducible number is a prime (Theorem 1.4.3).

The proof is by contradiction. Assume that a certain a has (at least) two essentially different decompositions into the product of irreducible elements:

(1.5.1) $$a = p_1 p_2 \dots p_r = q_1 q_2 \dots q_s.$$

If some p_i is an associate of a q_j, e.g. $p_1 = \varepsilon q_1$ where ε is a unit, then cancellation by q_1 yields

$$a' = \frac{a}{q_1} = (\varepsilon p_2) p_3 \dots p_r = q_2 q_3 \dots q_s,$$

hence also a' has two essentially different decompositions into the product of irreducible elements.

Continuing the process, we get finally an integer where the two decompositions do not share associate factors. Without loss of generality, we may assume that this is the case in (1.5.1), i.e. $p_i \neq \varepsilon q_j$.

Using (1.5.1), we have $p_1 \mid q_1 q_2 \dots q_s$. Since p_1 is irreducible, therefore it is a prime by Theorem 1.4.3; thus p_1 must divide at least one of the factors q_j.

If $p_1 \mid q_j$, then the irreducibility of q_j implies that p_1 is a unit or it is an associate of q_j, and both are impossible. $\qquad\square$

Second proof of uniqueness. This proof uses induction on $|a|$.

Since associates behave equivalently in every divisibility relation, we may restrict ourselves to the decompositions of positive integers into positive irreducible numbers.

For $a = 2$, the uniqueness holds as 2 is irreducible.

Assuming now that every integer $1 < a < n$ has a unique decomposition into the product of irreducible numbers, we show that then the decomposition of $a = n$ is unique. If not, then n has (at least) two essentially different decompositions into the product of irreducible numbers:

$$(1.5.2) \qquad\qquad n = p_1 p_2 \ldots p_r = q_1 q_2 \ldots q_s.$$

Clearly, $r \geq 2$, $s \geq 2$ and further $p_i \neq q_j$ since if e.g. $p_1 = q_1$, then also the number $1 < n/p_1 < n$ would have two different decompositions contradicting the induction hypothesis.

Suppose $p_1 < q_1$ and consider $n_1 = n - p_1 q_2 \ldots q_s$. We show that

$$(1.5.3) \qquad\qquad 1 < n_1 < n$$

and

$$(1.5.4) \qquad\qquad n_1 \text{ has two different decompositions,}$$

which is a contradiction.

By (1.5.2), the expression $n_1 = n - p_1 q_2 \ldots q_s$ can be rewritten as

$$(1.5.5) \qquad n_1 = p_1(p_2 \ldots p_r - q_2 \ldots q_s) = q_2 \ldots q_s(q_1 - p_1).$$

Clearly $n_1 < n$ and $p_1 < q_1$ implies

$$n_1 = q_2 \ldots q_s(q_1 - p_1) \geq q_2 \cdot 1 = q_2 > 1,$$

thus verifying (1.5.3).

Now, write the last factors in both decompositions in (1.5.5) as a product of irreducible numbers:

$$p_2 \ldots p_r - q_2 \ldots q_s = u_1 \ldots u_k \quad \text{and} \quad q_1 - p_1 = v_1 \ldots v_m.$$

Then n_1 has the following representations as a product of irreducible elements:

$$(1.5.6) \qquad\qquad n_1 = p_1 u_1 \ldots u_k = q_2 \ldots q_s v_1 \ldots v_m.$$

(If eventually $q_1 - p_1 = 1$, then the factors v_i are missing in which case the argument will be even more valid.)

We show that the two decompositions in (1.5.6) are essentially different. The first one contains p_1. But p_1 is missing from the second one, since on the one hand $p_1 \neq q_j$, and on the other hand, if $p_1 = v_i$ for some i, then

$$p_1 \mid v_1 \ldots v_m = q_1 - p_1 \implies p_1 \mid q_1,$$

which is impossible. Thus (1.5.4) is proven. □

Remarks: (1) Analyzing the first proof of uniqueness, we find that the division algorithm served as its basis, after all. It made possible the Euclidean algorithm, yielding the existence of a special common divisor based on which we showed (via Theorem 1.3.9) that an irreducible number is always a prime, giving the key step to the proof.

It is true also generally that if in some number sets (or integral domains) we can perform the division algorithm, then the Fundamental Theorem of Arithmetic holds there. Our proof of uniqueness remains valid literally also for the general case, whereas the decomposability may require some more refined arguments in certain sets. We shall see such examples in Chapters 7 and 10. In Section 11.3, using ideals, we shall give a unified proof for the general case that division algorithm always implies the Fundamental Theorem (both decomposability and uniqueness).

We note that the relation between the division algorithm and the Fundamental Theorem is not symmetric; there exist sets of numbers where the Fundamental Theorem is true but there do not exist division algorithms of any kind. We shall see an example in Chapter 10.

(2) The second proof of uniqueness did not rely on the theorems of Sections 1.3 and 1.4. Thus we can give new proofs for some of those theorems using the Fundamental Theorem. We emphasize two important results: the existence of a special common divisor (Theorem 1.3.3) and that every irreducible number is a prime (the "harder" part of Theorem 1.4.3). To derive these from the Fundamental Theorem, consult the proof of Theorem 1.6.4 for the first one, and Exercise 1.5.8 for the second one.

Exercises 1.5

1. Verify that the number of irreducible factors in the decomposition of a is at most $\log_2 |a|$.

2. Consider the set of even numbers.

   (a) Which numbers have an essentially unique decomposition into the product of irreducible elements?

   (b) Find a number that has exactly 1000 essentially distinct decompositions.

3. Analyze the reason why our proofs of uniqueness fail for the even numbers.

4. Demonstrate that the Fundamental Theorem is false among the integers divisible by 10 and there exist elements with decompositions not even having the same number of irreducible factors.

5. Consider the set F of finite decimal fractions.

   (a) Determine the units and the irreducible elements.

   (b) Prove that the Fundamental Theorem is valid in F.

 * (c) Verify that we can perform a division algorithm in F, i.e. we can assign to every $c \in F$ a non-negative integer $f(c)$ where $f(c) = 0$ if and only if $c = 0$ and to every a and $b \in F$, $b \neq 0$, there exist q and $r \in F$ satisfying $a = bq + r$ and $f(r) < f(b)$.

6. There are many variants of the second proof of uniqueness. Elaborate the argument if we work with $n_1 = n - p_1 q_2$.

7. Compute the number of decompositions of a given integer into the product of irreducible elements if we count separately those that differ only in the order of factors and/or in associates.

S 8. Derive from the Fundamental Theorem that every irreducible number is a prime.

9. Find all (not necessarily positive and not necessarily distinct) primes (among the integers) satisfying

$$\frac{1}{p_1 - p_2 - p_3} = \frac{1}{p_2} + \frac{1}{p_3}.$$

S* 10. Determine all positive primes (among the integers) a power of which (with positive integer exponent) is the sum of the cubes of two positive integers.

1.6. Standard Form

In the sequel, we shall deal only with positive divisors of positive integers, and a prime will always mean a positive irreducible number. Then the Fundamental Theorem reads as follows: Every integer $n > 1$ is the product of finitely many primes and this decomposition is unique apart from the order of the factors. (Units play no role now due to positivity.)

Combining the product of the same primes into a power, we can write n as the product of powers of distinct primes. This yields the following form of the Fundamental Theorem:

Theorem 1.6.1. *Every integer $n > 1$ can be decomposed as*

$$n = p_1^{\alpha_1} p_2^{\alpha_2} \dots p_r^{\alpha_r} = \prod_{i=1}^{r} p_i^{\alpha_i}$$

where $p_1, \dots, p_r$ are distinct (positive) primes and each $\alpha_i > 0$ is an integer. This form is unique apart from the order of the prime power factors $p_i^{\alpha_i}$. ♣

We call this decomposition the *standard form* (or *canonical representation*) of n.

We shall see that sometimes (e.g. when studying more numbers simultaneously) it is more convenient to allow 0 as an exponent for some primes. Then uniqueness is understood apart from these (eventually fictitious) primes, of course. This allows us to assign a standard form also to 1 (here all primes have exponent 0).

We shall always indicate when we need to allow exponent 0 in the standard form, and in all other cases we shall assume automatically that each exponent is positive.

First, we describe how the standard form helps us to characterize the divisors of an integer, the number of these divisors, and the greatest common divisor and the least common multiple of two integers.

Theorem 1.6.2. *A positive integer d divides the number n of standard form*

$$n = p_1^{\alpha_1} p_2^{\alpha_2} \dots p_r^{\alpha_r}$$

if and only if d has standard form

$$d = p_1^{\beta_1} p_2^{\beta_2} \dots p_r^{\beta_r}, \quad where \quad 0 \le \beta_i \le \alpha_i, \quad i = 1, 2, \dots, r.$$ ♣

We used the modified standard form for the divisors.

We obtain the trivial divisors 1 and n when $\beta_i = 0$ and $\beta_i = \alpha_i$, resp., for every i.

Proof. To verify sufficiency, assume that d is of the above form. Then

$$q = p_1^{\alpha_1 - \beta_1} p_2^{\alpha_2 - \beta_2} \ldots p_r^{\alpha_r - \beta_r}$$

is an integer since $\alpha_i \geq \beta_i$ and $n = dq$, i.e $d \mid n$. (This part did not need the uniqueness of the standard form, and we did not even use that the numbers p_i are primes.)

To prove the necessity, assume $d \mid n$, i.e. $n = dq$ for some (positive) integer q. Then we obtain the standard form of n by multiplying the standard forms of d and q. This means that every prime divisor of d occurs in the standard form of n, moreover with an exponent at least as big as in d, i.e. $\alpha_i \geq \beta_i$. $\qquad\square$

We denote the number of positive divisors of an integer $n > 0$ by $d(n)$.

Example. $d(1) = 1, d(10) = 4, d(n) = 2$ if and only if n is a prime.

Theorem 1.6.3. *The number of positive divisors of n with standard form*

$$n = p_1^{\alpha_1} p_2^{\alpha_2} \ldots p_r^{\alpha_r}$$

is

$$d(n) = (\alpha_1 + 1)(\alpha_2 + 1) \ldots (\alpha_r + 1). \qquad \clubsuit$$

Proof. By Theorem 1.6.2, we obtain all positive divisors of n if the exponents $\beta_1, \beta_2, \ldots, \beta_r$ in the standard form of

$$d = p_1^{\beta_1} p_2^{\beta_2} \ldots p_r^{\beta_r}$$

assume independently the values

$$\beta_1 = 0, 1, \ldots, \alpha_1, \quad \beta_2 = 0, 1, \ldots, \alpha_2, \quad \ldots, \quad \beta_r = 0, 1, \ldots, \alpha_r.$$

Hence, the exponent β_i can be chosen in $\alpha_i + 1$ ways and thus there are altogether

$$(1.6.1) \qquad\qquad (\alpha_1 + 1)(\alpha_2 + 1) \ldots (\alpha_r + 1)$$

options to choose the exponents $\beta_1, \ldots, \beta_r$ independently. Since every positive divisor has only one such decomposition (due to the uniqueness of its prime factorization), (1.6.1) yields the number of positive divisors of n. $\qquad\square$

Now, we turn to the standard form of the greatest common divisor of two integers. We use the modified standard form again: we include in the standard forms of both numbers also those primes that divide only one of our integers (these occur with exponent 0 in the standard form of the other integer, of course).

Theorem 1.6.4. *Let the standard forms of the positive integers a and b be*

$$a = p_1^{\alpha_1} p_2^{\alpha_2} \ldots p_r^{\alpha_r} \quad and \quad b = p_1^{\beta_1} p_2^{\beta_2} \ldots p_r^{\beta_r} \quad where \quad \alpha_i \geq 0, \beta_j \geq 0.$$

Then

$$(a, b) = p_1^{\min(\alpha_1, \beta_1)} p_2^{\min(\alpha_2, \beta_2)} \ldots p_r^{\min(\alpha_r, \beta_r)}$$

(where $\min(\alpha_i, \beta_i)$ means the smaller number of α_i and β_i if $\alpha_i \neq \beta_i$, and their common value if $\alpha_i = \beta_i$). $\qquad \clubsuit$

Proof. Consider

$$d = \prod_{i=1}^{r} p_i^{\min(\alpha_i, \beta_i)}.$$

We shall show that d is a common divisor of a and b and is a multiple of every common divisor. We shall rely on Theorem 1.6.2.

Since $\min(\alpha_i, \beta_i) \leq \alpha_i$ and $\min(\alpha_i, \beta_i) \leq \beta_i$, $d \mid a$ and $d \mid b$, so d is a common divisor.

Let c be an arbitrary common divisor of a and b. Then

$$c = \prod_{i=1}^{r} p_i^{\gamma_i} \quad \text{where} \quad \gamma_i \leq \alpha_i, \gamma_i \leq \beta_i.$$

This means that $\gamma_i \leq \min(\alpha_i, \beta_i)$, hence $c \mid d$.   □

Example. Compute the greatest common divisor of 4840 and 2156.

The standard forms of the numbers are $4840 = 2^3 \cdot 5 \cdot 11^2$ and $2156 = 2^2 \cdot 7^2 \cdot 11$. Thus $(4840, 2156) = 2^2 \cdot 5^0 \cdot 7^0 \cdot 11 = 44$.

Remark: This method seems to be very convenient to compute the gcd, but unfortunately it cannot be applied for large numbers, since we do not know a quick way to exhibit their standard forms. On the other hand, the Euclidean algorithm determines the gcd quickly even for very large integers. We shall investigate these problems (with applications) in Sections 5.7 and 5.8.

We turn now to the least common multiple, or shortly lcm. According to its name, this means the smallest positive element among the common multiples.

Definition 1.6.5. The *least common multiple* of integers a and b is the positive integer k if

(i) $a \mid k, b \mid k$

(ii) if $a \mid c$ and $b \mid c$ for some $c > 0$, then $c \geq k$.   ♣

We denote the least common multiple of a and b by $[a, b]$ (or $\operatorname{lcm}(a, b)$).

Since the product ab is clearly a common multiple of a and b, we can determine $[a, b]$ by checking the finitely many positive integers not greater than ab, seeing which is the smallest among the common multiples of a and b. Thus the existence and uniqueness of the least common multiple are obvious.

Analogously to the greatest common divisor, we can replace the minimality of the least common multiple by a more important special divisibility property: the least common multiple divides every common multiple (the lcm is often defined by this feature directly). We summarize this and further basic facts concerning the lcm in the next theorem.

Theorem 1.6.6. (I) *If the standard forms of the positive integers a and b are*

$$a = p_1^{\alpha_1} p_2^{\alpha_2} \dots p_r^{\alpha_r} \quad \text{and} \quad b = p_1^{\beta_1} p_2^{\beta_2} \dots p_r^{\beta_r} \quad \text{where} \quad \alpha_i \geq 0, \beta_j \geq 0,$$

then

$$[a, b] = p_1^{\max(\alpha_1, \beta_1)} p_2^{\max(\alpha_2, \beta_2)} \dots p_r^{\max(\alpha_r, \beta_r)}$$

(where $\max(\alpha_i, \beta_i)$ *denotes the larger number of* α_i *and* β_i *if* $\alpha_i \neq \beta_i$, *and their common value if* $\alpha_i = \beta_i$).

(II) $a \mid c, b \mid c$ *if and only if* $[a, b] \mid c$.

(III) $(a, b)[a, b] = ab$. ♣

Proof. I and II. A positive integer c is a common multiple of a and b if and only if both $a \mid c$ and $b \mid c$. This means that the exponent γ_i of any prime p_i in the standard form of c satisfies $\gamma_i \geq \alpha_i$ and $\gamma_i \geq \beta_i$, which is equivalent to $\gamma_i \geq \max(\alpha_i, \beta_i)$.

We obtain the smallest such c when $\gamma_i = \max(\alpha_i, \beta_i)$ $(i = 1, 2, \dots, r)$ and c is not divisible by any other primes than the p_i. This proves I.

We also obtained that the exponents of the primes p_i in the standard forms of all common multiples c are greater than or equal to the exponent in $[a, b]$ and there may also occur other primes in their standard forms. This means that the common multiples c are the same as the multiples of $[a, b]$. This proves II.

III. We show that every prime p_i occurs with the same exponent in the standard forms of $(a, b)[a, b]$ and ab, i.e.

$$\min(\alpha_i, \beta_i) + \max(\alpha_i, \beta_i) = \alpha_i + \beta_i, \qquad i = 1, 2, \dots, r.$$

If e.g $\alpha_i \leq \beta_i$, then the left-hand side is $\alpha_i + \beta_i$ which is the same as the right-hand side. □

Remarks: (1) An important consequence of III is that $ab = [a, b]$ if and only if $(a, b) = 1$.

  (2) Note that $a \mid c$ and $b \mid c$ do **not** imply $ab \mid c$, e.g. $4 \mid 36$, and $6 \mid 36$, but $24 \nmid 36$. The correct implication is given by II:

$$a \mid c, b \mid c \Longrightarrow [a, b] \mid c.$$

If a and b are coprime, then, according to the previous remark, we have $[a, b] = ab$, and obtain the following important special case:

$$a \mid c, b \mid c, (a, b) = 1 \Longrightarrow ab \mid c.$$

So, to prove $72 \mid c$, it suffices to verify that c is divisible both by 8 and 9. Also in general, any divisibility problem can be reduced to divisibilities by *prime powers*: If the standard form of m is $m = \prod_{i=1}^{r} p_i^{\alpha_i}$ $(\alpha_i > 0)$, then

$$m \mid c$$

if and only if

$$p_i^{\alpha_i} \mid c, \quad i = 1, 2, \dots, r.$$

  (3) The notion and properties of the lcm can be generalized for more than two integers. We shall often use that the least common multiple of finitely many positive integers equals their product if and only if the integers are pairwise coprime. We also note that the equality in III has no simple direct generalization for more than two numbers (see Exercise 1.6.15).

We infer from the Fundamental Theorem that two integers are coprime if and only if they share no common *prime* divisors. This implies the following theorem immediately:

Theorem 1.6.7.

$$(c, ab) = 1 \iff (c, a) = 1 \quad and \quad (c, b) = 1. \qquad \clubsuit$$

Therefore, if two positive integers are coprime, then generally it is best to exhibit their standard forms without common primes, as

$$a = \prod_{i=1}^{r} p_i^{\alpha_i}, \quad b = \prod_{j=1}^{s} q_j^{\beta_j}, \quad p_i \neq q_j.$$

Finally, we describe the standard form of $n!$:

Theorem 1.6.8 (Legendre's formula). *The standard form of $n!$ is*

$$n! = \prod_{p \leq n} p^{\alpha_p}, \quad where \quad \alpha_p = \sum_{k=1}^{\infty} \left\lfloor \frac{n}{p^k} \right\rfloor. \qquad \clubsuit$$

In the formula, $\lfloor x \rfloor$ is the floor or (lower) integer part of x and p under the product sign means a (positive) prime, so we have to form the product for *all primes p* satisfying $p \leq n$. We shall often meet similar notations later, as well,

$$\sum_{p \leq n} \frac{1}{p}, \quad \prod_{p \leq n} p, \quad \sum_{p \mid n} 1$$

mean the sum of reciprocals of primes not greater than n, the product of primes not greater than n, and the number of distinct prime divisors of n.

Observe that in Theorem 1.6.8, it is sufficient to consider only finitely many terms in the sum defining α_p since we have $\lfloor n/p^k \rfloor = 0$ for $p^k > n$ (hence the number of non-zero terms is $\lfloor \log_p n \rfloor$).

Proof. Since every factor in $n! = 1 \cdot 2 \cdots \cdot n$ is at most n, no primes greater than n can occur in the standard form of $n!$.

Let $p \leq n$ be a fixed prime and let α_p be the exponent of p in the standard form of $n!$. We have to verify

$$(1.6.2) \qquad \alpha_p = \sum_{k=1}^{\infty} \left\lfloor \frac{n}{p^k} \right\rfloor.$$

To determine α_p, we decompose the numbers $1, 2, \ldots, n$ into the product of primes and count how many times p will appear.

The multiples of p contain at least one p; we consider these first. The following numbers are divisible by p:

$$p, 2p, \ldots, tp \quad where \quad tp \leq n < (t+1)p.$$

Hence

$$t \leq \frac{n}{p} < t+1, \quad so \quad t = \left\lfloor \frac{n}{p} \right\rfloor.$$

This means that there are $\lfloor n/p \rfloor$ numbers divisible by p among the integers $1, 2, \ldots, n$.

The multiples of p^2 contain at least two copies of p, but we considered only one of these so far. Thus every multiple of p^2 yields a new p. The number of these newcomers is $\lfloor n/p^2 \rfloor$, similar to the previous case.

We can continue similarly. Every multiple of p^3 gives rise to a new p since there are at least three factors of p in them and we took only two of them into consideration in the first two steps of our argument. This means a further $\lfloor n/p^3 \rfloor$ copies of p, etc.

The procedure terminates in finitely many steps, since if $p^k > n$, then none of the numbers $1, 2, \ldots, n$ is divisible by p^k.

This method counted every prime occurring in $n!$ exactly once, hence α_p is equal to the sum in (1.6.2). $\qquad \square$

Exercises 1.6

(We always mean a positive integer by number, divisor, prime, etc. in the exercises.)

1. How can we see from the standard form that a given integer is a square, a cube, or in general, a kth power (of a positive integer)?

2. (a) Demonstrate that if the product of two coprime integers is a kth power, then the factors are kth powers.

   (b) How should we modify this assertion if we consider all integers (instead of positive numbers)?

   (c) How does the statement generalize for more factors?

S 3. Prove that the product of

   (a) 2

   (b) 3

   * (c) 4

   consecutive (positive) integers is never a power of an integer with exponent greater than one.

   Remark: It is true in general that the product of consecutive integers is never a power. This long-standing conjecture of Catalan was proven by Paul Erdős and John Selfridge in 1975.

S 4. For which primes p is $(2^{p-1} - 1)/p$ a square?

5. (a) Prove that $c \mid ab$ if and only if $c = a_1 b_1$ where $a_1 \mid a$ and $b_1 \mid b$.

   (b) Show that if $(a, b) = 1$, then the above a_1 and b_1 are unique.

   (c) Verify that if $(a, b) \neq 1$, then there exists a divisor $c \mid ab$ that can expressed in more than one way as $c = a_1 b_1$.

   (d) Prove that any $c \mid ab$ has at most $d((a, b))$ representations in the form $c = a_1 b_1$.

   (e) Which divisors $c \mid ab$ have $d((a, b))$ representations as $c = a_1 b_1$?

6. Assume that $a^k \mid b^{k+100}$ holds for every k. Prove that $a \mid b$.

7. Which is the smallest positive integer having exactly

    (a) 31

    (b) 33

    (c) 32

    (positive) divisors?

8. For which values of n is $d(n)$ odd?

9. A cruel lord keeps 400 prisoners in 400 separate dark cells. Turning the lock on the door of a cell once, the door opens, and turning it once again, the door closes, etc. At present, all doors are closed, of course. The lord decides to be generous on his birthday and orders a guard to turn the locks on each door once. Then he changes his mind and sends a second guard to turn the locks on each second door once. This guard is followed by a third one who turns the locks on each third door once, etc., and finally the four hundredth guard changes the position of the lock on the four hundredth door. Those prisoners get free whose door is open now. How many people were released by the lord?

S 10. A positive integer is *squarefree* if it has no square divisors greater than 1. E.g. 1 and 30 are squarefree, but 12 is not squarefree. Let $A(n)$ and $B(n)$ be the number of (positive) squarefree divisors and the number of square divisors of n.

    (a) Prove that $A(n)B(n) \geq d(n)$ for every n.

    (b) When do we have equality?

11. Prove

    (a) $d(n) \leq n/2 + 1$

    (b) $d(n) \leq n/3 + 2$

    (c) $d(n) \leq 2\sqrt{n}$.

12. Exhibit a simple formula for the product of the (positive) divisors of n.

13. What is the maximal number of divisors of 10^n where no one divides another one?

14. (a) For which pairs of integers a, b can we find positive integers having a as their gcd and b as their lcm?

    (b) How many such pairs exist if $a = 5$ and $b = 35000$?

    (c) Determine the number of such pairs in general for arbitrary a and b.

15. Verify the following assertions.

    (a) $(a, b, c)[a, b, c] \mid abc$ but equality does not hold in general.

    (b) $(a, b, c)[a, b, c] = abc$ if and only if a, b, c are pairwise coprime.

    (c) $(a, b, c)[ab, bc, ac] = abc$.

16. True or false?

    (a) $(a, b) = (a + b, ab)$.

    (b) $(a, b) = 1$ if and only if $(a + b, ab) = 1$.

(c) $(a, bc) = (a, b)(a, c)$.

(d) $(a^3, b^3) = (a, b)^3$.

17. Prove the following propositions.

   (a) $[a, b] \mid a + b$ if and only if $a = b$.

   (b) $a + b \mid [a, b]$ never holds.

   (c) There exist infinitely many pairs a, b, $a \neq b$, satisfying $a + b \mid ab$.

   (d) $a + b \mid ab$ if and only if $a + b \mid (a, b)^2$.

18. Show that if $(a, b^2) = (a^2, b)$, then $(a^7, b^{1000}) = (a^{1000}, b^7)$.

19. Verify the following distributive laws.

   (a) $[a, (b, c)] = ([a, b], [a, c])$.

   (b) $(a, [b, c]) = [(a, b), (a, c)]$.

20. (a) Prove that for given positive integers a, b, and c, there exist integers x, y, and z satisfying

$$(x, y) = a, \quad (y, z) = b, \quad \text{and} \quad (z, x) = c$$

   if and only if $(a, b) = (b, c) = (c, a)$.

   (b) Find the number of such triples x, y, z (for given a, b, and c).

   (c) Examine the dual problem for least common multiples instead of greatest common divisors.

21. Verify the divisibility $240 \mid p^4 - 1$ for every prime $p > 5$.

22. Prove that $504 \mid a^6 - b^6$ if $(ab, 42) = 1$.

23. Show that $a^6 + 85a^4 + 994a^2$ is divisible by 360 for any a.

24. Prove that $26^{101} - 33^{101} + 7^{101}$ is divisible by 606606.

25. How many digits 0 occur at the end of (a) 1111! (b) $\binom{125}{60}$?

26. (a) Prove that $c^n \nmid n!$ for $c > 1$.

   (b) Find all values of $n > 1$ and $c > 1$ satisfying $c^{n-1} \mid n!$.

27. Let $n \geq 2$ and $1 \leq k \leq n - 1$.

   (a) Show that if k and n are coprime, then $n \mid \binom{n}{k}$.

   (b) Is the converse also true?

   (c) Determine all values of n such that

   (c1) $n \mid \binom{n}{k}$ (c2) $\binom{n}{k}$ is even   (c3) $\binom{n}{k}$ is odd

   for every $1 \leq k \leq n - 1$.

   (d) Do there exist n and k, $1 \leq k \leq n - 1$, when n and $\binom{n}{k}$ are coprime?

S* 28. Finitely many monkeys sit around a round table and play the following game. In front of each monkey there is a dime on the table. At a command, each monkey checks the coin of her right neighbor: if it shows head, then she turns her own coin; if it shows tail, then she leaves her coin as it was. They repeat this procedure till all coins show tails. What can the number of monkeys be if the game terminates for every initial position of the coins?

S* 29. Show that each of the integers $n! + 1, \ldots, n! + n$ has a prime divisor that divides none of the other $n - 1$ numbers.

S 30. Consider 5000 distinct positive integers where any ten of them have the same lcm. At most how many of them can be pairwise coprime?

31. Which positive integers n satisfy $n \mid k^2 \implies n \mid k$ (i.e. n can divide a square of a number only if it divides the number itself)?

32. Show that the difference of two kth powers never divides their sum (for $k > 1$).

33. Prove that (a) $\sqrt[5]{100}$ (b) $\log_6 18$ are irrational numbers.

S* 34. Given a positive integer m, consider all sets of integers $a_1 < a_2 < \cdots < a_t$ where $a_1 = m$ and $a_1 a_2 \ldots a_t$ is a square ($t = 1$ is allowed). We denote the smallest possible value of a_t by $S(m)$. For example, $S(1) = 1$, $S(2) = 6$ since the product $2 \cdot 3 \cdot 6$ is the best choice for $m = 2$, $S(3) = 8$, $S(4) = 4$, etc.

Prove that the sequence $S(2), S(3), S(4), \ldots$ contains exactly the positive composite numbers and each of them occurs exactly once.

S* 35. (a) Can distinct powers form an infinite arithmetic progression?

(b) Can distinct powers form finite arithmetic progressions of arbitrary length?

Chapter 2

Congruences

We study the basic facts concerning congruences in this chapter. After introducing the notion of congruence, we investigate residue classes, residue systems, and Euler's function φ. We prove the theorems of Euler–Fermat and Wilson, using linear congruences for the latter one. Related to linear congruences, we treat also simultaneous systems of congruences. We shall learn more about congruences in Chapters 3 and 4.

2.1. Elementary Properties

We often see in divisibility problems that only the remainder matters, i.e. two integers behave identically if their remainders are the same. This (too) underlines the introduction of the notion below:

Definition 2.1.1. Let a and b be integers and m a positive integer. We say that a is *congruent* to b modulo m if $m \mid a - b$. ♣

Notation: $a \equiv b \pmod{m}$ or just $a \equiv b \, (m)$. The number m is called the *modulus* and is kept fixed, in general. As $m \mid a - b$ if and only if $m \mid b - a$, therefore

$$a \equiv b \pmod{m} \iff b \equiv a \pmod{m},$$

and so we may say also that "a and b are congruent modulo m". (Instead of "modulo m", we can use the expressions "mod m," or "with respect to the modulus m," or "related to the modulus m," as well.)

Clearly, a and b are congruent modulo m if and only if a and b give the same (least non-negative) remainder when they are divided by m. (The same holds for the remainder of least absolute value.)

If a and b are not congruent modulo m, we write $a \not\equiv b \pmod{m}$, and we say that a and b are *incongruent* modulo m (or a is incongruent to b modulo m).

Example. $11 \equiv 5 \pmod{3}$, $32 \equiv -1 \pmod{11}$, $21 \not\equiv 6 \pmod{10}$.

Clearly, any two integers are congruent with respect to the modulus $m = 1$.

The definition of congruence can trivially be extended for $m < 0$, but we can ignore it since $m \mid a - b$ if and only if $-m \mid a - b$.

Theorem 2.1.2.   (i) $a \equiv a \pmod{m}$ *for every* a.

(ii) $a \equiv b \pmod{m} \Longrightarrow b \equiv a \pmod{m}$.

(iii) $a \equiv b \pmod{m}$ *and* $b \equiv c \pmod{m} \Longrightarrow a \equiv c \pmod{m}$.

(iv) $a \equiv b \pmod{m}$ *and* $c \equiv d \pmod{m} \Longrightarrow a + c \equiv b + d \pmod{m}$ *and* $a - c \equiv b - d$ $\pmod{m}$.

(v) $a \equiv b \pmod{m}$ *and* $c \equiv d \pmod{m} \Longrightarrow ac \equiv bd \pmod{m}$.   ♣

Proof. All the assertions follow easily from the definition of congruence and the elementary properties of divisibility, hence we verify only property (v) as an illustration.

We rewrite the assumptions as $m \mid a - b$ and $m \mid c - d$ which imply

$$m \mid c(a - b) + b(c - d) = ac - bd, \quad \text{so} \quad ac \equiv bd \pmod{m}. \qquad \square$$

Properties (i), (ii), and (iii) express that congruence is *reflexive, symmmetric,* and *transitive,* hence it is an *equivalence relation.* We can thus divide the integers into (pairwise) disjoint sets of numbers congruent to each other, i.e. those that give the same remainder when divided by m. (Properties (i)–(iii) guarantee that the expression "congruent to each other" makes sense.) These sets are called *residue classes* modulo m. We shall study them in Section 2.2.

By (iv) and (v), congruences (with the same modulus) can be added, subtracted, and multiplied. This implies immediately that we can add the same number to both sides of a congruence, and this holds also for subtraction and multiplication. Further, a congruence can be multiplied by itself arbitrarily many times, so we may raise a congruence to a power with a positive integer exponent:

(vi) $a \equiv b \pmod{m} \Longrightarrow a + c \equiv b + c \pmod{m}$ and $a - c \equiv b - c \pmod{m}$.

(vii) $a \equiv b \pmod{m} \Longrightarrow ac \equiv bc \pmod{m}$.

(viii) $a \equiv b \pmod{m} \Longrightarrow a^n \equiv b^n \pmod{m}$.

The repeated application of these relations yields the useful law:

(ix) Let f be a polynomial with integer coefficients. Then

$$a \equiv b \pmod{m} \Longrightarrow f(a) \equiv f(b) \pmod{m}.$$

We illustrate the efficiency of the above rules with a few examples.

Examples.   **E1** Demonstrate that any natural number n satisfies

$$17 \mid 3^{3n+1}5^{2n+1} + 2^{5n+1}11^n.$$

Solution: We have to show

$$3^{3n+1}5^{2n+1} + 2^{5n+1}11^n \equiv 0 \pmod{17}.$$

We replace the left-hand side with congruent expressions till we obtain 0:

$$3^{3n+1}5^{2n+1} + 2^{5n+1}11^n = 3 \cdot 27^n \cdot 5 \cdot 25^n + 2 \cdot 32^n \cdot 11^n \equiv$$
$$\equiv 15(-7)^n 8^n + 2(-2)^n(-6)^n =$$
$$= 15(-56)^n + 2(12)^n \equiv 15(-5)^n + 2(-5)^n =$$
$$= 17(-5)^n \equiv 0 \pmod{17}.$$

E2 Give a new proof for the divisibility $a - b \mid a^n - b^n$.

Solution: Clearly, we can restrict ourselves to the case $a - b > 0$. Applying (viii), we have

$$a \equiv b \pmod{a - b} \implies a^n \equiv b^n \pmod{a - b}.$$

E3 Verify that $2^{32} + 1$ is a composite number. (Cf. with Exercise 1.4.4 and Section 5.2.)

Solution: We establish the divisibility $641 \mid 2^{32} + 1$ relying on

$$641 = 5^4 + 2^4 = 5 \cdot 2^7 + 1.$$

We infer

$$-1 \equiv 5 \cdot 2^7 \pmod{641} \quad \text{and} \quad 5^4 \equiv -2^4 \pmod{641}.$$

Raising the first congruence to the fourth power and substituting the result into the second one, we obtain

$$1 = (-1)^4 \equiv 5^4 \cdot 2^{28} \equiv -2^4 \cdot 2^{28} = -2^{32} \pmod{641},$$

so $641 \mid 2^{32} + 1$.

We have seen that concerning addition, subtraction, and multiplication, congruences behave like equalities. There is a big difference for division, however; two congruences **must not** be divided. First of all, the results of the divisions are not always integers, and then the congruence between the fractional quotients makes no sense since only integers can appear in congruences. But even if the quotients are integers, the congruence obtained after the division will not necessarily be true. For example,

$$28 \equiv 46 \pmod{6} \quad \text{and} \quad 2 \equiv 2 \pmod{6} \quad \text{but} \quad 14 \not\equiv 23 \pmod{6}.$$

Concerning division of congruences, we should be aware that also a fraction means a division. Therefore we must not replace the numerator or denominator of a fraction with an integer value even when the new fraction is an integer. E.g.

$$45 \equiv 35 \pmod{10} \quad \text{and} \quad 15 \equiv 5 \pmod{10} \quad \text{but} \quad 3 = \frac{45}{15} \not\equiv \frac{35}{5} = 7 \pmod{10}.$$

After clarifying what is forbidden, let us see what we are allowed to do. We shall deal only with the special case when division is just cancellation. The following theorem states that in performing the cancellation, we *have to change the modulus*:

Theorem 2.1.3. *Let $d = (c, m)$. Then $ac \equiv bc \pmod{m}$ if and only if $a \equiv b \left(\bmod \frac{m}{d}\right)$.*

♣

Proof. By the definition of congruence, we have

$$ac \equiv bc \pmod{m} \iff m \mid (a - b)c,$$

which is equivalent to the divisibility

(2.1.1) $$\frac{m}{d} \mid (a - b)\frac{c}{d}.$$

Since $(m/d, c/d) = 1$, (2.1.1) holds if and only if

$$\frac{m}{d} \mid a - b, \quad \text{i.e.} \quad a \equiv b \left(\text{mod } \frac{m}{d}\right). \qquad \square$$

An important special case of Theorem 2.1.3 is when c and the modulus are coprime. Then the congruence remains valid with the same modulus after cancellation by c:

Theorem 2.1.3A.

$$ac \equiv bc \pmod{m}, \ (c, m) = 1 \implies a \equiv b \pmod{m}.$$

Exercises 2.1

1. Prove $23 \mid 61^{k+1} + 11^k 7^{2k} 3^{3k} 2^{5k+3}$.

2. What are the last three digits of $999^{777^{888}}$ (in decimal representation)?

3. Give a new proof using congruences for the divisibility rules by 9 and 11 (Exercise 1.1.14) and for their generalizations in other number systems (Exercise 1.2.14).

4. True or false?

   (a) $k \mid n, \ a \equiv b \pmod{n} \implies a \equiv b \pmod{k}$.
   (b) $k \mid n, \ a \equiv b \pmod{k} \implies a \equiv b \pmod{n}$.
   (c) $a \equiv b \pmod{n}, a \equiv b \pmod{k} \iff a \equiv b \pmod{kn}$.
   (d) $a \equiv b \pmod{n}, a \equiv b \pmod{k} \iff a \equiv b \pmod{[k, n]}$.
   (e) $a \equiv b \pmod{n} \iff ka \equiv kb \pmod{kn}$.
   (f) $a \equiv b \pmod{n}, c \equiv d \pmod{k} \implies ac \equiv bd \pmod{kn}$.
   (g) $a^2 \equiv b^2 \pmod{n} \implies a \equiv \pm b \pmod{n}$.
   (h) $a^2 \equiv b^2 \pmod{101} \implies a \equiv \pm b \pmod{101}$.

5. There are several digits that can not be the last one in the decimal representation of a square. How many such digits can be found in the number system of base 101?

6. Comment on the following "theorem" and "proof" of Professor Donkey Monkey:

   "Theorem: For any integer $n > 3$, we have $\binom{n}{4} \equiv \binom{n+1}{4} \pmod{4}$.

Proof: Since $n + 1 \equiv n - 3 \pmod 4$ holds for every n,

$$\binom{n}{4} = \frac{n(n-1)(n-2)(n-3)}{1 \cdot 2 \cdot 3 \cdot 4} \equiv$$

$$\equiv \frac{n(n-1)(n-2)(n+1)}{1 \cdot 2 \cdot 3 \cdot 4} = \binom{n+1}{4} \pmod 4 ."$$

7. Verify: $m \mid a - b \implies m^2 \mid a^m - b^m$.

8. Assuming $3 \nmid a$ and $(6, n) = 1$, prove $a^n \equiv b^n \pmod{3^n} \implies a \equiv b \pmod{3^n}$.

9. Let $p > 2$ be a prime and $1 \leq k \leq p-1$. Verify the following congruences modulo p:

   (a) $\binom{p}{k} \equiv 0$

   (b) $\binom{p-1}{k} \equiv (-1)^k$

   (c) $\binom{p-2}{k} \equiv (-1)^k (k + 1)$.

10. Determine all primes p for which the remainder of $\binom{3p}{p}$ when divided by p is $p - 2$.

* 11. Let p be a prime. Prove the following congruences modulo p:

   (a) $\binom{n}{p} \equiv \left\lfloor \frac{n}{p} \right\rfloor$

   (b) $\binom{n}{kp} \equiv \binom{\lfloor n/p \rfloor}{k}$

   (c) $\binom{n}{p^k} \equiv \left\lfloor \frac{n}{p^k} \right\rfloor$.

2.2. Residue Systems and Residue Classes

We mentioned the notion of a residue class modulo m after Theorem 2.1.2: it is the set of all integers giving the same remainder when divided by m.

Definition 2.2.1. Given the modulus m, the set of integers congruent to a is called the *residue class* represented by a. ♣

Notation: $(a)_m$. If there is no ambiguity, we can omit the index m referring to the modulus.

Thus, the residue class $(a)_m$ is an infinite arithmetic progression in both directions with difference m and a being one of its elements. There are m residue classes mod m, and each contains infinitely many numbers. By the definition, $(a)_m = (c)_m$ if and only if $a \equiv c \pmod m$.

Example. $(23)_7 = \{\ldots, -5, 2, 9, 16, 23, 30, \ldots\} = (100)_7$.

Definition 2.2.2. Given the modulus m, choosing one element from each residue class, we obtain a *complete residue system* modulo m. ♣

Example. $\{33, -5, 11, -11, -8\}$ is a complete residue system modulo 5.

We use mostly the following complete residue systems:

(A) Least non-negative residues: $0, 1, \ldots, m - 1$.

(B) Residues of least absolute value:

$$0, \pm 1, \pm 2, \ldots, \pm \frac{m-1}{2}, \qquad \text{for } m \text{ odd}$$

and

$$0, \pm 1, \pm 2, \ldots, \pm \frac{m-2}{2}, \frac{m}{2}, \qquad \text{for } m \text{ even}$$

(in the latter, $m/2$ can be replaced by $-m/2$).

We can apply the following simple criterion to check whether or not given numbers form a complete residue system:

Theorem 2.2.3. *A set of integers forms a complete residue system modulo m if and only if*

(i) *their number is m and*

(ii) *they are pairwise incongruent modulo m.*   ♣

Proof. Let C_m be a complete residue system modulo m. Since there are m residue classes and we picked one element from each class, C_m contains exactly m numbers. Further, we took each number from a different residue class, hence the elements of C_m are pairwise incongruent modulo m.

Conversely, consider m integers pairwise incongruent modulo m. Then they belong to distinct residue classes. Since their number is m, they represent m residue classes, i.e. all classes are represented. Thus, these integers form a complete residue system modulo m.   □

Multiplying a complete residue system by an integer coprime to the modulus and then adding an arbitrary integer yields a complete residue system again:

Theorem 2.2.4. *If $r_1, r_2, \ldots, r_m$ is a complete residue system modulo m, $(a, m) = 1$, and b is any integer, then*

$$ar_1 + b, ar_2 + b, \ldots, ar_m + b$$

is a complete residue system modulo m.   ♣

Proof. Since the new system has m elements, it is enough to show, by Theorem 2.2.3, that the elements are pairwise incongruent mod m. We have to prove that $ar_i + b \equiv ar_j + b \pmod{m}$ implies $i = j$. Subtracting b from both sides, we obtain $ar_i \equiv ar_j \pmod{m}$. Since $(a, m) = 1$, by Theorem 2.1.3A, we can cancel a: $r_i \equiv r_j \pmod{m}$, and so $i = j$, indeed.   □

Note that for $(a, m) \neq 1$, the integers $ar_i + b$ never form a complete residue system; see Exercise 2.2.11.

We examine now the distribution of the integers coprime to the modulus in the residue classes. It turns out that in a residue class, either all elements, or no elements are coprime to the modulus:

Let $a \equiv b \pmod{m}$. Then $(a, m) = 1$ if and only if $(b, m) = 1$.

We prove a stronger assertion in the next theorem:

Theorem 2.2.5.
$$a \equiv b \pmod{m} \implies (a, m) = (b, m).$$
♣

Proof. By the assumption, $b = a + mc$ for some integer c.

On the right-hand side, both a and m are divisible by (a, m), hence $(a, m) \mid b$. This means that (a, m) is a common divisor of b and m, hence $(a, m) \mid (b, m)$.

We get the converse divisibility $(b, m) \mid (a, m)$ similarly, and so $(a, m) = (b, m)$. □

The residue classes with elements coprime to the modulus play an important role in the sequel:

Definition 2.2.6. A residue class $(a)_m$ is called a *reduced* residue class (mod m) if $(a, m) = 1$. ♣

As mentioned previously, Theorem 2.2.5 implies that if some element of a residue class is coprime to the modulus, then every element in the residue class has this property. Therefore Definition 2.2.6 does not depend on which number was picked to represent the residue class $(a)_m$.

We introduce now one of the most important functions in number theory:

Definition 2.2.7 (Euler's function φ). For n given, $\varphi(n)$ counts how many integers of $1, 2, \ldots, n$ are coprime to n. ♣

Example. $\varphi(1) = 1$, $\varphi(10) = 4$, $\varphi(n) = n - 1$ if and only if n is a prime.

Clearly, $\varphi(n)$ is also the number of reduced residue classes modulo n.

We can easily compute $\varphi(n)$ from the standard form of n; we shall discuss this formula in Section 2.3.

Next, we define the notion of a reduced residue system analogously to the complete residue system:

Definition 2.2.8. Given the modulus m, choosing one element from each *reduced* residue class, we obtain a *reduced residue system* modulo m. ♣

Example. $\{17, -5, 11, -11\}$ is a reduced residue system modulo 12.

The simplest way to obtain a reduced residue system is to select the elements coprime to the modulus from the least non-negative remainders or from the remainders of least absolute value.

Now, we prove the analogues of Theorems 2.2.3 and 2.2.4 for reduced residue systems.

Theorem 2.2.9. *A set of integers forms a reduced residue system modulo m if and only if*

 (i) *their number is $\varphi(m)$*

 (ii) *they are pairwise incongruent modulo m and*

(iii) *each of them is coprime to m.* ♣

Proof. Let R_m be a reduced residue system modulo m. Since there are $\varphi(m)$ reduced residue classes and we picked one element from each, R_m contains exactly $\varphi(m)$ elements. Further, because we took each element from a different residue class, the elements of R_m are pairwise incongruent modulo m. Finally, every element of R_m is coprime to m, since they were chosen from reduced residue classes.

Conversely, consider $\varphi(m)$ pairwise incongruent integers modulo m that are coprime to m. The pairwise incongruence and the relative primeness guarantee that they belong to distinct reduced residue classes. Since their number is $\varphi(m)$, they represent $\varphi(m)$ reduced residue classes, i.e. all classes are represented. Thus, these integers form a reduced residue system modulo m.   □

Theorem 2.2.10. *If $r_1, r_2, \ldots, r_{\varphi(m)}$ is a reduced residue system modulo m and $(a, m) = 1$, then*

$$ar_1, ar_2, \ldots, ar_m$$

is also a reduced residue system modulo m.   ♣

Proof. We check criteria (i)–(iii) of Theorem 2.2.9.

(i) The new system has $\varphi(m)$ elements.

(ii) $ar_i \equiv ar_j \pmod{m}, (a, m) = 1 \Longrightarrow r_i \equiv r_j \pmod{m} \Longrightarrow i = j$.

(iii) $(a, m) = 1, (r_i, m) = 1 \Longrightarrow (ar_i, m) = 1$.   □

Note that for $(a, m) \neq 1$, the integers ar_i never form a reduced residue system, and moreover none of them is coprime to m.

Adding an integer b to the elements of a reduced residue system will not, in general, yield a reduced residue system, a significant difference from the complete residue systems. See Exercise 2.2.12.

Exercises 2.2

We assume everywhere that the modulus $m \geq 2$.

1. Determine the modulus m knowing that the integers below are elements of a reduced residue system:

    (a) 2 and 14

    (b) 18, 78, and 178

    (c) a and $-a$.

2. In how many (a) complete (b) reduced residue systems does every element a_i satisfy $0 \leq a_i \leq 5m + 1$?

3. Given m, characterize those arithmetic progressions that are infinite in both directions and contain modulo m

    (a) a residue class

    (b) a complete residue system?

4. For which $m \geq 2$ can we find a complete residue system consisting of

    (a) odd numbers

    (b) composite numbers

    (c) squares

    (d) integers ending with 1357 (in decimal representation)

    (e) consecutive elements of a geometric series

S* (f) repunits (i.e. every digit is 1 in decimal system)

S* (g) powers?

5. For which $m \geq 2$ can we find a reduced residue system consisting of

    (a) multiples of 15

    (b) numbers not divisible by 15

    (c) squares

    (d) integers ending with 1357 (in decimal representation)

    (e) powers?

6. True or false?

    (a) If $r_1, r_2, \ldots, r_k$ is a reduced residue system modulo 7, then it is a reduced residue system modulo 14.

    (b) If $r_1, r_2, \ldots, r_k$ is a reduced residue system modulo 14, then it is a reduced residue system modulo 7.

7. (a) What is the remainder of the sum of elements of a complete residue system modulo m?

    (b) Let m be even, and $a_1, a_2, \ldots, a_m$ and $b_1, b_2, \ldots, b_m$ be two complete residue systems modulo m. Prove that $a_1 + b_1, \ldots, a_m + b_m$ *never* is a complete residue system modulo m. What can we say for m odd?

    (c) Examine the analogous questions for reduced residue systems instead of complete residue systems.

S 8. (a) There are m trees around a circular clearing with a squirrel in each tree. The squirrels want to get together in one tree, but they are allowed to move only the following way: every minute, any two squirrels may jump to an adjacent tree. For which values of m can they gather in one tree?

    (b) What happens if we modify the admissible step so that the two squirrels must jump to the adjacent trees in opposite directions (i.e. one of them clockwise, and the other counterclockwise).

* 9. (a) Determine all m for which $0, 0+1, 0+1+2, \ldots, 0+1+2+\cdots+(m-1)$ form a complete residue system mod m.

    (b) For which m does there exist a complete residue system $a_1, \ldots, a_m$ mod m so that $a_1, a_1 + a_2, a_1 + a_2 + a_3, \ldots, a_1 + a_2 + a_3 + \cdots + a_m$ is also a complete residue system mod m?

10. Let $k \mid m$. True or false?

    (a) Every residue class mod k is the union of residue classes mod m.

    (b) Every reduced residue class mod k is the union of reduced residue classes mod m.

    * (c) Every reduced residue class mod k contains a subset that is a reduced residue class mod m.

    (d) Every reduced residue system mod k can be extended to a reduced residue system mod m.

    * (e) Every reduced residue system mod m contains a reduced residue system mod k.

11. Let $r_1, r_2, \ldots, r_m$ be a complete residue system modulo m, $(a, m) \neq 1$, and b arbitrary.

    (a) Prove that $ar_1 + b, \ldots, ar_m + b$ is never a complete residue system modulo m.

    (b) How many residue classes modulo m are represented by the elements $ar_1 + b$, $\ldots, ar_m + b$ altogether?

S* 12. Let $r_1, r_2, \ldots, r_{\varphi(m)}$ be a reduced residue system modulo m.

    (a) Determine all integers a such that the numbers $ar_1, \ldots, ar_{\varphi(m)}$ are pairwise incongruent modulo m.

    (b) Find all integers b such that the numbers $r_1 + b, \ldots, r_{\varphi(m)} + b$ form a reduced residue system modulo m.

S* 13. For which integers m and k do there exist a complete residue system $a_1, \ldots, a_m$ modulo m and a complete residue system $b_1, \ldots, b_k$ modulo k so that the numbers $a_i b_j$ form a complete residue system modulo mk?

S 14. Let a and b be positive integers.

    (a) Prove that

$$T = \{ ib + ja \mid i = 1, 2, \ldots, a, j = 1, 2, \ldots, b \}$$

    is a complete residue system modulo ab if and only if $(a, b) = 1$.

    (b) Let $r_1, \ldots, r_{\varphi(a)}$ and $s_1, \ldots, s_{\varphi(b)}$ be reduced residue systems modulo a and modulo b. Prove that

$$R = \{ r_i b + s_j a \mid i = 1, 2, \ldots, \varphi(a), j = 1, 2, \ldots, \varphi(b) \}$$

    is a reduced residue system modulo ab if and only if $(a, b) = 1$.

    (c) Demonstrate that if $(a, b) = 1$, then $\varphi(ab) = \varphi(a)\varphi(b)$.

2.3. Euler's Function φ

We introduced Euler's function φ in Definition 2.2.7: If n is a positive integer, then $\varphi(n)$ is the number of integers coprime to n among the integers $1, 2, \ldots, n$.

This implies immediately that there are $\varphi(m)$ reduced residue classes modulo m and a reduced residue system consists of $\varphi(m)$ integers.

We prove now a formula for $\varphi(n)$ from the standard form of n:

Theorem 2.3.1. *Let the standard form of n be*

$$n = p_1^{\alpha_1} p_2^{\alpha_2} \dots p_r^{\alpha_r} = \prod_{i=1}^{r} p_i^{\alpha_i}, \qquad \text{where} \qquad \alpha_i > 0.$$

Then

$$\varphi(n) = \left(p_1^{\alpha_1} - p_1^{\alpha_1-1}\right) \dots \left(p_r^{\alpha_r} - p_r^{\alpha_r-1}\right) = \prod_{i=1}^{r} \left(p_i^{\alpha_i} - p_i^{\alpha_i-1}\right). \qquad \clubsuit$$

This formula for $\varphi(n)$ is valid only if the exponents α_i in the standard form of n are positive (in contrast e.g. to the formula for $d(n)$ in Theorem 1.6.3 which remains valid even if we allow 0 to occur among the exponents α_i). Some equivalent forms of the formula are:

$$\varphi(n) = \prod_{i=1}^{r} p_i^{\alpha_i-1}(p_i - 1) = n \prod_{i=1}^{r} \left(1 - \frac{1}{p_i}\right) = n \prod_{\substack{p|n \\ p \text{ prime}}} \left(1 - \frac{1}{p}\right).$$

We give two proofs of Theorem 2.3.1. A third one can be derived from Exercise 6.5.4b. Also, Exercises 2.2.14 and 2.6.10 contain two further verifications of assertion II which is the key step in the first proof.

First proof. We infer the theorem from the two propositions below:

(I) If p is a prime (and $\alpha > 0$), then $\varphi(p^\alpha) = p^\alpha - p^{\alpha-1}$.

(II) If $(a, b) = 1$, then $\varphi(ab) = \varphi(a)\varphi(b)$.

These imply the theorem: It follows from II by induction on the number of factors that if the integers $a_1, \dots, a_r$ are *pairwise* coprime, then $\varphi(a_1 \dots a_r) = \varphi(a_1) \dots \varphi(a_r)$. Applying this for $a_i = p_i^{\alpha_i}$ and substituting the value for $\varphi(p_i^{\alpha_i})$ obtained in I, we arrive at the desired formula.

We start with the verification of I. An integer is coprime to p^α if and only if it is not divisible by p. Hence, we obtain the coprime integers to p^α among $1, 2, \dots, p^\alpha$, if we discard the multiples of p. We thus discard $p, 2p, \dots, p^{\alpha-1}p$, which are $p^\alpha/p = p^{\alpha-1}$ numbers. This implies that $\varphi(p^\alpha) = p^\alpha - p^{\alpha-1}$ integers remain.

Now, we turn to the proof of II. (As indicated earlier, two other methods are available in Exercises 2.2.14 and 2.6.10.)

The number $\varphi(ab)$ is the number of positive integers not greater than ab that are coprime to ab, i.e. are relatively prime to both a and b.

Denoting the smallest positive elements of the reduced residue classes modulo a by $r_1, r_2, \dots, r_{\varphi(a)}$, we enumerate all positive integers not greater than ab and coprime to a:

$$
\begin{array}{cccc}
r_1 & r_2 & \cdots & r_{\varphi(a)} \\
a + r_1 & a + r_2 & \cdots & a + r_{\varphi(a)} \\
2a + r_1 & 2a + r_2 & \cdots & 2a + r_{\varphi(a)} \\
\vdots & \vdots & & \vdots \\
(b-1)a + r_1 & (b-1)a + r_2 & \cdots & (b-1)a + r_{\varphi(a)}
\end{array}
$$

(2.3.1)

We have to select those numbers from (2.3.1) that are coprime also to b.

Consider an arbitrary column of the table. For example, the integers in column i are

(2.3.2) $$r_i, a + r_i, 2a + r_i, \ldots, (b-1)a + r_i.$$

These numbers were obtained from the complete residue system $0, 1, \ldots, b-1$ modulo b by multiplying the elements by a coprime to b and then adding r_i. By Theorem 2.2.4, (2.3.2) is a complete residue system modulo b, so every column of table (2.3.1) is a complete residue system modulo b.

Since a complete residue system modulo b contains $\varphi(b)$ elements coprime to b, there are $\varphi(b)$ numbers relatively prime to b in each column of (2.3.1).

The number of columns in (2.3.1) is $\varphi(a)$, so the table has altogether $\varphi(a)\varphi(b)$ elements coprime to b.

This means that there are $\varphi(a)\varphi(b)$ numbers among the positive integers not greater than ab that are coprime both to a and b, i.e. to ab. By definition, this value equals $\varphi(ab)$, hence $\varphi(ab) = \varphi(a)\varphi(b)$, indeed. $\qquad\square$

Second proof. We use the Inclusion and Exclusion formula.

We have to determine, how many numbers are coprime to n among $1, 2, \ldots, n$, that is, how many are divisible by none of the primes $p_1, p_2, \ldots, p_r$.

Thus we have to delete those "bad" numbers from $1, 2, \ldots, n$ which are divisible by one or more primes p_j.

Consider first those elements that are multiples of a given p_j (disregarding whether or not they are divisible by some other prime factors of n). Clearly, there are n/p_j such integers.

Now we count those numbers that are divisible by a given set of primes p_j (not caring again whether or not they are multiples of some other prime factors of n). An integer is divisible by both of two (distinct) primes if and only if it is divisible by their product. Hence, $n/(p_1 p_2)$ elements are divisible by both p_1 and p_2, $n/(p_1 p_3 p_7)$ elements are divisible by each of p_1, p_3, and p_7, etc.

Thus, the Inclusion and Exclusion formula yields

(2.3.3) $$\varphi(n) = n - \frac{n}{p_1} - \frac{n}{p_2} - \cdots - \frac{n}{p_r} + \frac{n}{p_1 p_2} + \frac{n}{p_1 p_3} + \cdots + \frac{n}{p_{r-1} p_r} - \frac{n}{p_1 p_2 p_3} - \cdots$$

A simple direct calculation verifies that the right-hand side of (2.3.3) is equal to the product

$$n \prod_{i=1}^{r} \left(1 - \frac{1}{p_i}\right),$$

which is an alternative version of the formula in the theorem. □

Exercises 2.3

1. Verify that $\varphi(n)$ is even for every $n > 2$.

2. Find all values of n for which $\varphi(n)$ is (a) 2 (b) 4 (c) 14 (d) 60.

3. Which is the smallest n for which $\varphi(n)$ is divisible by

   (a) 2^{10}

   (b) 3^{10}?

4. Determine all possible values of $\varphi(100n)/\varphi(n)$ for n a positive integer.

5. Prove the following propositions.

   (a) $k \mid n \Longrightarrow \varphi(k) \mid \varphi(n)$.

   (b) $\varphi((a,b)) \mid (\varphi(a), \varphi(b))$ and $[\varphi(a), \varphi(b)] \mid \varphi([a,b])$.

   (c) $\varphi((a,b)) = (\varphi(a), \varphi(b)) \iff [\varphi(a), \varphi(b)] = \varphi([a,b])$.

6. Show that $\varphi(a)/\varphi(b) = a/b$ holds if and only if a and b have exactly the same prime factors.

7. Let $n > 2$. True or false?

   (a) If $(n, \varphi(n)) = 1$, then n is an odd squarefree number.

   (b) If n is an odd squarefree number, then $(n, \varphi(n)) = 1$.

* 8. Prove that for every positive integer k there exists an n satisfying $(n, \varphi(n)) = k$.

9. Verify that $\varphi(n) + d(n) \leq n + 1$ holds for every n. When do we have equality?

10. (a) Demonstrate that if $(a, b) \neq 1$, then $\varphi(ab) > \varphi(a)\varphi(b)$ (thus equality is never true in this case).

    (b) In the first proof of Theorem 2.3.1, the key step was the verification of II, i.e. of $(a, b) = 1 \Longrightarrow \varphi(ab) = \varphi(a)\varphi(b)$. Where does the argument fail if a and b are not coprime?

    (c) Show that
    $$\varphi(ab)\varphi((a,b)) = (a,b)\varphi(a)\varphi(b)$$
    holds for every a and b.

11. (a) Prove that $n - \varphi(n) \geq \sqrt{n}$ if n is composite. When is equality true?

    (b) Find those n for which $n - \varphi(n)$ is

        (b1) 1

        (b2) 6

(b3) 7

(b4) 10.

12. Which *integers* occur in the range of the function $n/\varphi(n)$?

13. Prove that $\varphi(n^2) = \varphi(k^2)$ holds only for $n = k$.

14. Verify $\sum_{d|n} \varphi(d) = n$.

15. Show that $\varphi(n) \to \infty$ if $n \to \infty$.

  * 16. Demonstrate that for every positive integer k there exists an n satisfying $\varphi(n) = \varphi(n + k)$.

  * 17. Exhibit 1000 distinct integers where the function φ assumes the same value.

S* 18. Determine all n satisfying $\varphi(n!) = k!$ for some k.

S* 19. For which m can a reduced residue system mod m form an arithmetic progression?

2.4. The Euler–Fermat Theorem

Theorem 2.4.1 (Euler–Fermat Theorem).

$$(a, m) = 1 \implies a^{\varphi(m)} \equiv 1 \ (\text{mod } m).$$ ♣

Proof. Let $r_1, r_2, \ldots, r_{\varphi(m)}$ be a reduced residue system modulo m.

Since $(a, m) = 1$, $ar_1, \ldots, ar_{\varphi(m)}$ is also a reduced residue system modulo m.

This means that to every $1 \leq i \leq \varphi(m)$, there exists exactly one $1 \leq j \leq \varphi(m)$ satisfying $ar_i \equiv r_j \ (\text{mod } m)$. Denote this r_j by s_i:

$$
\begin{aligned}
ar_1 &\equiv s_1 & (\text{mod } m), \\
ar_2 &\equiv s_2 & (\text{mod } m), \\
&\ \ \vdots \\
ar_{\varphi(m)} &\equiv s_{\varphi(m)} & (\text{mod } m).
\end{aligned}
$$

(2.4.1)

Here $s_1, \ldots, s_{\varphi(m)}$ is a permutation of the numbers $r_1, \ldots, r_{\varphi(m)}$.

Multiplying the congruences in (2.4.1), we obtain

$$a^{\varphi(m)} r_1 r_2 \ldots r_{\varphi(m)} \equiv s_1 s_2 \ldots s_{\varphi(m)} \ (\text{mod } m),$$

or

(2.4.2) $\qquad a^{\varphi(m)} r_1 r_2 \ldots r_{\varphi(m)} \equiv r_1 r_2 \ldots r_{\varphi(m)} \ (\text{mod } m).$

We can cancel every r_i in (2.4.2), since $(r_i, m) = 1$, which yields the desired congruence $a^{\varphi(m)} \equiv 1 \ (\text{mod } m)$. □

An important special case is when the modulus is a prime p. Then $\varphi(p) = p - 1$ and we obtain:

Theorem 2.4.1A (First form of Fermat's Little Theorem). *If p is a prime and $(a, p) = 1$, then $a^{p-1} \equiv 1 \ (\text{mod } p)$.*

Note that for a prime p, the conditions $(a, p) = 1$, $p \nmid a$, and $a \not\equiv 0 \pmod{p}$ are equivalent.

From Theorem 2.4.1A, it is easy to get a congruence valid for every a:

Theorem 2.4.1B (Second form of Fermat's Little Theorem). *If p is a prime, then $a^p \equiv a$ (mod p) holds for every a.*

Proof. If $p \nmid a$, then $a^{p-1} \equiv 1 \pmod{p}$ by Theorem 2.4.1A. Multiplying this congruence by a, we obtain the desired $a^p \equiv a \pmod{p}$.

If $p \mid a$, then $a \equiv 0 \pmod{p}$. Raising this to the pth power (or multiplying it by a^{p-1}), we get $a^p \equiv 0 \pmod{p}$, hence also $a^p \equiv a \pmod{p}$ holds. $\square$

Remarks: (1) The converse of the Euler–Fermat Theorem (Theorem 2.4.1) is also true, i.e. $(a, m) = 1$ is not only a sufficient, but also a *necesssary* condition for $a^{\varphi(m)} \equiv 1 \pmod{m}$. In fact, the following stronger proposition holds: There exists an exponent $k > 0$ such that $a^k \equiv 1 \pmod{m}$ *only* if a and m are coprime. Namely, $a^k \equiv 1 \pmod{m}$ implies $(a^k, m) = (1, m) = 1$ by Theorem 2.2.5, hence also $(a, m) = 1$ must hold.

(2) The second form of Fermat's Little Theorem (Theorem 2.4.1B) has no natural generalization for arbitrary modulus m, i.e. there exists no simple variant of the general Euler–Fermat Theorem that would be valid for every a (see Exercise 2.4.15).

(3) As their names indicate, Theorems 2.4.1A and B are due to Fermat. Both variants can be verified directly, without relying on Theorem 2.4.1. Form B can be proven by induction (on a), and form A follows easily (see Exercise 2.4.16). Theorem 2.4.1 was found by Euler as a generalization of Fermat's Little Theorem.

(4) The adjective "little" serves to distinguish this result from Fermat's Last Theorem which is a very famous and only recently solved problem of mathematics. We shall treat this topic in Chapter 7.

Exercises 2.4

1. Prove $n \mid 2^{n!} - 1$ for any odd n.

2. Determine the last two digits of 1793^{8642} (in decimal representation).

3. Verify that $n^{20} + 4n^{44} + 8n^{80}$ is a multiple of 13 for every n.

4. Show that if n is any integer, then at least one of $n^6 + 13$ and $n^2 + 21$ is a composite number.

5. Prove $1703601900 \mid a^{62} - a^2$ for every a.

6. Verify the following propositions:

   (a) $11 \mid a^{30} + b^{30} + c^{30} \implies 11^{30} \mid a^{30} + b^{30} + c^{30}$.

   (b) $9 \mid a^{30} + b^{30} + c^{30} \implies 9^{15} \mid a^{30} + b^{30} + c^{30}$.

7. Show that $a^{88} - b^{88}$ is *not* divisible by 23 if and only if exactly one of a and b is divisible by 23.

8. Let p be a prime and $r_1, \ldots, r_p$ be a complete residue system mod p. Prove that also $r_1^{2p-3}, \ldots, r_p^{2p-3}$ is a complete residue system mod p.

9. (a) Let p be a prime, a an integer, and i and j positive integers satisfying $i \equiv j$ (mod $p - 1$). Prove $a^i \equiv a^j$ (mod p).

   (b) How can we generalize the assertion in (a) for arbitrary m (instead of primes)?

10. True or false? (With decimal notation and powers with positive integer exponents.)

    (a) Infinitely many powers of 133 terminate with the string 133.

    (b) Infinitely many powers of 134 terminate with the string 134.

    (c) Infinitely many powers of 136 terminate with the string 136.

11. Show that an infinite arithmetic progression of distinct positive integers $a, a+d, \ldots,$ $a + kd, \ldots$ contains infinitely many powers of a (with positive integer exponents) if and only if $d/(a, d)$ and a are coprime.

12. Give a new solution to Exercise 1.3.12a using the Euler–Fermat Theorem.

13. Verify that every positive odd divisor of $n^2 + 1$ is of the form $4k + 1$.

14. Assume that 19 divides $a^{40} + b^{40}$. Show that then 19 must divide both a and b, as well.

15. Verify the following propositions and investigate their relation to Fermat's Little Theorem.

    (a) $a^{\varphi(m)+1} \equiv a$ (mod m) holds for every a if and only if m is squarefree.

    (b) $a^m \equiv a^{m-\varphi(m)}$ (mod m) holds for every m and a.

    (c) $a^{1729} \equiv a$ (mod 1729) holds for every a.

16. Give a direct proof of both versions of Fermat's Little Theorem: First verify Theorem 2.4.1B by induction and then deduce Theorem 2.4.1A.

2.5. Linear Congruences

This section deals with the simplest type of congruences with variables (or congruence equations), the linear congruences.

Definition 2.5.1. Let a and b be integers and m a positive integer. The congruence $ax \equiv b$ (mod m) is called a *linear congruence*, and by a *solution* of it we mean an integer s which substituted into x makes the congruence valid.   ♣

   Clearly, if s is a solution, then every other element of the residue class $(s)_m$ is a solution, too. Hence, to find all solutions, it is enough to check a complete residue system to see which elements of it satisfy the congruence; then all solutions are the integers congruent to them.

   Therefore the number of solutions of a linear congruence is defined as how many *pairwise incongruent* integers satisfy the congruence, i.e. what is the number of *residue classes* the solutions come from, or (again in a slightly different formulation) how many elements of a complete residue system make the congruence valid. The same applies for congruences of higher degree as well, thus we define this convention immediately for the general case.

Definition 2.5.2. Let f be a polynomial with integer coefficients. The *number of solutions* of the congruence $f(x) \equiv 0 \pmod{m}$ is how many elements s of a complete residue system modulo m satisfy $f(s) \equiv 0 \pmod{m}$.   ♣

Since $u \equiv v \pmod{m} \implies f(u) \equiv f(v) \pmod{m}$, this notion does not depend on which complete residue system modulo m we considered.

Returning to linear congruences, we want to answer the following questions arising for equations in general:

(i) What is a necessary and sufficient condition for solvability?

(ii) How many solutions do we have?

(iii) How can we describe or characterize all solutions?

(iv) Which methods yield these solutions?

We discuss solvability first.

Theorem 2.5.3. *The congruence $ax \equiv b \pmod{m}$ is solvable if and only if $(a, m) \mid b$.*   ♣

Proof. The solvability of $ax \equiv b \pmod{m}$ means that $as \equiv b \pmod{m}$ for some s.

This is equivalent to the existence of an integer t satisfying $as + mt = b$, i.e. s and t are a solution of the linear Diophantine equation $ax + my = b$.

Hence, the linear congruence $ax \equiv b \pmod{m}$ is solvable if and only if the linear Diophantine equation $ax + my = b$ is solvable.

The necessary and sufficient condition for the solvability of the latter is $(a, m) \mid b$, by Theorem 1.3.6. Thus the same criterion applies for the solvability of $ax \equiv b \pmod{m}$.   □

We see from the proof that the linear congruence $ax \equiv b \pmod{m}$ and the linear Diophantine equation $ax + my = b$ can be deduced from each other. (Moreover, the linear Diophantine equation $ax + my = b$ can also be transformed into the linear congruence $my \equiv b \pmod{|a|}$ if $a \neq 0$.)

Based on this, every result obtained for linear congruences can be used also for linear Diophantine equations and vice versa.

We should be aware, however, of the significant differences: The solutions of a linear congruence are integers (or rather residue classes), whereas the solutions of a linear Diophantine equation are *pairs* of integers; the number of solutions of a congruence is finite, but a linear Diophantine equation has infinitely many solutions, etc.

In the next theorem, we determine the number of solutions of a linear congruence, and also see how we can get all solutions from a given one.

Theorem 2.5.4.   (I) *If $ax \equiv b$ (mod m) is solvable, then there are (a, m) solutions.*

(II) *Let $(a, m) = d$, $m = dm_1$, and s be a solution of $ax \equiv b$ (mod m). Then*

(2.5.1) $$s, \quad s + m_1, \quad s + 2m_1, \quad \ldots, \quad s + (d-1)m_1$$

are pairwise incongruent modulo m, satisfy the congruence, and every solution is congruent to one of them modulo m.   ♣

Proof. We verify the two assertions simultaneously.

We assumed that s was a solution, so

(2.5.2) $$as \equiv b \ (\text{mod } m).$$

An integer t is a solution if and only if

(2.5.3) $$at \equiv b \ (\text{mod } m).$$

Using (2.5.2), formula (2.5.3) is equivalent to

(2.5.4) $$at \equiv as \ (\text{mod } m).$$

By Theorem 2.1.3, (2.5.4) is equivalent to

$$t \equiv s \ \left(\text{mod } \frac{m}{(m, a)}\right) \quad \text{or} \quad t \equiv s \ (\text{mod } m_1).$$

We can rewrite this as

(2.5.5) $$t = s + km_1,$$

with some integer k.

This means that the numbers t in (2.5.5) give all solutions of $ax \equiv b$ (mod m).

Thus, we have to prove that these integers t in (2.5.5) belong to d distinct residue classes and (2.5.1) lists a representative from each class.

When do two such t fall into the same residue class modulo m? Let

$$t' = s + k'm_1 \quad \text{and} \quad t'' = s + k''m_1.$$

Then

(2.5.6) $$t' \equiv t'' \ (\text{mod } m) \iff k'm_1 \equiv k''m_1 \ (\text{mod } m) \iff k' \equiv k'' \ (\text{mod } d).$$

Here, we first subtracted s from $t' \equiv t''$ (mod m), then cancelled m_1 and changed the modulus to $m/(m_1, m) = m/m_1 = d$, according to Theorem 2.1.3.

Implication (2.5.6) means that two integers t fall into the same residue class modulo m if and only if the relevant two integers k are congruent modulo d.

Thus, if k assumes the values $0, 1, \ldots, d - 1$, then the integers

$$t = s + km_1, \quad \text{or} \quad s, s + m_1, \ldots, s + (d-1)m_1$$

occurring in (2.5.1) are just the representatives of the relevant residue classes modulo m.   □

The most important special case of the linear congruence $ax \equiv b$ (mod m) is when $(a, m) = 1$. Then $(a, m) \mid b$ holds automatically, so the congruence is solvable, by Theorem 2.5.3, and it has $(a, m) = 1$ (pairwise incongruent) solutions, by Theorem 2.5.4.

We state this important result as a theorem:

Theorem 2.5.5. *If* $(a, m) = 1$, *then the congruence* $ax \equiv b$ (mod m) *is solvable for every* b *and the number of solutions is* 1. ♣

We make some general preliminary remarks concerning methods for finding the solutions.

(A) In general, it is advisable to check by the criterion of Theorem 2.5.3 whether the congruence is solvable at all.

(B) If $(a, m) = 1$, then the congruence is satisfied by the elements of just one residue class, so if we find somehow a solution, then we are done. Also, in the general case, it is sufficient to guess a single solution because we can easily obtain all solutions by Theorem 2.5.4/II.

(C) In most cases, the best start is to reduce the original linear congruence to one where the coefficient of x and the modulus are coprime. We can do this as follows.

If $ax \equiv b$ (mod m) is solvable, then $(a, m) \mid b$. Let $d = (a, m)$, then

$$a = da_1, \quad m = dm_1, \quad b = db_1, \quad \text{and} \quad (a_1, m_1) = 1.$$

Hence, we can divide the congruence by d (including also the modulus): $ax \equiv b$ (mod m) is equivalent to $a_1 x \equiv b_1$ (mod m_1) and here $(a_1, m_1) = 1$. (Looking at the corresponding Diophantine equations, this just means that $ax + my = b$ is divided by d to yield $a_1 x + m_1 y = b_1$.)

The word "equivalent" in the previous paragraph should remind us that though the two congruences are satisfied by the same integers, we have to group them into residue classes of different moduli: mod m at the first congruence and mod m_1 at the second one. As a consequence, the two congruences will differ also in the number of solutions (for $d > 1$).

We turn now to the detailed discussion of a few methods for finding the solutions of a linear congruence. Each will be illustrated by an example.

M1 *Trial.* We check each element of a complete residue system modulo m to see if it satisfies the congruence. (This should be applied only for very small moduli.)

E1 $23x \equiv 11$ (mod 5). To make calculations simpler, it is worthwhile to replace the coefficients with congruent numbers having smaller (absolute) value before substituting into x: $3x \equiv 1$ (mod 5) or $-2x \equiv 1$ (mod 5). Testing the numbers 0, 1, 2, 3, 4 (or 0, ±1, ±2), we obtain that the residue class $x \equiv 2$ (mod 5) is the only solution. (Since $(23, 5) = 1$ implies that there is only one solution, after finding it we do not have to check more numbers.)

M2 *Diophantine equation.* We reduce the linear congruence to a Diophantine equation as seen in the proof of Theorem 2.5.3, and then reconstitute its solutions into solutions of the congruence.

E2 $18x \equiv 38$ (mod 28). The corresponding Diophantine equation is $18x + 28y = 38$. Dividing by 2, we obtain $9x + 14y = 19$. Following the proof of Theorem 1.3.6, we write the gcd of 9 and 14 in form $9u + 14v$. From the Euclidean algorithm or after a few trials, we have $9 \cdot (-3) + 14 \cdot 2 = 1$. Multiplying by 19, we obtain $9 \cdot (-57) + 14 \cdot 38 = 19$, so $x = -57, y = 38$ is a solution of the equation $9x + 14y = 19$.

Returning to the congruence $18x \equiv 38 \pmod{28}$, this means that $x = -57$ is a solution. We find all solutions by Theorem 2.5.4/II: $x \equiv -57 \pmod{28}$ and $x \equiv -43 \pmod{28}$. (The representatives -57 and -43 can be replaced by any others, e.g. by -1 and 13.)

Note that to solve a linear Diophantine equation, it is more convenient to apply the procedure described in Section 7.1 that characterizes all solutions immediately in a parametric form. (Actually, also this is a variant of the Euclidean algorithm.)

M3 *Euler–Fermat Theorem.* We reduce the congruence $ax \equiv b \pmod{m}$ to $a_1 x \equiv b_1$ $\pmod{m_1}$ where $(a_1, m_1) = 1$, as seen in remark (C). T hen $a_1^{\varphi(m_1)} \equiv 1 \pmod{m_1}$ by the Euler–Fermat Theorem. Therefore $x = a_1^{\varphi(m_1)-1} b_1$ is a solution:

$$a_1 \cdot a_1^{\varphi(m_1)-1} b_1 = a_1^{\varphi(m_1)} b_1 \equiv b_1 \pmod{m_1}.$$

Hence, $x = a_1^{\varphi(m_1)-1} b_1$ is a solution of the original congruence, too. Finally, we can obtain all solutions from Theorem 2.5.4/II.

E3 $36x \equiv 81 \pmod{21}$. Here $(36, 21) = 3$, hence we can reduce the problem to the congruence $12x \equiv 27 \pmod{7}$. Decreasing the coefficients, we obtain $-2x \equiv -1 \pmod{7}$. Its solution is $x = (-2)^{6-1}(-1) \equiv 4 \pmod{7}$. Thus, all solutions of the original congruence are $x \equiv 4, 11, 18 \pmod{21}$.

Reducing the coefficients in the congruence $12x \equiv 27 \pmod{7}$, we may choose the least non-negative remainders instead of the ones with least absolute value. Then we get $5x \equiv 6 \pmod{7}$ and $x \equiv 5^5 \cdot 6 \pmod{7}$.

Since $(12, 7) = 1$, $12x \equiv 27 \pmod{7}$ has a unique solution modulo 7, i.e. $5^5 \cdot 6 \equiv 4 \pmod{7}$ For a direct verification, one should not compute the actual value of 5^5 but rather take the remainders modulo 7 while raising to powers:

$$5^2 = 25 \equiv 4 \pmod{7}, \quad 5^4 \equiv 4^2 \equiv 2 \pmod{7}, \quad 5^5 \equiv 5 \cdot 2 \equiv 3 \pmod{7},$$

hence $6 \cdot 5^5 \equiv 6 \cdot 3 \equiv 4 \pmod{7}$.

M4 *Tricks.* Multiplying or dividing the congruence by well-chosen integers coprime to the modulus, we get equivalent congruences till finally we can easily read the solution(s).

E4 Consider $80x \equiv 32 \pmod{108}$. Here $(80, 108) = 4$, so we can reduce the problem to solve $20x \equiv 8 \pmod{27}$.

As $(4, 27) = 1$, cancelling 4 yields an equivalent congruence: $5x \equiv 2 \pmod{27}$.

We show two methods of how to get rid of the coefficient 5 in $5x \equiv 2 \pmod{27}$.

I. Division: We can replace 2 on the right-hand side by -25: $5x \equiv -25 \pmod{27}$. Since $(5, 27) = 1$, we can cancel the 5: $x \equiv -5 \pmod{27}$.

II. Multiplication: We multiply by a suitable number to change the coefficient of x into an integer congruent to 1 (or -1) modulo 27. (This multiplier is then automatically coprime to 27 guaranteeing equivalence.) We can multiply our congruence $5x \equiv 2$

(mod 27) by 11: $55x \equiv 22$ (mod 27) and since $55 \equiv 1$ (mod 27) we obtain $x \equiv 22(\equiv -5)$ (mod 27).

So the solutions of the original congruence are $x \equiv -5, 22, 49, 76$ (mod 108).

Comparing the above methods, M3 or M4 could seem to be the easiest to apply at first sight. It turns out, however, that only M2 works for large moduli. This will be treated in Section 5.7.

Exercises 2.5

1. Solve Examples E1–E4 with every method M2–M4.

2. Solve the following congruences:

   (a) $24x \equiv 60$ (mod 51)

   (b) $100x \equiv 88$ (mod 116)

   (c) $555x \equiv 5555$ (mod 55555)

   (d) $(2^k + 1)x \equiv 2^{k+1} + 1 \left(\text{mod } 2^{k+2} + 1\right)$

   (e) $10x^{39} + 8x^{20} + 9x^3 + 7x \equiv 0$ (mod 19)

   (f) $13x^{41} \equiv 27$ (mod 100).

3. Determine the two smallest positive integers which when multiplied by 13 will have last digit 3 and next to last digit 4 in the number system of base seven.

4. Compute the last two digits of 3^{279} (in decimal representation).

5. Check (each of) the following conditions to see if they are sufficient for the solvability of the congruence $ax \equiv b$ (mod m).

   (a) $(a, m) \mid (a, b)$

   (b) $(a, b) \mid (a, m)$

   (c) a, m, b is an arithmetic progression

   (d) a, m, b is a geometric series

   (e) a, b, m is an arithmetic progression

   (f) a, b, m is a geometric series.

6. True or false?

   (a) The number of solutions of $ax \equiv b$ (mod m) is at most b if $b > 0$.

   (b) If $ax \equiv b$ (mod m) is solvable, then $a^2x \equiv b^2$ $\left(\text{mod } m^2\right)$ is solvable.

   (c) If both $a_1x \equiv b_1$ (mod m_1) and $a_2x \equiv b_2$ (mod m_2) are solvable, then $a_1a_2x \equiv b_1b_2$ (mod m_1m_2) is solvable.

S 7. Let a and m be fixed and denote the number of solutions of $ax \equiv b$ (mod m) by $f(b)$. Compute $\sum_{b=1}^{m} f(b)$.

2.6. Simultaneous Systems of Congruences

A simultaneous system of congruences means that several congruence conditions with different moduli are imposed on the same variable:

$$f_1(x) \equiv 0 \ (\text{mod } m_1), \quad f_2(x) \equiv 0 \ (\text{mod } m_2), \quad \ldots, \quad f_k(x) \equiv 0 \ (\text{mod } m_k)$$

where $f_1, \ldots, f_k$ are polynomials with integer coefficients.

Clearly, a necessary condition for the solvability of such a system is that each congruence should be solvable. Thus, after solving the individual congruences, we have to study only the (special linear) systems of the form

$$x \equiv c_1 \ (\text{mod } m_1), \quad x \equiv c_2 \ (\text{mod } m_2), \quad \ldots, \quad x \equiv c_k \ (\text{mod } m_k).$$

We consider first systems with two congruences.

Theorem 2.6.1. (I) *The simultaneous system of congruences*

(2.6.1)
$$x \equiv c_1 \ (\text{mod } m_1)$$
$$x \equiv c_2 \ (\text{mod } m_2)$$

is solvable if and only if

$$(m_1, m_2) \mid c_1 - c_2.$$

(II) *If solvable, the solutions form a residue class modulo* $[m_1, m_2]$. *Or, putting it into another form: if s is a solution, then all solutions t are given by*

$$t \equiv s \ (\text{mod } [m_1, m_2]), \quad \textit{or} \quad t = s + k[m_1, m_2], \quad \textit{where k is an integer.} \quad \clubsuit$$

The proof will yield a method for finding the solutions; one has to solve a linear Diophantine equation (or, equivalently, a linear congruence).

Proof. I. By the definition of congruences, (2.6.1) can be transformed into

(2.6.2)
$$x = c_1 + z_1 m_1, \quad x = c_2 + z_2 m_2$$

where z_1 and z_2 are integers.

Condition (2.6.2) is equivalent to

(2.6.3)
$$c_1 + z_1 m_1 = c_2 + z_2 m_2.$$

Rearranging (2.6.3), we obtain

(2.6.4)
$$c_1 - c_2 = z_2 m_2 - z_1 m_1.$$

This means that the system of congruences (2.6.1) can be reduced to the linear Diophantine equation (2.6.4).

By Theorem 1.3.6, it is solvable if and only if $(m_1, m_2) \mid c_1 - c_2$, hence the same applies for (2.6.1).

As we indicated before the proof, we also obtained a method of finding the solutions: we have to solve Diophantine equation (2.6.4) or a corresponding congruence.

II. Let s be a solution so

(2.6.5)
$$s \equiv c_1 \ (\text{mod } m_1),$$
$$s \equiv c_2 \ (\text{mod } m_2).$$

An integer t is a solution if and only if

(2.6.6)
$$t \equiv c_1 \pmod{m_1},$$
$$t \equiv c_2 \pmod{m_2}.$$

Using (2.6.5), condition (2.6.6) is equivalent to

(2.6.7)
$$t \equiv s \pmod{m_1}$$
$$t \equiv s \pmod{m_2}.$$

Rewrite (2.6.7) as divisibilities and apply the properties of lcm (Theorem 1.6.6/II):

$$\left.\begin{array}{l} m_1 \mid t - s \\ m_2 \mid t - s \end{array}\right\} \iff [m_1, m_2] \mid t - s \iff t \equiv s \pmod{[m_1, m_2]}. \qquad \square$$

The most importamt special case is when the moduli m_1 and m_2 in system (2.6.1) are coprime. Then $(m_1, m_2) \mid c_1 - c_2$ holds automatically, so the system of congruences is solvable and the solutions form a unique residue class modulo $m_1 m_2$. We state this important result as a theorem:

Theorem 2.6.1A. *If* $(m_1, m_2) = 1$, *then the simultaneous system of congruences*

$$x \equiv c_1 \pmod{m_1}$$
$$x \equiv c_2 \pmod{m_2}$$

is solvable for arbitrary c_1 *and* c_2, *and the solutions form a single residue class modulo* $m_1 m_2$.

Theorem 2.6.1A implies that if m_1 and m_2 are coprime, then the remainder of a number when divided by m_1 is independent of its remainder mod m_2. For example, the last digits of an integer give its remainder modulo a power of 10 and they provide no information on the remainder, say, modulo 3, 7, or 13, since these moduli are coprime to 10.

Turning to systems consisting of more than two congruences, we deal only with the case when the moduli are *pairwise* coprime (see Exercise 2.6.13 for the general case). This result was known by the Chinese mathematician Sun Tsu about 2000(!) years ago, therefore it is generally referred to as the Chinese Remainder Theorem.

Theorem 2.6.2 (Chinese Remainder Theorem). *Let* $m_1, \ldots, m_k$ *be pairwise coprime. Then the system of congruences*

(2.6.8)
$$x \equiv c_1 \pmod{m_1}$$
$$x \equiv c_2 \pmod{m_2}$$
$$\vdots$$
$$x \equiv c_k \pmod{m_k}$$

is solvable for any integers $c_1, \ldots, c_k$, *and the solutions form one residue class modulo* $m_1 m_2 \ldots m_k$. ♣

First proof. We can easily obtain the result from Theorem 2.6.1A by induction on k.

The case $k = 2$ is just Theorem 2.6.1A.

Assume now that the statement is true for systems of $k - 1$ congruences, and consider the system (2.6.8) of k congruences. The integers satsifying the first $k - 1$ congruences constitute one residue class modulo $m_1 m_2 \ldots m_{k-1}$ by the induction hypothesis, so we can replace the first $k - 1$ congruences by the congruence $x \equiv c$ (mod $m_1 m_2 \ldots m_{k-1}$) with a suitable integer c. Thus, (2.6.8) is equivalent to the system

(2.6.9)
$$x \equiv c \quad (\text{mod } m_1 m_2 \ldots m_{k-1})$$
$$x \equiv c_k \ (\text{mod } m_k)$$

Applying Theorem 2.6.1A to (2.6.9), we obtain just the statement for k. $\square$

Second proof. We show a new argument for solvability and we produce a solution in an explicit form (in a certain sense).

The procedure reminds us somewhat of the construction of the interpolation polynomials by Lagrange.

We consider first the special case of (2.6.8) when one c_i is 1 and all other c_j are 0, and then use this result to solve the general case.

Let us see the details. Let

$$M = m_1 \ldots m_k \quad \text{and} \quad M_i = \frac{M}{m_i}, \quad i = 1, 2, \ldots, k.$$

Since the moduli $m_1, \ldots, m_k$ are pairwise coprime,

(2.6.10) $(M_i, m_i) = 1, \quad i = 1, 2, \ldots, k.$

I. We fix an index $1 \le i \le n$ and solve the problem in the special case when $c_i = 1$ and $c_j = 0$ for $j \ne i$ in (2.6.8).

The congruences $x \equiv 0 \ (\text{mod } m_j)$ mean that x is a multiple of every m_j with $j \ne i$. The moduli m_j are pairwise coprime, hence equivalently x is a multiple of the product M_i of the numbers m_j: $x = M_i z$.

Substituting this in the remaining congruence $x \equiv 1 \ (\text{mod } m_i)$, we obtain

(2.6.11) $M_i z \equiv 1 \ (\text{mod } m_i).$

This a linear congruence for z that is solvable by (2.6.10).

Let b_i be a solution of (2.6.11). Then $x = b_i M_i$ is a solution of (2.6.8).

II. We consider now the general case with arbitrary c_i in (2.6.8). We show that

(2.6.12) $x = c_1 b_1 M_1 + \cdots + c_k b_k M_k$ (where $M_i b_i \equiv 1 \ (\text{mod } m_i), i = 1, \ldots, k$)

is a solution of (2.6.8).

Let us check for example the congruence $x \equiv c_3 \ (\text{mod } m_3)$. Since $b_3 M_3 \equiv 1$ (mod m_3) and all the other M_j are divisible by m_3, therefore the right-hand side of (2.6.12)

$$c_1 b_1 M_1 + \cdots + c_k b_k M_k \equiv c_3 b_3 M_3 \equiv c_3 \ (\text{mod } m_3).$$ $\square$

An important corollary of Theorem 2.6.2 is that any congruence with a composite modulus can be reduced to congruences with prime power moduli. If the standard form of m is $m = p_1^{\alpha_1} \dots p_r^{\alpha_r}$, then the congruence

(2.6.13) $f(x) \equiv 0 \pmod{m}$

is equivalent to the system

$$f(x) \equiv 0 \pmod{p_1^{\alpha_1}}$$
$$f(x) \equiv 0 \pmod{p_2^{\alpha_2}}$$

(2.6.14) $\vdots$

$$f(x) \equiv 0 \pmod{p_r^{\alpha_r}}.$$

We solve every congruence of (2.6.14) separately. If some of them are not solvable, then (2.6.13) is not solvable either. If all of them are solvable, then consider a solution of each, say $h_1, \dots, h_r$. Now, solving the system

$$x \equiv h_1 \pmod{p_1^{\alpha_1}}$$
$$x \equiv h_2 \pmod{p_2^{\alpha_2}}$$

$$\vdots$$

$$x \equiv h_r \pmod{p_r^{\alpha_r}},$$

we get a solution of the original congruence (2.6.13). We obtain all solutions by considering all possible solution systems $h_1, \dots, h_r$ for the congruences (2.6.14).

Example E1. Solve the congruence

(2.6.15) $10x^{84} + 3x + 7 \equiv 0 \pmod{245}.$

By the above, (2.6.15) is equivalent to the system

(2.6.16) $10x^{84} + 3x + 7 \equiv 0 \pmod{5}$

(2.6.17) $10x^{84} + 3x + 7 \equiv 0 \pmod{49}.$

(2.6.16) is identical to $3x + 7 \equiv 0 \pmod{5}$ since $10 \equiv 0 \pmod{5}$. The only solution of this linear congruence is

(2.6.16a) $x \equiv 1 \pmod{5}.$

In looking for the solutions of (2.6.17), we distinguish two cases:

(i) $(x, 49) = 1$

(ii) $(x, 49) \neq 1.$

In case (i),

$$x^{84} = x^{2\varphi(49)} \equiv 1 \pmod{49},$$

by the Euler–Fermat Theorem. Thus (2.6.17) is equivalent to $3x + 17 \equiv 0 \pmod{49}$ in this case. This has one solution

(2.6.17a) $x \equiv -22 \pmod{49}.$

In case (ii), $7 \mid x$. Then $x^{84} \equiv 0 \pmod{49}$. Thus (2.6.17) is equivalent to $3x + 7 \equiv 0 \pmod{49}$ in this case. The only solution (satisfying also the condition $7 \mid x$) is

(2.6.17b) $x \equiv 14 \pmod{49}.$

Thus, the solutions of (2.6.15) are obtained from the systems

(2.6.16a) $x \equiv 1$ (mod 5).

(2.6.17a) $x \equiv -22$ (mod 49).

and

(2.6.16a) $x \equiv 1$ (mod 5).

(2.6.17b) $x \equiv 14$ (mod 49).

To determine the solutions, we can use the procedure in the proof of Theorem 2.6.1, but it is often more convenient to apply the following method.

From the congruence (2.6.17a)—using the larger modulus—-we have:

(2.6.18) $x = 49z - 22.$

Substituting (2.6.18) into (2.6.16a), we get

$$49z - 22 \equiv 1 \ (\text{mod } 5).$$

We find that

(2.6.19) $z \equiv 2$ (mod 5)   so $z = 5w + 2.$

Substituting (2.6.19) back into (2.6.18), we obtain $x = 245w + 76$. Thus the solution of the first system of congruences is $x \equiv 76$ (mod 245).

Proceeding similarly, the solution of the second system is $x \equiv 161$ (mod 245).

Thus all solutions of (2.6.15) are

$$x \equiv 76 \ (\text{mod } 245) \quad \text{and} \quad x \equiv 161 \ (\text{mod } 245).$$

Finally, we discuss an application of the Chinese Remainder Theorem in computer science. Many operations in computers are composed of a sequence of additions, subtractions, and multiplications of integers. Therefore, it is essential to know how quickly these basic steps can be performed.

Consider e.g. addition. Using the usual representation in a number system, the addition of digits cannot be done independently since overflows influence the result significantly. In the so-called *remainder number systems*, however, we can perform the operations with the "digits." i.e. remainders, absolutely independently. This is mostly used if there are many parallel processors available.

The main point of the method is the following. Assume that only integers with absolute value less than N can occur during the operations. (This is no restriction since every computer can display and work with numbers only up to a given limit.) Let $m = p_1 \ldots p_r$ be the product of the first r (positive) primes, and choose r to satisfy $m > 2N$.

Then every integer with absolute value less than N is equal to its remainder of least absolute value modulo m. And this can be represented by the system of remainders modulo p_i, which will be the digits in the remainder number system.

The digits actually are a simultaneous system of congruences where the moduli p_i are pairwise coprime, hence the remainder modulo m, i.e. the original number itself, can be uniquely reconstructed.

Adding or multiplying two numbers, we have to add or multiply the corresponding remainders (i.e. digits), there is no overflow, and the operations can be performed independently for the various moduli. From the system of the remainders modulo p_i thus obtained, we have to determine the remainder modulo m, i.e. the number itself.

Example E2. As an illustration, let $N = 1000$, and we execute the multiplication $27 \cdot 34$ in the remainder number system.

We can take
$$m = 2 \cdot 3 \cdot 5 \cdot 7 \cdot 11 = 2310.$$
The remainders of 27 when divided by the primes 2, 3, 5, 7, and 11 are 1, 0, 2, 6, and 5, so the representation of 27 in the remainder number system is
$$27 = (1, 0, 2, 6, 5).$$
Similarly,
$$34 = (0, 1, 4, 6, 1).$$
To do the multiplication $27 \cdot 34$, we multiply the corresponding digits (there is no overflow), reduce the products modulo p_i, and solve the resulting system of congruences:
$$27 \cdot 34 = (1 \cdot 0, 0 \cdot 1, 2 \cdot 4, 6 \cdot 6, 5 \cdot 1) = (0, 0, 3, 1, 5).$$
The solution of the system
$$x \equiv 0 \ (\mathrm{mod} \ 2)$$
$$x \equiv 0 \ (\mathrm{mod} \ 3)$$
$$x \equiv 3 \ (\mathrm{mod} \ 5)$$
$$x \equiv 1 \ (\mathrm{mod} \ 7)$$
$$x \equiv 5 \ (\mathrm{mod} \ 11)$$
is
$$x \equiv 918 \ (\mathrm{mod} \ 2310).$$
Thus, $27 \cdot 34 = 918$.

If we perform more operations, we can keep working with the form in the remainder number system and convert only the final result into the usual representation of numbers.

We mention that systems of congruences can similarly be applied also to solve systems of linear equations (with rational coefficients). The main point of the method is that the system of equations is handled modulo various prime moduli, and from the solutions obtained we determine the solution modulo the product of these primes. This yields the solution wanted if certain conditions are satisfied and sufficiently many moduli are used. The advantage of the method in contrast with the traditional Gaussian elimination is that no too large (or too small) numbers can occur here, and thus there is no danger of overflow.

Exercises 2.6

(We use decimal representation unless stated otherwise.)

1. (a) A centipede wants to count its feet knowing that their number does not exceed 250. Counting them in elevens and in fifteens, 5 and 3 are left out. How many feet has the centipede?

   (b) Another centipede tries this method, too. It counts its feet by twelves and fifteens and finds that 4 and 8 are left out. Prove that it made a miscalculation.

2. The last digit of an integer in number system with base 20 is "eleven". What can be its last digit with base (a) 9 (b) 8?

3. Solve the congruences:

   (a) $2x^{20} + 3x + 4 \equiv 0 \pmod{176}$

   (b) $21x^{66} + 16x^{30} + 11x + 6 \equiv 0 \pmod{333}$

   (c) $3x^9 + 5x + 7 \equiv 0 \pmod{105}$.

4. Let a, b, and c be pairwise coprime integers greater than 1. What is the remainder

   (a) of $a^{\varphi(b)} + b^{\varphi(a)}$ modulo ab

   (b) of $a^{\varphi(bc)} + b^{\varphi(ac)} + c^{\varphi(ab)}$ modulo abc?

5. Determine the last three digits of 1234^{9876}.

6. I thought of an integer between 200 and 2000. Adding its 501st and 201st power to the original number, the sum will terminate in 998. Which number did I think of?

7. Which are those (a) two digit (b) three digit positive integers whose squares terminate in the same two and three digits, respectively?

8. (a) How many 21-digit positive integers have the property that every power of them terminates with the same 20 digits as the original number?

   (b) How many 21-digit positive integers have the property that every *odd* power of them terminates with the same 20 digits as the original number?

S 9. What will be the exact time (in hours and minutes) $39^{38^{37}}$ minutes after midnight?

10. (a) Let $(a, b) = 1$, and $r_1, \ldots, r_{\varphi(a)}$ and $s_1, \ldots, s_{\varphi(b)}$ be reduced residue systems modulo a and modulo b. For $i = 1, \ldots, \varphi(a)$, $j = 1, \ldots, \varphi(b)$, denote by c_{ij} a solution of the system

$$x \equiv r_i \pmod{a}$$
$$x \equiv s_j \pmod{b}.$$

   Show that the c_{ij} form a reduced residue system modulo ab. Use only the *definition* of the reduced residue system (Definition 2.2.8) during the proof, and do not rely on Theorem 2.2.9 or on part (b) of this exercise.

   (b) Give a new proof for $(a, b) = 1 \implies \varphi(ab) = \varphi(a)\varphi(b)$.

11. Verify that there are arbitrarily large gaps in the sequence of squarefree numbers. That is, for any K, there exist K consecutive positive integers none of which is squarefree.

* 12. (a) Prove that the following two systems are solvable for any positive integers a, b, and c.

   (a1) $x \equiv a + b \pmod{c}$

   $x \equiv b + c \pmod{a}$

   $x \equiv c + a \pmod{b}$

   (a2) $x \equiv ab \pmod{c}$

   $x \equiv bc \pmod{a}$

   $x \equiv ca \pmod{b}.$

   (b) Show that

   $$x \equiv b \pmod{c}, \qquad x \equiv c \pmod{a}, \qquad x \equiv a \pmod{b}$$

   is solvable if and only if $(a, b) = (b, c) = (c, a)$.

* 13. Demonstrate that the system

   $$x \equiv c_1 \pmod{m_1}, \quad x \equiv c_2 \pmod{m_2}, \quad \ldots, \quad x \equiv c_k \pmod{m_k}$$

   (where the moduli m_i are not necessarily pairwise coprime) is solvable if and only if $(m_i, m_j) \mid c_i - c_j$ for every $1 \leq i < j \leq k$.

14. Does there exist a polynomial $f(x)$ with integer coefficients for which the congruence $f(x) \equiv 0 \pmod{30}$ has exactly 14 solutions?

15. (a) Prove that there exist integers forming both a complete residue system modulo n and a reduced residue system modulo k if and only if $\varphi(k) = n$ and $(k, n) = 1$.

   ** (b) Prove that there exist integers forming a reduced residue system both modulo n and modulo k if and only if $\varphi(n) = \varphi(k)$.

16.* (a) Verify that for any distinct integers a_1, a_2, and a_3, there exist infinitely many positive numbers n such that $a_1 + n$, $a_2 + n$, and $a_3 + n$ are pairwise coprime.

   (b) Find distinct integers a_1, a_2, a_3, and a_4 such that the numbers $a_i + n$, $i = 1, 2, 3, 4$ are not pairwise coprime for any n.

   * (c) Demonstrate that for any distinct integers a_1, a_2, a_3, and a_4, there exist infinitely many positive numbers n such that $(a_i + n, a_j + n) \leq 2$ for every $i \neq j$.

   * (d) Verify that for any distinct integers a_1, a_2, a_3, and a_4, there exist infinitely many positive numbers n such that $(a_i + n, a_j + n, a_k + n) = 1$ for all $1 \leq i < j < k \leq 4$.

   * (e) Do the statements in (c) and (d) remain valid if we increase the number of integers a_i from four to five or six?

2.7. Wilson's Theorem

Theorem 2.7.1 (Wilson's Theorem). *If p is a (positive) prime, then $(p-1)! \equiv -1 \pmod{p}$.*

♣

Since the numbers $1, 2, \ldots, p-1$ form a reduced residue system modulo p and the product of the elements of every reduced residue system gives the same remainder modulo p, we can rewrite Wilson's Theorem in the following form:

If p is a (positive) prime, then the product of the elements of a reduced residue system is congruent to -1 modulo p.

We discuss generalizations for composite moduli and connections with group theory in Exercise 2.7.1 and in Section 2.8.

Proof. The theorem is clearly true for $p = 2$ and $p = 3$.

We show that for $p \geq 5$, the numbers $2, 3, \ldots, p-2$ can be paired so that the product of the two elements in every pair is congruent to 1 modulo p. This implies the theorem since then $2 \cdot 3 \cdot \cdots \cdot (p-2) \equiv 1 \pmod{p}$, hence

$$(p-1)! = 2 \cdot 3 \cdot \cdots \cdot (p-2) \cdot 1 \cdot (p-1) \equiv 1 \cdot 1 \cdot (p-1) \equiv -1 \pmod{p}.$$

We illustrate the pairing first for $p = 11$. The mate of 2 is obtained from the congruence $2x \equiv 1 \pmod{11}$. Its only solution is $x \equiv 6 \pmod{11}$, so 2 is matched with 6. Here, 2 and 6 correspond to each other mutually as $2 \cdot 6 = 6 \cdot 2 \equiv 1 \pmod{11}$.

Continuing similarly, we obtain the pairs 3–4, 5–9, and 7–8. Thus

$$10! = (2 \cdot 6) \cdot (3 \cdot 4) \cdot (5 \cdot 9) \cdot (7 \cdot 8) \cdot 1 \cdot 10 \equiv 1 \cdot 1 \cdot 1 \cdot 1 \cdot 1 \cdot (-1) = -1 \pmod{11}.$$

Let us see how this works in general. We have to verify the following facts to obtain a perfect match:

(i) To every integer $2 \leq a \leq p-2$, there exists exactly one $b = f(a)$ satisfying

$$ab \equiv 1 \pmod{p} \quad \text{and} \quad 2 \leq b \leq p-2.$$

(ii) If $f(a) = b$, then $f(b) = a$, so a and b are assigned mutually to each other.

(iii) $f(a) \neq a$, so no element is the partner of itself.

(i) Since $(a, p) = 1$, the congruence $ax \equiv 1 \pmod{p}$ is solvable and has exactly one solution b in the complete residue system $0, 1, 2, \ldots, p-1$. If $x = 0, 1$, or $p-1$, then $ax \equiv 0, a$, or $-a \pmod{p}$, thus $ax \not\equiv 1 \pmod{p}$ for these values of x. Hence, b falls in the interval $2 \leq b \leq p-2$, as required.

(ii) The condition $f(a) = b$ means $ab \equiv 1 \pmod{p}$. The value of $f(b)$ is the solution of the congruence $by \equiv 1 \pmod{p}$. Clearly, $y = a$ is a solution and we know from (i) that there is exactly one solution in the interval $2 \leq y \leq p-2$. Hence, necessarily $f(b) = a$.

(iii) The condition $b = a$ would mean $a^2 \equiv 1 \pmod{p}$. Considering the corresponding divisibility and using the prime property of p, we obtain

$$p \mid (a-1)(a+1) \implies p \mid a-1 \text{ or } p \mid a+1 \implies a \equiv \pm 1 \pmod{p}.$$

This, however, contradicts the condition $2 \leq a \leq p-2$. $\qquad\square$

For further proofs of Wilson's Theorem, see the note after Theorem 3.1.2 and Exercise 3.3.6.

Exercises 2.7

(Primes are assumed to be positive.)

1. *Generalizations of Wilson's Theorem for composite moduli.* Let m be composite. What is the remainder modulo m of

   (a) $(m - 1)!$

   * (b) $(\varphi(m))!$

   * (c) the product of all elements of a reduced residue system?

2. Which integers $m > 6$ satisfy $(m - 6)! \equiv 1 \pmod{m}$?

3. Let $a_1, \ldots, a_m$ and $b_1, \ldots, b_m$ be any two permutations of $1, 2, \ldots, m$.

   (a) Show that if $m > 2$ is a prime, then there exist i and j, $i \neq j$ satisfying

   $$m \mid a_i b_i - a_j b_j.$$

   * (b) Prove the same assertion if m is composite.

4. Let p be a prime of the form $4k - 1$. Prove

   $$\left(\frac{p-1}{2}\right)! \equiv \pm 1 \pmod{p}.$$

5. Verify

   $$p^p \mid (p^2 - 1)! - p^{p-1}$$

   for any prime p.

6. Let $p > 3$ be a prime. What is the remainder of $3(p - 3)!$ modulo p?

7. What is the remainder of 99! when divided by 10100?

8. Compute the possible values of $(n! + 3, (n + 2)! + 6)$ if n is a positive integer.

9. For which m does there exist a (a) complete (b) reduced residue system of numbers of the form $k!$?

10. Let $a_1, \ldots, a_{30}$ be a reduced residue system modulo 31. Prove

    $$31 \mid (a_1 a_2 a_3)^3 + (a_4 a_5 \ldots a_{30})^{27}.$$

11. Let $p > 2$ be a prime and construct an arithmetic progression of $p - 1$ integers. What can the remainder of the product of its elements modulo p be?

12. Solve the congruence $x! \, (z - x)! \equiv 1 \pmod{z}$ where $0 < x < z$ are integers.

* 13. For which primes p is $(p - 1)! + 1$ a power of p (with positive integer exponents)?

2.8. Operations with Residue Classes

We define an addition and a multiplication for residue classes modulo m and investigate their properties. We assume throughout that the modulus $m > 1$ is fixed.

Definition 2.8.1. The *sum* and *product* of the residue classes $(a)_m$ and $(b)_m$ are the residue classes $(a + b)_m$ and $(ab)_m$, i.e.

$$(a)_m + (b)_m = (a + b)_m \quad \text{and} \quad (a)_m(b)_m = (ab)_m. \qquad \clubsuit$$

We have to verify that we have defined the operations so that both addition and multiplication assign a *unique* residue class to any two given residue classes.

The difficulty is that addition and multiplication of residue classes were defined using representatives, thus we have to clarify that the resulting residue classes do not depend on which representatives in the initial two classes were chosen.

Consider addition. We have to show that if $(a)_m = (a')_m$ and $(b)_m = (b')_m$, then $(a + b)_m = (a' + b')_m$. This holds since

$$\left. \begin{array}{lll} (a)_m = (a')_m & \Longrightarrow & a \equiv a' \pmod{m} \\ (b)_m = (b')_m & \Longrightarrow & b \equiv b' \pmod{m} \end{array} \right\} \Longrightarrow a + b \equiv a' + b' \pmod{m}$$

$$\Longrightarrow (a + b)_m = (a' + b')_m.$$

We can argue similarly about multiplication.

We must be aware that there are many operations on the integers that cannot be defined for residue classes using representatives. We illustrate this by an example; for some further examples see Exercise 2.8.6.

Let a and b be integers and denote by $\max(a, b)$ the larger one (or their common value if $a = b$). This maximum assigns a unique integer to any two integers, so it is a well defined operation on the integers.

Among the residue classes modulo m, however, the specification $\max((a)_m, (b)_m)$ $= \big(\max(a, b)\big)_m$ does not define an operation, since the right-hand side of the equality (may) give different residue classes if we represent $(a)_m$ and/or $(b)_m$ with another element. For example, let the modulus be $m = 9$ and consider the two residue classes $A = (3)_9 = (12)_9$ and $B = (10)_9 = (1)_9$. Then $\max(A, B)$ would be $\big(\max(3, 10)\big)_9 = (10)_9$ on the one hand and $\big(\max(12, 1)\big)_9 = (12)_9$ on the other hand but $(10)_9 \neq (12)_9$.

We turn now to study the most important properties of addition and multiplication defined on the residue classes.

We can easily derive that most properties valid among the integers hold also for the residue classes:

Theorem 2.8.2. *Among the residue classes modulo m,*

- *addition is associative and commutative*
- $(0)_m$ *is a zero element, i.e.* $(0)_m + (a)_m = (a)_m + (0)_m = (a)_m$ *holds for every* $(a)_m$
- *the negative of* $(a)_m$ *is* $(-a)_m$, *i.e.* $(-a)_m + (a)_m = (a)_m + (-a)_m = (0)_m$
- *multiplication is associative and commutative*

- $(1)_m$ *is an* identity element, *i.e.* $(1)_m(a)_m = (a)_m(1)_m = (a)_m$ *holds for every* $(a)_m$
- *the* distributive law *is valid.* ♣

Proof. Each statement follows immediately from the definition of the operations and from the corresponding property of the integers. We illustrate this for the commutative law for addition:

$$(a)_m + (b)_m = (a + b)_m = (b + a)_m = (b)_m + (a)_m$$

(we applied the definition of addition for residue classes in the first and third equalities and the commutative law for the addition of integers in the second equality). □

Summarizing the properties listed in Theorem 2.8.2, the residue classes modulo m form a *commutative ring with identity element* with respect to addition and multiplication.

We mention that—as in every ring—also *subtraction* can be performed for residue classes, i.e. to any $(a)_m$ and $(b)_m$, there exists exactly one $(c)_m$ satisfying $(a)_m = (b)_m + (c)_m$; we obtain this $(c)_m$ as $(a)_m + (-b)_m$. (We can verify the existence of subtraction also by relying on subtraction among the integers; then we have $(c)_m = (a - b)_m$.)

We examine now which residue classes have a *multiplicative inverse* (or "reciprocal"), i.e. for which $(a)_m$ does there exist a residue class $(c)_m$ satisfying

(2.8.1) $$(c)_m(a)_m = (a)_m(c)_m = (1)_m?$$

Condition (2.8.1) is equivalent to $(ac)_m = (1)_m$, i.e. to $ac \equiv 1 \pmod{m}$ which means that the linear congruence $ax \equiv 1 \pmod{m}$ is solvable. By Theorem 2.5.3, this holds if and only if $(a, m) \mid 1$, or $(a, m) = 1$. This is exactly the case when $(a)_m$ is a reduced residue class. Thus, we have proved:

Theorem 2.8.3. *Among the residue classes modulo m, exactly the reduced residue classes have a multiplicative inverse.* ♣

We note that for any associative operation, every element can have only one inverse. Thus, the inverse of a reduced residue class is unique, as well. (This follows also from Theorem 2.5.5.)

A *field* is a commutative ring (with at least two elements) that has an identity element and every non-zero element has an inverse. By Theorem 2.8.3, the residue classes satisfy these requirements if and only if every non-zero residue class is reduced, i.e. m is a prime. This gives the result:

Theorem 2.8.4. *The residue classes modulo m form a field if and only if m is a prime.* ♣

It can occur that the product of two non-zero residue classes is the zero residue class, e.g. $(5)_{10}(4)_{10} = (0)_{10}$. A residue class $(a)_m \neq (0)_m$ is called a *zero divisor* if

(2.8.2) there exists some $(b)_m \neq (0)_m$ satisfying $(a)_m(b)_m = (0)_m$.

Thus, $(4)_{10}$ and $(5)_{10}$ are zero divisors in the previous example.

Theorem 2.8.5. *A residue class $(a)_m \neq (0)_m$ is a zero divisor if and only if $(a)_m$ is not a reduced residue class, i.e. $(a, m) \neq 1$.* ♣

The condition $(a)_m \neq (0)_m$ means $m \nmid a$ or $(a, m) < m$ for the representative a.

Proof. Rephrasing the definition in (2.8.2), the residue class $(a)_m \neq (0)_m$ is a zero divisor if and only if

(2.8.3)   there exists some $b \not\equiv 0$ (mod m) satisfying $ab \equiv 0$ (mod m).

Since $x \equiv 0$ (mod m) is always a solution of $ax \equiv 0$ (mod m), (2.8.3) means that $ax \equiv 0$ (mod m) has more solutions. The number of solutions is (a, m), hence $(a)_m \neq (0)_m$ is a zero divisor if and only if $(a, m) > 1$.   □

We see from Theorem 2.8.5 that residue classes modulo m contain a zero divisor if and only if m is composite.

Finally, we touch briefly some group theoretic connections of the residue classes.

A set G is called a *group* if an associative operation with an identity element is defined on G and every element has an inverse. If the operation is commutative we have a *commutative* or *Abelian* group.

Thus, the residue classes modulo m form a commutative group under addition, and the same is true for the reduced residue classes with respect to multiplication (this follows from the fact that the product of two reduced classes and the inverse of a reduced class is a reduced class again).

The Euler–Fermat Theorem can be considered as a special case of a general theorem for groups: For any element a of a finite group G, $a^{|G|}$ is the identity element (where $|G|$ denotes the number of elements in the group). This general result can be verified similarly to the Euler–Fermat Theorem for commutative groups (see Exercise 2.8.7) and follows from *Lagrange's Theorem* for arbitrary G.

Generalizing Wilson's theorem, we can ask which element of a finite commutative group will be equal to the product of all its elements (see Exercise 2.8.8).

Exercises 2.8

1. For which m does there exist a non-zero residue class that is the negative of itself?

2. Consider the ring of the residue classes modulo 100.

    (a) What is the multiplicative inverse of the residue class (13)?

    (b) What is the number of zero divisors?

    (c) How many zero divisor pairs belong to (40), i.e. how many residue classes $(b) \neq (0)$ satisfy $(40)(b) = (0)$?

    (d) Does there exist a residue class (c) satisfying $(35)(c) = (90)$?

3. How many residue classes modulo m are their own multiplicative inverses if m is

    (a) 47

    (b) 30

    (c) 800

  * (d) arbitrary?

4. Consider the ring of residue classes modulo a composite m.

   (a) Show that if (a) is a zero divisor, then $(a)(c)$ is a zero divisor or (0) for any (c).

   (b) Demonstrate that if $(a)(c)$ is a zero divisor, then at least one of (a) and (c) is a zero divisor.

   (c) Determine all m where the sum of any two zero divisors is a zero divisor or (0).

   (d) Compute the sum and product of all zero divisors.

   (e) For which m does there exist an $(a) \neq (0)$ satisfying $(a)^2 = (0)$?

5. (a) Let H be the set of those residue classes modulo 20 that are "divisible" by 4, i.e.
   $$H = \{(0)_{20}, (4)_{20}, (8)_{20}, (12)_{20}, (16)_{20}\}.$$
   Prove that H is a field under the addition and multiplication of residue classes.

   (b) Let K be the set of those residue classes modulo 40 that are divisible by 4, i.e.
   $$K = \{(0)_{40}, (4)_{40}, \ldots, (36)_{40}\}.$$
   Verify that K is a commutative ring under the addition and multiplication of residue classes, but it is not a field, it has no identity element, and every non-zero element is a zero divisor.

   S* (c) Generalize the problem (as far as possible).

6. Examine in detail whether it is possible to define the following operations for residue classes modulo m using their positive representatives.

   (a) Gcd: $\gcd\big((a)_m, (b)_m\big) = \big(\gcd(a, b)\big)_m$

   (b) Third power: $(a)_m^3 = (a^3)_m$

   (c) Cube root: $\sqrt[3]{(a)_m} = (\sqrt[3]{a})_m$

   (d) Arithmetic mean: $\big((a)_m + (b)_m\big)/2 = \big((a+b)/2\big)_m$

   (e) Exponentiation: $(a)_m^{(b)_m} = (a^b)_m$.

7. *Generalization of the Euler–Fermat Theorem.* In a finite commutative group G, let $|G|$ denote the number of elements and e be the identity element. Prove that $a^{|G|} = e$ holds for any $a \in G$.

* 8. *Generalization of Wilson's Theorem.* In a finite commutative group G, let e be the identity element and P the product of all elements. Show that if G contains exactly one element $c \neq e$ satisfying $c^2 = e$, then $P = c$, and $P = e$ in all other cases.

Chapter 3

Congruences of Higher Degree

We start with a few general remarks concerning congruences modulo a prime. Next, we discuss the most important properties of order, primitive roots, and discrete logarithms. Applying these, we "take roots" modulo p, i.e. examine binomial congruences. We will include an interesting theorem by Kőnig and Rados and another one by Chevalley. Finally, we show how congruences with composite moduli can be reduced to those with prime moduli.

3.1. Number of Solutions and Reduction

Let m be a fixed positive integer and f a polynomial with integer coefficients. We want to find the solutions of the congruence $f(x) \equiv 0 \pmod{m}$.

As with linear congruences, by a *solution* we mean an integer s which substituted for x makes the congruence valid. It is clear also in this general case that if an integer s is a solution, then every element of the residue class $(s)_m$ is a solution as well since $s \equiv r$ $(\bmod\ m)$ implies $f(s) \equiv f(r) \pmod{m}$. Therefore the *number of solutions* is defined as the number of the *pairwise incongruent* solutions, i.e. how many *residue classes* yield the solutions (see Definition 2.5.2).

Obviously, also it is only the residue class of the coefficients of f that matters.

By the above, it is often more convenient and more natural to handle both the coefficients and the solutions as residue classes modulo m (instead of integers). This means that f is considered as a polynomial over the ring $\mathbf{Z}_m$ of these residue classes and the solutions of the congruence $f(x) \equiv 0 \pmod{m}$ are the roots of f in $\mathbf{Z}_m$. We adopt this view also when defining the degree of a polynomial modulo m:

Definition 3.1.1. The *degree* of a polynomial $f = a_0 + a_1 x + \cdots + a_n x^n$ *modulo m* is k if $a_k \not\equiv 0 \pmod{m}$, but $a_i \equiv 0 \pmod{m}$ for every $i > k$. If $a_i \equiv 0 \pmod{m}$ for every i, so every coefficient of f is $0 \pmod{m}$, then f has no degree modulo m. ♣

Example. The polynomial $f = 6 + 12x + 15x^2 + 21x^3$ has degree 3 modulo 5, 2 modulo 7, and has no degree modulo 3.

The rest of this section deals with congruences with prime moduli.

Theorem 3.1.2. *If p is a prime and the degree of f modulo p is k, then the congruence $f(x) \equiv 0 \pmod{p}$ has at most k solutions.* ♣

Proof. According to the preliminary remarks, we consider f as a polynomial over the ring $\mathbf{Z}_p$ of the residue classes modulo p. Then the number of solutions is the number of roots of f in $\mathbf{Z}_p$.

Since $\mathbf{Z}_p$ is a field by Theorem 2.8.4, Theorem 3.1.2 follows immediately from a well-known basic result in classical algebra: If the degree of a polynomial over a field F is k, then f can have at most k roots in F. □

The statement of Theorem 3.1.2 is false for composite moduli. For example, the linear congruence

$$10x - 15 \equiv 0 \pmod{25}$$

has 5 solutions, the congruence

$$x(x-1)(x-2)(x-3) \equiv 0 \pmod{24}$$

of degree 4 has 24 solutions, etc.

Using Theorem 3.1.2, we can get a new proof for Wilson's Theorem (Theorem 2.7.1): If p is a prime, then $(p-1)! \equiv -1 \pmod{p}$.

This is obvious for $p = 2$. Let $p > 2$ and consider the polynomial

$$f = x^{p-1} - 1 - (x-1)(x-2)\dots(x-(p-1)) = a_0 + a_1 x + \cdots + a_{p-2}x^{p-2}.$$

By Fermat's Little Theorem, each of the (pairwise incongruent) numbers $x = 1, 2, \dots,$ $p - 1$ satisfies the congruence $f(x) \equiv 0 \pmod{p}$, hence the number of solutions is at least $p - 1$. If f had a degree modulo p, then this degree could be at most $p - 2$ contradicting Theorem 3.1.2. Therefore f has no degree modulo p, i.e. every coefficient a_i is $0 \pmod{p}$. Hence,

$$a_0 = -1 - (-1)^{p-1}(p-1)! = -1 - (p-1)! \equiv 0 \pmod{p},$$

thus proving Wilson's Theorem. □

Since a congruence modulo m can have at most m solutions, the statement of Theorem 3.1.2 becomes empty if the degree of f modulo p is p or larger. In this case, we can reduce the congruence $f(x) \equiv 0 \pmod{p}$ to a congruence of degree at most $p - 1$, in the following sense:

Theorem 3.1.3. *To every prime p and polynomial f with integer coefficients, there exists a polynomial g with integer coefficients such that*

(i) *the degree of g modulo p is at most $p - 1$ or every coefficient of g is $0 \pmod{p}$*

(ii) *$f(c) \equiv g(c) \pmod{p}$ for every integer c.* ♣

In other words, Theorem 3.1.3 asserts that to every polynomial over the field $\mathbf{Z}_p$, we can find a polynomial g of degree at most $p-1$ (allowing also the zero polynomial) such that the two polynomials define the same *function*.

The theorem clearly implies that the congruences $f(x) \equiv 0$ (mod p) and $g(x) \equiv 0$ (mod p) have exactly the same solutions, hence the number of solutions is at most the degree of g modulo p by Theorem 3.1.2.

First proof. Replace x^p by x everywhere in f as long as this is possible. We arrive at a polynomial g of degree at most $p-1$ modulo p or with all coefficients 0 (mod p). By Fermat's Little Theorem, $c^p \equiv c$ (mod p) for every c, hence also $f(c) \equiv g(c)$ (mod p) holds. $\square$

Second proof. Divide f by $x^p - x$. Since the leading coefficient of $x^p - x$ is 1, the quotient and the remainder will have integer coefficients. We show that the remainder serves as g. Indeed,

$$f = (x^p - x)h + g,$$

where the degree of g is at most $p-1$ or g is the zero polynomial. Then

$$f(c) = (c^p - c)h(c) + g(c) \equiv 0 + g(c) = g(c) \pmod{p}$$

for every integer c. $\square$

Remarks: (1) In the second proof, we can divide the polynomials over the field $\mathbf{Z}_p$, but this is slightly more complicated.

(2) Both proofs yield also an algorithm to find g (in fact, they are two interpretations of the same procedure).

(3) A third proof is obtained using the *interpolation polynomials*, but this is not really suitable for getting g in an explicit form (see Exercise 3.1.9).

(4) Note that the polynomial g meeting the requirements of Theorem 3.1.2 is unique (modulo p, see Exercise 3.1.8).

Exercises 3.1

1. What is the number of solutions of the following congruences?

   (a) $x^{100} + x \equiv 0$ (mod 101)

   (b) $x^{100} + x \equiv 0$ (mod 100)

   (c) $21x^9 + 18x^6 + 15 \equiv 0$ (mod 77)

   (d) $x(x^2 - 1)(x^2 - 4) \equiv 0$ (mod 60).

2. Prove that c is a solution of the congruence $f(x) \equiv 0$ (mod m) if and only if there exists a polynomial h with integer coefficients such that every coefficient of the polynomial $f - (x - c)h$ is a multiple of m.

3. Let $N(f, m)$ denote the number of solutions of the congruence $f(x) \equiv 0$ (mod m). True or false?

    (a) $N(fg, m) \leq N(f, m) + N(g, m)$.

    (b) $N(fg, m) \leq N(f, m) + N(g, m) + 1000$.

    (c) $N(fg, 13) \leq N(f, 13) + N(g, 13)$.

    (d) $N(fg, 13) = N(f, 13) + N(g, 13)$.

4. (a) Exhibit a polynomial f of degree 13 modulo 37 such that the congruence $f(x) \equiv 0 \pmod{37}$ has 12 solutions.

    (b) How many f satisfy the conditions in (a) if every coefficient is taken from the numbers $1, 2, \ldots, 37$?

5. Let p be a prime and denote the number of solutions of $f(x) \equiv 0 \pmod{p}$ by r. Prove

$$r \equiv - \sum_{i=1}^{p} f(i)^{p-1} \pmod{p}.$$

6. Let $p > 2$ be a prime and $1 \leq j \leq p - 2$. Show that the sum of all products with j distinct factors taken from the numbers $1, 2, \ldots, p - 1$ is divisible by p.

7. Let $p > 2$ be a prime and

$$f = a_0 + a_1 x + \cdots + a_n x^n \qquad \text{where} \qquad a_0 \not\equiv 0 \pmod{p}.$$

Prove that $f(x) \equiv 0 \pmod{p}$ can be reduced to a congruence of degree at most $p - 2$ in the following sense: We can find a polynomial h of degree at most $p - 2$ modulo p or with all coefficients $0 \pmod{p}$ satisfying $f(c) \equiv h(c) \pmod{p}$ for every $(c, p) = 1$.

8. Prove the existence of a polynomial g occurring in Theorem 3.1.3 using the interpolation polynomials by Lagrange or Newton.

9. Prove that the polynomial g satisfying the requirements of Theorem 3.1.3 is unique over $\mathbf{Z}_p$, i.e. its coefficients are uniquely determined modulo p.

10. Demonstrate that Theorem 3.1.3 remains valid also for composite moduli.

3.2. Order

It follows from the Euler–Fermat Theorem that if $(a, m) = 1$, then $a^t \equiv 1 \pmod{m}$ for some positive integer t and $\varphi(m)$ or any multiple of it can be taken as t. The *minimal* positive integer t with this property plays a distinguished role in the further investigations:

Definition 3.2.1. Let $(a, m) = 1$. The positive integer k is called the *order* of a modulo m, if $a^k \equiv 1 \pmod{m}$ and $a^i \not\equiv 1 \pmod{m}$ for every $0 < i < k$.   ♣

We denote the order of a by $o_m(a)$. For example, $o_7(2) = 3$, $o_{10}(3) = 4$, etc. If there is no ambiguity, we can omit the index referring to the modulus.

The Euler–Fermat Theorem implies that every a coprime to m has an order and $o_m(a) \leq \varphi(m)$.

The order can be defined only for $(a, m) = 1$: If $(a, m) \neq 1$, then there is no $k > 0$ satisfying $a^k \equiv 1 \pmod{m}$ (see the first remark after Theorem 2.4.1B).

It is clear from the definition of the order that

$$a \equiv b \pmod{m} \Longrightarrow o_m(a) = o_m(b),$$

thus all elements of a reduced residue class have the same order.

The next theorem summarizes the most important properties of the order.

Theorem 3.2.2. *Let t, u, and v be non-negative integers and* $(a, m) = 1$.

(i) $a^t \equiv 1 \pmod{m} \iff o_m(a) \mid t$.

(ii) $a^u \equiv a^v \pmod{m} \iff u \equiv v \pmod{}$).

(iii) *a has* $o_m(a)$ *pairwise incongruent powers with positive integer exponents modulo m.*

(iv) $o_m(a) \mid \varphi(m)$. ♣

Proof. (i) If $t = q o_m(a)$, then

$$a^t = \left(a^{o_m(a)}\right)^q \equiv 1^q = 1 \pmod{m}, .$$

For the converse, we apply the division algorithm for t by $o_m(a)$: $t = q o_m(a) + r$ where $0 \le r < o_m(a)$. Then

$$1 \equiv a^t = \left(a^{o_m(a)}\right)^q \cdot a^r \equiv 1 \cdot a^r = a^r \pmod{m}.$$

Since $r < o_m(a)$, only $r = 0$ is possible by the definition of the order, so $o_m(a) \mid t$.

(ii) Let $u \ge v$. Then, using $(a, m) = 1$ and (i), we obtain

$$a^u \equiv a^v \pmod{m} \iff a^{u-v} \equiv 1 \pmod{m}$$
$$\iff o_m(a) \mid u - v$$
$$\iff u \equiv v \pmod{o_m(a)}.$$

(iii) This is a direct consequence of (ii).

(iv) By the Euler–Fermat Theorem, $a^{\varphi(m)} \equiv 1 \pmod{m}$, hence (i) implies $o_m(a) \mid \varphi(m)$. □

Example. Compute the order of 13 modulo 59.

As 13 and 59 are coprime, $o_{59}(13)$ makes sense. Since $o_{59}(13) \mid \varphi(59)$ and $\varphi(59) = 58$,

$$o_{59}(13) = 1, 2, 29, \text{ or } 58.$$

Clearly, $13 \not\equiv 1 \pmod{59}$ and $13^2 \not\equiv 1 \pmod{59}$, so $o_{59}(13)$ can be only 29 or 58. This means that if $13^{29} \equiv 1 \pmod{59}$, then the order is 29, whereas if $13^{29} \not\equiv 1 \pmod{59}$, then the order is 58.

The remainder of 13^{29} modulo 59 can be determined using repeated squarings:

$$
\begin{aligned}
13^2 &= 169 \equiv -8 \pmod{59} \\
13^4 &\equiv (-8)^2 \equiv 5 \pmod{59} \\
13^8 &\equiv 5^2 = 25 \pmod{59} \\
13^{16} &\equiv 25^2 \equiv -24 \pmod{59}
\end{aligned}
$$

thus

$$13^{29} = 13^{16} \cdot 13^8 \cdot 13^4 \cdot 13$$
$$\equiv (-24) \cdot 25 \cdot 5 \cdot 13$$
$$= (-600) \cdot 65$$
$$\equiv (-10) \cdot 6$$
$$\equiv -1 \pmod{59}.$$

Hence $o_{59}(13) = 58$. (There is no need to compute the remainder of 13^{58} modulo 59; we know from the Euler–Fermat Theorem that it must be 1.)

Finally, we mention that Definition 3.2.1 is a special case of the order of an element in a group, and also the analog of Theorem 3.2.2 is true in arbitrary groups.

Exercises 3.2

(The notation $o(a)$ without an index refers to an arbitrary modulus unless a modulus m or p was specified in the exercise.)

1. Compute:

    (a) $o_{77}(155)$

    (b) $o_{100}(199)$

    (c) $o_{65}(2)$

    (d) $o_{47}(43)$.

2. Does there exist an a with $o_m(a) = 4$ if m is (a) 11 (b) 12 (c) 17?

3. Which moduli m satisfy $o_m(2) = 6$?

4. Let $(a, m) = 1$, $o_m(a) = k$, and $i \geq 0$. Prove

    (a) $o_m(a^i) \mid k$

    (b) $o_m(a^i) = k \iff (i, k) = 1$

    (c) $o_m(a^i) = k/(i, k)$.

5. What are the possible values of $o(a)$ if $o(a^3)$ is (a) 10 (b) 12?

S 6. Let $p > 2$ be a prime. Verify $o_p(a) = o_p(-a) \iff 4 \mid o_p(a)$.

7. Assume that a^5, a^{13}, and a^{21} belong to exactly two reduced residue classes modulo m. Compute $o_m(a)$.

8. Let p be a prime and $o_p(a) = 3$.

    (a) Show that $1 + a + a^2 \equiv 0 \pmod{p}$.

    (b) Determine $o_p(1 + a)$.

S 9. Assume that $p > 5$ is a prime and $a^{2p-10} \equiv -1 \pmod{p}$. Compute $o_p(a)$.

10. (a) Show that both congruences $a^n \equiv 1 \pmod{m}$ and $a^k \equiv 1 \pmod{m}$ hold simultaneously if and only if $a^{(n,k)} \equiv 1 \pmod{m}$.

    (b) Using (a), find a new proof for Exercise 1.3.13.

11. Show that $(a^n - 1, a^k + 1) \le 2$ if n is odd.

12. Let $p > 2$ be a prime and $(a, p) = 1$. Verify that $a^s \equiv -1 \pmod{p}$ is true for some s if and only if $o_p(a)$ is even. What happens if we replace p by a composite modulus m?

13. Prove

    (a) $(a, m) = 1, d \mid m \Longrightarrow o_d(a) \mid o_m(a)$

    (b) $(a, mn) = 1 \Longrightarrow o_{[m,n]}(a) = [o_m(a), o_n(a)]$.

14. How many of the integers $1, 2, \ldots, 999$ have order 2 modulo 1000?

* 15. Let $o(a) = u$ and $o(b) = v$. Verify

    (a) $o(ab) = uv \iff (u, v) = 1$

    (b) $\frac{[u,v]}{(u,v)} \mid o(ab)$ and $o(ab) \mid [u, v]$.

* 16. Assume $a^{o(b)} \equiv b^{o(a)} \pmod{m}$. Prove $o(a) = o(b)$.

17. Show that $n \mid \varphi(a^n - 1)$ holds for every $a > 1$ and $n > 0$.

* 18. Let $a_1, \ldots, a_{\varphi(m)}$ be a reduced residue system modulo m. Show that $\sum_{i=1}^{\varphi(m)} o_m(a_i)$ is always an odd number.

19. Let p be a prime and $(a, p) = 1$. What is the remainder of the sum and product below mod p?

    (a) $a + a^2 + \cdots + a^{o(a)}$

    (b) $a \cdot a^2 \cdot \cdots \cdot a^{o(a)}$.

20. *Decimal fractions.* We deal only with the digits following the decimal point, hence it is sufficient to consider numbers α with $0 < \alpha < 1$. We exclude those decimal fractions that end with infinitely many 9s. A decimal fraction is *finite* if it has only finitely many digits. We write these in their shortest form, so the last digit is not zero. An infinite decimal fraction is *periodic* if the sequence of the digits is eventually periodic. The periodicity is *pure* or *mixed* depending on whether or not the first period starts immediately after the decimal point. For fractions a/b we assume $b > 0$ and $(a, b) = 1$. Prove the following characterizations.

    (a) The decimal fraction of a real number α is finite or periodic if and only if α is rational.

    (b) The decimal fraction of the rational number a/b is finite if and only if the standard form of b contains no other primes than 2 and 5: $b = 2^r 5^s$. In this case, the number of digits after the decimal point is $k = \max(r, s)$, i.e. $b \mid 10^k$ but $b \nmid 10^{k-1}$.

    (c) The decimal fraction of the rational number a/b has a pure periodicity if and only if $(b, 10) = 1$. In this case, the length of the (smallest) period is $o_b(10)$.

(d) The decimal fraction of the rational number a/b has a mixed periodicity if and only if $(b, 10) > 1$ but b has also a prime divisor different from 2 and 5: $b = 2^r 5^s t$ where $(t, 10) = 1$, $t > 1$, and $k = \max(r, s) > 0$. Then the (first) period starts with the $(k + 1)$st digit after the decimal point and its length is $o_t(10)$.

3.3. Primitive Roots

We obtained from the Euler–Fermat Theorem that $o_m(a) \leq \varphi(m)$ for every $(a, m) = 1$. A special case is when this holds with equality.

Definition 3.3.1. A number g is a *primitive root* modulo m if $o_m(g) = \varphi(m)$. ♣

It is clear from the definition that a primitive root must be coprime to the modulus m, and in a reduced residue class either all elements are primitive roots, or there are no such elements at all.

Examples. **E1** 3 is a primitive root modulo 10 as $o_{10}(3) = \varphi(10) = 4$.

E2 2 is not a primitive root modulo 31 since $o_{31}(2) = 5 < \varphi(31) = 30$.

E3 There are no primitive roots modulo 12. It is enough to check the orders in the reduced residue system $\{\pm 1, \pm 5\}$: 1 has order 1 and all other elements have order 2, so every order is less than $\varphi(12) = 4$.

To determine whether or not a number a (coprime to m) is a primitive root modulo m, there is no need to check $a^{\varphi(m)} \equiv 1 \pmod{m}$ since this follows from the Euler–Fermat Theorem. Using $o_m(a) \mid \varphi(m)$, we have to test whether $a^d \equiv 1 \pmod{m}$ holds for some divisor $d < \varphi(m)$ of $\varphi(m)$; a is a primitive root if and only if there is no such d. In fact, it is enough to consider the maximal proper divisors of $\varphi(m)$, those of the form $\varphi(m)/q$ where q is a prime.

The applications of primitive roots are mostly based on the following property:

Theorem 3.3.2. *A number g is a primitive root modulo m if and only if* 1, g, g^2, ..., $g^{\varphi(m)-1}$ *form a reduced residue system modulo m.* ♣

Proof. Assume that g is a primitive root, so $o_m(g) = \varphi(m)$. Then the $\varphi(m)$ numbers $1, g, g^2, \ldots, g^{\varphi(m)-1}$ are pairwise incongruent modulo m by part (ii) of Theorem 3.2.2, and $(g, m) = 1$ implies that they are all coprime to m. Thus they constitute a reduced residue system mod m by Theorem 2.2.9.

For the converse, assume that the powers of g form a reduced residue system mod m. Then $(g, m) = 1$ implies that $o_m(g)$ exists and $o_m(g) \leq \varphi(m)$ by the Euler–Fermat Theorem. Further, the pairwise incongruence guarantees that none of g, g^2, ..., $g^{\varphi(m)-1}$ can be congruent to the element 1 in this reduced residue system. Hence, $o_m(g) = \varphi(m)$. □

Now we examine, for which moduli m a primitive root exists. Or, in a group theoretic formulation, for which m is the multiplicative group of the reduced residue classes cyclic.

We prove first that all prime moduli have a primitive root.

Theorem 3.3.3. *If p is a prime, then there exists a primitive root modulo p.* ♣

As a generalization of Theorem 3.3.3, it can be shown that any finite field contains an element whose powers supply all non-zero elements in the field. Theorem 3.3.3 is the special case for the field $\mathbf{Z}_p$ of the residue classes mod p.

We give two proofs for Theorem 3.3.3 and a third argument is sketched in Exercise 3.3.14. All three proofs can be adapted with suitable modifications to verify the more general proposition mentioned above.

First proof. If $p = 2$, then $g = 1$ (or any odd number) is a primitive root.

For $p > 2$, let $q_1, \ldots, q_s$ be the distinct prime divisors of $p - 1$.

To prove by contradiction, assume that there is no primitive root, so $o_p(i) = d_i < p - 1$ for every $1 \le i \le p - 1$. Since $d_i \mid p - 1$, therefore $d_i \mid (p - 1)/q$ for some prime divisor q of $p - 1$. This implies $d_i^{(p-1)/q} \equiv 1$ (mod p). Hence, every element of a reduced residue system is a root of the congruence $f(x) \equiv 0$ (mod p) where

$$(3.3.1) \qquad f = \left(x^{(p-1)/q_1} - 1\right)\left(x^{(p-1)/q_2} - 1\right)\ldots\left(x^{(p-1)/q_s} - 1\right).$$

Further, $f(0) = (-1)^s \not\equiv 0$ (mod p), thus the congruence has exactly $p - 1$ solutions.

Performing the multiplications in (3.3.1), we obtain f as a sum of monomials $\pm x^k$ where the exponent

$$(3.3.2) \qquad k \text{ is the sum of (one or more) terms } (p - 1)/q_j \text{ with distinct } q_j$$

or $k = 0$. Apply now the reduction in the first proof of Theorem 3.1.3: replace x^p by x as long as possible. The resulting polynomial g has degree at most $p - 1$ and $f(c) \equiv g(c)$ (mod p) for every c. This means that the congruence $g(x) \equiv 0$ (mod p) has exactly $p - 1$ solutions and therefore the degree of g modulo p must be equal to $p - 1$ by Theorem 3.1.2.

Then g must contain a term x^{p-1}. During the reduction, this was obtained from terms x^s in f where

$$(3.3.3) \qquad \text{the exponent } s \text{ is of the form } s = (p - 1)t \text{ with } t > 0,$$

since the exponents (greater than $p - 1$) were always decreased by $p - 1$ in each step.

Combining (3.3.2) and (3.3.3), we have (say)

$$(3.3.4) \qquad t = \frac{1}{q_1} + \frac{1}{q_2} + \cdots + \frac{1}{q_r}.$$

Multiplying (3.3.4) by $q_2 \ldots q_r$ makes all terms integers except the first term on the right-hand side. This is a contradiction. □

Second proof. Let $h(d)$ be the number of those integers among $1, 2, \ldots, p - 1$ that have order d modulo p. Clearly, $h(d) = 0$ if $d \nmid p - 1$ and

$$(3.3.5) \qquad \sum_{d \mid p-1} h(d) = p - 1.$$

We show that

$$(3.3.6) \qquad h(d) \le \varphi(d)$$

holds for every d.

If there is no element of order d, then (3.3.6) is true trivially since $0 = h(d) < \varphi(d)$.

Thus, we can assume that $o_p(a) = d$ for some a. Then $a, a^2, \ldots, a^d$ are pairwise incongruent modulo p, and by $(a^t)^d = (a^d)^t \equiv 1 \pmod{p}$, they are all solutions of the congruence $x^d \equiv 1 \pmod{p}$.

Since this congruence cannot have more than d solutions, every c satisfying $c^d \equiv 1 \pmod{p}$ must be congruent to one of the numbers $a, a^2, \ldots, a^d$.

Every integer of order d is a solution of $x^d \equiv 1 \pmod{p}$, hence it must be congruent to one of $a, a^2, \ldots, a^d$. By Exercise 3.2.4b, $o_p(a^j) = o_p(a) = d$ holds if and only if $(j, d) = 1$. Therefore exactly $\varphi(d)$ numbers will have order d among $a, a^2, \ldots, a^d$, i.e. $h(d) = \varphi(d)$. Thus we have verified (3.3.6).

Using (3.3.5), (3.3.6), and the equality $\sum_{d|p-1} \varphi(d) = p - 1$ from Exercise 2.3.14, we obtain

$$p - 1 = \sum_{d|p-1} h(d) \leq \sum_{d|p-1} \varphi(d) = p - 1.$$

This can hold only if $h(d) = \varphi(d)$ for every $d \mid p - 1$.

This proves that in a reduced residue system mod p, exactly $\varphi(d)$ elements have order d. For $d = p - 1$ this means that the number of primitive roots is $\varphi(p - 1)$ (implying the existence of primitive roots). $\qquad \square$

Remark: The second proof yielded a (seemingly) stronger result: Besides guaranteeing a primitive root, we obtained also the number of (pairwise incongruent) primitive roots, and even more generally, the number of elements of order d for every given d. This surplus, however, easily follows merely from the existence of a single primitive root (whichever proof produced it) using Theorem 3.3.2 and Exercises 3.2.4b and 3.2.4c (see Exercise 3.3.9).

We formulate these important results as a theorem:

Theorem 3.3.4. *Let the modulus be a prime p.*

  (i) *The ith power of a primitive root is a primitive root if and only if $(i, p - 1) = 1$.*

 (ii) *The number of pairwise incongruent primitive roots is $\varphi(p - 1)$.*

(iii) *In general, the number of elements of order d in a reduced residue system mod p is $\varphi(d)$ if $d \mid p - 1$.* ♣

In the next theorem we characterize the moduli that have primitive roots.

Theorem 3.3.5. *There exists a primitive root modulo $m > 1$ if and only if $m = p^\alpha, 2p^\alpha, 2$, or 4 where $p > 2$ is a prime and $\alpha > 0$.* ♣

Proof. The cases $m = p$ and $m = 2$ were verified in Theorem 3.3.3. Also, $g = 3$ is a primitive root for $m = 4$. For the other moduli, we perform the proof in the following steps:

(Y1) Yes, there exists a primitive root modulo p^2.

(Y2) Yes, there exists a primitive root modulo p^α for every $\alpha > 2$.

(Y3) Yes, there exists a primitive root modulo $2p^\alpha$ for every $\alpha > 0$.

(N1) No, there is no primitive root modulo m if m has an odd prime divisor and is divisible by 4, or it has at least two distinct odd prime divisors.

(N2) No, there is no primitive root modulo 2^α with $\alpha > 2$.

(Y1) Let g be a primitive root modulo p. We show that at least one of g and $g + p$ is a primitive root modulo p^2.

We know that

$$o_{p^2}(g) \mid \varphi(p^2)$$

and

$$o_p(g) \mid o_{p^2}(g)$$

by Exercise 3.2.13a. Substituting $\varphi(p^2) = p(p-1)$ and $o_p(g) = p - 1$, we get

$$p - 1 \mid o_{p^2}(g) \quad \text{and} \quad o_{p^2}(g) \mid p(p-1).$$

Hence, $o_{p^2}(g) = p - 1$ or $o_{p^2}(g) = p(p-1)$.

In the second case, g is a primitive root modulo p^2 (by definition).

We show that if $o_{p^2}(g) = p - 1$, then $g + p$ is a primitive root mod p^2.

Repeating the previous argument with $g + p$ instead of g, we find that $o_{p^2}(g + p)$ can equal only $p - 1$ or $p(p-1)$. Thus, it is sufficient to verify $(g+p)^{p-1} \not\equiv 1 \pmod{p^2}$. By the binomial theorem,

$$(g + p)^{p-1} \equiv g^{p-1} + (p-1)pg^{p-2} + \binom{p-1}{2}p^2 g^{p-3} + \dots.$$

The first term on the right-hand side is $1 \pmod{p^2}$ by our assumption and every other term is divisible by p^2 except the second one. Hence,

$$(g + p)^{p-1} \equiv g^{p-1} + (p-1)pg^{p-2} \equiv 1 - pg^{p-2} \not\equiv 1 \pmod{p^2}.$$

(Y2) We show that if g is a primitive root modulo p^2, then it is a primitive root modulo p^α for any $\alpha > 2$. As in (Y1), it is enough to check

$$g^{p^{\alpha-2}(p-1)} \not\equiv 1 \pmod{p^\alpha}.$$

We shall verify this in the form

(3.3.7) $$g^{p^{\alpha-2}(p-1)} = 1 + t_\alpha p^{\alpha-1} \quad \text{where} \quad p \nmid t_\alpha.$$

We prove (3.3.7) by induction on α.

For $\alpha = 2$, we have $g^{p-1} = 1 + t_2 p$ by Fermat's Little Theorem and here $p \nmid t_2$ as g is a primitive root modulo p^2.

Assume now that (3.3.7) holds for some $\alpha(\geq 2)$. We show that it holds also for $\alpha + 1$ (instead of α). We raise (3.3.7) to the pth power:

(3.3.8) $$g^{p^{\alpha-1}(p-1)} = (1 + t_\alpha p^{\alpha-1})^p = 1 + \binom{p}{1} t_\alpha p^{\alpha-1} + \binom{p}{2}(t_\alpha p^{\alpha-1})^2 + \dots.$$

Here the exponent of p in the third term is $1 + 2(\alpha - 1) \geq \alpha + 1$ and later terms have exponents at least that big. Therefore

$$g^{p^{\alpha-1}(p-1)} = 1 + t_\alpha p^\alpha + s p^{\alpha+1} = 1 + t_{\alpha+1} p^\alpha \quad \text{where} \quad p \nmid t_{\alpha+1}.$$

This means that (3.3.7) holds also for $\alpha + 1$.

(Y3) Let g be a primitive root modulo p^α. One of g and $g + p^\alpha$ is odd; denoting it by h, we show that h is a primitive root modulo $2p^\alpha$.

Since $h^i \equiv 1 \pmod 2$ for every i,

$$h^r \equiv 1 \pmod{p^\alpha} \iff h^r \equiv 1 \pmod{2p^\alpha}.$$

This means that

$$o_{2p^\alpha}(h) = o_{p^\alpha}(h) = \varphi(p^\alpha) = \varphi(2p^\alpha).$$

(N1) We show that if $(a, m) = 1$, then $a^r \equiv 1 \pmod m$ for some $0 < r < \varphi(m)$, hence a cannot be a primitive root.

The moduli can be written as $m = uv$ with suitable $u > 2$ and $v > 2$ where $(u, v) = 1$. We claim that the exponent $r = [\varphi(u), \varphi(v)]$ works.

Both $\varphi(u)$ and $\varphi(v)$ are even since $u > 2$ and $v > 2$ (see Exercise 2.3.1), thus $(\varphi(u), \varphi(v)) \geq 2$. Hence,

$$r = [\varphi(u), \varphi(v)] \leq \frac{\varphi(u)\varphi(v)}{2} = \frac{\varphi(m)}{2}.$$

Further, $\varphi(u) \mid r$ implies $a^r \equiv 1 \pmod u$ and the same holds also mod v. Therefore, $a^r \equiv 1 \pmod m$ is true as well.

(N2) We show by induction on α that

(3.3.9) $a^{2^{\alpha-2}} \equiv 1 \pmod{2^\alpha}$ so $o_{2^\alpha}(a) \leq 2^{\alpha-2} < \varphi(2^\alpha)$

if $\alpha \geq 3$ and a is odd. For $\alpha = 3$,

$$2^3 = 8 \mid a^2 - 1 = (a - 1)(a + 1).$$

Assume now that (3.3.9) holds for some α; we prove that it is also true for $\alpha + 1$. Considering

$$a^{2^{\alpha-1}} - 1 = \left(a^{2^{\alpha-2}} - 1\right)\left(a^{2^{\alpha-2}} + 1\right),$$

the first factor is divisible by 2^α by the induction hypothesis and the second factor is divisible by 2. Thus, the product is divisible by $2^{\alpha+1}$. $\square$

Exercises 3.3

1. Determine all primitive roots modulo (a) 7 (b) 10 (c) 18.

2. Exhibit a number that is a primitive root both modulo 11 and 14.

3. Exhibit

   (a) a primitive root mod 625

   (b) a primitive root mod 5 which is not a primitive root mod 625.

4. True or false?

   (a) If g is a primitive root mod 11, then g is a primitive root mod 22.

   (b) If g is a primitive root mod 22, then g is a primitive root mod 11.

   (c) If g is a primitive root mod m, then g^3 is a primitive root mod m.

(d) If g^3 is a primitive root mod m, then g is a primitive root mod m.

(e) If g is a primitive root mod m, then $g^{2\varphi(m)-1}$ is a primitive root mod m.

(f) If $(a, 34) = 1$ and $a^8 \not\equiv 1 \pmod{34}$, then a is a primitive root mod 34.

(g) If $(a, 25) = 1$ and $a^{10} \not\equiv 1 \pmod{25}$, then a is a primitive root mod 25.

5. Let the modulus be an arbitrary but fixed prime $p > 2$.

   (a) Show that the product of two primitive roots is never a primitive root.

   (b) Prove that there exist three primitive roots whose product is a primitive root.

   (c) For which primes p is the product of any three primitive roots a primitive root again?

6. Give a new proof for Wilson's Theorem using primitive roots.

7. Let $p > 2$ be a prime. What is the remainder of $1^k + 2^k + \cdots + (p-1)^k$ mod p?

8. Let $p > 2$ be a prime. What is the remainder of the product of all (pairwise incongruent) primitive roots mod p? (For the sum of the primitive roots, see Exercise 6.5.9c.)

9. (a) Let p be a prime, $d \mid p - 1$, g a primitive root mod p, and $(a, p) = 1$. Prove

$$o_p(a) = d \iff a \equiv g^j \pmod{p}, \quad \text{where } j = \frac{t(p-1)}{d} \text{ and } (t, d) = 1.$$

   (b) Using (a), determine the number of elements of order d in a reduced residue system modulo p.

S* 10. Let $p > 2$ be a prime and $(a, p) = (b, p) = 1$. Prove that $o_p(a) = o_p(b)$ holds if and only if $a \equiv b^r \pmod{p}$ and $b \equiv a^s \pmod{p}$ for some positive integers r and s.

11. How can Theorem 3.3.4 be generalized for composite moduli possessing a primitive root?

12. Let $m = 2^\alpha$ where $\alpha \geq 3$. Verify:

   (a) $o_m(5) = 2^{\alpha-2}$.

   (b) The congruence $5^x \equiv -1 \pmod{m}$ is not solvable.

   (c) The numbers $\pm 5^k, 0 \leq k < \varphi(m)/2$ form a reduced residue system modulo m. *Remark*: We know from Theorem 3.3.5 that there is no primitive root for $m = 2^\alpha$ if $\alpha \geq 3$. We obtain from (c) that 5 is "nearly" a primitive root for these moduli.

* 13. Let the standard form of the odd integer $m > 1$ be $m = p_1^{\alpha_1} \ldots p_r^{\alpha_r}$. Show that for suitable integers $u_1, \ldots, u_r$, the numbers

$$u_1^{k_1} \ldots u_r^{k_r}, \qquad 0 \leq k_i < \varphi(p_i^{\alpha_i}), \qquad i = 1, 2, \ldots, r$$

form a reduced residue system modulo m. Formulate and prove the analogous statement for even integers m.

14. Let $p > 2$ be a prime. Give a new proof for the existence of a primitive root modulo p following the argument below.

(a) Show that if a polynomial f with integer coefficients divides $x^{p-1} - 1$, then the number of solutions of $f(x) \equiv 0 \pmod{p}$ is equal to the degree of f.

(b) Assume that $q^\beta \mid p-1$ for some prime q and $\beta > 0$. Verify for the polynomials

$$f_1 = x^{q^\beta} - 1 \quad \text{and} \quad f_2 = x^{q^{\beta-1}} - 1$$

that the congruences $f_1(x) \equiv 0 \pmod{p}$ and $f_2(x) \equiv 0 \pmod{p}$ have exactly q^β and $q^{\beta-1}$ solutions.

(c) Using the notations and result of (b), exhibit the existence of a c satisfying $o_p(c) = q^\beta$.

(d) Using (c) and Exercise 3.2.15a, verify the existence of elements of order d for every $d \mid p - 1$.

3.4. Discrete Logarithm (Index)

In this and the next section, we assume that the modulus is a prime p. We note that the notions and results can be generalized to every modulus that possesses a primitive root.

Let g be a primitive root mod p. By Theorem 3.3.2, the integers 1, g, ..., g^{p-2} form a reduced residue system mod p, thus to any a coprime to p, there is a unique exponent $0 \leq k \leq p - 2$ satisfying $a \equiv g^k \pmod{p}$. This makes possible to introduce the logarithm.

Definition 3.4.1. Let g be a primitive root mod p and $(a, p) = 1$. Then the *discrete logarithm* or *index* of a with base g is the (unique) integer $0 \leq k \leq p - 2$ satisfying $a \equiv g^k \pmod{p}$. ♣

Notation: $\mathrm{ind}_{p,g}(a)$. Since the modulus p is fixed in most cases, we can write just $\mathrm{ind}_g a$ in general. If there is no ambiguity concerning the primitive root, then $\mathrm{ind}\, a$ is sufficient.

By the preliminary remark, $\mathrm{ind}_g a$ exists and is unique for any $(a, p) = 1$. Of course, the discrete logarithm of a number a depends on which primitive root g was chosen as base.

If $a \equiv b \pmod{p}$, then clearly $\mathrm{ind}_g a = \mathrm{ind}_g b$, thus all elements in a reduced residue class have the same discrete logarithm (with g fixed).

We shall often use the fact

$$g^s \equiv g^t \pmod{p} \iff s \equiv t \pmod{p - 1}$$

(which follows from assertion (ii) in Theorem 3.2.2 with $m = p$, $a = g$, and $o_p(g) = p - 1$).

According to this, all integers $j \geq 0$ satisfying $g^j \equiv a \pmod{p}$ are just the non-negative elements of a residue class modulo $p - 1$, i.e.

$$g^j \equiv a \pmod{p} \iff j \equiv \mathrm{ind}_g a \pmod{p - 1}.$$

(Therefore, as an alternative definition, the discrete logarithm of a with base g could mean this entire residue class mod $p - 1$.)

The analogs of the logarithmic identities are valid also for the discrete log (see Exercises 3.4.3 and 3.4.4).

The discrete logarithm will be the key for taking roots modulo p in the next section. An application in cryptography will be mentioned in Exercise 5.8.6.

As an illustration, we attach an exponential and a logarithmic table for the modulus $p = 13$ and the primitive root $g = 2$.

j	0	1	2	3	4	5	6	7	8	9	10	11
$2^j \pmod{13}$	1	2	4	8	3	6	12	11	9	5	10	7

a	1	2	3	4	5	6	7	8	9	10	11	12
$\mathrm{ind}_2\, a$	0	1	4	2	9	5	11	3	8	10	7	6

Exercises 3.4

Throughout the exercises, g and h denote primitive roots modulo a prime $p > 2$, a and b are coprime to p, and the index refers to base g unless indicated otherwise.

1. For which primes p is $\mathrm{ind}_{p,7}(2) = 3$?

2. Compute the discrete logarithms.

   (a) $\mathrm{ind}_g 1$

   (b) $\mathrm{ind}_g(-1)$

   (c) $\mathrm{ind}_g(-g)$.

3. Verify the logarithmic identities.

   (a) $\mathrm{ind}(ab) \equiv \mathrm{ind}\, a + \mathrm{ind}\, b \pmod{p-1}$

   (b) $\mathrm{ind}(a^k) \equiv k \cdot \mathrm{ind}\, a \pmod{p-1}$.

4. Demonstrate the law for switching between logarithms from base g to base h

   (a) $\mathrm{ind}_g h \cdot \mathrm{ind}_h g \equiv 1 \pmod{p-1}$

   (b) $\mathrm{ind}_h a \equiv \mathrm{ind}_h g \cdot \mathrm{ind}_g a \pmod{p-1}$.

5. Determine the smallest positive integer s satisfying $p - 1 \mid s \cdot \mathrm{ind}\, a$.

6. Prove that a is a primitive root mod p if and only if $(\mathrm{ind}_g a, p - 1) = 1$.

7. Verify the propositions.

   (a) $(\mathrm{ind}_g a, p - 1) = 1 \iff (\mathrm{ind}_h a, p - 1) = 1$.

   (b) $(\mathrm{ind}_g a, p - 1) = (\mathrm{ind}_h a, p - 1)$.

8. Let a, b, and c be arbitrary primitive roots modulo p. Show that
$$a^{\mathrm{ind}_b c}$$
   is a primitive root mod p.

S* 9. Show that $o_p(a) = o_p(b)$ holds if and only if $\mathrm{ind}_g a = \mathrm{ind}_h b$ for some primitive roots g and h.

10. Find the smallest positive primitive roots for the primes below and prepare the corresponding tables of indices (a) 7 (b) 11 (c) 17.

* 11. Prove that for any prime p and integer a, there are infinitely many positive integers satisfying $a \equiv k^k \pmod{p}$.

3.5. Binomial Congruences

To get the kth root of a positive real number, we divide its logarithm by k and the quotient yields the logarithm of the root (this is how calculators work). We can use the discrete logarithm similarly to take roots modulo p, i.e. to solve the congruence $x^k \equiv a$ (mod p) where p is a prime. As it has just two terms, it is called a *binomial* congruence. The general binomial congruence $cx^k \equiv d \pmod{p}$ with $c \not\equiv 0 \pmod{p}$ can be reduced to $x^k \equiv a \pmod{p}$ where a is the unique solution of the linear congruence $cy \equiv d \pmod{p}$.

If $(a, p) \neq 1$, then $a \equiv 0 \pmod{p}$, i.e. we get the congruence $x^k \equiv 0 \pmod{p}$. Using the prime property of p, it follows that $x \equiv 0 \pmod{p}$ is the only solution.

Thus, we assume $(a, p) = 1$ from now on.

Theorem 3.5.1. *Let p be a prime and $(a, p) = 1$. The congruence*

(3.5.1) $$x^k \equiv a \pmod{p}$$

is solvable if and only if

(3.5.2) $$a^{\frac{p-1}{(k,p-1)}} \equiv 1 \pmod{p}.$$

If it is solvable, then there are $(k, p - 1)$ (pairwise incongruent) solutions.

Condition (3.5.2) is equivalent to

(3.5.3) $$(k, p - 1) \mid \operatorname{ind}_g a$$

where g denotes an arbitrary primitive root modulo p.   ♣

Proof. We use the discrete logarithm with base g.

We look for solutions in the form $x \equiv g^{\operatorname{ind} x} \pmod{p}$. Then (3.5.1) can be written as

(3.5.4) $$g^{k \cdot \operatorname{ind} x} \equiv g^{\operatorname{ind} a} \pmod{p}.$$

Applying $g^s \equiv g^t \pmod{p} \iff s \equiv t \pmod{p-1}$, (3.5.4) is equivalent to

(3.5.5) $$k \cdot \operatorname{ind} x \equiv \operatorname{ind} a \pmod{p-1}.$$

(3.5.5) is a linear congruence for $\operatorname{ind} x$. By Theorem 2.5.3, it is solvable if and only if (3.5.3) holds, hence the same applies for the solvability of (3.5.1).

There is a one-to-one correspondence between the pairwise incongruent solutions modulo $p - 1$ of (3.5.5) and the pairwise incongruent solutions modulo p of (3.5.1), therefore the number of solutions of the two congruences is the same. By Theorem 2.5.4, it is $(k, p - 1)$.

We show the equivalence of (3.5.2) and (3.5.3). Since

$$(3.5.6) \qquad a^{\frac{p-1}{(k,p-1)}} \equiv \left(g^{\operatorname{ind} a}\right)^{\frac{p-1}{(k,p-1)}} = g^{(p-1)\frac{\operatorname{ind} a}{(k,p-1)}} \pmod{p},$$

$a^{(p-1)/(k,p-1)} \equiv 1 \pmod{p}$ is true if and only if the exponent of g in the last term of (3.5.6) is a multiple of $p-1$, i.e. $(k, p-1) \mid \operatorname{ind} a$. $\qquad\square$

Remarks: (1) The proof provides a method for obtaining the solutions assuming that we have a table of indices for some primitive root.

(2) The values of $\operatorname{ind}_g a$ (may) vary depending on the choice of the primitive root g. However, as the solvability of (3.5.1) does not depend on g, the condition in (3.5.3) has to be independent of g; it holds either for all, or for none of the primitive roots. (This follows from Exercise 3.4.7b.)

Example. Solve the congruence $5x^{22} \equiv 6 \pmod{13}$.

The (only) solution of $5y \equiv 6 \pmod{13}$ is $y \equiv 9 \pmod{13}$. This reduces our task to solving the congruence $x^{22} \equiv 9 \pmod{13}$.

According to the proof of Theorem 3.5.1, this is equivalent to

$$22 \cdot \operatorname{ind} x \equiv \operatorname{ind} 9 \pmod{12}.$$

Note that 2 is a primitive root modulo 13 and the relevant exponential and logarithmic tables are at the end of Section 3.4. From the logarithmic table, we have $\operatorname{ind} 9 = 8$.

The linear congruence

$$22 \cdot \operatorname{ind} x \equiv 8 \pmod{12}$$

is solvable since $(22, 12) \mid 8$ and it has $(22, 12) = 2$ (pairwise incongruent) solutions (mod 12). They are

$$\operatorname{ind} x \equiv 2 \pmod{12} \quad \text{and} \quad \operatorname{ind} x \equiv 8 \pmod{12}.$$

Using the exponential table, we obtain

$$x \equiv 4 \pmod{13} \quad \text{and} \quad x \equiv 9 \pmod{13}.$$

We note that it is unnecessary to first solve the congruence $5y \equiv 6 \pmod{13}$ as we can switch immediately to the indices:

$$\operatorname{ind} 5 + 22 \cdot \operatorname{ind} x \equiv \operatorname{ind} 6 \pmod{12} \quad \text{or} \quad 9 + 22 \cdot \operatorname{ind} x \equiv 5 \pmod{12}.$$

Thus, we arrived in a single step at the linear congruence $22 \cdot \operatorname{ind} x \equiv 8 \pmod{12}$.

Definition 3.5.2. Let p be a prime and $(a, p) = 1$. The number a is a kth *power residue* mod p if the congruence $x^k \equiv a \pmod{p}$ is solvable and it is a kth *power non-residue* mod p if the congruence $x^k \equiv a \pmod{p}$ is not solvable. $\qquad\spadesuit$

Theorem 3.5.3. *Let p be a prime and $(a, p) = 1$. The integer a is a kth power residue mod p if and only if*

$$a^{\frac{p-1}{(k,p-1)}} \equiv 1 \pmod{p} \quad i.e. \quad (k, p-1) \mid \operatorname{ind}_g a$$

where g is an arbitrary primitive root modulo p.

The number of (pairwise incongruent) kth power residues is $(p-1)/(k, p-1)$. $\qquad\clubsuit$

Proof. The first assertion is just a reformulation of (a part of) Theorem 3.5.1.

To prove the second assertion, note that by Theorem 3.5.1, the kth power residues are exactly the solutions of the congruence

$$z^{\frac{(p-1)}{(k,p-1)}} \equiv 1 \;(\text{mod } p)$$

and the number of solutions is

$$\left(\frac{p-1}{(k,p-1)}, p-1\right) = \frac{p-1}{(k,p-1)}. \qquad \qquad \square$$

Exercises 3.5

In the exercises, the modulus is a prime $p > 2$.

1. Solve the following congruences. (For moduli 11, 13, and 17 use the relevant tables of indices at the end of Section 3.4 and the hint to Exercise 3.4.10.)

   (a) $3x^{50} \equiv 2 \;(\text{mod } 101)$.

   (b) $x^{99} \equiv 2 \;(\text{mod } 101)$.

   (c) $x^{46} \equiv 50 \;(\text{mod } 23)$.

   (d) $5x^{14} \equiv 14x^2 \;(\text{mod } 17)$.

   (e) $4x^7 + 7x^4 \equiv 0 \;(\text{mod } 13)$.

   (f) $4x^{27} + 5x^{20} + 7x^{17} + 9x^8 + 3 \equiv 0 \;(\text{mod } 11)$.

2. Determine the number of solutions.

   (a) $(x^{30} - 1)(x^{45} - 1) \equiv 0 \;(\text{mod } 73)$.

   (b) $1 + x + x^2 + \cdots + x^k \equiv 0 \;(\text{mod } 31)$.

3. For which a is
   $$1 + x + \cdots + x^{p-2} \equiv a \;(\text{mod } p)$$
   solvable?

4. Show that if g is a primitive root, then the congruence $x^k \equiv g \;(\text{mod } p)$ can have at most one solution.

5. Denote by $b_1, \ldots, b_r$ the (pairwise incongruent) solutions of $x^k \equiv 1 \;(\text{mod } p)$. Let $(a, p) = 1$ and c be a solution of $x^k \equiv a \;(\text{mod } p)$. How can we describe all solutions of $x^k \equiv a$?

6. Determine the power residues of exponent

   (a) $p - 1$

   (b) $(p - 1)/2$.

7. For which k can we take unique kth roots mod p, i.e. when does the congruence $x^k \equiv a \;(\text{mod } p)$ have exactly one solution for any a?

8. For which primes can we form a complete residue system purely from cubes?

9. Prove the following assertions.

   (a) The product of two kth power residues is always a kth power residue.

   (b) The product of a kth power residue and a kth power non-residue is always a kth power non-residue.

10. How can we characterize in terms of p and k that there exist kth power residues and the product of any two of them is again a kth power residue?

11. What is the remainder mod p of (a) the sum (b) the product of all (pairwise incongruent) kth power residues?

S 12. Prove that a is both a 20th and 50th power residue modulo p if and only if it is a 100th power residue. Investigate also the generalized problem.

3.6. Chevalley's Theorem, Kőnig–Rados Theorem

We discuss two famous theorems concerning congruences with prime modulus. We consider first a system of congruences

(3.6.1) $$f_i(x_1, x_2, \ldots, x_t) \equiv 0 \pmod{p}, \qquad i = 1, 2, \ldots, k$$

where p is a prime, $k \geq 1$, and

$$f_i(x_1, x_2, \ldots, x_t), \qquad i = 1, 2, \ldots, k$$

are polynomials in t variables with integer coefficients and constant terms 0, i.e.

(3.6.2) $$f_i(0, 0, \ldots, 0) = 0, \qquad i = 1, 2, \ldots, k.$$

(3.6.2) implies that $x_1 \equiv x_2 \equiv \cdots \equiv x_t \equiv 0 \pmod{p}$ satisfies (3.6.1). We call this a trivial solution.

   Chevalley's Theorem asserts that with suitable requirements for the degrees of f_i, (3.6.1) has a non-trivial solution, too. (The degree of a term $x_1^{n_1} \ldots x_t^{n_t}$ is $n_1 + \cdots + n_t$ and the degree of a polynomial is the maximal degree among its terms having non-zero coefficient.)

Theorem 3.6.1 (Chevalley's Theorem). *If the polynomials f_i in (3.6.1) satisfy (3.6.2) and the sum of their degrees is less than the number of variables, i.e.*

(3.6.3) $$\sum_{i=1}^{k} \deg f_i < t,$$

then (3.6.1) has a non-trivial solution.   ♣

Examples. The system of congruences

$$x_1 + 2x_2 + 3x_3 + 4x_4 + 5x_5 \equiv 0 \pmod{23}$$
$$x_1^3 + 2x_1 x_2 + 3x_2 x_3 + 4x_3 x_4^2 + 5x_5^2 \equiv 0 \pmod{23}$$

has a non-trivial solution with not all x_i multiples of 23. (Here $k = 2$ and $5 = t > 1 + 3 = \deg f_1 + \deg f_2$.)

We can apply Theorem 3.6.1 also for $k = 1$, just one polynomial. For example, the divisibility

$$p \mid x_1^3 + 3x_2^3 + 5x_3^3 + 7x_4^3 + 9x_1x_2 + 11x_3x_4$$

can be satisfied for any prime p so that not all x_i are divisible by p. (Now $t = 4$ and $\deg f = 3$.)

Proof. Assuming that there is only a trivial solution, we shall force a contradiction.

We define two new polynomials in t variables:

$$F(x_1, x_2, \ldots, x_t) = \prod_{i=1}^{k} \left(1 - f_i^{p-1}(x_1, x_2, \ldots, x_t)\right) \quad \text{and}$$

$$G(x_1, x_2, \ldots, x_t) = \prod_{j=1}^{t} (1 - x_j^{p-1}).$$

By Fermat's Little Theorem,

$$c_j \not\equiv 0 \pmod{p} \Longrightarrow c_j^{p-1} \equiv 1 \pmod{p}.$$

This implies that substituting arbitrary integers $c_1, \ldots, c_t$ into G, we get

$$(3.6.4) \qquad G(c_1, c_2, \ldots, c_t) \equiv \begin{cases} 1 \pmod{p}, & \text{if } c_1 \equiv \cdots \equiv c_t \equiv 0 \pmod{p} \\ 0 \pmod{p}, & \text{otherwise.} \end{cases}$$

We show that the same holds also for F, i.e.

$$(3.6.5) \qquad F(c_1, c_2, \ldots, c_t) \equiv \begin{cases} 1 \pmod{p}, & \text{if } c_1 \equiv \cdots \equiv c_t \equiv 0 \pmod{p} \\ 0 \pmod{p}, & \text{otherwise.} \end{cases}$$

We consider first

$$c_1 \equiv \cdots \equiv c_t \equiv 0 \pmod{p}.$$

By (3.6.2),

$$f(c_1, \ldots, c_t) \equiv 0 \pmod{p}$$

for every i, so every factor of $F(c_1, \ldots, c_t)$ and so $F(c_1, \ldots, c_t)$ itself is congruent to 1 modulo p.

Now, we turn to the other case when at least one of the integers $c_1, \ldots, c_t$ is not a multiple of p. We assumed that (3.6.1) has only a trivial solution, hence $c_1, \ldots, c_t$ is not a solution, so

$$f_i(c_1, \ldots, c_t) \not\equiv 0 \pmod{p}$$

for at least one i. Applying Fermat's Little Theorem again, this implies

$$f_i^{p-1}(c_1, c_2, \ldots, c_t) \equiv 1 \pmod{p}.$$

This means that a factor of $F(c_1, \ldots, c_t)$ and so $F(c_1, \ldots, c_t)$ itself is divisible by p. Herewith we have proven (3.6.5).

By (3.6.4) and (3.6.5),

$$(3.6.6) \qquad\qquad F(c_1, \ldots, c_t) \equiv G(c_1, \ldots, c_t) \pmod{p}$$

for arbitrary integers $c_1, \ldots, c_t$.

From now on, we shall consider all polynomials as polynomials in t variables over the modulo p field.

In this interpretation, (3.6.6) tells us that F and G assume the same values for every substitution (the same polynomial *functions* belong to F and G; however, this does not imply in general the equality of the polynomials themselves, that is, the equality of the coefficients in the case of a finite field).

Let H^* be the reduced form of the polynomial H obtained by replacing every x_i^p in H with x_i as long as possible. The exponents of x_i in the terms of H^* are at most $p - 1$, and H and H^* assume the same values everywhere. It can be easily proven by induction on the number of variables that if the polynomials H and K assume the same values everywhere, then the (formal) polynomials H^* and K^* are equal (so they have the same coefficients).

We saw that F and G assume the same values everywhere, therefore the polynomials F^* and G^* are equal. Hence, $\deg G^* = \deg F^*$. However, by $G = G^*$ and (3.6.3), this leads to a contradiction:

$$\deg G^* = \deg G = (p-1)t > (p-1)\left(\sum_{i=1}^{k} \deg f_i\right) = \deg F \geq \deg F^*. \qquad \square$$

In the second half of this section, we express the number of solutions of a congruence $f(x) \equiv 0 \pmod{p}$ in an exact formula with the help of the coefficients. This theorem by Kőnig and Rados is rather only of theoretical significance; it can be hardly applied for computing the number of solutions in practice.

Theorem 3.6.2 (Kőnig–Rados Theorem). *Let p be a prime and $f = a_0 + a_1x + \cdots + a_{p-2}x^{p-2}$ be a polynomial with integer coefficients having $a_0 \not\equiv 0 \pmod{p}$. Then the number of solutions of the congruence $f(x) \equiv 0 \pmod{p}$ is $p - 1 - r$ where $r = r(A)$ is the rank of the cyclic $(p-1) \times (p-1)$ matrix A over the modulo p field,*

$$A = \begin{pmatrix} a_0 & a_1 & \cdots & a_{p-2} \\ a_{p-2} & a_0 & \cdots & a_{p-3} \\ \vdots & \vdots & \ddots & \vdots \\ a_1 & a_2 & \cdots & a_0 \end{pmatrix}. \qquad \clubsuit$$

Remarks: (1) The theorem immediately implies that $f(x) \equiv 0 \pmod{p}$ is solvable if and only if the rank of A is less than $p - 1$, i.e. $\det A \equiv 0 \pmod{p}$.

(2) The requirements imposed on f are not serious restrictions; we can obtain the number of solutions for an arbitrary polynomial f by a simple reduction to the Kőnig–Rados Theorem, see Exercise 3.6.11.

Proof. We shall need the following elementary results from linear algebra. They all refer to $n \times n$ matrices over a field F where $r(B)$ denotes the rank of matrix B; in our case, $n = p - 1$ and F is the modulo p field.

(i) Let $t_1, t_2, \ldots, t_n$ be distinct elements in F. Then the Vandermonde matrix

$$V = V(t_1, t_2, \ldots, t_n) = \begin{pmatrix} 1 & 1 & 1 & \cdots & 1 \\ t_1 & t_2 & t_3 & \cdots & t_n \\ t_1^2 & t_2^2 & t_3^2 & \cdots & t_n^2 \\ \vdots & \vdots & \vdots & \ddots & \vdots \\ t_1^{n-1} & t_2^{n-1} & t_3^{n-1} & \cdots & t_n^{n-1} \end{pmatrix}$$

has rank $r(V) = n$.

(ii) If $r(B) = n$, so B has an inverse, then $r(CB) = r(C)$ for an arbitrary C.

Assertion (ii) follows from the inequality

$$r(MN) \leq \min\big(r(M), r(N)\big)$$

valid for arbitrary matrices M and N. On the one hand, $r(CB) \leq r(C)$, and on the other hand, $r(C) = r((CB)B^{-1}) \leq r(CB)$.

Turning to the proof of Theorem 3.6.2, let s denote the number of solutions of the congruence $f(x) \equiv 0 \pmod{p}$. Consider the matrix $D = AV$ where $V = V(1, 2, \ldots, p-1)$. By (i) and (ii),

(3.6.7) $r(D) = r(A) = r.$

Performing the multiplication AV, the jth element of the first row in D is

$$d_{1j} = a_0 + a_1 j + a_2 j^2 + \cdots + a_{p-2} j^{p-2} = f(j).$$

For a simple form of the jth element in the second row, we use also $j^{p-1} \equiv 1 \pmod{p}$:

$$d_{2j} = a_{p-2} + a_0 j + a_1 j^2 + \cdots + a_{p-3} j^{p-2} \equiv$$
$$\equiv a_{p-2} j^{p-1} + a_0 j + a_1 j^2 + \cdots + a_{p-3} j^{p-2} = j f(j) \pmod{p}.$$

Similarly, for the jth element in the ith row, we obtain

$$d_{ij} \equiv j^{i-1} f(j) \pmod{p}.$$

This means that (working with equality in the modulo p field instead of congruences)

$$D = AV = \begin{pmatrix} f(1) & f(2) & f(3) & \cdots & f(p-1) \\ f(1) & 2f(2) & 3f(3) & \cdots & (p-1)f(p-1) \\ f(1) & 2^2 f(2) & 3^2 f(3) & \cdots & (p-1)^2 f(p-1) \\ \vdots & \vdots & \vdots & & \vdots \\ f(1) & 2^{p-2} f(2) & 3^{p-2} f(3) & \cdots & (p-1)^{p-2} f(p-1) \end{pmatrix}$$

In column j of D every element is 0 if and only if $f(j) \equiv 0 \pmod{p}$. Thus, D has exactly s columns with all 0s. The other columns are distinct columns of V multiplied by a non-zero scalar, so they are linearly independent according to (i). Hence, $r(D) = p - 1 - s$. Combined with (3.6.7), this proves the theorem. $\qquad\square$

Exercises 3.6

1. Which well-known theorem is obtained as a special case of Chevalley's Theorem when each polynomial f_i has degree one?

2. Verify that the congruence $ax^2 + by^2 + cz^2 \equiv 0 \pmod{p}$ has a non-trivial solution for every prime p and any integers a, b, and c.

3. Prove.

    (a) For any $n > 1$, there exist three integers such that taking the sum s of their squares, $n \mid s$ but $n^2 \nmid s$.

    (b) Moreover, even $(n, s/n) = 1$ can be attained.

4. Show that every prime p has a positive multiple less than $p^4/4$ that can be written as the sum of at most five fourth powers.

* 5. (a) Let $q_1, \ldots, q_k$ be distinct primes and $c_1, \ldots, c_t$ distinct positive integers not divisible by any other primes than the q_i. Prove that if $t \geq 2k + 1$, then we can select some distinct numbers from the c_j (maybe just one, maybe all of them) so that their product is a cube.

    (b) Generalize (a) for pth powers where p is an arbitrary prime.
    Remark: The analogous result can be proven (using different methods) for mth powers where m is a prime power, but it is not proved for any other values of m.

* 6. Show that from any $2n - 1$ integers, we can select n whose sum is divisible by n.

7. Prove the following generalization of Chevalley's Theorem. We omit the assumption that the constant terms of the polynomials are 0 and leave the other conditions unchanged. Then the following hold for the system of congruences in question:

    (a) If it is solvable, then there are at least two solutions.

    * (b) The number of solutions is divisible by p.

8. Let $p > 2$ be a prime and $(ab, p) = 1$. As an illustration of the Kőnig–Rados Theorem, determine the number of solutions of the congruence $f(x) \equiv 0 \pmod{p}$ for the following polynomials f:

    (a) $ax - b$

    (b) $1 + x + \cdots + x^{p-2}$

    (c) $x^{p-2} - a$.

9. Deduce the solvability of the following congruences from the Kőnig–Rados Theorem:

    (a) $x^k \equiv 1 \pmod{p}$ where p is an odd prime and $1 \leq k \leq p - 2$

    (b) $x^2 \equiv -1 \pmod{p}$, where p is a prime of the form $4k + 1$.

10. Let $p > 3$ be a prime, $(a_0, p) = (a_1, p) = (a_{p-2}, p) = 1$ and

$$\begin{aligned}
f &= a_0 &+ a_1 x + \cdots + a_{p-3} x^{p-3} + a_{p-2} x^{p-2} \\
g &= a_1 &+ a_2 x + \cdots + a_{p-2} x^{p-3} + a_0 x^{p-2} \\
h &= a_{p-2} + a_{p-3} x + \cdots + a_1 x^{p-3} + a_0 x^{p-2}.
\end{aligned}$$

Prove that the congruences

$$f(x) \equiv 0 \ (\mathrm{mod}\ p), \quad g(x) \equiv 0 \ (\mathrm{mod}\ p), \quad \text{and} \quad h(x) \equiv 0 \ (\mathrm{mod}\ p)$$

have the same number of solutions.

11. Let $g = b_0 + b_1 x + \cdots + b_n x^n$ be an arbitrary polynomial with integer coefficients. To determine the number of solutions of $g(x) \equiv 0 \ (\mathrm{mod}\ p)$ in the case $n > p - 2$ and/or $b_0 \equiv 0 \ (\mathrm{mod}\ p)$, how can we reduce the problem to the Kőnig–Rados Theorem?

3.7. Congruences with Prime Power Moduli

We saw in Section 2.6 that a congruence with a composite modulus can be reduced to congruences with prime power moduli by the Chinese Remainder Theorem. Now, we examine how we can reduce the prime power modulus case to that of a prime modulus.

Let p be a prime, k a positive integer, f a polynomial with integer coefficients, and consider the congruence

$$(3.7.1) \qquad f(x) \equiv 0 \ \left(\mathrm{mod}\ p^k\right).$$

If c is a solution of (3.7.1), then c also satisfies

$$(3.7.2) \qquad f(x) \equiv 0 \ (\mathrm{mod}\ p).$$

Therefore, we shall start from the solutions of (3.7.2) to determine the solutions of (3.7.1).

Theorem 3.7.1. *Let c be a solution of (3.7.2), and assume that $f'(c) \not\equiv 0 \ (\mathrm{mod}\ p)$ where f' stands for the derivative of f. Then (3.7.1) has exactly one solution $x \equiv c_k \ (\mathrm{mod}\ p^k)$ satisfying $c_k \equiv c \ (\mathrm{mod}\ p)$.* ♣

The proof yields a procedure to produce c_k and clarifies the situation in the case $f'(c) \equiv 0 \ (\mathrm{mod}\ p)$ as well.

Proof. We shall use the observation

$$(3.7.3) \qquad j \geq 1 \Longrightarrow f(a + t p^j) \equiv f(a) + t p^j f'(a) \ (\mathrm{mod}\ p^{j+1}).$$

To verify (3.7.3), consider the representation of $f(a + t p^j)$ by Taylor's formula,

$$(3.7.4) \qquad f(a + t p^j) = f(a) + t p^j f'(a) + t^2 p^{2j} \frac{f''(a)}{2!} + \cdots + t^n p^{nj} \frac{f^{(n)}(a)}{n!}$$

where n is the degree of f. Every $f^{(r)}(a)/(r!)$ is an integer, since the rth derivative of a term x^s is $s(s-1)\ldots(s-r+1)x^{s-r}$ (for $s \geq r$) and the product of r consecutive integers is always divisible by $r!$ (see Exercise 1.1.17b). This implies that on the right-hand side of (3.7.4), every term except the first two is divisible by p^{j+1}, thus proving (3.7.3).

We prove Theorem 3.7.1 by induction on k. The case $k = 1$ is obvious (even without the assumption on the derivative).

Assume now that the assertion is true for $k - 1$. This means that the congruence

(3.7.5) $$f(x) \equiv 0 \ (\mathrm{mod}\ p^{k-1})$$

has exactly one solution $x \equiv c_{k-1} \ (\mathrm{mod}\ p^{k-1})$ with $c_{k-1} \equiv c \ (\mathrm{mod}\ p)$.

We want to find a solution of (3.7.1) satisfying also $c_k \equiv c \ (\mathrm{mod}\ p)$. Because (3.7.5) holds for c_k, $c_k \equiv c_{k-1} \ (\mathrm{mod}\ p^{k-1})$, so

(3.7.6) $$c_k = c_{k-1} + tp^{k-1}.$$

Substituting (3.7.6) into (3.7.1) and applying (3.7.3) (with $a = c_{k-1}$ and $j = k - 1$) we obtain

(3.7.7) $$f(c_k) = f(c_{k-1} + tp^{k-1}) \equiv f(c_{k-1}) + tp^{k-1}f'(c_{k-1}) \equiv 0 \ (\mathrm{mod}\ p^k).$$

Here, $p^{k-1} \mid f(c_{k-1})$ by the induction hypothesis. Cancelling p^{k-1} in (3.7.7) and using $c_{k-1} \equiv c \ (\mathrm{mod}\ p)$, we obtain

(3.7.8) $$\frac{f(c_{k-1})}{p^{k-1}} + tf'(c) \equiv 0 \ (\mathrm{mod}\ p).$$

This is a linear congruence for t that has exactly one solution $t \equiv t_0 \ (\mathrm{mod}\ p)$ due to $f'(c) \not\equiv 0 \ (\mathrm{mod}\ p)$. Hence, $t = t_0 + sp$. Substitution into (3.7.6) yields

$$c_k = c_{k-1} + t_0 p^{k-1} + sp^k \quad \text{so} \quad c_k \equiv c_{k-1} + t_0 p^{k-1} \ (\mathrm{mod}\ p^k).$$

Thus we have proven that c_k exists and is unique mod p^k. $\qquad\qquad\square$

Following the proof, we can build up the values $c_2, \ldots, c_k$ recursively starting from $c = c_1$. (We can even get a formula for c_k, see Exercise 3.7.4.)

If $f'(c) \equiv 0 \ (\mathrm{mod}\ p)$, then either every t, or no t is a solution of (3.7.8) depending on whether or not $p^k \mid f(c_{k-1})$. This means that a solution c_{k-1} of (3.7.5) either gives rise to p suitable c_k, or to none. In this case, the above recursion is much more complicated.

Example. Solve the congruence $x^3 + 2x \equiv 22 \ (\mathrm{mod}\ 125)$.

We solve first
$$f(x) = x^3 + 2x - 22 \equiv 0 \ (\mathrm{mod}\ 5).$$

Checking the elements of the complete residue system $0, \pm 1, \pm 2$ modulo 5, we get two solutions:

(i) $x \equiv 2 \ (\mathrm{mod}\ 5)$, and

(ii) $x \equiv -1 \ (\mathrm{mod}\ 5)$.

(i) If $x \equiv 2 \ (\mathrm{mod}\ 5)$, then
$$f'(2) \equiv 3 \cdot 2^2 + 2 \equiv -1 \ (\mathrm{mod}\ 5),$$

hence we can apply Theorem 3.7.1.

Substituting $x = 2 + 5t$ into $x^3 + 2x - 22 \equiv 0 \ (\mathrm{mod}\ 25)$, we obtain

$$-10 + (5t) \cdot 14 \equiv 0 \ (\mathrm{mod}\ 25) \quad \text{so} \quad -2 - t \equiv 0 \ (\mathrm{mod}\ 5).$$

So $t \equiv -2 \ (\mathrm{mod}\ 5)$ and $t = 5s - 2$. Then

$$x = 2 + 5t = 2 + 5(5s - 2) = -8 + 25s.$$

Thus $x \equiv -8 \pmod{25}$ is the only solution of

$$x^3 + 2x - 22 \equiv 0 \pmod{25}$$

satisfying $x \equiv 2 \pmod 5$.

We proceed similarly from modulus 5^2 to 5^3. Writing $x = -8 + 25s$ in

$$x^3 + 2x - 22 \equiv 0 \pmod{125}$$

we get

$$-50 + (25s) \cdot 194 \equiv 0 \pmod{125}.$$

Thus $s \equiv -2 \pmod 5$, hence

$$x = -8 + 25s = -58 + 125r \quad \text{or} \quad x \equiv -58 \pmod{125}.$$

(ii) If $x \equiv -1 \pmod 5$, then $f'(-1) \equiv 0 \pmod 5$. According to the remark after the proof, we have to check in each step whether or not the value $f(c_{k-1})$ in (3.7.8) is divisible by p^k.

Since $f(-1) \equiv 0 \pmod{25}$, every $x \equiv -1 \pmod 5$ satisfies $x^3 + 2x - 22 \equiv 0 \pmod{25}$. Hence, the solutions are

$$x \equiv -1, 4, 9, 14, \text{ and } 19 \pmod{25}.$$

Out of these, only the last two will make $f(x)$ a multiple of 125, so

$$x \equiv 14 \pmod{25} \quad \text{and} \quad x \equiv 19 \pmod{25}$$

will be the solutions of $x^3 + 2x - 22 \equiv 0 \pmod{125}$ (these form $2 \cdot 5 = 10$ residue classes modulo 125).

Summarizing, all solutions of the congruence $x^3 + 2x \equiv 22 \pmod{125}$ are the following eleven residue classes modulo 125:

$$-58, \qquad 14 + 25j, \quad \text{and} \quad 19 + 25j \qquad \text{where} \qquad 0 \le j \le 4.$$

Exercises 3.7

1. What is the number of solutions of the following congruences?

   (a) $x^{80} + x^3 \equiv 8 \pmod{3^{20}}$

   (b) $x^{99} + x^3 \equiv 8 \pmod{3^{20}}$

   (c) $x^{60} \equiv 1 \pmod{73^{20}}$

   (d) $x^{73} \equiv 1 \pmod{73^{20}}$

   (e) $x(x - 1)(x - 2) \equiv 0 \pmod{10^{20}}$.

2. Let p be a prime and a and n positive integers not divisible by p. Prove that if $x^n \equiv a \pmod p$ is solvable, then $x^n \equiv a \pmod{p^k}$ is solvable for every k.

3. For which a coprime to the modulus are the following congruences solvable? Determine the number of solutions, too.

   (a) $x^{10} \equiv a \pmod{11^{50}}$

   S* (b) $x^2 \equiv a \pmod{2^{50}}$.

4. Assume that the conditions of Theorem 3.7.1 hold, and let u satisfy $uf'(c) \equiv 1$ (mod p). Prove that the values c_k obey the recursion

$$c_1 = c \quad \text{and} \quad c_k = c_{k-1} - uf(c_{k-1}) \quad \text{for } k > 1.$$

5. Solve $x^6 + 4x \equiv d \pmod{7^3}$ where d is

(a) 3

(b) 2

(c) 72.

Chapter 4

Legendre and Jacobi Symbols

Legendre symbol is the principal tool for handling quadratic congruences. Besides its basic properties, we shall prove Gauss's Lemma and the celebrated Quadratic Reciprocity Law, among other theorems. At the end of the chapter, we see that the Jacobi symbol provides a useful generalization of the Legendre symbol.

4.1. Quadratic Congruences

We assume throughout this section that $p > 2$ is a prime and $(a, p) = 1$.

As a special case $k = 2$ of Definition 3.5.2, we introduce the quadratic residues and non-residues.

Definition 4.1.1. Let $p > 2$ be a prime and $(a, p) = 1$. An integer a is a *quadratic residue* or a *quadratic non-residue* modulo p depending on whether the congruence $x^2 \equiv a \pmod{p}$ is solvable or not. ♣

The numbers $a \equiv 0 \pmod{p}$ are neither quadratic residues nor quadratic non-residues.

Theorem 4.1.2. (i) *An integer a is a quadratic residue mod p if and only if $a^{(p-1)/2} \equiv 1 \pmod{p}$. An equivalent condition is that the index (with base to any primitive root) of a is even.*

(ii) *An integer a is a quadratic non-residue mod p if and only if $a^{(p-1)/2} \equiv -1 \pmod{p}$. An equivalent condition is that the index (with base to any primitive root) of a is odd.*

(iii) *The number of (pairwise incongruent) quadratic residues is $(p-1)/2$ and the same holds for the number of non-residues.*

(iv) *If a is a quadratic residue, then the congruence $x^2 \equiv a \pmod{p}$ has two (pairwise incongruent) solutions.* ♣

Proof. We obtain (i) and (iii) from Theorem 3.5.3 and (iv) from an assertion of Theorem 3.5.1 as special cases of $k = 2$.

By (i), a is a quadratic non-residue if and only if $a^{(p-1)/2} \not\equiv 1 \pmod{p}$, or equivalently, the index of a is odd. Thus, to prove (ii), we need only the equivalence

(4.1.1) $\qquad\qquad a^{\frac{p-1}{2}} \not\equiv 1 \pmod{p} \iff a^{\frac{p-1}{2}} \equiv -1 \pmod{p}.$

Since $(a^{(p-1)/2})^2 = a^{p-1} \equiv 1 \pmod{p}$ and p is a prime, only $a^{(p-1)/2} \equiv \pm 1 \pmod{p}$ are possible. Also, $1 \not\equiv -1 \pmod{p}$ for $p > 2$, therefore (4.1.1) holds. $\qquad\square$

Definition 4.1.3. The *Legendre symbol* $\left(\dfrac{a}{p}\right)$ is defined by

$$\left(\frac{a}{p}\right) = \begin{cases} 1, & \text{if } a \text{ is a quadratic residue mod } p \\ -1, & \text{if } a \text{ is a quadratic non-residue mod } p. \end{cases} \qquad \clubsuit$$

Remark: It is sometimes useful to extend the Legendre symbol to the case $p \mid a$ as $\left(\dfrac{a}{p}\right) = 0$ (see Exercise 4.1.15). We restrict ourselves, however, to the condition $(a, p) = 1$ unless stated otherwise.

Example. $\left(\dfrac{2}{7}\right) = 1$ since $x^2 \equiv 2 \pmod{7}$ is solvable: $x \equiv 3 \pmod{7}$ is a solution. We can verify the solvability also by checking

$$2^{\frac{7-1}{2}} = 2^3 \equiv 1 \pmod{7}.$$

Combining the definition of the Legendre symbol with Theorem 4.1.2, we obtain

(4.1.2) $\qquad\qquad\qquad\qquad a^{\frac{p-1}{2}} \equiv \left(\frac{a}{p}\right) \pmod{p}$

for every a.

We summarize some basic properties of the Legendre symbol in the next theorem.

Theorem 4.1.4.   (i) $a \equiv b \pmod{p} \implies \left(\dfrac{a}{p}\right) = \left(\dfrac{b}{p}\right).$

(ii) $\left(\dfrac{ab}{p}\right) = \left(\dfrac{a}{p}\right)\left(\dfrac{b}{p}\right).$

(iii)

$$\left(\frac{-1}{p}\right) = \begin{cases} 1, & \text{if } p \equiv 1 \pmod{4} \\ -1, & \text{if } p \equiv -1 \pmod{4}. \end{cases} \qquad \clubsuit$$

Proof. Each assertion follows from (4.1.2) immediately; we describe the details only for (ii):

$$\left(\frac{ab}{p}\right) \equiv (ab)^{\frac{p-1}{2}} = a^{\frac{p-1}{2}} b^{\frac{p-1}{2}} \equiv \left(\frac{a}{p}\right)\left(\frac{b}{p}\right) \pmod{p}.$$

Thus

$$K = \left(\frac{ab}{p}\right) - \left(\frac{a}{p}\right)\left(\frac{b}{p}\right)$$

is divisible by $p > 2$. As K can assume no other values than 0 and ± 2, only $K = 0$ is possible. $\qquad\square$

By Theorem 4.1.4, we can reduce the calculation of a Legendre symbol to the determination of $\left(\frac{2}{p}\right)$ and $\left(\frac{q}{p}\right)$ where $q > 2$ is a prime different from p. We discuss these results in the next section.

Exercises 4.1

(The symbol p always represents an odd prime.)

1. Verify by three different methods that c^2 is a quadratic residue mod p if $(c, p) = 1$.

2. Compute the Legendre symbols

   (a) $\left(\dfrac{39}{19}\right)$

   (b) $\left(\dfrac{37}{19}\right)$

   (c) $\left(\dfrac{-100}{19}\right)$.

3. Compute the sum and product of the Legendre symbols

   $$\left(\frac{1}{p}\right), \quad \left(\frac{2}{p}\right), \quad \ldots, \quad \left(\frac{p-1}{p}\right).$$

4. Demonstrate that every quadratic residue is congruent to exactly one of the numbers

   $$1^2, 2^2, \ldots, \left(\frac{p-1}{2}\right)^2.$$

5. Show that if $a^2 + b^2$ is a multiple of 77, then it is divisible by 5929.

6. Let p be a prime of the form $4k + 1$. Prove that the solutions of $x^2 \equiv -1 \pmod{p}$ are

   $$x \equiv \pm\left(\frac{p-1}{2}\right)! \pmod{p}.$$

7. Let p be a prime of the form $4k + 3$ and a a quadratic residue mod p. Verify that the solutions of $x^2 \equiv a \pmod{p}$ are

   $$x \equiv \pm a^{\frac{p+1}{4}} \pmod{p}.$$

8. (a) Show that if $o_p(a)$ is odd, then a is a quadratic residue mod p.

   (b) For which primes p is the converse true?

9. (a) Prove that every primitive root is a quadratic non-residue modulo p.

   * (b) For which primes p is the converse true?

10. Assume that $(c, 97) = 1$ and c is neither a quadratic residue nor a primitive root mod 97. Compute $o_{97}(c)$.

11. Prove that if

    (a) $p = 4k - 1$

    (b) $p = 4k + 1$,

    then $x^2 \equiv k \pmod{p}$ is solvable.

12. Show that for every p, at least one of the congruences

$$x^2 \equiv 30, \quad x^2 \equiv 33, \quad x^2 \equiv 70, \quad x^2 \equiv 105, \quad \text{and} \quad x^2 \equiv 165$$

has a solution mod p.

13. Solve the congruences

    S (a) $3x^2 + 5x + 5 \equiv 0 \pmod{13}$
       (b) $7x^2 + 8x \equiv 5 \pmod{17}$
       (c) $6x^{25} + x^5 + 5x \equiv 0 \pmod{23}$
       (d) $2x^{17} + 5x + 1 \equiv 0 \pmod{19}$.

14. Denote by $n(p)$ the smallest positive integer that is a quadratic non-residue mod p. For example, $n(5) = 2$ and $n(7) = 3$. Prove that

    (a) $n(p)$ is always a prime;
  ** (b) $n(p) < 1 + \sqrt{p}$.

15. Extend the definition of the Legendre symbol for $p \mid a$ as $(\frac{a}{p}) = 0$. Verify the following assertions for

$$S(a, p) = \sum_{i=1}^{p} \left(\frac{i(i + a)}{p} \right).$$

    (a) $S(0, p) = p - 1$
  * (b) $(a, p) = 1 \Longrightarrow S(a, p) = S(1, p)$
    (c) $\sum_{a=0}^{p-1} S(a, p) = 0$
    (d) $S(1, p) = -1$.

16. Let $M(p)$ be the number of those integers a, $1 \le a \le p - 2$, for which both a and $a + 1$ are quadratic residues mod p.

    (a) Prove

$$4M(p) = \sum_{a=1}^{p-2} \left(\left(\frac{a}{p} \right) + 1 \right) \left(\left(\frac{a+1}{p} \right) + 1 \right).$$

    (b) Show that $M(p)$ is approximately $p/4$: if $p = 4k \pm 1$, then $M(p) = k - 1$.

4.2. Quadratic Reciprocity

We assume in this section too that $p > 2$ is a prime. We shall discuss theorems concerning the Legendre symbols $\left(\frac{2}{p} \right)$ and $\left(\frac{q}{p} \right)$ where $q > 2$ is a prime. Both results will require the following lemma:

Theorem 4.2.1 (Gauss's Lemma). *Let $(a, p) = 1$ and consider the least positive remainders of a, $2a$, $\ldots$, $\frac{p-1}{2}a$ modulo p. Let v denote how many of them are greater than $\frac{p}{2}$. Then*

$$\left(\frac{a}{p} \right) = (-1)^v.$$

♣

Proof. Taking the least positive remainders of the given $\frac{p-1}{2}$ numbers, let $r_1, \ldots, r_u$ be the ones smaller than $\frac{p}{2}$ and $p - s_1, \ldots, p - s_v$ the ones greater than $\frac{p}{2}$ ($u + v = \frac{p-1}{2}$). Then for every $1 \le t \le \frac{p-1}{2}$

$$(4.2.1) \qquad ta \equiv r_i \quad \text{or} \quad ta \equiv p - s_j \pmod{p}$$

with a suitable i or j. Note that every r_i and s_j is one of the integers $1, 2, \ldots, \frac{p-1}{2}$.

We show that r_i and s_j are distinct, therefore they are the same as the numbers $1, 2, \ldots, \frac{p-1}{2}$ in some order.

Assuming $r_i = r_k$ for some $i \ne k$, we have

$$\lambda a \equiv r_i = r_k \equiv \mu a \pmod{p}$$

with suitable numbers $\lambda, \mu, 1 \le \lambda < \mu \le \frac{p-1}{2}$. Since $(a, p) = 1$, cancelling a gives $\lambda \equiv \mu$ $\pmod{p}$, which is a contradiction.

We get a contradiction similarly assuming the equality of two s_j.

Finally, if $r_i = s_j$, then

$$\lambda a \equiv r_i = s_j \equiv -\mu a \pmod{p},$$

so $p \mid a(\lambda + \mu)$. However, $(a, p) = 1$ and $0 < \lambda + \mu < p$, hence none of the factors is divisible by p, which contradicts the prime property of p.

Multiplying the congruences (4.2.1) for $t = 1, 2, \ldots, (p-1)/2$, we obtain

$$(4.2.2) \qquad \left(\frac{p-1}{2}\right)! \, a^{\frac{p-1}{2}} \equiv r_1 \ldots r_u (p - s_1) \ldots (p - s_v) \equiv$$
$$\equiv (-1)^v r_1 \ldots r_u s_1 \ldots s_v = (-1)^v \left(\frac{p-1}{2}\right)! \pmod{p}.$$

Cancelling $\left(\frac{p-1}{2}\right)!$ in (4.2.2), we arrive at

$$a^{\frac{p-1}{2}} \equiv (-1)^v \pmod{p} \quad \text{or} \quad \left(\frac{a}{p}\right) = (-1)^v. \qquad \square$$

As a simple application of Gauss's Lemma, we determine which primes have 2 as a quadratic residue.

Theorem 4.2.2.

$$\left(\frac{2}{p}\right) = \begin{cases} 1, & \text{if } p \equiv \pm 1 \pmod 8 \\ -1, & \text{if } p \equiv \pm 3 \pmod 8. \end{cases} \qquad \clubsuit$$

Proof. To apply Gauss's Lemma for $a = 2$, we count how many of the numbers $2, 4, 6, \ldots, p - 1$ exceed $\frac{p}{2}$.

There are altogether $\frac{p-1}{2}$ numbers, $\lfloor \frac{p-1}{4} \rfloor$ of which are less than $\frac{p}{2}$, hence the v to be computed is

$$v = \frac{p-1}{2} - \left\lfloor \frac{p-1}{4} \right\rfloor.$$

If $p = 8k + 1$, then $v = 4k - 2k = 2k$, so $\left(\frac{2}{p}\right) = (-1)^{2k} = 1$.

We get the propositions for $p = 8k \pm 3$ and $8k - 1$ similarly. $\qquad \square$

It is easy to check that an equivalent form of Theorem 4.2.2 is

$$\left(\frac{2}{p}\right) = (-1)^{(p^2-1)/8}.$$

Now, we turn to the most important result concerning the Legendre symbol.

Theorem 4.2.3 (Quadratic Reciprocity Law). *If $p > 2$ and $q > 2$ are two distinct primes, then*

(4.2.3) $$\left(\frac{q}{p}\right)\left(\frac{p}{q}\right) = (-1)^{\frac{p-1}{2}\cdot\frac{q-1}{2}},$$

or

$$\left(\frac{q}{p}\right) = \begin{cases} -\left(\frac{p}{q}\right), & \text{if } p \equiv q \equiv -1 \pmod{4} \\ \left(\frac{p}{q}\right), & \text{otherwise.} \end{cases}$$ ♣

Proof. We shall verify two claims:

(A) If $(a, p) = 1$ and a is odd, then

(4.2.4) $$\left(\frac{a}{p}\right) = (-1)^w \qquad \text{where} \qquad w = \sum_{t=1}^{\frac{p-1}{2}}\left\lfloor\frac{ta}{p}\right\rfloor.$$

(B) If b and c are coprime odd numbers greater than 1, then

(4.2.5) $$\sum_{\mu=1}^{(c-1)/2}\left\lfloor\frac{\mu b}{c}\right\rfloor + \sum_{\nu=1}^{(b-1)/2}\left\lfloor\frac{\nu c}{b}\right\rfloor = \frac{b-1}{2}\cdot\frac{c-1}{2}.$$

These imply Theorem 4.2.3: by (4.2.4),

$$\left(\frac{q}{p}\right)\left(\frac{p}{q}\right) = (-1)^z \qquad \text{where} \qquad z = \sum_{\mu=1}^{(p-1)/2}\left\lfloor\frac{\mu q}{p}\right\rfloor + \sum_{\nu=1}^{(q-1)/2}\left\lfloor\frac{\nu p}{q}\right\rfloor,$$

and since

$$z = \frac{p-1}{2}\cdot\frac{q-1}{2}$$

by (4.2.5), (4.2.3) holds.

To prove (A), we apply Gauss's Lemma (Theorem 4.2.1). Keeping the previous notations, it is sufficient to show

(4.2.6) $$w = \sum_{t=1}^{(p-1)/2}\left\lfloor\frac{ta}{p}\right\rfloor \equiv v \pmod{2}.$$

We rewrite the congruences in (4.2.1) as equalities obtained from the division algorithm:

(4.2.7) $$ta = \left\lfloor\frac{ta}{p}\right\rfloor p + \begin{cases} \text{either} & r_i \\ \text{or} & p - s_j. \end{cases}$$

Taking the sum of the equalities (4.2.7) for $t = 1, 2, \ldots, \frac{p-1}{2}$, we obtain

$$\left(1 + 2 + \cdots + \frac{p-1}{2}\right)a = p \sum_{t=1}^{(p-1)/2} \left\lfloor \frac{ta}{p} \right\rfloor + \sum_{i=1}^{u} r_i + \sum_{j=1}^{v} (p - s_j).$$

Since $r_1, \ldots, r_u, s_1, \ldots, s_v$ is a permutation of $1, 2, \ldots, \frac{p-1}{2}$, we get, after ordering

$$(4.2.8) \qquad \left(1 + 2 + \cdots + \frac{p-1}{2}\right)(a - 1) + 2 \sum_{j=1}^{v} s_j = p \left(\sum_{t=1}^{(p-1)/2} \left\lfloor \frac{ta}{p} \right\rfloor + v \right).$$

As a is odd, the left-hand side of (4.2.8) is even. Since $p > 2$, (4.2.6) holds.

To verify (B), consider a rectangle R in the plane with vertices

$$A = (0, 0), \quad B = \left(\frac{b}{2}, 0\right), \quad C = \left(\frac{b}{2}, \frac{c}{2}\right), \quad \text{and} \quad D = \left(0, \frac{c}{2}\right).$$

The right-hand side of (4.2.5) is the number of points with integer coordinates (*lattice points*) inside R.

We show that also the left-hand side of (4.2.5) is the number of these lattice points. We halve the rectangle R along the diagonal $y = \frac{c}{b}x$ connecting A and C. The diagonal itself does not contain lattice points since $(b, c) = 1$.

Now, we count the number n of lattice points inside the lower triangle ABC. Consider such a lattice point on the vertical line $x = v$. Its first coordinate is v and its second coordinate y satisfies $1 \leq y < \frac{c}{b}v$. Thus, there are $\lfloor \frac{vc}{b} \rfloor$ lattice points on this vertical segment. To obtain the total number of lattice points inside the triangle ABC, we sum these values $\lfloor \frac{vc}{b} \rfloor$ for $v = 1, 2, \ldots, \frac{b-1}{2}$:

$$n = \sum_{v=1}^{(b-1)/2} \left\lfloor \frac{vc}{b} \right\rfloor.$$

This is just the second sum on the left-hand side of (4.2.5).

We can verify the same way that counting the lattice points inside the upper triangle ACD by the horizontal lines $y = \mu$, we get the first sum on the left-hand side of (4.2.5). Thus (4.2.5) is proven and so we have completed the proof of Theorem 4.2.3.
□

The next example illustrates how Theorems 4.1.4, 4.2.2, and 4.2.3 can be used to compute a Legendre symbol.

Example. Is the congruence $x^2 \equiv 198 \pmod{1997}$ solvable? (1997 is a prime.)

The standard form of 198 is $198 = 2 \cdot 3^2 \cdot 11$, therefore

$$\left(\frac{198}{1997}\right) = \left(\frac{2}{1997}\right) \left(\frac{3}{1997}\right)^2 \left(\frac{11}{1997}\right).$$

$1997 \equiv -3 \pmod 8$, thus $\left(\frac{2}{1997}\right) = -1$ by Theorem 4.2.2.

$1997 \equiv 1 \pmod 4$, so using Theorem 4.2.3, then $1997 \equiv -5 \pmod{11}$, etc.,

$$\left(\frac{11}{1997}\right) = \left(\frac{1997}{11}\right) = \left(\frac{-5}{11}\right) = \left(\frac{-1}{11}\right)\left(\frac{5}{11}\right) = (-1)\left(\frac{11}{5}\right) = (-1)\left(\frac{1}{5}\right) = -1.$$

Hence,

$$\left(\frac{198}{1997}\right) = (-1) \cdot 1 \cdot (-1) = 1,$$

so $x^2 \equiv 198 \pmod{1997}$ is solvable.

For very large numbers, a problem arises. We have to factor the "numerators" of the Legendre symbols and no fast algorithm is known for that. We shall see in the next section how the Jacobi symbol eliminates this difficulty.

Exercises 4.2

1. Which congruences are solvable?

    (a) $x^2 \equiv 66 \pmod{191}$

    (b) $x^2 \equiv 7! \pmod{83}$

    (c) $x^2 \equiv 94! \pmod{101}$

    (d) $x^2 \equiv 30 \pmod{77}$

    (e) $x^2 \equiv 38 \pmod{187}$

    (f) $2x^2 + 3x + 5 \equiv 0 \pmod{101}$.

2. For which primes $p > 2$ are the following congruences solvable?

    (a) $x^2 \equiv -2 \pmod{p}$

    (b) $x^2 \equiv 3 \pmod{p}$

    (c) $x^2 \equiv -3 \pmod{p}$

    (d) $x^2 \equiv 5 \pmod{p}$

    (e) $x^4 \equiv 4 \pmod{p}$

  * (f) $x^4 \equiv -4 \pmod{p}$

  * (g) $x^8 \equiv 16 \pmod{p}$

  * (h) $x^8 \equiv 81 \pmod{p}$.

3. Verify that if $1999 \mid a^2 + 2b^2$, then $1999 \mid a$ and $1999 \mid b$.

* 4. Prove that $43^{100} \mid 2c^8 + 1$ for some integer c.

5. Demonstrate the following propositions ($c \neq 0$).

    (a) Every prime divisor of $8c^2 - 1$ is of the form $8k \pm 1$ and at least one of them is of the form $8k - 1$.

    (b) Every prime divisor of $12c^2 - 1$ is of the form $12k \pm 1$ and at least one of them is of the form $12k - 1$.

    (c) An odd number $c^2 + 4$ has a prime divisor of the form $8k + 5$ and for $3 \nmid c$ also a prime divisor of the form $12k + 5$ (these two prime divisors may coincide).

6. Let p_1, p_2, p_3, p_4, p_5 be distinct odd primes, $P = p_1 \ldots p_5$, and $a_i = P/p_i$, $i = 1, 2, 3, 4, 5$.

(a) Verify that among the congruences

$$x_i^2 \equiv a_i \pmod{p_i}, \qquad i = 1, 2, 3, 4, 5,$$

an even number is solvable if and only if

$$\sum_{i=1}^{5} \left(\frac{-1}{p_i} \right) = \pm 1.$$

(b) Assume that each congruence

$$z_i^2 \equiv p_i \pmod{a_i}, \qquad i = 1, 2, 3, 4, 5$$

is solvable. Show that

$$\sum_{i=1}^{5} \left(\frac{-1}{p_i} \right) \geq 3.$$

7. (a) Prove that the sum of the squares of 19 consecutive integers is never a power.

* (b) Show that 19 can be replaced by any prime of the form $12k \pm 5$.

S** 8. Construct a polynomial f with integer coefficients so that the equation $f(x) = 0$ has no rational roots but the congruence $f(x) \equiv 0 \pmod{m}$ is solvable for every m.

4.3. Jacobi Symbol

Definition 4.3.1. Let the odd number $m > 1$ have the factorization $m = p_1 \ldots p_r$ into (not necessarily distinct) positive primes. For $(a, m) = 1$, we define the *Jacobi symbol* $\left(\frac{a}{m} \right)$ as the product of the Legendre symbols $\left(\frac{a}{p_i} \right)$:

$$\left(\frac{a}{m} \right) = \left(\frac{a}{p_1} \right) \cdots \left(\frac{a}{p_r} \right). \qquad \clubsuit$$

Example. $\left(\frac{7}{45} \right) = \left(\frac{7}{3} \right)^2 \left(\frac{7}{5} \right) = \left(\frac{2}{5} \right) = -1.$

For m prime, the Jacobi symbol equals the Legendre symbol. Therefore, no ambiguity can arise if we use the same notation for both.

In contrast to prime moduli, the solvability of $x^2 \equiv a \pmod{m}$ cannot be characterized with the Jacobi symbol $\left(\frac{a}{m} \right)$ for composite m (see Exercise 4.3.2).

On the other hand, the Jacobi symbol inherits the properties of the Legendre symbol listed in Theorems 4.1.4, 4.2.2, and 4.2.3.

Theorem 4.3.2. *Assume that the Jacobi symbols below make sense, i.e. every "denominator" is an odd number greater than 1 coprime to the "numerator" (thus e.g. in (v), m and n are coprime odd integers greater than 1).*

(i) $a \equiv b \pmod{m} \implies \left(\frac{a}{m} \right) = \left(\frac{b}{m} \right).$

(ii) $\left(\frac{ab}{m} \right) = \left(\frac{a}{m} \right)\left(\frac{b}{m} \right), \qquad \left(\frac{a}{mn} \right) = \left(\frac{a}{n} \right)\left(\frac{a}{m} \right).$

(iii) $\left(\frac{-1}{m} \right) = \begin{cases} 1, & \text{if } m \equiv 1 \pmod 4 \\ -1, & \text{if } m \equiv -1 \pmod 4. \end{cases}$

(iv) $\left(\frac{2}{m}\right) = \begin{cases} 1, & \text{if } m \equiv \pm 1 \pmod 8 \\ -1, & \text{if } m \equiv \pm 3 \pmod 8. \end{cases}$

(v) $\left(\frac{m}{n}\right) = \begin{cases} -\left(\frac{n}{m}\right), & \text{if } n \equiv m \equiv -1 \pmod 4 \\ \left(\frac{n}{m}\right), & \text{otherwise.} \end{cases}$ ♣

Proof. Each property follows from the definition of the Jacobi symbol and from the corresponding property of the Legendre symbol. We verify this for (v) (i.e. for reciprocity) in detail; the others can be proven similarly.

Let $m = p_1 \dots p_r$ and $n = q_1 \dots q_s$ (where $p_i \neq q_j$). The definition of the Jacobi symbol and the multiplicativity of the Legendre symbol (or properties (ii) of the present theorem) imply

(4.3.1) $$\left(\frac{m}{n}\right) = \prod_{\substack{1 \leq i \leq r \\ 1 \leq j \leq s}} \left(\frac{p_i}{q_j}\right) \quad \text{and} \quad \left(\frac{n}{m}\right) = \prod_{\substack{1 \leq i \leq r \\ 1 \leq j \leq s}} \left(\frac{q_j}{p_i}\right).$$

Denote by u and v the number of primes of the form $4k - 1$ among the p_i and the q_j. Then $\left(\frac{p_i}{q_j}\right) = -\left(\frac{q_j}{p_i}\right)$ for these uv pairs p_i, q_j, and $\left(\frac{p_i}{q_j}\right) = \left(\frac{q_j}{p_i}\right)$ for all other pairs. Hence, by (4.3.1),

$$\left(\frac{m}{n}\right) = -\left(\frac{n}{m}\right) \iff uv \text{ is odd}$$
$$\iff u \text{ and } v \text{ are odd}$$
$$\iff m \equiv n \equiv -1 \pmod 4. \qquad \Box$$

Example. Is the congruence $x^2 \equiv 2342 \pmod{11239}$ solvable? (11239 is a prime.)

We compute the Legendre symbol $\left(\frac{2342}{11239}\right)$ as a Jacobi symbol using Theorem 4.3.2. We have to separate only the largest power of two from the actual numerator, and we can apply the reciprocity directly for the remaining odd part without factoring it.

$$\left(\frac{2342}{11239}\right) = \left(\frac{2}{11239}\right)\left(\frac{1171}{11239}\right) = 1(-1)\left(\frac{11239}{1171}\right) = -\left(\frac{-471}{1171}\right) =$$

$$= -\left(\frac{-1}{1171}\right)\left(\frac{471}{1171}\right) = -(-1)(-1)\left(\frac{1171}{471}\right) = -\left(\frac{229}{471}\right) =$$

$$= -\left(\frac{471}{229}\right) = -\left(\frac{13}{229}\right) = -\left(\frac{229}{13}\right) = -\left(\frac{8}{13}\right) = -\left(\frac{2}{13}\right)^3 = 1.$$

Thus, the congruence has a solution.

The procedure is a variant of the Euclidean algorithm.

The Jacobi symbol plays an important role also in primality testing (see Theorem 5.7.4).

Exercises 4.3

1. Compute the following Jacobi symbols:

   (a) $\left(\dfrac{1234567}{225}\right)$

   (b) $\left(\dfrac{31}{95}\right)$

   (c) $\left(\dfrac{589}{1999}\right)$

   (d) $\left(\dfrac{1113}{11131}\right)$.

2. Let $m > 1$ be an odd number and $(a, m) = 1$.

   (a) Show that if $x^2 \equiv a \pmod{m}$ is solvable, then $\left(\dfrac{a}{m}\right) = 1$.

   (b) Demonstrate with an example that the converse of (a) is false.

   * (c) For which m is the converse of (a) true?

3. Prove that if p is a prime and $p = a^2 + b^2$, then at least one of the congruences

   $$x^2 \equiv a \pmod{p} \quad \text{and} \quad x^2 \equiv b \pmod{p}$$

   is solvable.

4. Compute the sums of Jacobi symbols:

   (a) $\displaystyle\sum_{k=1}^{111} \left(\dfrac{2}{2k+1}\right)$

   (b) $\displaystyle\sum_{k=1}^{111} \left(\dfrac{k}{2k+1}\right)$.

5. Let a, m, and n be greater than 1, m and n odd, and $(a, m) = (a, n) = 1$.

   (a) Prove that if $a \equiv 0$ or $1 \pmod 4$, then

   $$m \equiv n \pmod a \Longrightarrow \left(\dfrac{a}{m}\right) = \left(\dfrac{a}{n}\right).$$

   (b) Show that for $a \equiv 2$ or $3 \pmod 4$ we can find m and n with

   $$m \equiv n \pmod a \quad \text{but} \quad \left(\dfrac{a}{m}\right) \neq \left(\dfrac{a}{n}\right).$$

6. Let $m > 1$ be odd. Compute the sum and product of Jacobi symbols:

   (a) $\displaystyle\sum_{\substack{1 \leq r \leq m \\ (r,m)=1}} \left(\dfrac{r}{m}\right)$

   (b) $\displaystyle\prod_{\substack{1 \leq r \leq m \\ (r,m)=1}} \left(\dfrac{r}{m}\right)$.

7. (a) Determine all odd numbers $m > 1$ satisfying $\left(\dfrac{a}{m}\right) = 1$ for every a coprime to m.

   S* (b) Determine all integers a satisfying $\left(\dfrac{a}{m}\right) = 1$ for every odd $m > 1$ coprime to a.

Chapter 5

Prime Numbers

The notion of primes is very simple, but they form perhaps the most mysterious sequence in mathematics. Euclid's *Elements* contains a proof that there are infinitely many of them, but we do not know whether the same holds for twin primes. After introducing some other similar famous, innocent looking but hopelessly difficult unsolved problems, we shall deal with primes of special forms such as Mersenne and Fermat primes and with primes in arithmetic progressions. Concerning the distribution of primes, we shall establish lower and upper bounds for the number of primes not exceeding x and investigate the sum of reciprocals of the primes. Finally, we shall study how we can determine practically whether a large number is prime or not (primality testing), and how we can factor a large composite number (prime factorization). The amount of time needed to solve these two types of problems differ dramatically (at least according to our present knowledge), and we shall discuss the RSA scheme, the widely applied public key cryptosystem based on this discrepancy.

5.1. Classical Problems

Throughout this chapter, by prime we shall always mean a positive prime number (generally in the sense of a positive irreducible integer) and p will always denote a (positive) prime (so $\prod_{p \leq n} p$ stands for the product of the primes in the interval $(0, n]$).

First we discuss two remarkable results of ancient Greek mathematics.

Theorem 5.1.1. *There are infinitely many primes.* ♣

Proof. Assume the converse, i.e. there exist only finitely many primes, $p_1(= 2)$, ..., p_r. Consider the number $A = p_1 \dots p_r + 1$.

Clearly, A is not divisible by any of the primes $p_1, \dots, p_r$.

As with every integer greater than 1, A has a prime divisor. It must differ from the primes $p_1, \ldots, p_r$, which contradicts the assumption that these were the only primes. $\qquad\square$

Remark: The proof yields also an upper bound

$$p_n < 2^{2^n},$$

where p_n denotes the nth prime number (Exercise 5.1.9a). A much better upper bound will be established in Section 5.4.

Now we present the *sieve of Eratosthenes*. This procedure generates all primes up to a given limit N.

Theorem 5.1.2 (Sieve of Eratosthenes). *We list all integers from 2 to N. In the first step we mark the number 2 and delete all multiples of 2 greater than 2: 4, 6, 8, ... Then we mark the smallest integer not yet marked or deleted; this is the number 3, and then we delete all its multiples greater than itself: 6, 9, ... (6, 12, etc. are deleted the second time).*

We repeat the above process always with the smallest integer not yet marked or deleted as long as this number does not exceed $\sqrt{N}$. If every number up to $\sqrt{N}$ is either marked or deleted, then we stop.

At this point, the remaining numbers (i.e. the marked and the unmarked but undeleted integers together) form all primes not greater than N (the marked ones are the primes not greater than $\sqrt{N}$, whereas those unmarked but undeleted are the primes between $\sqrt{N}$ and N). ♣

Proof. The deleted numbers are clearly composite since they have a proper divisor greater than 1.

We show by induction that the marked numbers are primes. The first marked number, 2 is irreducible. Let now $s \leq \sqrt{N}$ be the kth marked integer, and assume that the first $k - 1$ marked integers constitute all irreducible elements less than s. None of them divides s (since s was not deleted), i.e. s is not divisible by any irreducible element less than s, hence s must be irreducible itself.

Finally, let t be any other undeleted (and unmarked) integer ($\sqrt{N} < t \leq N$). If t were composite, then (e.g. by Exercise 1.4.7a-b) t would have an irreducible factor $p \leq \sqrt{t} \leq \sqrt{N}$. This is a contradiction, however, since t was not divisible by any marked number, i.e. by any irreducible integer up to $\sqrt{N}$. $\qquad\square$

Now we mention a few famous unsolved problems about prime numbers. We shall deal with some of them more in detail in later sections of this chapter.

Twin primes. $\{3, 5\}, \{5, 7\}, \{11, 13\}, \{17, 19\}, \ldots$: Does it occur infinitely often that two consecutive odd integers are both primes?

Remarks: (1) As of Feb. 2019 the largest known twin primes are $2996863034895 \cdot 2^{1290000} \pm 1$ (these numbers have 388342 digits in decimal system).

(2) Replacing 2 by any other even number $2k$, it is unknown whether there exist infinitely many pairs of primes with a difference of $2k$. It was a major break-through, however, when, improving the recent results and ideas of Goldston, Pintz, and Yildirim, Zhang proved in 2013 that there exists such a number $2k < 70000000$. The Polymath8 group led by Terence Tao obtained the presently known best bound $2k \leq 246$ in 2014.

(3) As further generalizations, one can investigate prime triples, quadruples, etc. It is easy to check that each of $n, n + 2$, and $n + 4$ is prime only if $n = 3$, but it is conceivable that n, $n + 2$, and $n + 6$, or even n, $n + 2$, $n + 6$, and $n + 8$ are all primes for infinitely many n, etc. (Cf. with Exercises 1.4.1 and 5.1.1.)

(4) The twin prime problem asks whether the difference of two consecutive primes is very small infinitely often. Another famous conjecture in the opposite direction is that there is always a prime between any two consecutive squares, so the difference of consecutive primes cannot grow too fast. We investigate the gaps between consecutive primes in more detail in Section 5.5.

(5) The twin primes (even if there are infinitely many of them) are very rare among the primes. The sum of their reciprocals converges, whereas the sum of recipro-cals of all primes diverges (see Section 5.6).

(6) Another interesting result is that there exist infinitely many primes p where $p + 2$ is either prime or the product of two primes (i.e. just one step is missing from the solution of the twin prime problem).

Goldbach conjecture. Notice that $4 = 2 + 2, 6 = 3 + 3, 8 = 5 + 3, 10 = 7 + 3$, $12 = 7 + 5, \dots$. Is every even number greater than 2 the sum of two primes?

Remarks: (1) This problem is often called the *even* Goldbach conjecture to distinguish it from the *odd* (or ternary or weak) Goldbach conjecture stating that every odd integer greater than 5 is the sum of three primes. This latter statement immediately follows from the even conjecture (see Exercise 5.1.2), and in contrast to its still unsolved even brother, has been settled completely. The first step was done by Vinogradov in 1937 who showed that every sufficiently large odd integer is the sum of three primes. The proof also yielded an upper bound from where this type of representation holds for the odd numbers, so the remaining task was "just" to check this property for the finitely many odd integers below the bound. Unfortunately, the bound was so huge that the check could not be done till recently even using computers and the newer results decreasing the bound. Finally, Helfgott proved the odd Goldbach conjecture completely in 2013.

(2) Some partial results concerning the (even) Goldbach conjecture:

(A) Every even integer is the sum of at most four primes. This is a direct consequence of the odd Goldbach conjecture: It clearly holds for $2k \leq 8$, and otherwise $2k = 3 + (2k - 3)$ where $2k - 3$ is the sum of three primes. (The first result in this direction was obtained by Schnirelmann in 1930 with a few thousand summands instead of four.)

(B) Every sufficiently large even integer can be written in the form $p + m$ where p is a prime and m is either a prime, or the product of two primes. (The first result in this direction where m is the product of at most k primes with some fixed k was found by Rényi in 1947.)

(C) The even integers possibly not representable as the sum of two primes occur as very rare exceptions (in a precisely defined sense). Unfortunately "rare" cannot be replaced yet by "finitely many".

Long arithmetic progressions. $\{3, 5, 7\}, \{5, 11, 17, 23, 29\}, \{7, 37, 67, 97, 127, 157\},$ ...: Are there arbitrarily long (nonconstant) arithmetic progressions consisting purely of primes? It was a great surprise when Ben Green and Terence Tao proved in 2004 that the answer is yes.

Remarks: (1) It is very hard to exhibit such long arithmetic progressions explicitly. The record length as of February 2019 is 26; one of the record-holders is

$$43142746595714191 + 23681770 \cdot 223092870k, \qquad k = 0, 1, \ldots, 25.$$

Here 223092870 is the product of all primes less than 26, which necessarily divides the difference of any such arithmetic progression (see Exercise 5.1.5).

(2) An infinite arithmetic progression cannot consist purely of primes (see Exercise 1.4.2), but there are infinitely many primes in it if its first (or any other) term and the difference are coprime (Dirichlet's Theorem, see Section 5.3).

Primes of special form.

- Are there infinitely many primes of the form $2^k - 1$ and $2^k + 1$ (Mersenne and Fermat primes, see Section 5.2)?
- Are there infinitely many primes of the form $n^2 + 1$ (cf. Exercise 1.4.6)?
- Are there infinitely many primes among the repunits (having all digits 1 in decimal system), among the integers of the form $333\ldots31$, among the Fibonacci numbers, etc.?

Formulas for primes. Can we establish a formula of practical value that yields the nth prime for every n, or at least an effectively computable function defined on the natural numbers that assumes only prime values (among its infinitely many values)?

Remarks: (1) It is generally agreed that there is no real hope of finding such a function. The formulas in Exercises 5.1.9b and 5.5.9b do not meet the requirement of practical computability.

(2) As noticed by Euler, $n^2 + n + 41$ is a prime for every $0 \le n \le 39$ (but it is composite for $n = 40$). This immediately implies that

$$(n - 40)^2 + (n - 40) + 41 = n^2 - 79n + 1601$$

is a prime for every $0 \le n \le 79$. If we allow polynomials with rational coefficients, then we can construct arbitrarily long such sequences of primes (Exercise 5.1.7). However, a (nonconstant) polynomial cannot yield a general formula for primes since it cannot assume prime values at every integer (Exercise 5.1.8).

(3) On the other hand, we have the following surprising result (also of theoretical significance only): There are polynomials in several variables where on substituting all *non-negative* integers into the variables, the set of *positive* values is exactly the set of all (positive) primes. (Such a polynomial may assume the same prime values at different places and it assumes negative values as well.)

The existence of such a polynomial was first shown by Matiyasevich in 1970 as a by-product when he (crowning the work of many other mathematicians) provided a negative answer to Hilbert's tenth problem: he disproved the existence of a general algorithm that could decide for every Diophantine equation whether or not it has a(n integer) solution. The present records for such polynomials are the following: (i) the minimal degree is 5 with 42 variables; (ii) the minimal number of variables is 10, but then the degree is about $1.6 \cdot 10^{45}$.

Exercises 5.1

See also Exercises 1.4.1–1.4.7.

1. Consider integers $r_1, \ldots, r_k$, and assume that $n + r_1, \ldots, n + r_k$ are all primes for infinitely many integers n. Show that the numbers $r_1, \ldots, r_k$ cannot contain a complete residue system modulo m for any $m > 1$.

2. Prove that

    (a) the odd Goldbach conjecture follows from the even one; and

    (b) the even Goldbach conjecture is equivalent to the proposition that every integer $n \geq 6$ is the sum of three primes.

3. Which even integers are the sums of two (positive) composite numbers? And of two *odd* (positive) composite numbers?

4. Determine all prime pairs where both their sum and difference are primes.

5. Consider an arithmetic progression of primes with n terms and difference $d > 0$. Verify that

    (a) $n = 4$ implies $6 \mid d$

    (b) $n = 6$ implies $30 \mid d$

    S (c) in general, d is divisible by all primes less than n.

6. Show that there are infinitely many composite numbers among each type of numbers listed above in "Primes of special form".

7. Let k be any positive integer. Prove that there exists a polynomial f with rational coefficients where $f(i)$ is the ith prime for every $1 \leq i \leq k$.

8. (a) Let f be any nonconstant polynomial in one variable with integer coefficients. Show that $f(n)$ cannot be a prime for every natural number n.

    (b) Verify the same property also for polynomials

        (i) with rational coefficients

(ii) with complex coefficients

(iii) of several variables.

9. Let p_n denote the nth prime.

   (a) Prove $p_n < 2^{2^n}$.

   (b) Consider

   $$c = \sum_{n=1}^{\infty} \frac{p_n}{10^{2^{2^n}}} = 0.000200000000000000300\ldots,$$

   where the digits in the decimal fraction c are obtained from the decimal expansions of the primes written one after the other and separated by sufficiently many 0 digits to avoid collision. Show that

   $$p_n = \left\lfloor 10^{2^{2^n}} c \right\rfloor - 10^{2^{2^n} - 2^{2^{n-1}}} \cdot \left\lfloor 10^{2^{2^{n-1}}} c \right\rfloor.$$

   (c) Why is the formula in (b) not suitable to determine p_n effectively?

10. Find a number K so that:

   An integer c in the range $10^4 \le c \le 10^8$ is prime if and only if $(c, K) = 1$.

5.2. Fermat and Mersenne Primes

In this section we investigate the primes of the form $2^k + 1$ and $2^k - 1$; they are called *Fermat* and *Mersenne* primes, respectively. As mentioned in the previous section, it is unknown whether or not there exist infinitely many Fermat or Mersenne primes.

In Exercise 1.4.4 we have seen that if $2^k + 1$ is a prime, then k is necessarily a power of two, whereas if $2^k - 1$ is a prime, then k itself must be a prime. Thus it is enough to investigate the Fermat numbers $F_n = 2^{2^n} + 1$ and the Mersenne numbers $M_p = 2^p - 1$ (where p is a prime).

We consider Fermat numbers first. Fermat believed that F_n was always a prime (this is *not* the famous Fermat's Last Theorem to be discussed in Chapter 7). For $0 \le n \le 4$ these are primes (3, 5, 17, 257, and 65537), but Euler showed that $F_5 = 2^{32} + 1$ is composite, since it is divisible by 641.

As of February 2019 we know that F_n is composite for $5 \le n \le 32$ and also for some larger values of n. The record is $F_{3329780}$ (with more than $10^{1000000}$ decimal digits!) having a factor $193 \cdot 2^{3329782} + 1$. No other Fermat primes have been found other than the F_n with $n \le 4$. We have no information about F_{33}. No factors of F_{20} or F_{24} are known (though they are known to be composite). The factorization of F_5, F_6, and F_7 can be found in the table of Fermat numbers at the end of this book (the complete factorization of F_n is known only for $n \le 11$).

The Fermat primes play a central role in the Euclidean constructibility of regular polygons: Gauss's theorem states that a regular N-gon is constructible if and only if the standard form of N, $N \ge 3$ is $N = 2^{\alpha} p_1 \ldots p_r$ where $\alpha \ge 0$, $r \ge 0$, and the numbers p_i are distinct Fermat primes. The first few values are $N = 3, 4, 5, 6, 8, 10, 12, 15, 16, 17, 20, \ldots$.

The following two theorems give practical tools for investigating the Fermat numbers. Theorem 5.2.1 is an effective help in finding their prime divisors and Theorem 5.2.2 yields a (relatively) fast algorithm to test whether a given Fermat number is prime or composite.

Theorem 5.2.1. *Any (positive) divisor of F_n is of the form $k2^{n+1} + 1$, and for $n \geq 2$ it is of the form $r2^{n+2} + 1$.* ♣

Presumably Euler used this theorem for proving that F_5 is composite: the prime divisors of F_5 can only be primes of form $128k + 1$. The first two of these are 257 and 641, and the latter one divides F_5.

Proof. First we verify the statement if the divisor is a prime p. Then $p \mid F_n$ means

(5.2.1) $$2^{2^n} \equiv -1 \ (\text{mod } p).$$

Squaring both sides, we obtain

(5.2.2) $$2^{2^{n+1}} \equiv 1 \ (\text{mod } p).$$

By Theorem 3.2.2(i),

$$2^j \equiv 1 \ (\text{mod } p) \qquad \Longleftrightarrow \qquad o_p(2) \mid j.$$

Hence (5.2.2) implies

$$o_p(2) \mid 2^{n+1},$$

and by (5.2.1), we have

$$o_p(2) \nmid 2^n,$$

since clearly $p > 2$, and thus $-1 \not\equiv 1 \ (\text{mod } p)$. It follows that

$$o_p(2) = 2^{n+1}.$$

Using $o_p(2) \mid p - 1$, we obtain $2^{n+1} \mid p - 1$, so $p = k2^{n+1} + 1$ for a suitable integer k.

If $n \geq 2$, then this implies $p = 8s + 1$, so

$$\left(\frac{2}{p}\right) = 1, \qquad \text{hence} \qquad 2^{\frac{p-1}{2}} \equiv 1 \ (\text{mod } p).$$

Therefore

$$o_p(2) = 2^{n+1} \ \bigg| \ \frac{p-1}{2},$$

so $p = r2^{n+2} + 1$ for a suitable integer r.

These results can be written also as $p \equiv 1 \ (\text{mod } 2^{n+1})$, and for $n \geq 2$, as $p \equiv 1 \ (\text{mod } 2^{n+2})$.

Consider an arbitrary divisor $d \mid F_n$. Write d as the product of (not necessarily distinct) primes (if $d > 1$): $d = p_1 \ldots p_s$. We have just proven that $p_i \equiv 1 \ (\text{mod } 2^{n+1})$ for every i. Multiplying these congruences, we see that also $d \equiv 1 \ (\text{mod } 2^{n+1})$ holds. We can use the same argument also for the modulus 2^{n+2}. □

Theorem 5.2.2 (Pepin's test). *Let $n \geq 1$. Then F_n is prime if and only if*

(5.2.3) $$3^{(F_n-1)/2} \equiv -1 \ (\text{mod } F_n).$$ ♣

Proof. Assume first that F_n is a prime. Then (5.2.3) means that 3 is a quadratic non-residue modulo F_n, i.e.

$$\left(\frac{3}{F_n}\right) = -1.$$

To verify this, we use that $n \geq 1$ yields $2^{2^n} = 4^t$, hence

$$F_n \equiv 1 \pmod{4}, \quad \text{and} \quad F_n = 4^t + 1 \equiv -1 \pmod{3}.$$

Applying quadratic reciprocity, we obtain

$$\left(\frac{3}{F_n}\right) = \left(\frac{F_n}{3}\right) = \left(\frac{-1}{3}\right) = -1.$$

To prove the converse, assume that (5.2.3) holds. Squaring both sides, we get

(5.2.4) $$3^{F_n-1} \equiv 1 \pmod{F_n}.$$

Congruences (5.2.4) and (5.2.3) imply

$$o_{F_n}(3) \mid F_n - 1 \quad \text{and} \quad o_{F_n}(3) \nmid \frac{F_n - 1}{2}.$$

Since $F_n - 1$ is a power of two, we infer

$$o_{F_n}(3) = F_n - 1.$$

Therefore $F_n - 1 \mid \varphi(F_n)$ follows. Clearly $\varphi(F_n) \leq F_n - 1$, therefore $F_n - 1 = \varphi(F_n)$, or equivalently, F_n is a prime.   □

Using Theorem 5.2.2, we can show the compositeness of $F_5 = 2^{32} + 1$ by computing the residue of $3^{2^{31}}$ modulo F_5 by 31 squarings and reducing modulo F_5 in every step. It turns out that this residue is not -1. Moreover, even Fermat's Little Theorem is sufficient for our purposes: 32 such steps of squaring and reduction reveal

$$3^{F_5-1} = 3^{2^{32}} \not\equiv 1 \pmod{F_5},$$

hence F_5 cannot be a prime. So Fermat could have disproved his conjecture about Fermat primes with his own theorem (the computations would have been no obstacle since even more lengthy calculations were regularly done in those times).

Theorem 5.2.2 is an efficient tool in general for determining whether a Fermat number is prime or composite: we can check the validity of (5.2.3) quickly by repeated squarings (and reducing modulo F_n); we need altogether $2^{n-1} \approx \log_2 F_n$ such steps. Unfortunately, the practical application is limited by the fact that the Fermat numbers grow with enormous speed, $F_n \approx F_{n-1}^2$, therefore computers are unable to handle even relatively small values of n.

Now we turn to study the Mersenne numbers $M_p = 2^p - 1$ (where p is a prime). It is easy to see that not all of them are primes; the smallest composite number is obtained for $p = 11$:

$$2^{11} - 1 = 2047 = 23 \cdot 89.$$

The significance of Mersenne primes lies, partly, in their connection with the even perfect numbers (see Theorem 6.3.2). Mersenne was a superb scientific manager in the seventeenth century, corresponding intensively with Fermat, Descartes, and other leading scientists, and encouraged the search for such primes in the hope of finding new perfect numbers.

Mersenne was aware of the difficulty of determining whether a large integer is prime or composite. He wrote in his book in 1644: "To tell if a given number of 15 or 20 digits is prime or not, all time would not suffice for the test, whatever use is made of what is already known." A few pages later, however, we can read his claim that: $2^p - 1$ is a prime for $p = 2, 3, 5, 7, 13, 17, 19, 31, 67, 127, 257$, but for no other values of p below 257.

For more than two centuries, nobody knew whether Mersenne's list was correct or not. The first error was discovered in 1876(!) by another Frenchman, Édouard Lucas, who proved that $2^{67} - 1$ is composite. It is interesting that Lucas proved the compositeness of $2^{67} - 1$ without exhibiting any factors of it (based on Theorem 5.2.4 bearing also his name). The factorization

$$193707721 \cdot 761838257287$$

was found only in 1903(!) by the American mathematician F. N. Cole who spent three years of Sunday afternoons wrestling with the problem (remember, he had to work by hand without computers, since these were invented half a century later).

Later four other errors were discovered in Mersenne's list: the missing $2^{61} - 1$, $2^{89} - 1$, and $2^{107} - 1$ are primes and $2^{257} - 1$ is composite.

The presently (as of February 2019) known 51 Mersenne primes are $2^p - 1$ where $p = 2, 3, 5, 7, 13, 17, 19, 31, 61, 89, 107, 127, 521, 607, 1279, 2203, 2281, 3217, 4253,$ $4423, 9689, 9941, 11213, 19937, 21701, 23209, 44497, 86243, 110503, 132049, 216091,$ $756839, 859433, 1257787, 1398269, 2976221, 3021377, 6972593, 13466917, 20996011,$ $24036583, 25964951, 30402457, 32582657, 37156667, 42643801, 43112609, 57885161,$ $74207281, 77232917,$ and 82589933. The last number, $2^{82589933} - 1$ is the largest known prime—it has 24862048 decimal digits! It is a famous unsolved problem whether there are infinitely many Mersenne primes.

In the table of Mersenne numbers at the end of this book you can find the prime factorization of all composite Mersenne numbers for the (prime) exponents between 10 and 100.

Now we prove the analogues of Theorems 5.2.1 and 5.2.2 for Mersenne numbers.

Theorem 5.2.3. *Let $p > 2$ be a prime. Then any (positive) divisor of $M_p = 2^p - 1$ is of the forms $2kp + 1$ and $8r \pm 1$.* ♣

Example. Consider $p = 47$. Then for any prime divisor q of $M_{47} = 2^{47} - 1$, we have $q = 94k + 1 = 8r \pm 1$. Solving the system of simultaneous congruences

$$x \equiv 1 \pmod{94}, \qquad x \equiv \pm 1 \pmod 8$$

we obtain

$$x \equiv 1 \text{ or } 95 \pmod{376}.$$

The primes satisfying these conditions are

$$q = 1129, 1223, 2351, \ldots$$

We find that $2351 \mid M_{47}$, hence M_{47} is composite.

It is conceivable that also Mersenne found this divisor of M_{47}, and therefore he did not include $p = 47$ into his list (and the missing of this value is not just a lucky coincidence).

Proof. Similar to the argument seen at the Fermat numbers, it is sufficient to prove the statement for prime divisors.

Assume that a prime q satisfies

$$q \mid 2^p - 1, \quad \text{i.e.} \quad 2^p \equiv 1 \pmod{q}.$$

Then $o_q(2) \mid p$, and $o_q(2) \neq 1$, hence $o_q(2) = p$.

We infer $p \mid q - 1$, thus $q = tp + 1$. Since q and p are odd, therefore t is even, so $q = 2kp + 1$.

To verify $q = 8r \pm 1$, we have to show that 2 is a quadratic residue mod q. This follows from the congruence $2^p \equiv 1 \pmod{q}$ by the properties of the Legendre symbol using that p is odd:

$$\left(\frac{2}{q}\right) = \left(\frac{2}{q}\right)^p = \left(\frac{2^p}{q}\right) = \left(\frac{1}{q}\right) = 1. \qquad \square$$

Theorem 5.2.4 (Lucas–Lehmer-test). *Let $p > 2$ be a prime, $a_1 = 4$, and $a_{i+1} = a_i^2 - 2$ for $i \geq 1$. Then M_p is a prime if and only if*

$$(5.2.5) \qquad\qquad\qquad M_p \mid a_{p-1}. \qquad\qquad\qquad \clubsuit$$

Example. Put $p = 5$. Then

$$a_1 = 4, \quad a_2 = 14, \quad a_3 = 194 \equiv 8 \pmod{31}, \quad \text{and} \quad a_4 \equiv 62 \equiv 0 \pmod{31},$$

hence $M_5 = 31$ is a prime.

When checking (5.2.5), we compute the modulo M_p remainders of the a_i, which requires $p - 2 \approx \log_2 M_p$ steps of squaring (plus subtracting and reducing).

Proof. The numbers $a + b\sqrt{3}$ (where a, b are integers) form a (commutative) ring (with identity element and without zero divisors) for the usual operations; we denote this ring by H. In our proof we shall rely on the elementary properties of divisibility, congruences, and order in H (which hold exactly the same way as for the integers). Unique prime factorization is valid in H (see Theorem 10.3.6 and Exercise 10.3.1), but we shall not need this result in our argument.

I. We can easily verify by induction that

$$a_k = (2 + \sqrt{3})^{2^{k-1}} + (2 - \sqrt{3})^{2^{k-1}}$$

holds for every k. Hence (5.2.5) is equivalent to the divisibility

$$(5.2.6) \qquad\qquad M_p \mid (2 + \sqrt{3})^{2^{p-2}} + (2 - \sqrt{3})^{2^{p-2}}.$$

Factoring the right-hand side in (5.2.6), we obtain

$$(5.2.7) \qquad\qquad M_p \mid (2 - \sqrt{3})^{2^{p-2}}\left((2 + \sqrt{3})^{2^{p-1}} + 1\right).$$

We note that the divisibility in (5.2.7) holds among the integers if and only if it is valid in H (see Exercise 5.2.10), and $(2 - \sqrt{3})(2 + \sqrt{3}) = 1$ implies that $2 \pm \sqrt{3}$ raised to integer powers are units in H. Therefore (5.2.7) and thus (5.2.5) are equivalent to the congruence

(5.2.8) $$(2 + \sqrt{3})^{2^{p-1}} \equiv -1 \; (\text{mod } M_p).$$

We conclude that Theorem 5.2.4 can be reformulated as follows: M_p is a prime if and only if (5.2.8) holds.

II. We shall need the following lemma: For any prime $q > 3$, we have

(5.2.9) $$(a + b\sqrt{3})^q \equiv a + \left(\frac{3}{q}\right) b\sqrt{3} \; (\text{mod } q).$$

Proof of the lemma: Consider the binomial expansion

(5.2.10) $$(a + b\sqrt{3})^q = a^q + \binom{q}{1} a^{q-1} b\sqrt{3} + \binom{q}{2} a^{q-2} 3b^2 + \cdots + b^q 3^{(q-1)/2}\sqrt{3}.$$

By Fermat's Little Theorem,

$$a^q \equiv a \; (\text{mod } q) \quad \text{and} \quad b^q \equiv b \; (\text{mod } q),$$

further, each of

$$\binom{q}{1}, \binom{q}{2}, \ldots, \binom{q}{q-1}$$

is divisible by q, and

$$3^{(q-1)/2} \equiv \left(\frac{3}{q}\right) \; (\text{mod } q).$$

Substituting these into (5.2.10), we obtain (5.2.9) as stated.

III. Now we are in the position to show that (5.2.8) implies the primality of M_p. Squaring (5.2.8), we have

(5.2.11) $$(2 + \sqrt{3})^{2^p} \equiv 1 \; (\text{mod } M_p).$$

Let q be a prime divisor of M_p (clearly $q > 3$). Then (5.2.11) and (5.2.8) hold also for the modulus q instead of M_p. This yields (similar to the argument used in the proofs of Theorems 5.2.1 and 5.2.2) that $o_q(2 + \sqrt{3}) = 2^p$.

If $\left(\frac{3}{q}\right) = 1$, then by (5.2.9) we obtain

$$(2 + \sqrt{3})^{q-1} = (2 - \sqrt{3})(2 + \sqrt{3})^q \equiv (2 - \sqrt{3})(2 + \sqrt{3}) = 1 \; (\text{mod } q),$$

hence

$$o_q(2 + \sqrt{3}) = 2^p \le q - 1.$$

But this is impossible since $q \le M_p = 2^p - 1$.

If $\left(\frac{3}{q}\right) = -1$, then similarly

$$(2 + \sqrt{3})^{q+1} \equiv (2 - \sqrt{3})(2 + \sqrt{3}) = 1 \; (\text{mod } q),$$

thus

$$o_q(2 + \sqrt{3}) = 2^p \le q + 1.$$

Comparing this with $q \le M_p = 2^p - 1$, we have $q = M_p$, i.e. M_p is a prime.

IV. Finally, we prove that if M_p is a prime, then (5.2.8) must hold.

We shall use that $M_p \equiv -1 \pmod 8$ implies

(5.2.12) $$\left(\frac{2}{M_p}\right) = 1,$$

further, using $M_p \equiv 1 \pmod 3$, $M_p \equiv -1 \pmod 4$, and the law of reciprocity we get

(5.2.13) $$\left(\frac{3}{M_p}\right) = -\left(\frac{M_p}{3}\right) = -\left(\frac{1}{3}\right) = -1.$$

Starting from the equality

$$2(2 + \sqrt{3}) = (1 + \sqrt{3})^2,$$

we raise both sides to the power $(M_p + 1)/2 = 2^{p-1}$:

(5.2.14) $$2^{(M_p+1)/2} \cdot (2 + \sqrt{3})^{2^{p-1}} = (1 + \sqrt{3})^{M_p+1}.$$

For the first factor on the left-hand side of (5.2.14), using (5.2.12), we obtain

(5.2.15) $$2^{(M_p+1)/2} = 2 \cdot 2^{(M_p-1)/2} \equiv 2\left(\frac{2}{M_p}\right) = 2 \pmod{M_p}.$$

The right-hand side of (5.2.14) can be transformed as follows, applying (5.2.9) with $a + b\sqrt{3} = 1 + \sqrt{3}$ and $q = M_p$, and using (5.2.13):

$$(1 + \sqrt{3})^{M_p+1} = (1 + \sqrt{3})(1 + \sqrt{3})^{M_p}$$

(5.2.16) $$\equiv (1 + \sqrt{3})\left(1 + \left(\frac{3}{M_p}\right)\sqrt{3}\right)$$

$$= (1 + \sqrt{3})(1 - \sqrt{3}) = -2 \pmod{M_p}.$$

Substituting (5.2.15) and (5.2.16) into (5.2.14), we infer

(5.2.17) $$2(2 + \sqrt{3})^{2^{p-1}} \equiv -2 \pmod{M_p}.$$

Multiplying (5.2.17) by 2^{p-1} and using $2^p \equiv 1 \pmod{M_p}$, we obtain the desired congruence (5.2.8). $\qquad\square$

Exercises 5.2

1. (a) Verify $F_{n+1} = F_0 F_1 \dots F_n + 2$.

   (b) Demonstrate that the Fermat numbers are pairwise relatively prime (cf. Exercise 1.3.14).

   (c) Use part (b) to devise a new proof for the existence of infinitely many primes.

   (d) Give a new proof for the statement of Exercise 5.1.9a.

2. Show that Theorem 5.2.2 remains valid for $n \geq 2$ if 3 is replaced by 5 or 10 in formula (5.2.3).

3. Let $n \geq 2$. Prove that $K_n = 5 \cdot 2^n + 1$ is a prime if and only if

   $$3^{(K_n-1)/2} \equiv -1 \pmod{K_n}.$$

4. Verify that $\varphi(N)$ is a power of two if and only if $N = 2^\alpha p_1 \dots p_r$, where $\alpha \geq 0, r \geq 0$ and p_i are distinct Fermat primes.

S 5. For how many values of k can we construct a regular $(2^k - 1)$-gon?

6. Find the smallest prime divisors of the numbers:

   (a) $2^{23} - 1$
   (b) $2^{29} - 1$
   (c) $2^{37} - 1$
   (d) $2^{43} - 1$.

S 7. Prove that M_p is divisible by $2p+1$ if and only if $2p+1$ is a prime and $p \equiv 3 \pmod 4$. (Illustration: $11 \equiv 3 \pmod 4$, $2 \cdot 11 + 1 = 23$ is a prime, and also $23 \mid 2^{11} - 1$.)

8. Assume that for a prime q its square q^2 divides a Fermat number or a Mersenne number. Prove
$$2^{q-1} \equiv 1 \pmod{q^2}.$$

Remark: It is an unsolved problem whether the assumption of the exercise can hold at all; it may well happen that all Fermat and Mersenne numbers are squarefree. It is also unknown, how many primes q satisfy the above congruence; it is possible that the only such primes q are the presently known 1093 and 3511.

S 9. The pairs 8 and 9, 16 and 17, or 31 and 32 are adjacent prime powers (also primes are considered to be prime powers). Characterize all such pairs $n, n + 1$.

10. Let H denote the ring of the numbers $a + b\sqrt 3$ (where a, b are integers, see the proof of Theorem 5.2.4), and let k and n be integers. Show that the divisibility $k \mid n$ holds in H if and only if it is valid among the integers.

* 11. Another unsolved problem is whether there are infinitely many *composite* Fermat numbers. Similarly, we do not know if there are infinitely many primes or composite numbers $H_n = 6^{2^n} + 1$. Prove, however, that there must occur infinitely many composite numbers in at least one of the two sequences F_n and H_n.

5.3. Primes in Arithmetic Progressions

By arithmetic progression we mean an infinite arithmetic progression of integers with a positive difference:

$$a + kd, \qquad \text{where } d > 0 \text{ and } a \text{ are integers, } k = 0, 1, 2, \dots.$$

We saw in Section 5.1 that such a sequence cannot consist purely of primes. Also, if $(a, d) = t > 1$, then every element is divisible by t, thus the sequence can contain one (positive) prime at most. For $(a, d) = 1$, however, the sequence contains infinitely many primes:

Theorem 5.3.1 (Dirichlet's Theorem). *If the integers $d > 0$ and a are coprime, then there are infinitely many primes in the arithmetic progression $a + kd$, $k = 0, 1, 2, \dots$.* ♣

We do not prove this general theorem; we shall verify only a few special cases.

Theorem 5.3.2. *There are infinitely many primes of the form $4k + 3$.* ♣

Proof. We follow the Euclidean ideas seen in Theorem 5.1.1. For a proof by contradiction, we assume that there exist only finitely many primes of the form $4k + 3$. Let them be $p_1 = 3, \ldots, p_r$, and let $A = 4p_1 \ldots p_r - 1$.

Clearly, no p_i divides A.

We write A as a product of primes: $A = q_1 \ldots q_s$ (possibly $s = 1$ or $q_i = q_j$). Every $q_j > 2$, since A is odd. Further, all factors q_j cannot satisfy $q_j \equiv 1 \pmod 4$, because multiplying these congruences would yield $A \equiv 1 \pmod 4$ which is false. Therefore there must be a prime of the form $4k + 3$ among the q_j. This differs from the primes $p_1, \ldots, p_r$, providing thus a contradiction.   □

Theorem 5.3.3. *There are infinitely many primes of the form $4k + 1$.*   ♣

Proof. We need a further refinement of the Euclidean ideas. Again, we assume that there exist only finitely many such primes, $p_1 = 5, \ldots, p_r$. We consider now $A = (2p_1 \ldots p_r)^2 + 1$.

Clearly, no p_i divides A.

Let q be any prime divisor of A. Obviously, $q > 2$. We rewrite the divisibility $q \mid A$ as

$$(2p_1 \ldots p_r)^2 \equiv -1 \pmod q.$$

It follows that the congruence $x^2 \equiv -1 \pmod q$ is solvable, i.e. $q \equiv 1 \pmod 4$. Thus we found a new prime of the form $4k + 1$ which is a contradiction.   □

Using quadratic congruences, we can settle many other special cases of Dirichlet's Theorem, too, see Exercise 5.3.3.

Now we verify Dirichlet's Theorem for any arithmetic progression having 1 as its first term:

Theorem 5.3.4. *For any $m > 0$, there are infinitely many primes in the sequence $mk + 1$, $k = 0, 1, 2, \ldots$.*   ♣

Proof. We shall use the following facts about cyclotomic polynomials and multiple roots of polynomials:

(i) The mth cyclotomic polynomial Φ_m has leading coefficient 1 and its zeros are the complex mth primitive roots of unity. Thus the degree of Φ_m is $\varphi(m)$. Examples:

$$\Phi_4 = x^2 + 1, \qquad \Phi_{11} = x^{10} + x^9 + \cdots + 1.$$

It can be shown that Φ_m has integer coefficients, and

(5.3.1) $\Phi_m \mid x^m - 1$,   moreover, $x^m - 1 = \displaystyle\prod_{d \mid m} \Phi_d.$

(ii) Let F be any (commutative) field and let $f \in F[x]$. An element $\alpha \in F$ is called a multiple root of f if $(x - \alpha)^2 \mid f$. This holds if and only if $f(\alpha) = f'(\alpha) = 0$, where f' is the (formal) derivative of f.

Using the above notions and theorems, we prove first the following lemma of independent interest:

Let c be an integer and q a prime. Then

$$(5.3.2) \qquad o_q(c) = m \iff q \mid \Phi_m(c) \quad \text{and} \quad q \nmid m.$$

Proof of the lemma. Assume first $o_q(c) = m$. Then $m \mid q - 1$, hence $q \nmid m$.

Substitute c for x in (5.3.1):

$$(5.3.3) \qquad c^m - 1 = \prod_{d \mid m} \Phi_d(c).$$

Because $o_q(c) = m$ implies $c^m \equiv 1 \pmod{q}$, q divides the left-hand side of (5.3.3). Since q is a prime, a factor $\Phi_d(c)$ on the right-hand side must be a multiple of q. Due to $\Phi_d(c) \mid c^d - 1$ we get $c^d \equiv 1 \pmod{q}$ for some $d \mid m$. But $o_q(c) = m$, therefore only $d = m$ can occur, so $q \mid \Phi_m(c)$.

Turning to the converse, we assume $q \mid \Phi_m(c)$ and $q \nmid m$. Then $\Phi_m(c) \mid c^m - 1$ implies $c^m \equiv 1 \pmod{q}$. Assuming $o_q(c) = t < m$, we shall arrive at a contradiction. We have $t \mid m$ and $c^t \equiv 1 \pmod{q}$. Applying (5.3.3) for t instead of m, we obtain $q \mid \Phi_d(c)$ for some $d \mid t$. This means that at least two factors are divisible by q on the right-hand side of the original (5.3.3).

We shall consider the identity $x^m - 1 = \prod_{d \mid m} \Phi_d$ in (5.3.1) over the modulo q field $\mathbf{Z}_q$. Then the last sentence of the previous paragraph can be interpreted so that c (as an element of $\mathbf{Z}_q$) is a root of at least two factors in $\prod_{d \mid m} \Phi_d$. This product equals $x^m - 1$, hence c is a multiple root of the polynomial $f = x^m - 1 \in \mathbf{Z}_q[x]$. By (ii), we have $f'(c) = mc^{m-1} = 0$ (in $\mathbf{Z}_q$).

Since $q \nmid m$ and $q \nmid c$, i.e. $m \neq 0$ and $c \neq 0$ in the field $\mathbf{Z}_q$, therefore mc^{m-1} cannot be 0, which is a contradiction. This completes the proof of the lemma.

Turning to the proof of Theorem 5.3.4, we assume that there are only finitely many primes (possibly none) of the form $mk+1$, $p_1, \ldots, p_r$. Define c as $c = vmp_1 \ldots p_r$, where v is any positive integer ($c = vm$ if $r = 0$). If v is large enough, $\Phi_m(c) > 1$.

Let q be any prime divisor of $\Phi_m(c)$. Here $\Phi_m(c) \mid c^m - 1$ guarantees $(q, c) = 1$, hence $q \nmid m$. Thus $o_q(c) = m$, by the lemma.

Therefore $m \mid q - 1$, so q is of the form $q = mk + 1$. Finally, $(q, c) = 1$ implies $q \neq p_i$, which contradicts our assumption that $p_1, \ldots, p_r$ were all primes of the form $mk + 1$. $\qquad \square$

Exercises 5.3

1. How many modulo 9999 residue classes contain a positive prime?

2. Why can one not apply the proof of Theorem 5.3.2 to Theorem 5.3.3 directly, taking $A = 4p_1 \ldots p_r + 1$?

3. Prove without relying on the general form of Dirichlet's Theorem that the arithmetic progressions below contain infinitely many primes:

   (a) $6k + 5$

   (b) $8k + 3$

   (c) $8k + 5$

   (d) $8k + 7$

   (e) $10k + 9$

   (f) $12k + 5$

   (g) $12k + 7$

   (h) $12k + 11$.

4. How many primes have 4321 as last four digits in their decimal representation?

5. Write all primes one after the other following the decimal point. Show that the resulting number $0.235711131719\ldots$ is irrational.

6. For which positive integers a, b, c does the set of numbers $a + bk + cn$ contain infinitely many primes, where $k = 0, 1, 2, \ldots, n = 0, 1, 2, \ldots$?

7. (a) Show that every non-zero integer is a quadratic residue mod p for some suitable prime p.

   (b) Which integers are quadratic non-residues mod p for some suitable prime p?

8. Prove that for any $n > 1$, there exists a polynomial f with integer coefficients of degree n reducible over the rational field such that $f(v_i)$ is a positive prime for each of the suitably chosen integers $v_1, \ldots, v_n$.

9. Show (without relying on the general form of Dirichlet's Theorem), that if there exists a prime of the form $a + kd$ for every pair of coprime integers a and d, then there always exist infinitely many such primes. (This means that the main difficulty in proving Dirichlet's Theorem lies not in guaranteeing the infinitude of such primes but in showing that there exist such primes at all in every suitable arithmetic progression.)

5.4. How Big Is $\pi(x)$?

We denote the number of (positive) primes not exceeding x by $\pi(x)$. For example, $\pi(1) = 0$, $\pi(6.7) = 3$, $\pi(20) = 8$. It is sufficient to investigate the values of $\pi(x)$ for positive integers x.

Though the distribution of primes is very irregular, the asymptotic behavior of $\pi(x)$ can be well characterized by the so-called Prime Number Theorem stated below without proof:

Theorem 5.4.1 (Prime Number Theorem). *Let* log *stand for the natural logarithm. Then*

$$\lim_{x \to \infty} \frac{\pi(x)}{\frac{x}{\log x}} = 1,$$

i.e. $\pi(x)$ *is asymptotically equal to* $\frac{x}{\log x}$.   ♣

Remarks: (1) The Prime Number Theorem refers to the *ratio*, and not to the *difference* of $\pi(x)$ and $\frac{x}{\log x}$. In fact, $\lim_{x \to \infty} \pi(x) - \frac{x}{\log x} = \infty$.

(2) The Prime Number Theorem states that there are approximately $\frac{x}{\log x}$ primes not exceeding x. Whether this is much or few, depends on to which set it is compared. Compared to all positive integers, the primes are very scarce as

$$\lim_{x \to \infty} \frac{\pi(x)}{\lfloor x \rfloor} = \lim_{x \to \infty} \frac{\frac{x}{\log x}}{x} = \lim_{x \to \infty} \frac{1}{\log x} = 0.$$

At the same time, the primes occur much more densely than, for example, the squares since there are $\lfloor \sqrt{x} \rfloor$ squares up to x, and

$$\lim_{x \to \infty} \frac{\pi(x)}{\lfloor \sqrt{x} \rfloor} = \lim_{x \to \infty} \frac{\frac{x}{\log x}}{\sqrt{x}} = \lim_{x \to \infty} \frac{\sqrt{x}}{\log x} = \infty.$$

(3) The Prime Number Theorem was first conjectured at the end of the 18th century by Legendre and Gauss independently. Gauss was just 15 years old, and $x/\log x$ is replaced by the logarithmic integral

$$\mathrm{Li}(x) = \int_2^x \frac{dt}{\log t}$$

in his conjecture. Later it turned out that this integral approximates $\pi(x)$ much better than $x/\log x$. The way towards the proof of the Prime Number Theorem was devised some 70 years later by Riemann, and the first proofs were achieved in 1896 independently by de la Vallée Poussin and Hadamard. Erdős and Selberg found a so-called elementary proof (not relying on deep theorems from analysis) in 1949.

The Prime Number Theorem provides an asymptotic formula for the nth prime.

Theorem 5.4.2. *Let p_n be the nth prime. Then*

$$\lim_{n \to \infty} \frac{p_n}{n \log n} = 1. \tag{5.4.1}$$ ♣

Proof. Since $\pi(p_n) = n$, the Prime Number Theorem implies

$$\lim_{n \to \infty} \frac{\pi(p_n)}{\frac{p_n}{\log p_n}} = \lim_{n \to \infty} \frac{n \log p_n}{p_n} = 1. \tag{5.4.2}$$

The reciprocal of the sequence on the left-hand side of (5.4.1) can be written as

$$\frac{n \log n}{p_n} = \frac{n \log p_n}{p_n} \cdot \frac{\log n}{\log p_n}. \tag{5.4.3}$$

By (5.4.2) and (5.4.3), to prove (5.4.1) we have to show that the limit of the second fraction on the right-hand side of (5.4.3) is 1, i.e.

$$\lim_{n \to \infty} \frac{\log n}{\log p_n} = 1. \tag{5.4.4}$$

Taking the logarithm of (5.4.2) we obtain

$$(5.4.5) \qquad \lim_{n\to\infty} \log\left(\frac{n \log p_n}{p_n}\right) = \lim_{n\to\infty} \left(\log n + \log\log p_n - \log p_n\right) = 0.$$

As $1/(\log p_n)$ is bounded, (5.4.5) implies

$$(5.4.6) \qquad \lim_{n\to\infty} \left(\frac{\log n}{\log p_n} + \frac{\log\log p_n}{\log p_n} - 1\right) = 0.$$

Here

$$\lim_{n\to\infty} \frac{\log\log p_n}{\log p_n} = 0,$$

hence (5.4.4) and thus also (5.4.1) follow from (5.4.6). $\qquad\qquad\square$

In the remaining part of the section we prove a result weaker than the Prime Number Theorem:

Theorem 5.4.3. *There exist positive constants c_1 and c_2 and an x_0 such that every $x \geq x_0$ satisfies*

$$(5.4.7) \qquad c_1 \frac{x}{\log x} < \pi(x) < c_2 \frac{x}{\log x}. \qquad\qquad\clubsuit$$

Remarks: (1) Theorem 5.4.3 means that the order of magnitude of $\pi(x)$ is the same as that of $x/\log x$. This in itself is sufficient to answer several questions, e.g. the density comparisons in Remark 2 after Theorem 5.4.1.

(2) To parallel Theorems 5.4.1 and 5.4.3, the quotient of $\pi(x)$ and $x/\log x$ tends to 1 by the Prime Number Theorem, and stays between two positive constants (for x large enough) by Theorem 5.4.3. It immediately follows that (5.4.7) can hold only with constants $c_1 \leq 1$ and $c_2 \geq 1$. The Prime Number Theorem means that the estimates of Theorem 5.4.3 are valid for *any* constants $0 < c_1 < 1$ and $c_2 > 1$, i.e. there exists an x_0 for *any* constants $0 < c_1 < 1$ and $c_2 > 1$ so that (5.4.7) holds for every $x \geq x_0$. (Moreover, even $c_1 = 1$ is possible; see Remark 1 after Theorem 5.4.1.)

(3) In Theorem 5.4.3 even $x_0 = 2$ is possible (at the price of obtaining worse values for c_1 and c_2), see Exercise 5.4.2.

(4) Theorem 5.4.3 was first proven by Chebyshev in 1850. Below we present Erdős's proof for the lower bound and a joint proof by Erdős and Kalmár for the upper bound.

Proof. I. Lower bound for $\pi(x)$.

We need the following lemma:

Lemma 5.4.4. *Any prime power divisor of the binomial coefficient $\binom{n}{k}$ is less than or equal to n.* $\qquad\qquad\clubsuit$

Proof. Using the standard form of

$$(5.4.8) \qquad \binom{n}{k} = \frac{n!}{k!\,(n-k)!} = \prod_{p\leq n} p^{\beta_p},$$

we have to show $p^{\beta_p} \leq n$, i.e. $\beta_p \leq \lfloor \log_p n \rfloor$.

Consider a prime p, and denote $\lfloor \log_p n \rfloor$ by t. We determine the exponent of p in $n!$, $k!$, and $(n-k)!$ by Legendre's formula (Theorem 1.6.8). The exponent of p in $\binom{n}{k}$ is

$$\beta_p = \quad \left\lfloor \frac{n}{p} \right\rfloor + \left\lfloor \frac{n}{p^2} \right\rfloor + \cdots + \left\lfloor \frac{n}{p^t} \right\rfloor -$$

$$- \left\lfloor \frac{k}{p} \right\rfloor - \left\lfloor \frac{k}{p^2} \right\rfloor - \cdots - \left\lfloor \frac{k}{p^t} \right\rfloor -$$

$$- \left\lfloor \frac{n-k}{p} \right\rfloor - \left\lfloor \frac{n-k}{p^2} \right\rfloor - \cdots - \left\lfloor \frac{n-k}{p^t} \right\rfloor .$$

The sum of terms in each of the t columns is of the form $\lfloor a+b \rfloor - \lfloor a \rfloor - \lfloor b \rfloor$. It follows that each expression is always 0 or 1 (see Exercise 5.4.1), hence $\beta_p \leq t$. $\qquad\square$

Now we turn to the proof of the lower bound for $\pi(x)$. The right-hand side of (5.4.8) is the product of (at most) $\pi(n)$ prime powers, and each of these factors is less than or equal to n, by Lemma 5.4.4. This immediately implies

(5.4.9)
$$\binom{n}{k} = \prod_{p \leq n} p^{\beta_p} \leq n^{\pi(n)}.$$

Summing the inequalities (5.4.9) for $k = 0, 1, \ldots, n$, we get

$$2^n = \sum_{k=0}^{n} \binom{n}{k} \leq (n+1)n^{\pi(n)}.$$

Taking the logarithm, we obtain

$$n \log 2 \leq \log(n+1) + \pi(n) \log n.$$

This yields

(5.4.10)
$$\pi(n) \geq \log 2 \cdot \frac{n}{\log n} - \frac{\log(n+1)}{\log n}.$$

The second term on the right-hand side of (5.4.10) is bounded, hence it is less than (say) $0.01n/\log n$ for n large enough, so

$$\pi(n) > (\log 2 - 0.01)\frac{n}{\log n}.$$

II. Upper bound for $\pi(x)$.

Here again we need a lemma that provides an upper bound for the product of primes not exceeding n:

Lemma 5.4.5. *For any positive integer n, we have*

(5.4.11)
$$\prod_{\substack{p \leq n \\ (p \text{ prime})}} p < 4^n. \qquad\clubsuit$$

Proof. We proceed by induction.

Clearly, (5.4.11) holds for $n = 1, 2$, and 3.

We assume now that it holds for $n = 1, 2, \ldots, m$ (where $m \geq 3$), and show that it is valid also for $n = m + 1$.

If m is odd, then $m + 1 > 2$ is even, so it is composite. Applying the induction hypothesis for $n = m$, we get

$$\prod_{p \leq m+1} p = \prod_{p \leq m} p < 4^m < 4^{m+1}.$$

Let now m be even, $m = 2k$, so $m + 1 = 2k + 1$. We write our product as

$$(5.4.12) \qquad \prod_{p \leq 2k+1} p = \prod_{p \leq k+1} p \cdot \prod_{k+2 \leq p \leq 2k+1} p.$$

We apply the induction hypothesis for $n = k + 1$ to the first product on the right-hand side of (5.4.12):

$$(5.4.13) \qquad \prod_{p \leq k+1} p < 4^{k+1}.$$

We get an upper bound for the second product using the binomial coefficient

$$\binom{2k+1}{k} = \frac{(2k+1)(2k)\ldots(k+2)}{k!}.$$

Every prime $k + 2 \leq p \leq 2k + 1$ occurs in the numerator, but none of them divides the denominator, hence (the integer) $\binom{2k+1}{k}$ is divisible by each of them, so it is a multiple of their product, as well:

$$\prod_{k+2 \leq p \leq 2k+1} p \ \Big| \ \binom{2k+1}{k}.$$

Hence

$$(5.4.14) \qquad \prod_{k+2 \leq p \leq 2k+1} p \leq \binom{2k+1}{k}.$$

Further,

$$(5.4.15) \qquad \binom{2k+1}{k} = \frac{1}{2}\left(\binom{2k+1}{k} + \binom{2k+1}{k+1}\right) < \frac{1}{2} \cdot 2^{2k+1} = 4^k.$$

(5.4.14) and (5.4.15) imply

$$(5.4.16) \qquad \prod_{k+2 \leq p \leq 2k+1} p < 4^k.$$

Finally, substituting (5.4.13) and (5.4.16) into (5.4.12), we obtain the desired inequality

$$\prod_{p \leq 2k+1} p < 4^{2k+1}. \qquad\qquad \square$$

Now we turn to the proof of the upper bound for $\pi(x)$. There are $\pi(n)$ factors on the left-hand side of (5.4.11). To get an upper bound for $\pi(n)$, we try to replace every factor by the smallest prime, i.e. by 2. Unfortunately, this gives only

$$2^{\pi(n)} < \prod_{p \leq n} p < 4^n,$$

yielding $\pi(n) < 2n$ which is worse than the trivial upper bound n.

We refine the method of reducing the product on the left-hand side of (5.4.11) so that we omit the small primes and replace the other factors (roughly) by their minimum:

(5.4.17)
$$\prod_{p \le n} p \ge \prod_{\sqrt{n} < p \le n} p \ge \sqrt{n}^{\pi(n) - \pi(\sqrt{n})}.$$

Combining (5.4.17) and (5.4.11), we get

$$\sqrt{n}^{\pi(n) - \pi(\sqrt{n})} < 4^n.$$

Taking the logarithm gives

$$(\pi(n) - \pi(\sqrt{n})) \log(\sqrt{n}) < n \log 4.$$

Hence

(5.4.18)
$$\pi(n) < 2 \cdot \log 4 \cdot \frac{n}{\log n} + \pi(\sqrt{n}).$$

Finally, using $\pi(\sqrt{n}) < \sqrt{n}$ and

$$\lim_{n \to \infty} \frac{\sqrt{n}}{\frac{n}{\log n}} = \lim_{n \to \infty} \frac{\log n}{\sqrt{n}} = 0,$$

we obtain that $\pi(\sqrt{n})$ is less than (say) $0.01 n / \log n$ for n large enough, thus (5.4.18) implies

$$\pi(n) < (2 \log 4 + 0.01) \frac{n}{\log n}. \qquad \square$$

Exercises 5.4

p always denotes a prime, p_n stands for the nth prime, and $u_n \sim v_n$ means u_n is asymptotically equal to v_n, i.e. $\lim_{n \to \infty} u_n / v_n = 1$.

1. Verify that $\lfloor a + b \rfloor - \lfloor a \rfloor - \lfloor b \rfloor$ equals 0 or 1 for any real numbers a and b.

2. Show that Theorem 5.4.3 holds with $x_0 = 2$, i.e. the corresponding inequalities (5.4.7) are true with suitable positive constants c_1' and c_2' for every real number $x \ge 2$.

* 3. Which lower and upper bounds follow for p_n if (instead of Theorem 5.4.1) we rely on (the weaker) Theorem 5.4.3?

4. Verify the estimates below using the Prime Number Theorem.

   (a) $\sum_{p \le n} \log p \sim n$.

   (b) The product of all primes not exceeding n is approximately e^n in the following sense (cf. Lemma 5.4.5): To any $\varepsilon > 0$ there exists an n_0 such that

   $$e^{(1-\varepsilon)n} < \prod_{p \le n} p < e^{(1+\varepsilon)n}$$

   holds for every $n > n_0$.

* 5. Let $1 \leq a_1 < a_2 < \dots$ be an arbitrary subsequence of the positive integers and let $A(n)$ denote the number of its elements not greater than n, i.e. $A(n) = \sum_{a_i \leq n} 1$. Prove the equivalence of the following four statements.

(i) $A(n) \sim n/\log n$.

(ii) $a_n \sim n \log n$.

(iii) $\sum_{a_i \leq n} \log a_i \sim n$.

(iv) To any $\varepsilon > 0$ there exists an n_0 such that
$$e^{(1-\varepsilon)n} < \prod_{a_i \leq n} a_i < e^{(1+\varepsilon)n}$$
is valid for every $n > n_0$.

Remark: This shows that the statements of Theorems 5.4.1 and 5.4.2, and of Exercise 5.4.4 are strongly correlated for more general sequences than the sequence of primes.

* 6. Let $S(n)$ denote the sum of primes not exceeding n, i.e. $S(n) = \sum_{p \leq n} p$. Prove the estimates for $S(n)$.

(a) There exist positive constants c_3 and c_4 such that
$$c_3 \frac{n^2}{\log n} < S(n) < c_4 \frac{n^2}{\log n}$$
is true for every $n > 1$.

(b) $S(n) \sim n^2/(2 \log n)$.

7. (a) Show that to any K there exists an even integer that has at least K representations as the sum of two primes.

(b) Demonstrate the similar statement for differences instead of sums.

8. Verify
$$\pi(n) = \sum_{j=2}^{n} \left(\left\lfloor \frac{(j-1)!+1}{j} \right\rfloor - \left\lfloor \frac{(j-1)!}{j} \right\rfloor \right).$$
Is this formula suitable for the practical computation of $\pi(n)$?

5.5. Gaps between Consecutive Primes

We show first that there occur arbitrarily large gaps between consecutive primes:

Theorem 5.5.1. *For any positive integer K there exist K consecutive composite numbers.*

♣

Proof. Take any $N > K$, and consider the integers $a_i = N! + i$, $i = 2, 3, \dots, K + 1$. Clearly, $i \mid a_i$ and $a_i > i$, hence every a_i is composite. □

Remark: We can replace $N!$ in the proof by the product of primes not exceeding N.

Generalizing Theorem 5.5.1, we prove now that even both of two consecutive gaps can be arbitrarily large, i.e. there exist primes surrounded by many composite numbers from both sides (these are called *solitary* primes).

Theorem 5.5.2. *For any positive integer K there exists a prime p such that all numbers $p \pm 1, p \pm 2, \ldots, p \pm K$ are composite.* ♣

Proof. We choose a prime $q \geq K + 2$, and consider

$$d = 2 \cdot 3 \ldots (q-2)(q-1)(q+1)(q+2) \ldots (2q-2) = \frac{(2q-2)!}{q}.$$

Here $(q, d) = 1$, thus there exist (infinitely many) $k > 0$ for which $p = q + dk$ is a prime. We show that such a p meets the requirements. For any $1 \leq j \leq q - 2$ we have

$$p \pm j = q + kd \pm j = (q \pm j) + \frac{k(2q-2)!}{q} = (q \pm j)(1 + c_j),$$

where c_j is a positive integer. Thus every $p \pm j$ is composite. □

Now we prove Chebyshev's Theorem stating that there must occur a prime between any number and its double.

Theorem 5.5.3 (Chebyshev's Theorem). *For any integer $n \geq 1$ there exists a prime p satisfying $n < p \leq 2n$.* ♣

This obviously implies that the theorem remains valid for any real numbers $n \geq 1$ (instead of integers).

Another name for this result is *Bertrand's postulate*, because the conjecture was first formulated in 1845 by Bertrand in a slightly stronger form: To every $n > 3$ there is a prime p satisfying $n < p \leq 2n - 2$. (This version is true as well, and even much stronger results hold, see assertions (A) in Theorems 5.5.4 and 5.5.5.) Theorem 5.5.3 was proved by Chebyshev in 1852. The proof below was found by Erdős when he was 19 years old.

Proof. The basic idea is to observe that the product of primes between n and $2n$ is closely related to the binomial coefficient $\binom{2n}{n}$. We assume $n \geq 5$ from now on.

I. We write the standard form of $\binom{2n}{n}$ and break it into the product of three factors according to the size of the primes the following way:

(5.5.1) $$\binom{2n}{n} = \prod_{p \leq 2n} p^{\nu_p} = \prod_{p \leq \sqrt{2n}} p^{\nu_p} \cdot \prod_{\sqrt{2n} < p \leq n} p^{\nu_p} \cdot \prod_{n+1 \leq p \leq 2n} p^{\nu_p}.$$

We denote the three subproducts on the right-hand side of (5.5.1) by A, B, and C. It is sufficient to show $C > 1$, since then there must exist a prime p satisfying $n + 1 \leq p \leq 2n$. (It can be easily shown that every exponent ν_p in C is 1, so C equals the product of primes between n and $2n$, see Exercise 5.5.7a.)

To verify $C > 1$, we establish upper bounds for A and B, and a lower bound for $\binom{2n}{n}$.

II. Lower bound for $\binom{2n}{n}$: Since $\binom{2n}{k} \leq \binom{2n}{n}$ for every $0 \leq k \leq 2n$ (see Exercise 5.5.5)

$$(2n+1)\binom{2n}{n} > \sum_{k=0}^{2n} \binom{2n}{k} = 2^{2n},$$

So

(5.5.2)
$$\binom{2n}{n} > \frac{4^n}{2n+1}.$$

III. Upper bound for A: By Lemma 5.4.4, we have $p^{\nu_p} \leq 2n$, hence

(5.5.3)
$$A = \prod_{p \leq \sqrt{2n}} p^{\nu_p} \leq (2n)^{\pi(\sqrt{2n})} < (2n)^{\sqrt{2n}}.$$

IV. Upper bound for B: Again, $p^{\nu_p} \leq 2n$, by Lemma 5.4.4, and since $p > \sqrt{2n}$, this implies $\nu_p \leq 1$.

We show that $\nu_p = 0$ for ($p > 2$ and) $2n/3 < p \leq n$. Indeed, such a p occurs exactly to the first power both in the numerator and denominator of

$$\binom{2n}{n} = \frac{2n(2n-1)\ldots(n+1)}{n!},$$

it appears only in the factor $2p$ in the numerator, and in the factor p in the denominator.

Hence

(5.5.4)
$$B = \prod_{\sqrt{2n}<p\leq n} p^{\nu_p} = \prod_{\sqrt{2n}<p\leq 2n/3} p^{\nu_p} \leq \prod_{\sqrt{2n}<p\leq 2n/3} p.$$

This and Lemma 5.4.5 imply

(5.5.5)
$$B < \prod_{p\leq 2n/3} p < 4^{2n/3}.$$

V. Substituting (5.5.2), (5.5.3), and (5.5.5) into (5.5.1), and expressing C, we get

(5.5.6)
$$C > \frac{4^n}{(2n+1)(2n)^{\sqrt{2n}} \cdot 4^{2n/3}} > \frac{4^{n/3}}{(2n+1)^{1+\sqrt{2n}}}.$$

To prove $C > 1$ it is sufficient to verify that the logarithm of the expression s_n on the right-hand side of (5.5.6) is positive. Since

(5.5.7) $$\log s_n = \frac{n\log 4}{3} - (1+\sqrt{2n})\log(2n+1) \to \infty \quad \text{as } n \to \infty,$$

$\log s_n > 0$ if n is large enough. A calculation shows that $n > 511$ guarantees positivity, hence $C > 1$ for $n > 511$.

VI. Finally, we verify the statement directly for $n \leq 511$. This can be done by generating a sequence of primes starting with 2 where every element is less than the double of the previous element: 2, 3, 5, 7, 13, 23, 43, 83, 163, 317, 631 is such a sequence. (It is Chebyshev's Theorem which guarantees the existence of an *infinite* sequence with this property.)   □

Related to Chebyshev's Theorem the following more general problem arises concerning the "gap function":

For which functions $h(n)$ is it true that the open interval $(n, n + h(n))$ always contains a prime if n is large enough?

Chebyshev's Theorem asserts that $h(n) = n$ works, but according to Theorem 5.5.1, a constant $h(n)$ is not suitable, as the interval $(n, n + K)$ is primefree for infinitely many n however we fix K.

The order of magnitude of the best $h(n)$ is a famous unsolved problem. We state the related strongest results without proof:

Theorem 5.5.4. (A) *Let $\theta = 0.525$. Then the interval $(n, n + n^\theta)$ contains a prime for every n is large enough.*

(B) *There exists a constant $c > 0$ such that the interval*

$$\left(n, n + \frac{c \cdot \log n \cdot \log \log n \cdot \log \log \log \log n}{\log \log \log n} \right)$$

is primefree for infinitely many positive integers n.   ♣

Both assertions in Theorem 5.5.4 are very deep results (they are much sharper than the ones deducible from the Prime Number Theorem, see Theorem 5.5.5). There is, however, an enormous gulf between them: $h(n)$ can be chosen as n^θ, and cannot be chosen as a function not much bigger than the logarithm. Some probabilistic considerations suggest that the boundary should be around $(\log n)^2$.

It is interesting to note that (A) does not imply even the innocent looking conjecture mentioned in Section 5.1 claiming that every interval between two consecutive squares contains a prime. To prove this conjecture one has to reduce the exponent θ to 1/2 which could not be verified even assuming the famous unproved Riemann Hypothesis.

Another remarkable fact about the difficulties in this field is that the previous best result concerning primefree intervals was achieved in 1936(!), which differed from (B) just in the denominator being squared, and there was no progress at all for nearly 80(!) years, in spite of all efforts and a prize of 10000(!) US dollars offered by Erdős. The five authors of this slight improvement thus got the biggest prize ever given (now with the contribution of Ron Graham) for the solution of an Erdős problem.

In what follows, we show how the results of Theorems 5.5.3 and 5.5.1 can be sharpened using the Prime Number Theorem.

Theorem 5.5.5. (A) *For any $\varepsilon > 0$ there exists an n_0 (depending on ε) such that the interval $(n, (1 + \varepsilon)n)$ contains a prime for every $n > n_0$.*

(B) *For any $0 < \varepsilon < 1$ there exist infinitely many positive integers so that the interval $(n, n + (1 - \varepsilon) \log n)$ is primefree.*   ♣

Proof. To prove (A), we have to verify

(5.5.8) $$\pi\big((1 + \varepsilon)n\big) - \pi(n) > 0$$

for every n large enough. Using the Prime Number Theorem in two different directions, we get

(5.5.9) $$\pi(n) < \left(1 + \frac{\varepsilon}{4}\right) \cdot \frac{n}{\log n}$$

on the one hand, and

(5.5.10) $$\pi\big((1+\varepsilon)n\big) > \Big(1 - \frac{\varepsilon}{4}\Big) \cdot \frac{(1+\varepsilon)n}{\log\big((1+\varepsilon)n\big)}$$

on the other hand, if n is sufficiently large. Further,

(5.5.11) $$\log\big((1+\varepsilon)n\big) = \log(1+\varepsilon) + \log n < \Big(1 + \frac{\varepsilon}{4}\Big)\log n.$$

From (5.5.9), (5.5.10), and (5.5.11) we get

(5.5.12) $$\pi\big((1+\varepsilon)n\big) - \pi(n) > \left(\frac{\big(1 - \frac{\varepsilon}{4}\big)(1+\varepsilon)}{1 + \frac{\varepsilon}{4}} - \Big(1 + \frac{\varepsilon}{4}\Big)\right)\frac{n}{\log n}.$$

The coefficient of $n/\log n$ on the right-hand side of (5.5.12) is

$$\frac{\big(1 - \frac{\varepsilon}{4}\big)(1+\varepsilon) - \big(1 + \frac{\varepsilon}{4}\big)^2}{1 + \frac{\varepsilon}{4}} = \frac{\frac{\varepsilon}{4}\big(1 - \frac{5\varepsilon}{4}\big)}{1 + \frac{\varepsilon}{4}} > 0$$

(since we may assume $\varepsilon < 4/5$), hence (5.5.8) follows from (5.5.12).

We apply a proof by contradiction for (B): we assume that the interval $\big(n, n + (1-\varepsilon)\log n\big)$ contains a prime for every $n > n_0$ for some given $\varepsilon > 0$ and n_0.

We fix a large integer N and consider all primes between n_0 and N: $n_0 < p_r < p_{r+1} < \cdots < p_k \le N$. Using our assumption, we obtain the inequalities

(5.5.13)
$$\begin{aligned}
p_{r+1} &< & p_r + & (1-\varepsilon)\log p_r \\
p_{r+2} &< & p_{r+1} + & (1-\varepsilon)\log p_{r+1} \\
&\vdots \\
p_{k+1} &< & p_k + & (1-\varepsilon)\log p_k.
\end{aligned}$$

Summing the inequalities in (5.5.13), the terms $p_{r+1}, \ldots, p_k$ get cancelled, and we obtain

(5.5.14) $$p_{k+1} < p_r + (1-\varepsilon)\sum_{j=r}^{k}\log p_j.$$

By the definition of p_k, we have $p_{k+1} > N$, thus to get a contradiction it is sufficient to show that the right-hand side of (5.5.14) is less than N.

To achieve this, we estimate the right-hand side of (5.5.14) from above the following way:

(5.5.15) $$p_r + (1-\varepsilon)\sum_{j=r}^{k}\log p_j < p_r + (1-\varepsilon)\pi(N)\log N.$$

If N is large enough, then

(5.5.16) $$\pi(N) < \Big(1 + \frac{\varepsilon}{4}\Big)\frac{N}{\log N}$$

by the Prime Number Theorem and

(5.5.17) $$p_r < \frac{\varepsilon N}{4}$$

if N is large enough. Substituting (5.5.16) and (5.5.17) into (5.5.15), we obtain that the right-hand side of (5.5.14) is less than

$$\left((1-\varepsilon)\left(1+\frac{\varepsilon}{4}\right)+\frac{\varepsilon}{4}\right)N < \left(1-\frac{\varepsilon}{2}\right)N < N,$$

yielding the desired contradiction.   □

Exercises 5.5

1. Prove that $n!$ is not a perfect power if $n > 1$.

2. Verify that at least one of any two consecutive integers is representable as the sum of distinct primes (we allow sums consisting of a single term).

3. Demonstrate that infinitely many primes have

   (a) 1 as first digit

   (b) 4 as the first thousand digits in decimal system.

4. Prove that neither of the following sums is an integer for $1 \le k < n$:

   (a) $\displaystyle\sum_{j=1}^{n} \frac{1}{j}$

   (b) $\displaystyle\sum_{j=k}^{n} \frac{1}{j}$.

5. Show that $\binom{2n}{n}$ is the largest among the binomial coefficients $\binom{2n}{k}$, $0 \le k \le 2n$.

6. Give another proof for Theorem 5.5.2 on the following lines: Choose $2K$ primes greater than K, $p_1, \ldots, p_K, q_1, \ldots, q_K$, and consider the system of simultaneous congruences

   $$x \equiv j \pmod{p_j}, \qquad x \equiv -j \pmod{q_j}, \qquad j = 1, 2, \ldots, K.$$

   Show that the solutions contain (infinitely many) primes p and they meet the requirements of the theorem.

7. (a) Prove that $\binom{2n}{n}$ is divisible by exactly the first power of every prime $n + 1 \le p \le 2n$.

   (b) Show that if $p > 3$ is a prime and $2n/5 < p \le n/2$, then $\binom{2n}{n}$ is not divisible by p. How can we generalize this observation?

8. Show that the proof of Chebyshev's Theorem yields the following stronger result (for $n \ge 2$): There are more than $cn/\log n$ primes between n and $2n$ where c is a suitable positive constant.

S 9. (a) Using (A) in Theorem 5.5.4, verify that there is a prime between any two sufficiently large consecutive cubes.

* (b) Prove the existence of a real number $\alpha > 1$ such that $\lfloor \alpha^{3^n} \rfloor$ is a prime for every positive integer n.

  (c) Why can one not generate large primes practically with the formula in (b)?

10. Establish results similar to (B) in Theorem 5.5.5 using the following facts or methods instead of the Prime Number Theorem:

    (a) Theorem 5.4.3

    (b) the proof of Theorem 5.5.1

    (c) the Remark after the proof of Theorem 5.5.1.

* 11. (Cf. with Remark 2 on twin primes in Section 5.1.) Prove that for any $\varepsilon > 0$ there exist infinitely many positive integers n satisfying $p_{n+1} - p_n < (1 + \varepsilon) \log n$. (As usual, p_n denotes the nth prime.)

5.6. The Sum of Reciprocals of Primes

In this section we prove that the infinite series of the reciprocals of primes is divergent. This means that the reciprocals of primes decrease slowly, i.e. the primes themselves grow slowly, so they occur fairly densely among the positive integers. As a comparison, the infinite series of the reciprocals of squares converges, i.e. the squares form a rare subsequence of the natural numbers (cf. with Remark 2 after Theorem 5.4.1).

We present three proofs for the divergence of the sum of reciprocals of primes. The first one shows that this follows from the Prime Number Theorem (or even from the weaker Theorem 5.4.3). The second one is an ingenious proof by contradiction of Erdős. The third one is due to Euler who was the first to state and prove this theorem.

Finally, we show that the sum of reciprocals of primes not exceeding x can be approximated extremely well by the function $\log \log x$.

Theorem 5.6.1. *The infinite series of the reciprocals of primes diverges, i.e.*

$$\sum_p \frac{1}{p} = \infty. \qquad \clubsuit$$

First proof. We have to show

$$(5.6.1) \qquad \lim_{n \to \infty} \sum_{j=1}^{n} \frac{1}{p_j} = \infty,$$

where p_j denotes the jth prime.

By Theorem 5.4.2 (or Exercise 5.4.3), there exist c and n_0 such that $p_j < c j \log j$ for every $j \geq n_0$. Hence

$$(5.6.2) \qquad \sum_{j=1}^{n} \frac{1}{p_j} > \frac{1}{c} \sum_{j=n_0}^{n} \frac{1}{j \log j}.$$

For every integer $n_0 \leq j \leq n$, we draw a rectangle so that its base is the segment $[j, j+1]$ on the x-axis, and its height is $1/(j \log j)$. Then the sum of the areas of the rectangles is just the sum on the right-hand side of (5.6.2) (without the multiplier $1/c$).

As the function $1/(x \log x)$ is strictly decreasing for $x > 1$, in the interval $[n_0, n+1]$, its graph lies in the region formed by the rectangles. Hence, the area below the graph of the function is less than the total area of the rectangles, i.e.

$$(5.6.3) \qquad \sum_{j=n_0}^{n} \frac{1}{j \log j} > \int_{n_0}^{n+1} \frac{dx}{x \log x}.$$

Computing the integral on the right-hand side of (5.6.3), we obtain

$$(5.6.4) \qquad \int_{n_0}^{n+1} \frac{dx}{x \log x} = \left[\log \log x\right]_{n_0}^{n+1} = \log \log(n + 1) - \log \log n_0.$$

Combining (5.6.2), (5.6.3), and (5.6.4), we get

$$(5.6.5) \qquad \sum_{j=1}^{n} \frac{1}{p_j} > \frac{1}{c}\left(\log \log(n + 1) - \log \log n_0\right).$$

Since

$$\lim_{n \to \infty} \log \log n = \infty,$$

the right-hand side in (5.6.5) tends to infinity if $n \to \infty$. But then the same holds also for the left-hand side, so (5.6.1) is true. $\qquad \square$

Remark: The proof yields

$$(5.6.5a) \qquad \sum_{p \le n} \frac{1}{p} > c' \log \log n.$$

with a suitable positive constant c' if n is large enough. We can show similarly that

$$\sum_{p \le n} \frac{1}{p} < c'' \log \log n.$$

A slightly more refined use of the Prime Number Theorem (or equivalently, of Theorem 5.4.2) gives

$$\sum_{p \le n} \frac{1}{p} \sim \log \log n.$$

Much sharper estimates will be obtained in Theorem 5.6.2 (without relying on the Prime Number Theorem). Even (5.6.13) in our third proof of Theorem 5.6.1 is much better than (5.6.5a).

Second proof. For a proof by contradiction, assume that the sum of reciprocals of primes converges. Then

$$(5.6.6) \qquad \sum_{j=k+1}^{\infty} \frac{1}{p_j} < \frac{1}{2}$$

for some k. We fix k, and divide the positive integers into two groups: the first group consists of the numbers with a prime divisor greater than p_k, and the second group is formed by the numbers with all prime divisors less than or equal to p_k.

Let N be a (large) natural number, and consider the set $H = \{1, 2, \ldots, N\}$. We show that each of the two groups contains less than the half of the elements in H for N large enough, which is a contradiction.

We start with the first group. There are $\lfloor \frac{N}{p} \rfloor$ elements in H divisible by a prime p. This yields the following upper bound for the size of the first group:

$$\sum_{p_k < p \leq N} \left\lfloor \frac{N}{p} \right\rfloor \leq \sum_{p_k < p \leq N} \frac{N}{p} < N \sum_{j=k+1}^{\infty} \frac{1}{p_j} < \frac{N}{2}$$

(we used (5.6.6) in the last step). This means that fewer than half of the elements in H belong to the first group.

To investigate the second group, we shall use the fact that every positive integer has a (unique) representation as a product of a square and a squarefree number. This is a direct consequence of unique prime factorization: separating the even and odd exponents in the standard form of n,

$$n = q_1^{2\beta_1} \ldots q_r^{2\beta_r} q_{r+1}^{2\beta_{r+1}+1} \ldots q_s^{2\beta_s+1}$$

($r = 0$ or $r = s$ may occur), we obtain the required representation as

$$n = \left(q_1^{\beta_1} \ldots q_r^{\beta_r} q_{r+1}^{\beta_{r+1}} \ldots q_s^{\beta_s} \right)^2 \cdot (q_{r+1} \ldots q_s).$$

We write the elements of the second group in H in the form $a^2 b$ where b is squarefree. Then $1 \leq a \leq \lfloor \sqrt{N} \rfloor$, and b is the product of some of the primes $p_1, \ldots, p_k$ (possibly of all of them, or b can be also the empty product when $b = 1$).

Hence, a^2 can assume $\lfloor \sqrt{N} \rfloor$ values, and b can be chosen in 2^k ways (this is the number of subsets in the set $\{p_1, \ldots, p_k\}$). Thus there are at most $\sqrt{N} \cdot 2^k$ such products $a^2 b$. Since k is fixed, $2^k < \sqrt{N}/2$ for N large enough, so $\sqrt{N} \cdot 2^k < N/2$. This proves that fewer than half of the elements in H belong to the second group. $\square$

Third proof. We shall use the following theorems from analysis:

(i) $\sum_{j=1}^{n} \frac{1}{j} > \log n$

(ii) $\sum_{j=1}^{\infty} \frac{1}{j^2} < 2$

(iii) $\log \frac{1}{1-x} = x + \frac{x^2}{2} + \frac{x^3}{3} + \ldots \leq x + x^2$ if $0 \leq x \leq \frac{1}{2}$.

To prove our theorem, we consider the product

$$A_n = \prod_{p \leq n} \left(1 + \frac{1}{p} + \frac{1}{p^2} + \cdots + \frac{1}{p^{\nu_p}} \right),$$

where $n > 1$ is an integer and

$$p^{\nu_p} \leq n < p^{\nu_p+1}, \quad \text{so} \quad \nu_p = \lfloor \log_p n \rfloor.$$

We claim that

(5.6.7)
$$A_n \geq \sum_{j=1}^{n} \frac{1}{j}.$$

To illustrate the idea, we consider first $n = 10$, and write the factors of A_{10} in detail:

$$A_{10} = \left(1 + \frac{1}{2} + \frac{1}{2^2} + \frac{1}{2^3}\right)\left(1 + \frac{1}{3} + \frac{1}{3^2}\right)\left(1 + \frac{1}{5}\right)\left(1 + \frac{1}{7}\right).$$

For $j \leq 10$, the standard form of j may contain only the primes 2, 3, 5, and 7 with an exponent not greater than the ones in the corresponding factors of A_{10}. Therefore $j \leq 10$ is a (unique) product of these prime powers. This means that performing the multiplication in A_{10}, we shall obtain the reciprocals of all integers $j \leq 10$ (and of some others, too), thus $A_{10} \geq \sum_{j=1}^{10} 1/j$.

Applying the same argument for any n instead of 10, we obtain (5.6.7). Using (i), we infer

(5.6.8) $$A_n > \log n.$$

Now we establish an upper bound for A_n. The summation of the geometric series in the factors of A_n gives

(5.6.9) $$A_n = \prod_{p \leq n} \frac{1 - \left(\frac{1}{p}\right)^{v_p + 1}}{1 - \frac{1}{p}} < \prod_{p \leq n} \frac{1}{1 - \frac{1}{p}}.$$

By (5.6.8) and (5.6.9), we have

(5.6.10) $$\log n < \prod_{p \leq n} \frac{1}{1 - \frac{1}{p}}.$$

Taking the logarithm of (5.6.10), we get

(5.6.11) $$\log \log n < \sum_{p \leq n} \log \frac{1}{1 - \frac{1}{p}}.$$

We estimate the right-hand side of (5.6.11) by (iii):

(5.6.12) $$\log \log n < \sum_{p \leq n} \frac{1}{p} + \sum_{p \leq n} \frac{1}{p^2}.$$

Finally, the second sum on the right-hand side of (5.6.12) is less than 2 by (ii), hence

(5.6.13) $$\sum_{p \leq n} \frac{1}{p} > \log \log n - 2,$$

which implies the theorem. $\qquad\square$

We observe from the third proof that the sum of reciprocals of primes not greater than n cannot be much less than $\log \log n$ (see (5.6.13)). We sharpen this result by showing that the difference of this sum of reciprocals and of $\log \log n$ is bounded:

Theorem 5.6.2. *There exists a constant c such that*

(5.6.14) $$\left|\sum_{p \leq n} \frac{1}{p} - \log \log n\right| < c$$

holds for every integer $n \geq 3$. ♣

Proof. We shall need an estimate for the sum $\sum_{p \leq n} (\log p)/p$.

Theorem 5.6.3. *There exists a constant c' such that*

$$(5.6.15) \qquad \left| \sum_{p \leq n} \frac{\log p}{p} - \log n \right| < c'$$

holds for every integer $n \geq 2$.   ♣

Proof. We take the logarithm of the standard form of $n!$ (see Theorem 1.6.8):

$$(5.6.16) \qquad \log n! = \sum_{p \leq n} \log p \left(\left\lfloor \frac{n}{p} \right\rfloor + \left\lfloor \frac{n}{p^2} \right\rfloor + \left\lfloor \frac{n}{p^3} \right\rfloor + \dots \right).$$

We shall show that the left-hand side of (5.6.16) is about $n \log n$, and we can omit the floor in the multiplier of $\log p$ on the right-hand side and only the first term counts, i.e. the right-hand side is about $n \sum_{p \leq n} (\log p)/p$. Then dividing by n, we get (5.6.15).

Let us see the details. To estimate $\log n!$ on the left-hand side of (5.6.16), we use

$$\left(\frac{n}{e} \right)^n < n! < n^n$$

for $n \geq 2$. The upper bound is obvious, and the lower bound can be easily verified by induction. Taking the logarithm, we obtain

$$(5.6.17) \qquad n(\log n - 1) < \log n! < n \log n.$$

The sum $\lfloor n/p \rfloor + \lfloor n/p^2 \rfloor + \dots$ can be estimated as follows:

$$(5.6.18) \qquad \frac{n}{p} - 1 < \left\lfloor \frac{n}{p} \right\rfloor + \left\lfloor \frac{n}{p^2} \right\rfloor + \dots < \frac{n}{p} + \frac{n}{p^2} + \dots = \frac{n}{p} + \frac{n}{p(p-1)}.$$

Denoting the right-hand side of (5.6.16) by J, we get the following bounds from (5.6.18):

$$(5.6.19) \qquad n \sum_{p \leq n} \frac{\log p}{p} - \sum_{p \leq n} \log p < J < n \sum_{p \leq n} \frac{\log p}{p} + n \sum_{p \leq n} \frac{\log p}{p(p-1)}.$$

By Lemma 5.4.5,

$$(5.6.20) \qquad \sum_{p \leq n} \log p = \log \prod_{p \leq n} p < \log 4^n = n \log 4,$$

further

$$(5.6.21) \qquad \sum_{p \leq n} \frac{\log p}{p(p-1)} < \sum_{k=2}^{\infty} \frac{\log k}{k(k-1)},$$

where the infinite series on the right-hand side of (5.6.21) is convergent and it can be shown that its sum is less than 4. Using (5.6.20) and (5.6.21), we infer from (5.6.19) that

$$(5.6.22) \qquad \left| \frac{J}{n} - \sum_{p \leq n} \frac{\log p}{p} \right| < 4.$$

At the same time, $J = \log n!$ by (5.6.16), hence (5.6.17) implies

$$(5.6.23) \qquad \left| \frac{J}{n} - \log n \right| < 1.$$

Finally, (5.6.22) and (5.6.23) guarantee (5.6.15) (with $c' = 5$).   □

Turning to the proof of Theorem 5.6.2, it is more convenient to extend Theorem 5.6.3 from the integers to every real number $x \geq 2$. Observe that

$$\sum_{p \leq x} \frac{\log p}{p} = \sum_{p \leq \lfloor x \rfloor} \frac{\log p}{p} \quad \text{and} \quad |\log x - \log \lfloor x \rfloor| = \log \frac{x}{\lfloor x \rfloor} < \log \frac{3}{2},$$

thus (5.6.15) implies

$$\left| \sum_{p \leq x} \frac{\log p}{p} - \log x \right| \leq \left| \sum_{p \leq \lfloor x \rfloor} \frac{\log p}{p} - \log \lfloor x \rfloor \right| + |\log \lfloor x \rfloor - \log x| < c' + \log \frac{3}{2}.$$

Thus we verified that

(5.6.24) $$\left| \sum_{p \leq x} \frac{\log p}{p} - \log x \right| < 6$$

holds for every real number $x \geq 2$.

We define the following functions for every real number $x \geq 2$:

(5.6.25) $$f(x) = \sum_{p \leq x} \frac{\log p}{p}, \quad g(x) = \frac{1}{\log x}, \quad \text{and} \quad h(x) = f(x) - \log x.$$

Then $f(2)g(2) = 1/2$, and for any integer $k \geq 3$ we have

$$(f(k) - f(k-1))g(k) = \begin{cases} \frac{1}{k}, \text{if } k \text{ is a prime} \\ 0, \text{if } k \text{ is not a prime.} \end{cases}$$

This implies

(5.6.26) $$\sum_{p \leq n} \frac{1}{p} = f(2)g(2) + \sum_{k=3}^{n} (f(k) - f(k-1))g(k)$$

for every integer $n \geq 3$. Rewriting the right-hand side of (5.6.26) by Abel's partial summation, we obtain

(5.6.27) $$\sum_{p \leq n} \frac{1}{p} = f(2)(g(2) - g(3)) + f(3)(g(3) - g(4)) + \dots$$
$$\dots + f(n-1)(g(n-1) - g(n)) + f(n)g(n).$$

We show that a general term on the right-hand side of (5.6.27) (except the last one) can be transformed into

(5.6.28) $$f(k)(g(k) - g(k+1)) = -\int_{k}^{k+1} f(t)g'(t)\,dt.$$

Indeed, the function $f(t)$ assumes the constant value $f(k)$ on the interval $[k, k+1)$ (closed from the left and open from the right), further

$$\int_{k}^{k+1} g'(t)\,dt = g(k+1) - g(k)$$

by the Newton–Leibniz law.

Combining (5.6.27) and (5.6.28), we get

$$(5.6.29) \qquad \sum_{p \leq n} \frac{1}{p} = f(n)g(n) - \int_2^n f(t)g'(t)\,dt.$$

Now we compute the integral on the right-hand side of (5.6.29). Using

$$f(t) = \log t + h(t) \quad \text{and} \quad g'(t) = \left(\frac{1}{\log t}\right)' = \frac{-1}{t(\log t)^2}$$

we have

$$(5.6.30) \qquad -\int_2^n f(t)g'(t)\,dt = \int_2^n \frac{dt}{t \log t} + \int_2^n \frac{h(t)\,dt}{t(\log t)^2}.$$

The first integral on the right-hand side of (5.6.30) is

$$(5.6.31) \qquad \int_2^n \frac{dt}{t \log t} = \left[\log \log t\right]_2^n = \log \log n - \log \log 2.$$

To estimate the second integral on the right-hand side of (5.6.30), we rely on $|h(t)| < 6$ (which follows from (5.6.24) and (5.6.25)):

$$(5.6.32) \qquad \left|\int_2^n \frac{h(t)\,dt}{t(\log t)^2}\right| < 6 \int_2^n \frac{dt}{t(\log t)^2} = 6\left[\frac{-1}{\log t}\right]_2^n = \frac{6}{\log 2} - \frac{6}{\log n}.$$

Substituting (5.6.32) and (5.6.31) into (5.6.30), we obtain

$$(5.6.33) \qquad -\int_2^n f(t)g'(t)\,dt = \log \log n + s(n), \quad \text{where } s(n) \text{ is bounded.}$$

Now we verify that the product $f(n)g(n)$ on the right-hand side of (5.6.29) is bounded:

$$(5.6.34) \qquad |f(n)g(n)| = \left|\frac{\log n + h(n)}{\log n}\right| = \left|1 + \frac{h(n)}{\log n}\right| < 1 + 6 = 7.$$

Finally, combining (5.6.29), (5.6.33), and (5.6.34), we obtain the statement of Theorem 5.6.2. □

Remark: Repeating the estimate in (5.6.32) for the interval $[n, N]$ instead of $[2, n]$, it turns out that the second integral on the right-hand side of (5.6.30) has a limit as $n \to \infty$, and the difference between the integral and the limit is at most $6/\log n$ in absolute value. The same is obvious for $f(n)g(n)$. This proves that with suitable constants c_1 and c_2,

$$\left|\sum_{p \leq n} \frac{1}{p} - \log \log n - c_1\right| \leq \frac{c_2}{\log n}$$

is valid for every integer $n \geq 3$.

Exercises 5.6

S 1. Let L be a fixed positive integer. Consider the following sequences of positive integers and determine whether the series of the reciprocals of their elements converge or diverge:

    (a) the multiples of L

    (b) the perfect powers

    (c) the squarefree numbers

    (d) the integers with no prime divisor greater than L

    (e) the integers with no prime divisors less than L

    (f) the squareful numbers, i.e. the ones where no prime has exponent 1 in the standard form.

Examine in each case except for (c), about how many elements are in the sequence up to some large n; more precisely, find asymptotics or good estimates for the counting function $U(n) = \sum_{u_i \leq n} 1$ of the sequence $U = \{u_1 < u_2 < \dots\}$. (For the squarefree numbers see Exercise 6.7.2.)

2. Using the integral criterion seen in the first proof of Theorem 5.6.1, determine whether the following infinite series converge or diverge:

    (a) $\displaystyle\sum_{n=1}^{\infty} \frac{1}{n^{1.01}}$

    (b) $\displaystyle\sum_{n=2}^{\infty} \frac{1}{n(\log n)^2}$

    (c) $\displaystyle\sum_{n=2}^{\infty} \frac{1}{n \cdot \log n \cdot \log\log n}$.

3. In the infinite series below, the summation is over all primes. Investigate the question of convergence or divergence:

    (a) $\displaystyle\sum_{p} \frac{1}{p \log p}$

    (b) $\displaystyle\sum_{p} \frac{1}{p \log\log p}$.

4. Consider sequences $a_1 < a_2 < \dots$ of positive integers with the properties below. What can be asserted about the convergence/divergence of the infinite series of the reciprocals of their elements? (Possible answers: always convergent—always divergent—can be convergent, but can be divergent, as well.)

    (a) The elements a_n are pairwise coprime composite numbers.

    (b) The sum of exponents of primes in the standard form of a_n is at least $2 \log n$ for every n.

    (c) $a_{n+1} - a_n < 10^{1000}$ for every n.

    (d) $a_{n+1}/a_n < 1.00001$ for every n.

    (e) No two a_n have the same number of divisors.

5. If $\sum_{n=1}^{\infty} 1/a_n < \infty$ for the sequence $A = \{a_1 < a_2 < \dots\}$ of positive integers, this means that A is a rare subsequence of the natural numbers. Is it worth refining the notion of rarity according to the value of $\sum_{n=1}^{\infty} 1/a_n$?

S 6. The Riemann zeta function is defined as

$$(5.6.35) \qquad\qquad \zeta(s) = \sum_{n=1}^{\infty} \frac{1}{n^s}$$

for any real number $s > 1$. It is well known (or can be proven similarly to Exercise 5.6.2a) that the infinite series on the right-hand side of (5.6.35) converges for $s > 1$. E.g. $\zeta(2) = \pi^2/6$.

Now we define an infinite product (p ranges over the primes):

$$(5.6.36) \qquad\qquad \prod_{p} \frac{1}{1 - \frac{1}{p^s}} = \lim_{n \to \infty} \prod_{p \le n} \frac{1}{1 - \frac{1}{p^s}}.$$

Verify for $s > 1$ that the limit on the right-hand side of (5.6.36) exists and is equal to $\zeta(s)$.

7. Let $0 < a_j < 1$, $j = 1, 2, \dots$, and define the infinite product

$$\prod_{j=1}^{\infty} (1 - a_j) = \lim_{n \to \infty} \prod_{j=1}^{n} (1 - a_j).$$

Prove

$$\sum_{j=1}^{\infty} a_j = \infty \iff \prod_{j=1}^{\infty} (1 - a_j) = 0.$$

Remark: In general, an infinite product (with no zero elements) is called convergent, if its partial products tend to a *finite* limit *different from 0*.

* 8. In the third proof of Theorem 5.6.1 we demonstrated

$$\log n < \prod_{p \le n} \frac{1}{1 - \frac{1}{p}}$$

(see formula (5.6.10)). In the other direction, exhibit the following lower bound: There exists a constant c such that

$$c \log n > \prod_{p \le n} \frac{1}{1 - \frac{1}{p}}$$

for every $n \ge 2$.

9. For $n > 1$, let $p(n)$ and $P(n)$ denote the smallest and largest prime divisor of n. Determine whether the following infinite series converge or diverge:

    (a) $\displaystyle\sum_{n=2}^{\infty} \frac{1}{n p(n)}$

** (b) $\displaystyle\sum_{n=2}^{\infty} \frac{1}{nP(n)}$.

10. Give a new proof for Exercise 5.3.5 based on the following observation: If writing positive integers $a_1 < a_2 < \ldots$ one after the other following the decimal point, the resulting decimal fraction is rational, then $\sum_{i=1}^{\infty} 1/a_i < \infty$.

5.7. Primality Tests

Is it easy to find the prime factorization of an integer? Seemingly yes since we just have to check whether it is divisible by 2, 3, 5, etc. If we find a (prime) divisor, then we continue by factoring the quotient. And if there was no divisor up to the square root of the number, then it must be a prime (see Exercise 1.4.7a).

This way we can factor e.g. $143(= 11 \cdot 13)$, or can show that 197 is a prime (it is not divisible by any prime up to 13).

For large numbers we cannot manage to just try the primes as potential divisors since we do not have a list of them. We do not have to try all numbers, of course: we divide our number in question repeatedly by 2 till we get an odd number, and then we can restrict ourselves to divisibility by odd numbers. An improved version of this idea is when besides 2 we do the same with the powers of (say) 3 and 5, and then look only for divisors coprime to 30.

For really big numbers, however, trial division is absolutely useless: the time to perform the huge amount of trial divisions would require many billions of years even for the fastest computers. And the same holds for the improved versions of the method or for other factorization algorithms invented so far; they are all hopeless from a practical point of view: A composite number with 500 digits with no special property cannot be factored in the lifetime of Earth according to our present knowledge. (This might change in the future if quantum computers can be implemented effectively.)

At the same time, there are algorithms that can decide quickly (with absolute or nearly absolute certainty) whether a given large number is prime or composite (but cannot find factors in the latter case). These procedures are called *primality tests*.

The existence of quick primality tests seems to be very surprising at first hearing, especially compared to the task of detecting a non-trivial divisor which is harder than finding a needle in a haystack. These algorithms, however, instead of looking for divisors, check some quickly verifiable properties where the primes pass the test, but the composite numbers practically fail on it. (Here "practically" means that most methods run a minimal risk of error by allowing the possibility of some very rare exceptions.)

We have already proved primality tests for some special types of integers: see Theorems 5.2.2 and 5.2.4 for the tests of Fermat and Mersenne numbers.

Before discussing general primality tests, we show that there exist quick algorithms to solve some basic problems in number theory.

Theorem 5.7.1. *Let a, b, c, and m be integers where $b > 1$ and $m > 0$. Then we can compute*

I *the remainder of a^b modulo m*

II *the gcd of a and b*

III *the Jacobi symbol $\left(\frac{a}{b}\right)$ (for b odd and $(a, b) = 1$)*

IV *the solutions of the linear Diophantine equation $ax + by = c$ and*

V *the solutions of the linear congruence $ax \equiv c$ (mod b)*

in at most $5\log_2 b$ steps, where a step is an addition, subtraction, multiplication, or a division algorithm of two integers.   ♣

Thus, considering a large b with 500 digits, these computations can be executed in

$$5\log_2 b \approx 2500\log_2 10 < 9000$$

steps at most. They can be performed in a split second by a fast computer and the procedures can even be speeded up and automated by a more efficient organization.

Proof. I. The remainder of a^b modulo m can be computed by repeated squarings and reducing the result modulo m after each step. (This method occurred in the Example of Section 3.2 when we determined the residue of 13^{29} modulo 59, and in the test of the Fermat numbers, see the remarks after the proof of Theorem 5.2.2.)

We write the exponent b in binary system:

$$b = 2^{i_1} + 2^{i_2} + \cdots + 2^{i_s}, \qquad \text{where} \qquad 0 \le i_1 < i_2 < \cdots < i_s \le t = \lfloor \log_2 b \rfloor.$$

Probably b is stored in the computer in this form, but if necessary, the conversion from another base can be done in not more than $\log_2 b$ steps, since we obtain the digits from a sequence of division algorithms, by Theorem 1.2.2.

Then we compute the remainder of

$$a^2, a^4, a^8, \ldots, a^{2^t}$$

modulo m by repeated squarings (and reducing always mod m). Finally,

$$a^b = a^{2^{i_1}} a^{2^{i_2}} \ldots a^{2^{i_s}}$$

yields the desired residue.

For example, to determine 5^{1000} modulo m, we compute first the remainders of

$$5^2, 5^4, 5^8, \ldots, 5^{512}$$

modulo m, and then multiply the relevant ones (reducing modulo m in each step):

$$5^{1000} = 5^8 \cdot 5^{32} \cdot 5^{64} \cdot 5^{128} \cdot 5^{256} \cdot 5^{512}.$$

To determine the remainder of a^b modulo m, we do t squarings and not more than t further multiplications (and reductions modulo m). This requires at most $2t \le 2\log_2 b$ multiplications and reductions, i.e. division algorithms. Counting the representation of b in the binary system, we needed altogether at most $5\log_2 b$ steps (multiplications or division algorithms) to get the remainder of a^b modulo m.

II. To compute a greatest common divisor, we apply the Euclidean algorithm with remainders of least absolute value (i.e. allowing also negative remainders, but the absolute value of a remainder is at most the half of the absolute value of the divisor, see Theorem 1.2.1A):

$$a = bq_1 + r_1 \qquad \text{where} \qquad |r_1| \le \frac{b}{2}$$

$$b = r_1 q_2 + r_2 \qquad \text{where} \qquad |r_2| \le \frac{|r_1|}{2} \le \frac{b}{4}$$

$$r_1 = r_2 q_3 + r_3 \qquad \text{where} \qquad |r_3| \le \frac{|r_2|}{2} \le \frac{b}{8}$$

$$\vdots$$

$$r_{n-2} = r_{n-1} q_n + r_n \qquad \text{where} \qquad |r_n| \le \frac{|r_{n-1}|}{2} \le \frac{b}{2^n}$$

$$r_{n-1} = r_n q_{n+1} \qquad\qquad\qquad (r_{n+1} = 0).$$

The Euclidean algorithm consists of $n + 1$ steps in this case. Since

$$1 \le |r_n| \le \frac{b}{2^n},$$

therefore

$$2^n \le b \quad \text{or} \quad n \le \log_2 b.$$

This shows that the Euclidean algorithm requires $1 + \log_2 b$ steps at most (where each step is a division algorithm).

We note that also the usual Euclidean algorithm with least non-negative residues terminates in at most a constant times $\log b$ steps, see Exercise 5.7.1.

III. By Theorem 4.3.2, we can compute a Jacobi symbol by the repeated application of detaching the powers of two in the numerator (we call, in a mild abuse of language, the top and bottom of Jacobi and Legendre symbols the *numerator* and the *denominator*) and using the law of reciprocity (which is just a variant of the Euclidean algorithm, as in the Example after Theorem 4.3.2).

Let us see the details. To compute $\left(\frac{a}{b}\right)$, we first perform the division algorithm of a by b, and have

$$\left(\frac{a}{b}\right) = \left(\frac{r}{b}\right), \qquad \text{where} \qquad |r| < \frac{b}{2}.$$

If necessary, we can achieve $r > 0$ with the help of $\left(\frac{-1}{b}\right)$. If r is even, then we can halve the numerator by separating a factor of 2. If r is odd, then using the law of reciprocity, r gets transferred into the denominator, and the new numerator is the remainder s of b when divided by r. Thus $|s| < r/2$, and we can achieve $s > 0$ now, as well. This means that the numerator gets halved in each step, so no more than $\log_2 b$ steps occur. To compute $\left(\frac{-1}{v}\right)$ and $\left(\frac{2}{v}\right)$ we need the modulo 4 and modulo 8 residues of v, which can be obtained by a division algorithm or can be seen directly from the two or three last digits in the binary representation of v. It is likewise simple to check the parity for the numerator and to halve it if it is even.

The Jacobi symbol $\left(\frac{a}{b}\right)$ makes sense only for odd $b > 1$ and $(a, b) = 1$. This latter condition can be checked in advance by the Euclidean algorithm, but there is no need for that: If $(a, b) = d > 1$, then applying the procedure, we shall run into a situation where the numerator is d, and the denominator is a multiple of d (see Exercise 5.7.2). Thus we get stuck, and the Jacobi symbol $\left(\frac{a}{b}\right)$ does not exist. (This cannot occur for $(a, b) = 1$ because the last step is to compute a Jacobi symbol $\left(\frac{\pm 1}{v}\right)$ or $\left(\frac{\pm 2}{v}\right)$.)

IV–V. We saw in Section 2.5 that the two tasks are equivalent. Further, by Theorems 1.3.6 and 1.3.5 (or 7.1.1), we can find the solutions of a Diophantine equation $ax + by = c$ from the Euclidean algorithm, which gives the desired bound for the number of steps. □

Now we turn to the discussion of primality tests. The simplest general test is a direct consequence of Fermat's Little Theorem:

If $2^{n-1} \not\equiv 1 \pmod{n}$ for some $n > 2$, then n is composite.

This condition can be checked quickly, by Theorem 5.7.1. But what can we say about n if $2^{n-1} \equiv 1 \pmod{n}$? Unfortunately, we cannot be absolutely certain that n is a prime, since infinitely many *composite* n satisfy $2^{n-1} \equiv 1 \pmod{n}$, as well. They are called *pseudoprimes* of base 2 (the smallest one is 341).

It can be shown, however, that the pseudoprimes of base 2 occur very rarely compared to the primes: the ratio of the number of pseudoprimes up to x and $\pi(x)$ tends (very strongly) to 0 when $x \to \infty$. (As an illustration, up to 10^{10} there are 14887 pseudoprimes of base 2 and 455052511 primes, their ratio is roughly one to thirty thousand.)

Thus if a large number n satisfies $2^{n-1} \equiv 1 \pmod{n}$, then we can declare that n is a prime with very high probability. This assertion means that if we execute the test for many random integers n, then it will happen only very rarely (practically never) that the remainder of 2^{n-1} is 1, but n is composite.

We summarize the above in a theorem.

Theorem 5.7.2. *Let $n > 2$. If $2^{n-1} \not\equiv 1 \pmod{n}$, then n is necessarily composite. If $2^{n-1} \equiv 1 \pmod{n}$, then it is nearly sure that n is a prime.* ♣

The condition can be checked quickly if we compute the power by repeated squarings. We can improve the test by checking the residue of a^{n-1} modulo n not just for $a = 2$, but for (say) all primes less than 1000: if the residue is different from 1 for at least one a (and $n > 1000$), then n must be composite by Fermat's Little Theorem. It is even more efficient, if a is chosen randomly from the numbers not divisible by n (see Exercise 5.7.13).

If $a^{n-1} \equiv 1 \pmod{n}$ for every tested a, then n is even more probably prime, but we can still not be absolutely sure because there exist composite numbers n satisfying $a^{n-1} \equiv 1 \pmod{n}$ for every $(a, n) = 1$. For example, 1729 has this property (see Exercise 2.4.15c). These integers are called *universal pseudoprimes* or *Carmichael numbers*.

We summarize the types of pseudoprimes in the following definition:

Definition 5.7.3. If a composite integer n satisfies $a^{n-1} \equiv 1 \pmod{n}$, then n is a *pseudoprime of base a.*

If a composite integer n satisfies the above congruence for every $(a, n) = 1$, then n is a *universal pseudoprime* or *Carmichael number*. ♣

For some equivalent characterizations of Carmichael numbers see Exercise 5.7.7.

It has long been known that there are infinitely many pseudoprimes for any base $a > 1$ (see Exercise 5.7.5). In 1992 it was proved that the same holds for the universal pseudoprimes.

We present now two primality tests that detect also the pseudoprimes. Both use random numbers in the following sense: we consider a large but finite set of integers, and select elements one after the other so that all numbers occur with the same probability (like drawing balls from a box). For example, to generate a random number of 2000 binary digits, we write 1 as a first digit and determine the other digits by tossing a coin 1999 times. Of course, it is the computer who tosses the coin, or rather, it uses some pseudorandom number generator that produces a pseudorandom sequence of integers that is very similar to a truly random sequence.

Theorem 5.7.4 (Solovay–Strassen primality test). (A) *Let $n > 1$ be an odd integer and consider the congruence*

(5.7.1)
$$a^{\frac{n-1}{2}} \equiv \left(\frac{a}{n}\right) \pmod{n}$$

where $\left(\frac{a}{n}\right)$ denotes the Jacobi symbol.

If n is a prime, then every $a \not\equiv 0 \pmod{n}$ satisfies (5.7.1).

If n is composite, then (5.7.1) is satisfied for fewer than half of the elements in a complete residue system modulo n.

(B) *Using criterion (A), we can decide whether a large odd n is prime or composite as follows. We select (say) 1000 random numbers $a \not\equiv 0 \pmod{n}$ and check (5.7.1) for each of them. If (5.7.1) is not satisfied for at least one a, then n is necessarily composite. If every chosen a satisfies (5.7.1), then the probability of n being composite is less than 2^{-1000}.* ♣

Remarks: (1) For $(a, n) > 1$ the Jacobi symbol $\left(\frac{a}{n}\right)$ makes no sense, so (5.7.1) cannot hold.

(2) Condition (5.7.1) can be checked quickly (even for 1000 values of a) by Theorem 5.7.1.

(3) Even the Solovay–Strassen primality test cannot avoid the error of declaring a composite number as a prime. However, it is a great advance over the test in Theorem 5.7.2 from both the theoretical and the practical point of view.

The test in Theorem 5.7.2 is unable to detect the pseudoprimes of base 2, it fails completely in this case. So we are wrong when we think that a pseudoprime is a prime as suggested by the test (though this happens only very seldom as pseudoprimes are rare). Similarly, the improved version of checking (say) a million values of a is not suitable for detecting a large universal pseudoprime: we will falsely think that this composite n is a prime (except if some a was not coprime to n but the probability of this is practically zero).

At the same time, no composite integer can hide from the Solovay–Strassen test, there are no pseudoprimes related to it: there are lots of, so-called, witnesses who certify the compositeness of n. This means that the probability of error can be made arbitrarily small (independent of the tested integer) by checking sufficiently many values of a. (The error probability of 2^{-1000} in the case of a thousand trials provides a perfect practical security.)

Proof. (B) is a direct consequence of (A), thus it is sufficient to verify the latter.

For a prime n, we obtain (5.7.1) from Theorem 4.1.2 and the definition of the Legendre symbol (see formula (4.1.2) after Definition 4.1.3).

Let now n be composite. Since (5.7.1) can be valid only for a coprime to n, it is enough to show that (5.7.1) is satisfied by at most half of the elements in a reduced residue system modulo n.

Let us call a coprime to n a *witness* (for compositeness) if (5.7.1) is false, and an *accomplice* if (5.7.1) is true. Thus we have to prove that at least half of the elements in a reduced residue system are witnesses.

We start by showing that there exists a witness for any odd n.

Consider first the case when n is not squarefree, i.e. $q^2 \mid n$ for some prime q. Let $q = q_1, q_2, \ldots, q_s$ be the distinct prime divisors of n, let g be a primitive root modulo q^2, and let v be a solution of the system of congruences

$$x \equiv g \ (\mathrm{mod}\ q^2), \qquad x \equiv 1 \ (\mathrm{mod}\ q_i), \qquad 2 \leq i \leq s$$

(for $s = 1$, take $v = g$). We claim that v is a witness.

Since $(v, q_i) = 1$ for every i, $(v, n) = 1$. For a proof by contradiction, assume

(5.7.2) $$v^{\frac{n-1}{2}} \equiv \left(\frac{v}{n}\right) \ (\mathrm{mod}\ n).$$

Squaring (5.7.2), we obtain

(5.7.3) $$v^{n-1} \equiv \left(\frac{v}{n}\right)^2 = 1 \ (\mathrm{mod}\ n).$$

Since $q^2 \mid n$, (5.7.3) remains valid if we replace the modulus n by q^2. Using $v \equiv g \ (\mathrm{mod}\ q^2)$, this gives

(5.7.4) $$g^{n-1} \equiv 1 \ (\mathrm{mod}\ q^2).$$

As g is a primitive root mod q^2, its order is $\varphi(q^2) = q(q-1)$, so (5.7.4) implies $q(q-1) \mid n - 1$. But $q^2 \mid n$ so q divides both n and $n - 1$, which is a contradiction.

Now we turn to the case where n is squarefree, $n = q_1 \ldots q_s$ with distinct primes q_i and $s \geq 2$.

Let h be a quadratic non-residue modulo q_1, and let w be a solution of the system of congruences

(5.7.5) $$x \equiv h \ (\mathrm{mod}\ q_1), \qquad x \equiv 1 \ (\mathrm{mod}\ q_i), \qquad 2 \leq i \leq s.$$

We claim that w is a witness. Assume the converse, i.e. w satisfies (5.7.1). Then $(w, n) = 1$ and

$$\left(\frac{w}{n}\right) = \left(\frac{w}{q_1}\right)\left(\frac{w}{q_2}\right) \cdots \left(\frac{w}{q_s}\right) = \left(\frac{h}{q_1}\right)\left(\frac{1}{q_2}\right) \cdots \left(\frac{1}{q_s}\right) = -1.$$

By (5.7.1), we have

$$w^{\frac{n-1}{2}} \equiv -1 \ (\mathrm{mod}\ n).$$

Since $q_2 \mid n$ and $w \equiv 1 \ (\mathrm{mod}\ q_2)$ by (5.7.5), we infer

(5.7.6) $$-1 \equiv w^{\frac{n-1}{2}} \equiv 1 \ (\mathrm{mod}\ q_2),$$

which is a contradiction. Hence w is a witness.

We have proved that there exists a witness for any odd composite n.

Finally, we show that at least half of the elements in a reduced residue system are witnesses.

Let w be an arbitrary witness and let $a_1, a_2, \ldots, a_k$ be pairwise incongruent accomplices. We claim that $wc_1, \ldots, wc_k$ are pairwise incongruent witnesses.

From $(w, n) = (a_i, n) = 1$, $(wa_i, n) = 1$ and wa_i are pairwise incongruent modulo n. For a proof by contradiction, assume that some wa_i is an accomplice, i.e.

(5.7.7) $$(wa_i)^{\frac{n-1}{2}} \equiv \left(\frac{wa_i}{n}\right) \ (\mathrm{mod}\ n).$$

Since a_i is an accomplice,

(5.7.8) $$a_i^{\frac{n-1}{2}} \equiv \left(\frac{a_i}{n}\right) \ (\mathrm{mod}\ n).$$

Multiplying (5.7.7) and (5.7.8), we obtain

(5.7.9) $$w^{\frac{n-1}{2}} a_i^{n-1} \equiv \left(\frac{w}{n}\right)\left(\frac{a_i}{n}\right)^2 \ (\mathrm{mod}\ n).$$

Squaring (5.7.8), we have

$$a_i^{n-1} \equiv \left(\frac{a_i}{n}\right)^2 = 1 \ (\mathrm{mod}\ n)$$

which substituted into (5.7.9) yields

$$w^{\frac{n-1}{2}} \equiv \left(\frac{w}{n}\right) \ (\mathrm{mod}\ n).$$

This means that w is an accomplice, which is a contradiction.

Thus we verified that multiplying pairwise incongruent accomplices by a fixed witness gives pairwise incongruent accomplices. So the number of witnesses in a reduced residue system is at least as big as the number of accomplices: at least half of the elements are witnesses. $\square$

The next primality test is based on Fermat's Little Theorem and on the fact that if $u^2 \equiv 1 \ (\mathrm{mod}\ p)$ for a prime p, then $u \equiv \pm 1 \ (\mathrm{mod}\ p)$. This implies that for $p \nmid a$, the sequence of remainders of least absolute value of the numbers

$$a^{p-1}, a^{\frac{p-1}{2}}, a^{\frac{p-1}{4}}, \ldots$$

starts with 1 and either remains 1 to the very end, or the first remainder different from 1 must be -1. At the same time, replacing p by a composite n, the sequence of remainders will not obey this rule for many values of a. This gives the following primality test (for technical reasons, we state the above condition in a modified form, essentially for the inverted sequence):

Theorem 5.7.5 (Miller–Lenstra–Rabin primality test). *Let $n > 1$ be odd and $n-1 = 2^k r$ with r odd. The numbers*

(5.7.10) $$a^r, a^{2r}, a^{4r}, \ldots, a^{2^{k-2}r} = a^{\frac{n-1}{4}}, a^{2^{k-1}r} = a^{\frac{n-1}{2}}$$

form a good sequence if either -1 occurs among their residues of least absolute value modulo n, or the residue of a^r is 1.

For a prime n, (5.7.10) is a good sequence for every $a \not\equiv 0 \pmod{n}$.

For a composite n, (5.7.10) is a good sequence only for fewer than half of the elements of a complete residue system modulo n. ♣

This criterion can be checked quickly: we compute the remainder of a^r modulo n by repeated squarings and then continued squarings give the other elements of the sequence one by one.

Based on the criterion, we can formulate the concrete algorithm, similarly to part (B) in Theorem 5.7.4.

Outline of proof. We follow the ideas and the usage of "witness" and "accomplice" seen in Theorem 5.7.4 with suitable modifications.

If n is a prime, then we sketched before stating Theorem 5.7.5 that every $p \nmid a$ produces a good sequence.

If n is composite and is not squarefree, then we can construct a witness exactly as in the proof of Theorem 5.7.4.

If n is composite and squarefree, then consider the largest $0 \le j \le k - 1$ satisfying

(5.7.11) $$a^{2^j r} \not\equiv 1 \pmod{n}$$

for some a coprime to n. Since (5.7.11) holds with some j and a, e.g. with $j = 0$ and $a = -1$ (as $(-1)^r \not\equiv 1 \pmod{n}$), therefore a maximal j exists.

By (5.7.11),

$$a^{2^j r} \not\equiv 1 \pmod{q_1}$$

for some prime divisor q_1 of n. Then w obtained from the system of congruences (5.7.6) in the proof of Theorem 5.7.4 is a witness, since similarly to the argument seen there,

$$w^{2^j r} \not\equiv \pm 1 \pmod{n},$$

but

$$w^{2^{j+1} r} \equiv 1 \pmod{n}$$

by the maximal property of j (for $j < k - 1$).

Finally, multiplying this w, or v in the not squarefree case, by pairwise incongruent accomplices we obtain pairwise incongruent witnesses as seen in the proof of Theorem 5.7.4 (but w cannot be replaced now by an arbitrary witness). Thus we proved that if n is composite, then at least half of the elements in a reduced residue system are witnesses. □

Remarks: (1) The Miller–Lenstra-Rabin test is even more efficient than stated in Theorem 5.7.5: it can be shown by more refined methods that more than 75% of elements in a reduced residue system are witnesses.

(2) Comparing the Solovay–Strassen and Miller–Lenstra–Rabin tests, it turns out that the latter is more efficient in detecting composite numbers (see Exercise 5.7.17).

Agrawal, Kayal, and Saxena devised a quick primality test in 2002 that determines not with 99.99999999999% but 100% certainty whether n is prime or composite. The test starts with a polynomial version of Fermat's Little Theorem. We sketch the basic idea below.

For $(c, n) = 1$, we consider the polynomials $f_c = x^n - c$ and $g_c = (x - c)^n$ over $\mathbf{Z}_n$. If n is a prime, then $f_c = g_c$ (i.e. their coefficients are equal, which is a stronger statement than the equality of the corresponding values assumed by the functions). For the constant terms $-c$ and $(-c)^n$, this follows from Fermat's Little Theorem, the leading coefficients are 1, and the other coefficients $\binom{n}{k}(-c)^k$ in g_c are divisible by n as n is prime (see Exercise 2.1.9a), hence they are 0 in $\mathbf{Z}_n$. It is another simple observation that $\binom{n}{k}$ is not divisible by n for some $0 < k < n$ if n is composite, and as $(c, n) = 1$, the coefficient of x^{n-k} is not 0 in g_c implying $f_c \neq g_c$. Thus this is a perfect primality test (e.g. with $c = 1$), but unfortunately it is awfully slow since computing the coefficients of g requires many steps even using repeated squarings, due to the huge number of terms.

The ingenious idea of the AKS test is that instead of $f_c = g_c$ we check just the equality of remainders of f_c and g_c divided by a suitable polynomial $h \in \mathbf{Z}_n[x]$. If h is of sufficiently small degree (compared to n), then the computation can be carried out if during the repeated squarings we reduce also modulo h. This reduction is particularly simple if h is of the form $h = x^r - 1$, since then we just have to reduce the exponents in the powers of x mod r (i.e. we replace x^j by x^{j-r} as long as possible).

If n is a prime, then $f_c = g_c$ implies that the remainders are equal modulo any h. The main point in the AKS test is that choosing r appropriately, no composite integer satisfies this, so to any composite n there exists some $c \leq K$ (where K is very small compared to n), such that f_c and g_c do not yield the same remainder when divided by $x^r - 1$.

The corresponding algorithm hence selects a suitable r and then checks $f_c \equiv g_c$ (mod $x^r - 1$) for every $c = 1, 2, \ldots, K$. If this fails for some c, then n is composite (this follows from our initial considerations). On the other hand, if it holds for every c, then n is a prime for sure (this is the hard part in the proof of the test).

We have to select r as a not too big prime with some special properties, its existence is guaranteed by a deep theorem in number theory. To prove that after fixing this r, any composite number gets detected by checking a few values of c, we need some basic results about finite fields.

Exercises 5.7

1. Consider the usual Euclidean algorithm for the integers $a, b, a > b > 0$, where the remainders satisfy $b = r_0 > r_1 > r_2 > \ldots \geq 0$.

   (a) Verify $r_{k+2} < \frac{r_k}{2}$ for every k.

(b) Which upper bound follows from this for the number of steps in the algorithm?

* (c) Prove that if the algorithm requires exactly s steps, then the minimal possible value of b is φ_{s+1} where φ_j denotes the jth Fibonacci number (defined in Exercise 1.2.5).

  Remark: By the explicit formula

$$\varphi_j = \frac{1}{\sqrt{5}}\left(\left(\frac{1+\sqrt{5}}{2}\right)^j - \left(\frac{1-\sqrt{5}}{2}\right)^j\right)$$

for the Fibonacci numbers, (c) implies that the usual Euclidean algorithm requires at most $\log_\gamma b + \delta$ steps where $\gamma = (1+\sqrt{5})/2$ and δ is a suitable constant, and this bound is best possible.

2. Consider the procedure in part III of the proof of Theorem 5.7.1 for computing the Jacobi symbol $\left(\frac{a}{b}\right)$. Show that for $(a, b) = d > 1$, this leads to a situation when the numerator is d and the denominator is a multiple of d. (Thus the method reveals that the Jacobi symbol makes no sense in this case, and there is no need to check separately whether or not a and b are coprime.)

3. Show that 341 is a pseudoprime of base 2, but not of base 3.

S 4. Prove that if n is a pseudoprime of base 2, then so is $2^n - 1$.

5. Let $a > 1$. Show that if the prime $p > 2$ does not divide $a \pm 1$, then

$$n = \frac{a^{2p} - 1}{a^2 - 1}$$

is a pseudoprime of base a. (For $a = 2$ and $p = 5$ we obtain $n = 341$.)

6. Verify that 561 is a universal pseudoprime.

7. Prove the equivalence of the following conditions

(a) $a^{n-1} \equiv 1 \pmod{n}$ for any $(a, n) = 1$.

(b) n is squarefree and $p \mid n \implies p - 1 \mid n - 1$.

(c) $a^n \equiv a \pmod{n}$ for any a.

Remark: This means that in Definition 5.7.3 of universal pseudoprimes we could have chosen condition (c) (or (b)) instead of (a).

8. Show that a universal pseudoprime has at least three prime divisors.

9. (a) In the primality tests discussed, we check a condition, and there is no need to compute in advance whether a and n are coprime. What advantage is there if we compute (a, n)?

(b) If n is the product of two primes of hundred digits, then roughly what is the chance that a random a is not coprime to n?

10. Prove that if $a^2 \equiv 1 \pmod{n}$ but $a \not\equiv \pm 1 \pmod{n}$, then we can determine a nontrivial divisor of n.

* 11. Verify that if we know a (non-zero) multiple of $\varphi(n)$ besides n, then we can find the standard form of n quickly. (More precisely, there is a theoretical chance that we still cannot factor n, but this occurs practically never.)

12. Analyze whether Wilson's Theorem and its converse, i.e. checking whether or not n divides $(n-1)! + 1$, are suitable or not as a primality test.

13. (a) Show that if a composite number n is not a universal pseudoprime, then $a^{n-1} \equiv 1 \pmod{n}$ holds for fewer than half of the elements in a complete residue system modulo n.

    (b) Describe the concrete primality test based on part (a).

14. Prove that the following primality test can be performed quickly, and its probability of error can be reduced below an arbitrarily small bound (prescribed in advance).

    We want to decide whether an odd integer $n > 1$ is prime or composite. We check the remainder of $a^{(n-1)/2}$ modulo n for a fixed (but sufficiently large) amount of random integers a where $n \nmid a$. We declare n to be a prime if every such remainder is ± 1 with -1 occurring at least once among them.

15. Let $n = 2^k r + 1$ with $k \geq 1$, r odd, and $0 < r < 2^k$. Assume that

$$a^{\frac{n-1}{2}} \equiv -1 \pmod{n}$$

    holds for some integer a. Prove that n is a prime.

16. Let $n > 2$. Show that any of the following conditions imply that n is a prime.

    (a) There is an integer a satisfying $a^{n-1} \equiv 1 \pmod{n}$ and

    $$a^{\frac{n-1}{p_i}} \not\equiv 1 \pmod{n}$$

    for every prime divisor p_i of $n - 1$.

    * (b) To any prime divisor p_i of $n-1$ there exists an integer a_i satisfying

    $$a_i^{n-1} \equiv 1 \pmod{n} \quad \text{and} \quad a_i^{\frac{n-1}{p_i}} \not\equiv 1 \pmod{n}.$$

    * (c) There exists a divisor c of $n-1$ greater than $\sqrt{n}$ such that for any prime divisor p_i of c there exists an integer a_i satisfying

    $$a_i^{n-1} \equiv 1 \pmod{n} \quad \text{and} \quad \left(a_i^{\frac{n-1}{p_i}} - 1, n\right) = 1.$$

S 17. Show that the Miller–Lenstra–Rabin test is more efficient than the Solovay–Strassen test in the following sense. If a is a witness for n in the Solovay–Strassen test, then the same a is a witness in the Miller–Lenstra–Rabin test; i.e. if condition (5.7.1) of Theorem 5.7.4 is false for some a, then the set (5.7.10) in Theorem 5.7.5 cannot form a good sequence for this a.

5.8. Cryptography

In classical cryptography A and B agree in advance on an encoding key E (e.g. to write always the next letter instead of each letter in the alphabet). The inverse of E is the decoding key D (in the example above, this means to write the preceding letter). When communicating, (say) A encodes the plain text by E into a ciphertext and sends it to B who can decode it by D.

The keys may refer not just to letters but also to sequences of characters, and can be very complicated. In that case, computers do the encoding and decoding and the messages are sent electronically instead of by a messenger.

These schemes meet two basic requirements, that only B can understand the message of A, and no third party can send a false message in the name of A. There are, however, several disadvantages: the two parties have to agree on the keys in advance, which may be a difficult (and dangerous) task; no disputes between A and B can be resolved, since either party can falsify a message with the common keys in the other's name; and the bilateral communication of A with several parties (e.g. in business) requires a new pair of keys with each partner.

Diffie and Hellman suggested a cryptosystem based on a revolutionary new idea: we make the key E public and keep only D secret.

This sounds absurd at the first hearing, since if we know the procedure in one direction, then we can find it out in the opposite direction. Let the functions E and D be bijections of the set $\{1, 2, \ldots, N\}$ (we shall see that we can always assume this without loss of generality). If we want to determine (say) $D(5)$, then we compute $E(1)$, $E(2), \ldots$ with the help of the public key E till $E(k) = 5$ occurs, providing $D(5) = k$.

This sounds good in principle, but if N has (say) 500 digits, then it cannot be carried out in practice. A computer could determine only a negligible fraction of the values $E(1)$, $E(2), \ldots$ even in billions of years and so most probably would never find $D(5)$. (We illustrate the situation with an analogy. An English-French dictionary can be used as a French-English dictionary in principle: if we want to find the English equivalent of the French word "eau", then we go through the English words of the English-French dictionary (in alphabetic order) till we find "eau" among the French meanings. This will occur at the English word "water". So probably nobody would not also buy the French-English dictionary.)

Hence it is not inconceivable that E being public, D can still remain secret. We now discuss *public key cryptosystems* based on this idea.

Each party creates a pair of keys E and D which are inverses of each other, makes E public, but keeps D in secret. Let E_A and D_A be the keys of A, and E_B and D_B be the keys of B. Then A transforms the plain text u into the ciphertext $v = E_B(D_A(u))$ and sends it to B who can decode it as $u = E_A(D_B(v))$:

$$E_A(D_B(v)) = E_A(D_B(E_B(D_A(u)))) = E_A(D_A(u)) = u.$$

(To compute v, A uses his own function D_A and the public function E_B, and B can act similarly.)

This scheme meets the two basic requirements discussed above: only B can understand A's message, since no one else knows D_B needed for the decoding, and a third party cannot falsify a message in the name of A since only A knows D_A necessary for the encoding.

The method has several further important advantages. There is no need to agree about the keys in advance, and everybody can use the same keys with each partner. There cannot be any dispute about the message between A and B, since D_A cannot be falsified even by B, it acts as an electronic signature for A.

To implement the system, we have to construct pairs of keys E and D where the owner knows both keys but other persons cannot determine D even using the publicly accessible E.

We saw previously that the prime factorization of a large number can serve as such a secret known only by the person who formed the product of these primes. Based on this, Rivest, Shamir, and Adleman made a concrete realization of the Diffie–Hellman principle. Their procedure is called the RSA scheme from the initials of the discoverers (or inventors?).

Before discussing RSA, we show that any cryptosystem can be reduced to the case where E and D are permutations, i.e. bijections of the set $\{1, 2, \ldots, N\}$ where N is a sufficiently large integer. To see this, we encode (in a publicly known standard way) letters and other characters as numbers, thus transforming a message into a sequence of integers. Then we cut it into blocks of a given size, and consider each block as one (large) number with many digits. These numbers will constitute both the domain and the range of the functions E and D.

We can transform letters and other characters into numbers for example in the following way: A $\mapsto$ 01, B $\mapsto$ 02, $\ldots$, Z $\mapsto$ 26, comma $\mapsto$ 27, space $\mapsto$ 28, etc. and say that four such two-digit numbers should form a block. Then any message is converted into a sequence of integers between 1 and $10^8 - 1$ so $N = 10^8 - 1$.

Let us find the equivalent of the expression "number theory". N is converted into 14, U into 21, M into 13, etc., so we get the sequence

$$14211302|05182820|08051518|25.$$

Hence the blocks are 14211302, 05182820, 08051518, and 25999999 (the last block was completed with 9s). We apply the keys E and D to these four numbers. (We repeatedly emphasize that this conversion of the text into numbers is publicly known and its only purpose is to provide a unified and comfortable handling of the functions E and D.)

Now we turn to the construction of the keys E and D in the RSA.

Let $N = pq$ where p and q are two large primes. The holder of the key keeps p and q secret, but makes N public. Further, he/she chooses an integer $e > 1$ coprime to $\varphi(N)$, and declares publicly his/her key E:

(5.8.1) $E(r) = $ the least positive residue (mod N) of r^e, $\quad r = 1, 2, \ldots, N$.

How can we get $D = E^{-1}$? We try to find it in a similar form:

(5.8.2) $D(s) = $ the least positive residue (mod N) of s^d, $\quad r = 1, 2, \ldots, N$.

This meets the requirements if and only if for every r we have

$$r = ED(r) = DE(r) = \text{the least positive residue } (\text{mod } N) \text{ of } r^{ed},$$

i.e. if and only if

(5.8.3) $r^{ed} \equiv r \ (\text{mod } N)$

for every r. Using Fermat's Little Theorem for the primes p and q, we easily derive that

(5.8.4) $r^{1+k\varphi(N)} \equiv r \ (\text{mod } N)$

holds for any k and r (see Exercise 5.8.3a).

By (5.8.4), we obtain a suitable d in (5.8.3) (and thus in (5.8.2), as well) if we solve the linear Diophantine equation

(5.8.5) $de = 1 + k\varphi(N)$

for d (and k). Since $(e, \varphi(N)) = 1$, (5.8.5) is solvable and we can get a solution quickly using the Euclidean algorithm.

But all this can be done only by the holder of the key, as nobody else can compute $\varphi(N)$ for lack of the prime factors of N.

The holder of the key generates the primes p and q in the following way. He chooses odd random numbers with (say) 400 and 500 digits and checks (e.g. with one of the primality tests in Section 5.7) whether or not they are primes. He will get p and q fairly soon, since the primality tests are fast and there are many primes with 400 and 500 digits: according to the Prime Number Theorem, a random 500-digit odd number is prime with probability $1/\left(\log(10^{500})/2\right) \approx 1/576$.

Both $E(r)$ in (5.8.1) and $M(s)$ in (5.8.2) can be computed quickly by repeated squarings (of course, the latter can be done only by the holder of the key).

When selecting p, q, and e, a few safety measures have to be taken. If p and q are too close to each other, then N can be factored more easily, therefore we had to test random numbers of different sizes for p and q. Similar reasons require that $p - 1$ and $q - 1$ should have large prime factors. We do not discuss these and similar technical details.

How safe is this procedure? Presently it seems that (following the precautionary measures) we do not have to worry. It is not completely impossible, however, that somebody finds a quick method for factoring integers, and then can get D, too. It is also conceivable that one can exhibit D in some different form. But these are highly improbable.

We summarize the essential points of RSA in the the following theorem.

Theorem 5.8.1 (RSA scheme). *Let p and q be two large primes, $N = pq$, and $(e, \varphi(N)) = 1$. Define the pair of keys E and D by (5.8.1) and (5.8.2) where d satisfies (5.8.5). N, e, and E are public, but p, q, $\varphi(N)$, d, and D are secret. Then $D = E^{-1}$, and D cannot be determined even knowing E.*

The holder of the key generates the primes p and q by testing random numbers, and can find d quickly. The holder of the key can compute $D(s)$ quickly, and anyone can compute $E(r)$ quickly. ♣

Exercises 5.8

1. What type of problem can occur if A sends to B just $v' = E_B(u)$ instead of $v = E_B(D_A(u))$ in the Diffie–Hellman scheme?

2. Show that the function E defined in (5.8.1) is invertible if and only if $(e, \varphi(N)) = 1$.

3. Let $N = pq$ where p and q are distinct primes.

    (a) Verify $r^{1+k\varphi(N)} \equiv r \pmod{N}$ for every r.

    (b) Find all integers $v > 0$ satisfying $r^v \equiv r \pmod{N}$ for every r.

4. Assume that (due to the imperfection of primality testing) we use a universal pseudoprime p in RSA. Do we have to worry about this?

* 5. Show that RSA is not safe if the order of the exponent e is small modulo $\varphi(N)$.

6. Let p be a large prime and g a primitive root modulo p. At present we know no quick algorithm for computing the discrete logarithm, so it is not possible to determine $\mathrm{ind}_g\, a$. This means that we can compute the residue $a \bmod p$ of g^k, but no one else can find k from a.

    A and B choose exponents k_A and k_B, keeping them secret, but make the remainders of g^{k_A} and g^{k_B} modulo p public. Prove that both A and B can compute the remainder of $g^{k_A k_B}$ modulo p, but (hopefully) no one else can do this. (This means that A and B can find a common password without preliminary negotiations and without revealing their secret exponents k_A and k_B to each other).

7. It seemed for a while that also the following scheme, the so-called modular knapsack or subset problem could be used for public key cryptosystems, but later it turned out that it is not safe.

    (a) A sequence of positive integers $C = \{c_0, c_1, \dots, c_{k-1}\}$ is *sum injective*, if the sums of arbitrarily many distinct c_is are all distinct.
    Prove that if C is super-increasing, i.e.

(5.8.6)
$$c_i > \sum_{j=0}^{i-1} c_j, \qquad i = 1, 2, \dots, k-1,$$

   then C is sum injective.

    (b) Let C be sum injective, $m > \sum_{i=0}^{k-1} c_i$, $(r, m) = 1$, and

(5.8.7) $d_i =$ the least positive remainder $\pmod{m}$ of rc_i, $\quad 0 \le i \le k-1$.

   Verify that the sequence $d_0, \dots, d_{k-1}$ is sum injective.

    (c) Let $0 \le u < 2^k$, and write u in binary system:

$$u = \sum_{i=0}^{k-1} \delta_i 2^i, \quad \text{where} \quad \delta_i = 0 \text{ or } 1, \quad i = 0, 1, \dots, k-1.$$

Show that if H is sum injective, then u can be determined in principle from the number

$$v = \sum_{i=0}^{k-1} \delta_i h_i.$$

(d) Prove that u can be determined quickly also in practice for sequences of type (5.8.6) and their derivatives of type (5.8.7).

Based on the above considerations, we take a sequence C of type (5.8.6) and convert it into a sequence D of type (5.8.7). The sequence D will be public, but c_i, m, and r are kept secret. Then anybody can compute v from u quickly, and we can do this also backwards using C, m, and r. Since it is very hard to get u from v in practice for general sum injective sequences, it seemed reasonable that this should be the case also for sequences of type (5.8.7) without knowing c_i, m, and r. As mentioned earlier, this belief turned out to be false.

Chapter 6

Arithmetic Functions

An arithmetic function is a complex-valued function defined on the positive integers. We shall mostly deal with those reflecting some arithmetic properties of positive integers, such as $d(n)$, the number of positive divisors of n, or Euler's function $\varphi(n)$ indispensable for congruences, which appeared in Chapters 1 and 2. Some further important examples are $\sigma(n)$, the sum of positive divisors of n, related also to the perfect numbers, and the Möbius function $\mu(n)$ that plays an important role in the summation and inversion functions. Using $d(n)$ as an example, we illustrate how double-faced many arithmetic functions behave, as they assume hectically oscillating values on the one hand, but show a regular pattern if considered on average on the other hand. Applying convolution, we extend this type of investigation of mean values to $\sigma(n)$ and $\varphi(n)$. The latter result gives also the probability (in a precisely defined meaning) of two numbers being relatively prime. (This probability turns out to be surprisingly big: $6/\pi^2 \approx 0.61$.) The study of $\omega(n)$, denoting the number of distinct (positive) prime divisors of n, is of special interest, since (in contrast to $d(n)$) it assumes mostly values close to its mean value. We present Turán's simple proof for this famous theorem of Hardy and Ramanujan whose argument became the starting point of probabilistic number theory. Finally, we give a glimpse into a topic initiated by Erdős, namely which conditions can characterize the logarithm among the additive arithmetic functions.

6.1. Multiplicative and Additive Functions

Definition 6.1.1. An *arithmetic function* is a complex-valued function defined on the positive integers. ♣

Examples. $d(n)$ is the number of positive divisors of n (see Theorem 1.6.3)

Euler's function φ (see Definition 2.2.7 and Theorem 2.3.1)

$f(n) = (-1)^n$, $g(n) = \sqrt{n^2 + 5} + i \sin n$, etc.

We shall discuss some important arithmetic function in Section 6.2.

The following properties often play an important role:

Definition 6.1.2. An arithmetic function f is *multiplicative* if $f(ab) = f(a)f(b)$ for every coprime a and b. ♣

Definition 6.1.3. An arithmetic function f is *completely multiplicative* (or *totally multiplicative*), if $f(ab) = f(a)f(b)$ for every a and b. ♣

Examples. Euler's function φ is multiplicative (this was verified in the first proof of Theorem 2.3.1), but it is not completely multiplicative, as $\varphi(8) \neq \varphi(2)\varphi(4)$. The same holds for $d(n)$ (see Exercise 6.1.1).

If α is a fixed real number, then $f(n) = n^{\alpha}$ is completely multiplicative (hence it is multiplicative.

$g(n) = 3n - 2$ is not multiplicative, since $(2, 3) = 1$, but $g(6) \neq g(2)g(3)$.

Requiring similar conditions for the sum of the values instead of their product, we get the notion of additive and completely additive arithmetic functions, resp.:

Definition 6.1.4. An arithmetic function f is *additive* if $f(ab) = f(a) + f(b)$ for every coprime a and b. ♣

Definition 6.1.5. An arithmetic function f is *completely additive* (or *totally additive*), if $f(ab) = f(a) + f(b)$ for *every* a and b. ♣

The definitions both of additivity and complete additivity refer to the values of $f(ab)$ (and not of $f(a + b)$).

Examples. The logarithm function (with any base) is completely additive.

$f(n) = 1 + (-1)^n$ is additive, but not completely additive.

$g(n) = 1 + \log_2 n$ is not additive (hence it cannot be completely additive either).

The identically zero function $f = 0$ is both completely multiplicative and completely additive, but no other function can be both multiplicative and additive (this follows from Theorem 6.1.6).

We show first that additive and non-zero multiplicative functions can assume only special values at 1:

Theorem 6.1.6. *If f is multiplicative and $f \neq 0$, then $f(1) = 1$.*

If g is additive, then $g(1) = 0$. ♣

Proof. Let a be a positive integer satisfying $f(a) \neq 0$. Then $(a, 1) = 1$ implies $f(a) = f(a \cdot 1) = f(a)f(1)$, and dividing by $f(a) \neq 0$ we get $1 = f(1)$.

The other statement can be proved similarly. □

Theorem 6.1.6 gives a necessary (but not sufficient) condition for a function to be additive or multiplicative.

The definitions of additivity and multiplicativity imply that additive and ($\neq 0$) multiplicative functions are uniquely determined by their values at prime powers:

Theorem 6.1.7. *Let f be multiplicative, g additive, and $n = p_1^{\alpha_1} \dots p_r^{\alpha_r}$ be the standard form of $n > 1$. Then*

$$f(n) = f(p_1^{\alpha_1}) \dots f(p_r^{\alpha_r}) \quad and \quad g(n) = g(p_1^{\alpha_1}) + \dots + g(p_r^{\alpha_r}). \qquad \clubsuit$$

We used this fact deducing the formula for $\varphi(n)$ (in the first proof of Theorem 2.3.1).

Similarly, completely additive and ($\neq 0$) completely multiplicative functions are uniquely determined by their values at primes:

Theorem 6.1.8. *Let f be completely multiplicative, g completely additive, and $n = p_1^{\alpha_1} \dots p_r^{\alpha_r}$ be the standard form of $n > 1$. Then*

$$f(n) = f(p_1)^{\alpha_1} \dots f(p_r)^{\alpha_r} \quad and \quad g(n) = \alpha_1 g(p_1) + \dots + \alpha_r g(p_r). \qquad \clubsuit$$

We can add to Theorem 6.1.7 that additivity or multiplicativity does not impose any restrictions on the values assumed at prime powers, these can be chosen freely. This means that prescribing the values arbitrarily at prime powers, gives a multiplicative/additive function. An analogous statement holds with primes instead of prime powers for completely multiplicative/additive functions (see Exercise 6.1.4).

Exercises 6.1

1. Verify that $d(n)$ is multiplicative but not completely.

2. Which of the following functions are multiplicative, completely multiplicative, additive, and completely additive?

   (a) $f(n) = \begin{cases} 0, & \text{if } 6 \mid n \\ 1, & \text{if } 6 \nmid n. \end{cases}$

   (b) $g(n) = \begin{cases} 0, & \text{if } 3 \mid n \\ 1, & \text{if } 3 \nmid n. \end{cases}$

   (c) $h(n) = \begin{cases} 0, & \text{if } 3 \mid n \\ 2, & \text{if } 3 \nmid n. \end{cases}$

   (d) $k(n) = \begin{cases} 2, & \text{if } 3 \mid n \\ 0, & \text{if } 3 \nmid n. \end{cases}$

3. Does there exist an (a) additive (b) multiplicative function h satisfying $h(6) = 0$, $h(10) = 1$, and $h(15) = 3$?

4. Consider the sequence of primes $p_1, p_2, \dots = 2, 3, 5, 7, \dots$, the sequence of prime powers $q_1, q_2, \dots = 2, 3, 4, 5, 7, 8, 9, 11, \dots$, and let $c_1, c_2, \dots$ be arbitrary complex numbers.

   (a) Prove that there exists exactly one multiplicative function $f \neq 0$ and exactly one additive function g satisfying

   $$f(q_i) = g(q_i) = c_i, \qquad i = 1, 2, \dots.$$

(b) Prove that there exists exactly one completely multiplicative function $s \neq 0$ and exactly one completely additive function t satisfying

$$s(p_i) = t(p_i) = c_i, \qquad i = 1, 2, \dots.$$

5. If g can assume only positive integer values, then we can define the composite function $h(n) = (f \circ g)(n) = f(g(n))$ for any f. True or false?

   (a) If f and g are completely multiplicative, then h is completely multiplicative.

   (b) If f and g are completely additive, then h is completely additive.

   (c) If f is multiplicative and g is completely multiplicative, then h is multiplicative.

   (d) If f is completely multiplicative and g is multiplicative, then h is multiplicative.

6. (a) Let f be completely additive. For which positive integers k is the function $g(n) = f(kn)$ completely additive?

   (b) Solve the problem for the case when we prescribe only additivity instead of complete additivity (for both of f and g).

   (c) Investigate the variants for completely multiplicative and multiplicative functions.

S 7. (a) Show that if f is completely additive, then

(A.6.1) $f(a) + f(b) = f((a, b)) + f([a, b])$ holds for every a and b.

   (b) Prove (A.6.1) for any additive f.

   * (c) Determine all functions f satisfying (A.6.1).

   * (d) Investigate also the corresponding equation $f(a)f(b) = f((a, b))f([a, b])$.

8. Let f be real valued and $g(n) = 2^{f(n)}$. Demonstrate that g is multiplicative if and only if f is additive.

   Remark: This means that properties of additive functions assuming real values and of multiplicative functions assuming positive values can be mutually deduced from each other.

9. (a) Verify that both the sum and the difference of two additive functions are additive, and the same holds if "additive" is replaced by "completely additive."

   (b) Prove that the product of two completely additive functions is never completely additive except in the trivial case when at least one of the factors is the 0 function.

   (c) Give examples when the product of two $\neq 0$ additive functions is (c1) additive (c2) not additive.

   S* (d) Find all pairs of additive functions whose product is additive.

   (e) Show that the product of two multiplicative functions is multiplicative, and the same holds if "multiplicative" is replaced by "completely multiplicative."

   (f) Verify that neither the sum nor the difference of two distinct $\neq 0$ multiplicative functions can be multiplicative.

10. (a) Show that the arithmetic mean of two additive or completely additive functions has the same property.

   (b) Prove that if the arithmetic mean of two completely multiplicative functions is completely multiplicative then the two functions are equal. What happens if we require only multiplicativity instead of complete multiplicativity (for all three functions)?

11. Assume that f is multiplicative, g is additive, and $f + g$ is constant. Show that $f^{1000} + g^{1000}$ is multiplicative and $f^{1000}g^{1000}$ is additive.

* 12. Let h be an additive function.

   (a) Prove that if h is the difference of two multiplicative functions, then $h(a)h(b)h(c) = 0$ for any pairwise coprime integers a, b, and c.

   (b) If h has only the trivial representation $1 \cdot h = h$ as the product of a multiplicative and an additive function, then $h(a)h(b)h(c) = 0$ for any pairwise coprime integers a, b, and c.

S 13. (a) Assume that the range $R(f)$ of an additive function f is finite. Show that every $c \in R(f)$ occurs infinitely often, i.e., there are infinitely many positive integers b satisfying $f(b) = c$.

   (b) Give an example that shows that the same does not necessarily hold for multiplicative functions.

   (c) Assume that the range $R(f)$ of a multiplicative function f is finite and some $d \in R(f)$ occurs only finitely many times, i.e. $f(b) = d$ holds only for finitely many positive integers b. Prove that there exists a K such that $f(n) = 0$ for every n having a prime divisor greater than K.

14. True or false?

   (a) If f is additive and $f(ab) = f(a) + f(b)$ for some a and b not coprime, then f is completely additive.

   (b) If f is additive and $f(ab) = f(a) + f(b)$ for some a and b not coprime, then there exist infinitely many such a and b.

   (c) If f is additive but not completely, then $(a, b) \neq 1$ implies $f(ab) \neq f(a) + f(b)$.

   (d) If f is additive but not completely, then $f(ab) \neq f(a) + f(b)$ for infinitely many a and b.

   (e) If f is multiplicative but not completely, then $f(ab) \neq f(a)f(b)$ for infinitely many a and b.

S* 15. Let $\varphi_2(n)$ denote the number of integers $i \in \{1, 2, \ldots, n\}$ satisfying $(i, n) = (i + 1, n) = 1$. Give a formula for $\varphi_2(n)$ based on the standard form of n.

* 16. Prove
$$\sum_{\substack{1 \leq k \leq n \\ (k,n)=1}} (k - 1, n) = \varphi(n)d(n).$$

6.2. Some Important Functions

We introduce some basic functions in this section: $\sigma(n)$, $\mu(n)$, $\omega(n)$, $\Omega(n)$, and $d_k(n)$.

Definition 6.2.1. $\sigma(n)$ is the sum of positive divisors of n.   ♣

Example. $\sigma(1) = 1$, $\sigma(10) = 18$; $\sigma(n) = n + 1 \iff n$ is a prime.

A divisor will always mean a positive divisor in this chapter.

Theorem 6.2.2. *If the standard form of n is $n = p_1^{\alpha_1} \dots p_r^{\alpha_r}$, then*

$$\sigma(n) = \prod_{i=1}^{r} (1 + p_i + p_i^2 + \cdots + p_i^{\alpha_i}) = \prod_{i=1}^{r} \frac{p_i^{\alpha_i + 1} - 1}{p_i - 1}.$$ ♣

Proof. We follow the argument applied for deducing the formula for $d(n)$ (Theorem 1.6.3).

By Theorem 1.6.2, all divisors d of n are

(6.2.1) $d = p_1^{\beta_1} p_2^{\beta_2} \dots p_r^{\beta_r}$

where the exponents $\beta_1, \beta_2, \dots, \beta_r$ assume the values

$$\beta_1 = 0, 1, \dots, \alpha_1, \quad \beta_2 = 0, 1, \dots, \alpha_2, \quad \dots, \quad \beta_r = 0, 1, \dots, \alpha_r,$$

further, every divisor has a unique representation in that form. Accordingly, $\sigma(n)$ is the sum of all these values of d.

On the other hand, we get the same sum performing the multiplication

(6.2.2) $\prod_{i=1}^{r} (1 + p_i + p_i^2 + \cdots + p_i^{\alpha_i}) :$

product (6.2.1) occurs if we multiply $p_1^{\beta_1}$ from the first factor of (6.2.2), $p_2^{\beta_2}$ from the second factor, etc.

This proves the first equality stated in the theorem.

The second equality follows from the well-known summation formula for finite geometric series.   □

For another possible proof of Theorem 6.2.2, see Exercise 6.2.1.

Definition 6.2.3. The *Möbius function $\mu(n)$* is defined by

$$\mu(n) = \begin{cases} 1, & \text{if } n = 1 \\ (-1)^r, & \text{if } n = p_1 \dots p_r \text{ where } p_j \text{ are distinct primes} \\ 0, & \text{if } p^2 \mid n \text{ for some prime.} \end{cases}$$ ♣

Example. $\mu(10) = 1$, $\mu(20) = 0$, $\mu(30) = -1$.

The following property is the key to the important applications of the Möbius function μ:

Theorem 6.2.4.

$$\sum_{d|n} \mu(d) = \begin{cases} 1, & \text{if } n = 1 \\ 0, & \text{if } n > 1. \end{cases}$$ ♣

Proof. If $n = 1$, then $\sum_{d|1} \mu(d) = \mu(1) = 1$.

For $n > 1$, let $n = p_1^{\alpha_1} \dots p_r^{\alpha_r}$ be the standard form of n. As $\mu(k) = 0$ if k is not squarefree, it is sufficient to do the summation for the squarefree divisors of n. Hence,

$$\sum_{d|n} \mu(d) = \mu(1) + \mu(p_1) + \dots + \mu(p_r) +$$

$$+ \mu(p_1 p_2) + \mu(p_1 p_3) + \dots + \mu(p_{r-1} p_r) + \dots + \mu(p_1 p_2 \dots p_r) =$$

$$= 1 - r + \binom{r}{2} - \binom{r}{3} + \dots + (-1)^r \binom{r}{r} = (1-1)^r = 0.$$ □

Definition 6.2.5. $\omega(n)$ is the number of distinct (positive) prime divisors of n.

$\Omega(n)$ is the total number of (positive) prime divisors of n, i.e. we count the primes according to their multiplicity given by the exponent in the standard form of n.

So $\omega(1) = \Omega(1) = 0$, and if the standard form of n is

$$n = p_1^{\alpha_1} \dots p_r^{\alpha_r} \qquad \text{(where every } \alpha_i > 0),$$

then

$$\omega(n) = r \quad \text{and} \quad \Omega(n) = \alpha_1 + \dots + \alpha_r.$$ ♣

Example. $\omega(500) = 2$, $\Omega(500) = 5$; $\omega(n) = \Omega(n) \iff n$ is squarefree.

Definition 6.2.6. Let k be a fixed positive integer. Then $d_k(n)$ is the number of positive integer solutions of the equation $n = x_1 x_2 \dots x_k$ where two solutions are considered as distinct even if they differ only in the order of the factors. ♣

Clearly, $d_1(n) = 1$, $d_k(1) = 1$, and $d_2(n) = d(n)$ (thus $d_k(n)$ is a generalization of $d(n)$).

Theorem 6.2.7. *If the standard form of n is $n = p_1^{\alpha_1} \dots p_r^{\alpha_r}$, then*

$$d_k(n) = \prod_{i=1}^{r} \binom{\alpha_i + k - 1}{k - 1}.$$ ♣

Proof. No prime divisor can occur in x_i other than $p_1, \dots, p_r$, hence the standard form of x_i is

$$x_1 = p_1^{\beta_{11}} \dots p_r^{\beta_{r1}}, \qquad \dots, \qquad x_k = p_1^{\beta_{1k}} \dots p_r^{\beta_{rk}},$$

where

$$0 \le \beta_{ij} \le \alpha_i, \qquad i = 1, 2, \dots, r, \qquad j = 1, 2, \dots k.$$

(The first index in the exponents refers to the prime, the second index refers to the variable.)

Then $n = x_1 x_2 \dots x_k$ holds if and only if

(6.2.3) $$\alpha_1 = \beta_{11} + \beta_{12} + \dots + \beta_{1k}, \qquad \dots, \qquad \alpha_r = \beta_{r1} + \beta_{r2} + \dots + \beta_{rk}.$$

System (6.2.3) contains r equations of the type

(6.2.4) $\qquad \alpha = y_1 + y_2 + \cdots + y_k, \qquad y_i$ is a non-negative integer.

We establish a formula for the number of solutions of (6.2.4). That is, we determine in how many ways α can be written as the sum of k non-negative integers where the order of the terms counts, so two representations are considered as distinct even if they differ only in the order of the summands.

We take a segment of length α, and measure segments of length $y_1, \ldots, y_k$ (including the ones of length 0) onto it in this order. We can encode this process by writing y_1 pieces of 1s, then writing a sign $*$ to indicate the end of the first segment, then writing y_2 pieces of 1s followed again by a delimiter $*$, etc., and finally y_k pieces of 1s close the row.

E.g. if $\alpha = 7$ and $k = 4$, then the representation $7 = 4 + 0 + 1 + 2$ can be encoded as $1111 * *1 * 11$. Conversely, $*1111 * 111*$ stands for the representation $7 = 0 + 4 + 3 + 0$.

Thus the number of solutions of (6.2.4) is equal to the number of such sequences composed of 1s and $*$s. A sequence consists of α pieces of 1s and $k - 1$ pieces of $*$s, in arbitrary order. Hence the number of sequences is

(6.2.5) $\qquad \dbinom{\alpha + k - 1}{k - 1}.$

By (6.2.5), the equations in (6.2.3) have

(6.2.6) $\qquad \dbinom{\alpha_1 + k - 1}{k - 1}, \dbinom{\alpha_2 + k - 1}{k - 1}, \ldots, \dbinom{\alpha_r + k - 1}{k - 1}$

solutions. Since these equations are independent, the number of solutions of (6.2.3) is the product of the number of solutions of the individual equations, i.e. of the binomial coefficients listed in (6.2.6). $\qquad\qquad\square$

We note that the formulas for $\sigma(n)$, $\Omega(n)$, and $d_k(n)$ (hence also for $d(n)$) remain valid even if the standard form of n may contain some exponents $\alpha_i = 0$, but the formulas for $\varphi(n)$ and $\omega(n)$ are valid only if every exponent in the standard form is strictly positive.

Finally, we examine these functions from the point of view of multiplicativity and additivity.

Theorem 6.2.8. $\varphi(n)$, $\sigma(n)$, $\mu(n)$, and $d_k(n)$ are multiplicative, but not completely (apart from the trivial case $d_1(n) = 1$).

$\omega(n)$ is additive, but not completely.

$\Omega(n)$ is completely additive. $\qquad\qquad\qquad\qquad\qquad\qquad\qquad$ ♣

Proof. The multiplicativity of $\varphi(n)$ was shown in the first proof of Theorem 2.3.1 (and also in Exercises 2.2.14 and 2.6.10). Further,

$$6 = \varphi(9) \neq \varphi(3)\varphi(3) = 4,$$

thus $\varphi(n)$ is not completely multiplicative. (Moreover, $\varphi(ab) = \varphi(a)\varphi(b)$ never holds if a and b are not coprime, see Exercise 2.3.10a.)

To show that $\sigma(n)$ is multiplicative, we use the formula in Theorem 6.2.2 (another proof can be obtained based on Exercise 1.6.5a-b, see Exercise 6.2.1).

If $a = 1$ or $b = 1$, then $\sigma(1) = 1$ guarantees $\sigma(ab) = \sigma(a)\sigma(b)$.

If a and b are coprime and their standard forms are

$$a = p_1^{\alpha_1} \dots p_r^{\alpha_r} \quad \text{and} \quad b = q_1^{\beta_1} \dots q_s^{\beta_s},$$

where $p_i \neq q_j$ (due to $(a, b) = 1$), then the standard form of ab is

$$ab = p_1^{\alpha_1} \dots p_r^{\alpha_r} q_1^{\beta_1} \dots q_s^{\beta_s}.$$

Applying the formula of σ for a, b, and ab, we obtain

$$\sigma(a)\sigma(b) = \frac{p_1^{\alpha_1+1} - 1}{p_1 - 1} \cdot \dots \cdot \frac{p_r^{\alpha_r+1} - 1}{p_r - 1} \cdot \frac{q_1^{\beta_1+1} - 1}{q_1 - 1} \cdot \dots \cdot \frac{q_s^{\beta_s+1} - 1}{q_s - 1} = \sigma(ab).$$

Because

$$36 = \sigma(2)\sigma(6) \neq \sigma(12) = 28,$$

thus $\sigma(n)$ is not completely multiplicative. (Moreover, $\sigma(ab) = \sigma(a)\sigma(b)$ never holds if a and b are not coprime, see Exercise 6.2.2.)

We verify the multiplicativity of $\mu(n)$ using its Definition 6.2.3. If $a = 1$ or $b = 1$, then $\mu(ab) = \mu(a)\mu(b)$ since $\mu(1) = 1$. If at least one of a and b is not squarefree, then their product is not squarefree, so $\mu(ab) = \mu(a)\mu(b) = 0$. Finally if both a and b are squarefree and are coprime, then their product is squarefree:

$$a = p_1 \dots p_r, \qquad b = q_1 \dots q_s, \qquad ab = p_1 \dots p_r q_1 \dots q_s,$$

thus

$$\mu(a)\mu(b) = (-1)^r(-1)^s = (-1)^{r+s} = \mu(ab).$$

Because

$$-1 = \mu(5)\mu(15) \neq \mu(75) = 0,$$

hence $\mu(n)$ is not completely multiplicative.

(We note that—in contrast with the behavior of $d(n)$, $\varphi(n)$, and $\sigma(n)$—there are infinitely many pairs a and b with $(a, b) \neq 1$ for which $\mu(a)\mu(b) = \mu(ab)$; e.g. $a = 4$ and b is an arbitrary even number.)

For $d_k(n)$, we can proceed similarly as seen at $\sigma(n)$.

Finally, the statements for $\omega(n)$ and $\Omega(n)$ follow directly from Definition 6.2.5. $\quad\square$

Exercises 6.2

1. Prove the multiplicativity of $\sigma(n)$ via Exercise 1.6.5a-b, and deduce the formula for $\sigma(n)$ from the multiplicative property.

2. Show that if $(a, b) \neq 1$, then $\sigma(ab) < \sigma(a)\sigma(b)$ and $d_k(ab) < d_k(a)d_k(b)$ for $k > 1$.

3. Assume that $n\varphi(n)\sigma(n)$ is not divisible by 3. Verify that n must be a square.

4. Prove that to any n there exist infinitely many k satisfying $\sigma(n) \mid \sigma(n^k)$.

5. We divide the sum of divisors of n by the sum of reciprocals of the divisors of n. What is the quotient?

6. For which values of n is $\sigma(n)$ (a) odd (b) a power of 2?

S* 7. Show that infinitely many positive integers are missing from the range of $\sigma(n)$.

S* 8. Find all positive integers n for which $\sigma(n!) = k!$ with a suitable k.

9. Prove $\sigma(n) \geq n + \sqrt{n} + 1$ for any composite n. When does equality hold?

10. Consider the equation $\sigma(n) = n + c$ where n is the variable and c is a fixed positive integer.

    (a) Solve the equation if c is (a1) 1; (a2) 5; (a3) 8; (a4) 11.

    (b) For which c are there infinitely many solutions?

    (c) Assume that the even Goldbach conjecture holds in the following slightly stronger form: Every even number greater than 6 is the sum of two *distinct* primes. Show that the above equation has a solution for any odd number $c \neq 5$.

    Remark: It was a long-standing unsolved problem whether the equation has no solutions for infinitely many positive integers c. Finally, Erdős showed that there exist infinitely many such (even) c.

11. Consider the equation $\sigma(n) - \varphi(n) = c$ where n is the variable and c is a fixed positive integer.

    (a) Solve the equation if c is (a1) 2; (a2) 4; (a3) 5; (a4) 10.

    (b) For which c are there infinitely many solutions?

    (c) Assume that the even Goldbach conjecture holds in the following slightly stronger form: Every even number greater than 6 is the sum of two *distinct* primes. Find infinitely many c for which the equation has a solution.

12. How many pairs of composite integers $a \neq b$ satisfy

    (a) $a + \varphi(b) = b + \varphi(a)$

  * (b) $a + \sigma(b) = b + \sigma(a)$?

13. Show that the following inequalities hold for every n and determine the cases of equality.

    (a) $\sigma(n) \leq \dfrac{(n+1)d(n)}{2}$

    (b) $\sigma(n) \leq \dfrac{nd(n)}{2} + 1$

    (c) $\sigma(n) \geq n + 2d(n) - 3$.

* 14. Solve the equation $2\sigma(n) = nd(n)$.

15. (a) Show that the following inequalities hold for every n and determine the cases of equality.

    (a1) $\sigma(n)\varphi(n) \leq n^2 - 1$

    (a2) $\sigma(n) + \varphi(n) \geq 2n$.

   (b) Demonstrate

* (b1) $\sigma(n)\varphi(n) > \dfrac{n^2}{2}$

   (b2) $\inf \dfrac{\sigma(n)\varphi(n)}{n^2} = \dfrac{6}{\pi^2}$.

* 16. Prove
$$\varphi(n) \mid n\sigma(n) - 2 \iff n \text{ is a prime or } n = 1, 4, 6, 22.$$

17. What is the range of the following functions?

   (a) $f(n) = \mu(n) + \mu(2n) + \mu(5n) + \mu(10n)$

 S (b) $g(n) = \displaystyle\sum_{k \mid 100!} \mu(kn)$.

18. (a) How many consecutive integers are there such that $\mu(n)$ is zero for none of
       them?

   (b) How many consecutive integers are there such that $\mu(n)$ is zero for each of
       them?

* 19. Show that the sum of the nth primitive complex roots of unity is $\mu(n)$.

20. Give a simpler form for the function $\mu(n)(\Omega(n) - \omega(n))$.

21. (a) Prove
$$2^{\omega(n)} \le d(n) \le 2^{\Omega(n)}$$
       for every n. When do we have equality?

   (b) How can we generalize part (a) for $d_k(n)$ instead of $d(n)$?

22. True or false?

   (a) If n is a square, then $d(n) \mid d_3(n)$.

   (b) If $d(n) \mid d_3(n)$, then n is a square.

23. Let ν be an arbitrary real number and define $\sigma_\nu(n)$ to be the sum of νth powers of
    the divisors of n:
$$\sigma_\nu(n) = \sum_{d \mid n} d^\nu.$$
    In particular: $\sigma_1(n) = \sigma(n)$ and $\sigma_0(n) = d(n)$.

    Find a formula for $\sigma_\nu(n)$ and show that $\sigma_\nu(n)$ is multiplicative.

6.3. Perfect Numbers

Antique Greek numerology viewed the proper divisors of a number as parts of it (i.e. the
number itself was not regarded as a divisor), and called a number perfect if it can be
"assembled from its parts". E.g. $6 = 1+2+3$ and $28 = 1+2+4+7+14$ have this property.
The famous book *Elements* by Euclid provides the following general construction (with
proof!):

   If we form a geometric series of double proportion starting from the unity till the
sum will be a prime, and multiply the sum by the last term, then the product is a perfect
number.

In modern terminology, a number n is perfect if and only if $\sigma(n) = 2n$ (since we count also n itself as a divisor), and Euclid's theorem claims that

$$(1 + 2 + 2^2 + \cdots + 2^k)2^k = (2^{k+1} - 1)2^k$$

is perfect if $2^{k+1} - 1$ is a prime. For $k = 1$ and $k = 2$ we obtain 6 and 28.

Primes of the form $2^s - 1$ are the Mersenne primes (see Section 5.2), and then s is necessarily a prime, too. As mentioned in Section 5.2, Mersenne (and many other contemporaries) investigated these primes in search of large perfect numbers.

Euler proved that every even perfect number is given by Euclid's construction. This means that there are exactly as many even perfect numbers as Mersenne primes. It is unknown whether there exist infinitely many Mersenne primes, hence we do not know whether there are infinitely many even perfect numbers. Another unsolved problem is whether there are odd perfect numbers at all. These simply formulated questions, more than 2000 years old, are perhaps the most ancient unsolved problems in mathematics.

Now we repeat the definition of perfect numbers and prove the theorems of Euclid and Euler characterizing the even perfect numbers.

Definition 6.3.1. The positive integer n is a *perfect number* if $\sigma(n) = 2n$.   ♣

Theorem 6.3.2. *An even number n is perfect if and only if $n = 2^{p-1}(2^p - 1)$ where $2^p - 1$ is a (Mersenne) prime (and thus also p is a prime).*   ♣

Proof. First we show that these numbers are perfect. Since $2^p - 1$ is a prime, therefore n is given in its standard form, and

$$\sigma(n) = (1 + 2 + \cdots + 2^{p-1})(1 + (2^p - 1)) = (2^p - 1)2^p = 2n,$$

by Theorem 6.2.2.

For the converse, assume that n is even and perfect, i.e.

(6.3.1) $n = 2^k t,$ where $k \geq 1$ and t is odd, and $\sigma(n) = 2n$.

Since $(2^k, t) = 1$, we get

(6.3.2) $2^{k+1} t = 2n = \sigma(n) = \sigma(2^k)\sigma(t) = (2^{k+1} - 1)\sigma(t),$

using the multiplicativity of σ and the formula for $\sigma(2^k)$.

Subtracting $(2^{k+1} - 1)t$ from the first and last terms in (6.3.2), we can factor t as

(6.3.3) $t = (2^{k+1} - 1)(\sigma(t) - t).$

We observe from (6.3.3) that $\sigma(t) - t$ is a divisor of t. Also, $k \geq 1$ implies $2^{k+1} - 1 > 1$, thus $\sigma(t) - t \neq t$, by (6.3.3).

Since $\sigma(t) - t$ and t are distinct divisors of t, with sum $\sigma(t)$ which is the sum of all divisors of t, t has no other divisors. This means that t is a prime, so $\sigma(t) - t = 1$.

Substituting into (6.3.3) and (6.3.1), we obtain

$$n = 2^k(2^{k+1} - 1), \qquad \text{where } 2^{k+1} - 1 \text{ is a prime,}$$

which yields the desired form of n (after replacing $k + 1$ by p).   □

Exercises 6.3

1. Show that the last digit of an even perfect number is 6 or 8 (in the decimal system).

2. Prove that if there exists an odd perfect number n, then

   (a) $n = s^2 p$ where p is a prime of the form $4k + 1$

   (b) $n \equiv 1 \pmod{12}$ or $n \equiv 9 \pmod{36}$.

3. Following the ancient Greeks, we call a natural number *deficient* if it is greater than the sum of its proper divisors (i.e. the total of its parts is less than the number itself). A number is *abundant* if this sum is greater than the number (i.e. its parts together surpass the number). For example, 10 is deficient since $1 + 2 + 5 < 10$, but 12 is abundant as $1 + 2 + 3 + 4 + 6 > 12$.

   Verify the following statements.

   (a) Every prime power is deficient.

   (b) If an odd number has only two distinct prime divisors, then it is deficient.

   (c) For every $k \geq 3$ there are both infinitely many odd abundant numbers and infinitely many odd deficient numbers with exactly k distinct prime divisors.

   (d) Every multiple of an abundant number is abundant.

   (e) Every deficient number has both infinitely many abundant multiples and infinitely many deficient multiples.

* 4. If we disregard trivial divisors (1 and the number itself), and want to assemble a number from its other divisors, then we get the condition $\sigma(n) = 2n + 1$. Prove that n must be the square of an odd integer.

   Remark: These numbers are called *quasiperfect*. It is unknown whether there exist any quasiperfect numbers.

S* 5. A positive integer n is called *superperfect* if $\sigma(\sigma(n)) = 2n$. Prove the following assertions.

   (a) An even number n is superperfect if and only if $n = 2^{p-1}$ where $2^p - 1$ is a (Mersenne) prime.

   (b) An odd superperfect number must be a square.

   (c) An odd prime power cannot be superperfect.

   Remark: By part (a), there are as many even superperfect numbers as Mersenne primes, thus it is unknown whether there exist infinitely many even superperfect numbers. It is also unknown whether there are any odd superperfect numbers.

6. A positive integer n is a *harmonic number* (or *Ore number*) if the harmonic mean of its divisors is an integer. Verify the following propositions.

   (a) n is harmonic if and only if $\sigma(n) \mid n d(n)$.

   (b) Every perfect number is harmonic.

(c) No prime powers are harmonic.

(d) 6 is the only squarefree harmonic number.

Remark: Numbers that are not perfect can be harmonic, e.g. 1 and 140 are harmonic. It is unknown whether there are infinitely many harmonic numbers, and whether there exists an odd harmonic number greater than 1.

7. The positive integers $a \neq b$ form an *amicable pair* if $\sigma(a) = \sigma(b) = a + b$. E.g. 220 and 284 form an amicable pair.

(a) Show that every amicable pair consists of a deficient and an abundant number (see the definitions in Exercise 6.3.3).

(b) Verify that a power of two cannot be a member of an amicable pair.

Remark: The origin of this notion is the ancient Greek numerology, as well: Each of the two numbers can be assembled from the parts (i.e. from the proper divisors) of the other. It is unknown whether there are infinitely many amicable pairs, and whether there exists an amicable pair where the members are coprime or have opposite parity.

6.4. Behavior of $d(n)$

We show first that the values of $d(n)$ fluctuate capriciously, with arbitrarily deep canyons and arbitrarily high peaks in the graph of the function.

Theorem 6.4.1 (Canyon theorem). *Given any positive integer K, there are infinitely many n satisfying*

$$(6.4.1) \qquad d(n-1) - d(n) > K \quad and \quad d(n+1) - d(n) > K$$

simultaneously. ♣

Proof. We shall choose n as a suitable prime number, so $d(n) = 2$.

Then (6.4.1) requires that both $n - 1$ and $n + 1$ have at least $K + 3$ divisors. This is certainly true if e.g. $2^{K+2} \mid n - 1$ and $3^{K+2} \mid n + 1$, so n is a solution of the system of congruences

$$(6.4.2) \qquad x \equiv 1 \ (\mathrm{mod}\ 2^{K+2}), \qquad x \equiv -1 \ (\mathrm{mod}\ 3^{K+2}).$$

Since $(2^{K+2}, 3^{K+2}) = 1$, (6.4.2) is solvable and all (positive) solutions are of the form $x \equiv x_0 \ (\mathrm{mod}\ 6^{K+2})$, or

$$(6.4.3) \qquad x = x_0 + t6^{K+2}, \quad t = 0, 1, 2, \ldots.$$

We have to show that the arithmetic progression (6.4.3) contains infinitely many primes. By Dirichlet's Theorem (Theorem 5.3.1), this holds if x_0 and 6^{K+2} are coprime. Since x_0 is a solution of (6.4.2), x_0 is relatively prime to both 2 and 3, hence also to 6^{K+2}. □

Theorem 6.4.2 (Peak theorem). *Given any positive integer K, there are infinitely many n satisfying*

(6.4.4) $$d(n) - d(n-1) > K \quad and \quad d(n) - d(n+1) > K$$

simultaneously. ♣

Proof. We choose n as the product of the first r primes:

(6.4.5) $$n = p_1 \dots p_r \quad so \quad d(n) = 2^r.$$

We shall show

(6.4.6) $$d(n-1) \leq 2^{r-1} \quad and \quad d(n+1) \leq 2^{r-1}.$$

From (6.4.5) and (6.4.6)

$$d(n) - d(n-1) \geq 2^{r-1} \quad and \quad d(n) - d(n+1) \geq 2^{r-1},$$

so (6.4.4) is true if $2^{r-1} > K$.

We verify the second inequality in (6.4.6); the first one can be treated similarly.

Write $n + 1$ as a product of primes: $n + 1 = q_1 \dots q_s$ (now $q_i = q_j$ may occur, too). Since n is the product of the first r primes and $(n + 1, n) = 1$, $q_i > p_r$ for every i (where p_r is the rth prime).

Each divisor of $n + 1$ is the product of some of its prime divisors (e.g. 1 and $n + 1$ are obtained when no q_j or every q_j is taken). If the q_j are not all distinct, then some products may give the same divisor. Hence, $d(n + 1) \leq 2^s$.

Thus the second inequality $d(n + 1) \leq 2^{r-1}$ in (6.4.6) follows if we show $s \leq r - 1$.

For a proof by contradiction, assume $s \geq r$. Then we get a contradiction (for $r \geq 2$) from the chain of inequalities

$$n + 1 = q_1 \dots q_s \geq q_1 \dots q_r \geq p_r^r + 1 \geq p_1 \dots p_r + 2 = n + 2. \qquad \square$$

The canyon and peak theorems illustrate that the behavior of $d(n)$ is very irregular. Now we shall investigate the average of the first n values of the function. It turns out that this *mean value* function (or *average value* function) is already very nice.

Theorem 6.4.3. *Let*

$$D(n) = \sum_{i=1}^{n} d(i).$$

Then

(6.4.7) $$\left| \frac{D(n)}{n} - \log n \right| \leq 1$$

for every n. ♣

Proof. We shall use the fact that

(6.4.8) $$\log n < \sum_{j=1}^{n} \frac{1}{j} \leq 1 + \log n$$

for every n. (Inequalities (6.4.8) can be proved by comparing suitable areas and integrals similar to the method applied in the first proof of Theorem 5.6.1.)

We construct an $n \times n$ matrix where the jth element of the ith row a_{ij} is 1 or 0, depending on whether j divides i or not:

$$a_{ij} = \begin{cases} 1, & \text{if } j \mid i \\ 0, & \text{if } j \nmid i. \end{cases}$$

For example, we obtain the following matrix for $n = 6$:

$$\begin{pmatrix} 1 & 0 & 0 & 0 & 0 & 0 \\ 1 & 1 & 0 & 0 & 0 & 0 \\ 1 & 0 & 1 & 0 & 0 & 0 \\ 1 & 1 & 0 & 1 & 0 & 0 \\ 1 & 0 & 0 & 0 & 1 & 0 \\ 1 & 1 & 1 & 0 & 0 & 1 \end{pmatrix}.$$

The key idea of the proof is to determine the sum of all elements in the matrix (i.e. the number of 1s) in two different ways.

In row i there are 1s whenever $j \mid i$, so the sum of elements in row i is $d(i)$. Thus summing by rows, we obtain that the sum of all elements in the matrix is

(6.4.9) $$D(n) = \sum_{i=1}^{n} d(i).$$

In column j there are 1s exactly in places

$$j, 2j, \ldots, \left\lfloor \frac{n}{j} \right\rfloor j,$$

thus the sum of elements in column j is $\lfloor n/j \rfloor$. Summing by columns, we get that the sum of all elements in the matrix is

(6.4.10) $$\sum_{j=1}^{n} \left\lfloor \frac{n}{j} \right\rfloor.$$

Both (6.4.9) and (6.4.10) provide the sum of elements in the matrix, so

(6.4.11) $$D(n) = \sum_{j=1}^{n} \left\lfloor \frac{n}{j} \right\rfloor.$$

Using the inequalities

$$\frac{n}{j} - 1 < \left\lfloor \frac{n}{j} \right\rfloor \le \frac{n}{j}$$

and (6.4.8), we deduce from (6.4.11) that

(6.4.12a) $$D(n) \le \sum_{j=1}^{n} \frac{n}{j} = n \sum_{i=1}^{n} \frac{1}{j} \le n(1 + \log n)$$

and

(6.4.12b) $$D(n) > \sum_{j=1}^{n} \left(\frac{n}{j} - 1 \right) = \left(n \sum_{j=1}^{n} \frac{1}{j} \right) - n > n(-1 + \log n).$$

Dividing inequalities (6.4.12a) and (6.4.12b) by n, we obtain (6.4.7). $\square$

Theorem 6.4.3 can be written also in the form $|D(n) - n \log n| \leq n$. The next theorem gives a better estimate for the difference of $D(n)$ and $n \log n$ (i.e. we obtain a better bound for the error term).

We shall need a more precise estimate of the sum $\sum_{j=1}^{n} 1/j$ than that given by (6.4.8): The sequence $\sum_{j=1}^{n} 1/j - \log n$ converges, its limit is known as Euler's constant, $\gamma = 0.577\ldots$, and

$$(6.4.13) \qquad \left| \sum_{j=1}^{n} \frac{1}{j} - \log n - \gamma \right| \leq \frac{10}{n}$$

for every n.

Theorem 6.4.4. *There exists a constant c such that*

$$(6.4.14) \qquad |D(n) - n \log n - (2\gamma - 1)n| < c\sqrt{n}$$

for every n. ♣

Proof. $d(i)$ is the number of pairs of positive integers x and y satisfying $xy = i$ (where the order of x and y counts). Therefore $D(n) = \sum_{i=1}^{n} d(i)$ is the number of pairs of positive integers x and y satisfying $xy \leq n$.

This means that $D(n)$ is the number of lattice points (x, y) (with integer coordinates) in the region defined by the positive halves of the coordinate axes and the hyperbola $xy = n$, including the lattice points on the hyperbola but not the ones on the axes. Now we count these lattice points.

Let $A(n)$ be the number of lattice points (x, y) with $x \leq \sqrt{n}$. As lattice points are symmetric about the line $y = x$, the number of lattice points with $y \leq \sqrt{n}$ is also $A(n)$.

We took thus all lattice points into consideration, but counted twice the lattice points satisfying both $x \leq \sqrt{n}$ and $y \leq \sqrt{n}$. These are the lattice points in the square where one of the diagonals is the segment connecting the origin and $(\sqrt{n}, \sqrt{n})$, so there are $\lfloor \sqrt{n} \rfloor^2$ lattice points in this square.

Thus the total number of lattice points is

$$(6.4.15) \qquad D(n) = 2A(n) - \lfloor \sqrt{n} \rfloor^2.$$

Now we determine $A(n)$. There are $\lfloor n/j \rfloor$ lattice points with first coordinate j, so

$$(6.4.16) \qquad A(n) = \sum_{j=1}^{\lfloor \sqrt{n} \rfloor} \left\lfloor \frac{n}{j} \right\rfloor.$$

Estimating the sum on the right-hand side of (6.4.16) similar to the proof of Theorem 6.4.3, we obtain

$$(6.4.17) \qquad A(n) = n \sum_{j=1}^{\lfloor \sqrt{n} \rfloor} \frac{1}{j} + f(n), \qquad \text{where} \quad |f(n)| < \sqrt{n}.$$

We apply (6.4.13) for the sum in (6.4.17):

(6.4.18)
$$\sum_{j=1}^{\lfloor\sqrt{n}\rfloor} \frac{1}{j} = \log\lfloor\sqrt{n}\rfloor + \gamma + g(n) \quad \text{where } |g(n)| \le \frac{10}{\lfloor\sqrt{n}\rfloor}.$$

Substituting back into (6.4.17), we get

(6.4.19a)
$$A(n) = n\log\lfloor\sqrt{n}\rfloor + \gamma n + h(n)$$

where

(6.4.19b)
$$|h(n)| = |ng(n) + f(n)| < \frac{10n}{\lfloor\sqrt{n}\rfloor} + \sqrt{n} < \frac{10n}{\frac{\sqrt{n}}{2}} + \sqrt{n} = 21\sqrt{n}.$$

To replace $\log\lfloor\sqrt{n}\rfloor$ in (6.4.19a) by $\log\sqrt{n} = \frac{\log n}{2}$, we estimate the error term, the difference $\frac{\log n}{2} - \log\lfloor\sqrt{n}\rfloor$.

Applying the mean value theorem of Lagrange and $(\log x)' = 1/x$, to any $a > 1$ there exists some u satisfying $a - 1 < u < a$ and

$$\log a - \log(a-1) = \frac{\log a - \log(a-1)}{a - (a-1)} = \frac{1}{u} < \frac{1}{a-1}.$$

Therefore

(6.4.20)
$$0 \le \frac{\log n}{2} - \log\lfloor\sqrt{n}\rfloor < \log\sqrt{n} - \log(\sqrt{n}-1) < \frac{1}{\sqrt{n}-1} \le \frac{2}{\sqrt{n}}$$

for any $n \ge 4$.

By (6.4.20), we can rewrite (6.4.19a) and (6.4.19b) as

(6.4.21)
$$A(n) = \frac{n\log n}{2} + \gamma n + k(n), \quad \text{where} \quad |k(n)| < 23\sqrt{n}.$$

To eliminate the floor sign in (6.4.15) and to replace $\lfloor\sqrt{n}\rfloor^2$ by n, we estimate $n - \lfloor\sqrt{n}\rfloor^2$:

(6.4.22)
$$\begin{aligned}
0 &\le n - \lfloor\sqrt{n}\rfloor^2 \\
&= (\sqrt{n})^2 - \lfloor\sqrt{n}\rfloor^2 \\
&= (\sqrt{n} - \lfloor\sqrt{n}\rfloor)(\sqrt{n} + \lfloor\sqrt{n}\rfloor) \\
&< 1(\sqrt{n} + \sqrt{n}) \\
&= 2\sqrt{n}.
\end{aligned}$$

Finally, substituting (6.4.21) and (6.4.22) into (6.4.15), we obtain

$$D(n) = n\log n + (2\gamma - 1)n + \ell(n) \quad \text{where} \quad |\ell(n)| < 48\sqrt{n}. \qquad \square$$

Remarks: (1) Improving the bound (6.4.14) for the error term in Theorem 6.4.4 is called the divisor problem and has an extensive literature. It was shown that $\sqrt{n}$ can be replaced by $n^{0.32}$, but not by $n^{0.25}$.

(2) As
$$\log 1 + \log 2 + \cdots + \log n \sim n \log n$$
(the two functions are asymptotically equal, their ratio tends to 1), Theorem 6.4.3
(or 6.4.4) implies

(6.4.23) $$d(1) + d(2) + \cdots + d(n) \sim \log 1 + \log 2 + \cdots + \log n.$$

Relation (6.4.23) expresses that the average order of magnitude of $d(n)$ is $\log n$.

This does not mean, however, that a typical n has about $\log n$ divisors; we prove in
Section 6.7 (see Exercise 6.7.6), that the number of divisors is smaller in general:
$d(n)$ is about
$$(\log n)^{\log 2} = (\log n)^{0.69\cdots}$$
for most integers n. The bigger average $\log n$ is due to those rarely occurring numbers that have extremely many divisors.

Finally, we examine a few further properties of the range of $d(n)$.

Note that $d(n)$ assumes every value $k \geq 2$ infinitely often, since $d(p^{k-1}) = k$ for
any prime p.

As for upper bounds for $d(n)$ depending on n, we established some of them in
Exercise 1.6.11. The next theorem improves those results significantly:

Theorem 6.4.5. *For any fixed $\delta > 0$,*
$$\lim_{n \to \infty} \frac{d(n)}{n^\delta} = 0. \qquad \clubsuit$$

The proof relies on the following fact of independent interest:

Theorem 6.4.6. *Let*
$$\{q_1 < q_2 < \ldots\} = \{2, 3, 4, 5, 7, 8, 9, 11, \ldots\}$$
be the sequence of all prime powers and f an arbitrary multiplicative function. Then
$$\lim_{j \to \infty} f(q_j) = 0 \Longrightarrow \lim_{n \to \infty} f(n) = 0. \qquad \clubsuit$$

Proof. The condition implies

(6.4.24) $$|f(q_j)| \leq H \quad \text{for every } j \text{ and } |f(q_j)| \leq 1, \text{ for } j > k$$

with suitable values H and k.

First we show

(6.4.25) $$|f(m)| \leq H^k$$

for every m. If the standard form of m is $m = \prod_{i=1}^{r} p_i^{\alpha_i}$, then

(6.4.26) $$|f(m)| = \prod_{i=1}^{r} |f(p_i^{\alpha_i})|$$

since f is multiplicative. By (6.4.24), at most k factors on the right-hand side of (6.4.26)
are greater than 1, and each is less than or equal to H, thus (6.4.25) holds.

Let $\varepsilon > 0$ be arbitrary. We have to guarantee an $n_0 = n_0(\varepsilon)$ such that $|f(n)| < \varepsilon$ for every $n > n_0$.

By the condition, there exists an $s = s(\varepsilon)$ such that

$$(6.4.27) \qquad\qquad |f(q_j)| < \frac{\varepsilon}{H^k}, \quad \text{for every } j > s.$$

We claim that $q_1 \ldots q_s$ can be chosen as n_0.

If $n > q_1 \ldots q_s$, then there must occur a prime power q_j greater than q_s: $n = q_j m$ where $(q_j, m) = 1$.

By (6.4.27), $|f(q_j)| < \varepsilon/H^k$, and $|f(m)| \le H^k$, by (6.4.25), so

$$|f(n)| = |f(q_j)| \cdot |f(m)| < \frac{\varepsilon}{H^k} \cdot H^k = \varepsilon. \qquad\qquad \square$$

Proof of Theorem 6.4.5. We apply Theorem 6.4.6 for the function

$$f(n) = \frac{d(n)}{n^\delta}.$$

To do this, we have to show

$$(6.4.28) \qquad\qquad \lim_{j \to \infty} \frac{d(q_j)}{q_j^\delta} = 0.$$

Let $q_j = p^\alpha$ (where p is a prime). Then

$$d(q_j) = d(p^\alpha) = \alpha + 1 \le 2\alpha = \frac{2\log(p^\alpha)}{\log p} \le \frac{2\log q_j}{\log 2},$$

hence

$$(6.4.29) \qquad\qquad \frac{d(q_j)}{q_j^\delta} \le \frac{2}{\log 2} \cdot \frac{\log q_j}{q_j^\delta}.$$

Since

$$\lim_{x \to \infty} \frac{\log x}{x^\delta} = 0,$$

the right-hand side in (6.4.29) tends to 0, therefore this is true also for the left-hand side. $\qquad \square$

Remark: It can be shown that the maximal order of magnitude of $d(n)$ is approximately

$$n^{\frac{\log 2}{\log\log n}}.$$

The precise formulation is:

(i) For any $\varepsilon > 0$, there exists an $n_0 = n_0(\varepsilon)$ such that

$$d(n) < n^{\frac{(1+\varepsilon)\log 2}{\log\log n}}$$

for every $n > n_0$.

(ii) For any $\varepsilon > 0$ there exist infinitely many n satisfying

$$d(n) > n^{\frac{(1-\varepsilon)\log 2}{\log\log n}}.$$

The proof of (ii) is Exercise 6.4.3b.

Exercises 6.4

* 1. Show that the statements of Theorems 6.4.1 and 6.4.2 remain valid if $d(n)$ is replaced by $\sigma(n)$, $\varphi(n)$, $\Omega(n)$, $\omega(n)$, or $d_k(n)$ with $k > 1$.

2. Prove
$$\lim_{n \to \infty} \frac{d_k(n)}{n^\delta} = 0$$
for any fixed $\delta > 0$ and positive integer k.

3. Let $\varepsilon > 0$ be arbitrary. Find infinitely many n satisfying

   (a) $d(n) > (\log n)^{100}$

 * (b) $d(n) > n^{\frac{(1-\varepsilon)\log 2}{\log\log n}}$.

4. Prove $\Omega(n) \le \log_2 n$ for every n. When do we get equality?

* 5. Let $\varepsilon > 0$ be arbitrary. Prove the following statements.

   (a) If n is sufficiently large, then
   $$\omega(n) < \frac{(1+\varepsilon)\log n}{\log\log n}.$$

   (b) There are infinitely many n satisfying
   $$\omega(n) > \frac{(1-\varepsilon)\log n}{\log\log n}.$$

6. Show that if n is large enough, then

   (a) $\varphi(n) > n^{0.99}$

   (b) $\varphi(n) > \dfrac{n}{2\log n}$

 * (c) $\varphi(n) > \dfrac{n}{C\log\log n}$

   (d) $\sigma(n) < n^{1.01}$

   (e) $\sigma(n) < 2n\log n$

 * (f) $\sigma(n) < Cn\log\log n$

   where C is a suitable absolute constant in parts (c) and (f).

7. Verify.

   (a) The range of $\varphi(n)/n$ is everywhere dense in the interval $[0, 1]$.

   (b) The range of $\sigma(n)/n$ is everywhere dense in $[1, \infty]$.

* 8. Dirichlet's Theorem (Theorem 5.3.1) states that if the positive integers a and d are coprime, then the arithmetic progression $a + kd$, $k = 0, 1, 2, \ldots$ contains infinitely many primes. The following significantly stronger results hold as well:

   (i) The sum of reciprocals of these primes is divergent.

(ii) The number of such primes not greater than n (with a and d fixed) is asymptotically

$$\frac{n}{\varphi(d)\log n}$$

when $n \to \infty$.

(i) and (ii) are far-reaching generalizations of Theorems 5.6.1 and 5.4.1.

(a) Let k be a fixed positive integer. Apply (i) to show that $k \mid \varphi(n)$ holds for nearly every n. More precisely, let $F(N)$ be the number of integers $x \le N$ satisfying $k \mid \varphi(x)$; then $\lim_{N \to \infty} F(N)/N = 1$.

S (b) Prove that nearly all positive integers are missing from the range of $\varphi(n)$. (Similar to the previous interpretation, let $G(N)$ be the number of values $y \le N$ occurring in the range of $\varphi(n)$; then $\lim_{N \to \infty} G(N)/N = 0$.)

* 9. Show that the statements of the previous exercise remain valid if φ is replaced by σ.

6.5. Summation and Inversion Functions

Definition 6.5.1. The *summation function* with respect to divisors of the arithmetic function f is

$$f^+(n) = \sum_{d\mid n} f(d). \qquad \qquad \clubsuit$$

Examples. The summation function of $f(n) = 1$ is $f^+(n) = d(n)$, the one of $g(n) = n$ is $g^+(n) = \sigma(n)$.

By Exercise 2.3.14, $\varphi^+(n) = n$, and by Theorem 6.2.4, $\mu^+(n) = e(n)$ where

(6.5.1) $$e(n) = \begin{cases} 1, & \text{if } n = 1 \\ 0, & \text{if } n > 1. \end{cases}$$

Theorem 6.5.2. *To every arithmetic function f there exists exactly one function having f as its summation function. This uniquely determined function is called the* inversion function *of f and is denoted by $\tilde{f}$.* $\clubsuit$

Proof. We write the equalities

$$f(n) = \sum_{d\mid n} \tilde{f}(d)$$

required from the inversion function for every n:

$$f(1) = \tilde{f}(1)$$
$$f(2) = \tilde{f}(1) + \tilde{f}(2)$$
$$f(3) = \tilde{f}(1) + \tilde{f}(3)$$
$$f(4) = \tilde{f}(1) + \tilde{f}(2) + \tilde{f}(4)$$
$$f(5) = \tilde{f}(1) + \tilde{f}(5)$$
$$f(6) = \tilde{f}(1) + \tilde{f}(2) + \tilde{f}(3) + \tilde{f}(6)$$
$$\vdots$$

We have to show that this system consisting of infinitely many equations and containing infinitely many variables $\tilde{f}(1)$, $\tilde{f}(2)$, ... has a unique solution.

The first equation is satisfied if and only if

$$\tilde{f}(1) = f(1).$$

Both of the first two equations are valid if and only if $\tilde{f}(1)$ is the value obtained from the first equation and

$$\tilde{f}(2) = f(2) - \tilde{f}(1).$$

We can proceed similarly by induction. Assume that the system of the first $m - 1$ equations has exactly one solution $\tilde{f}(1), \ldots, \tilde{f}(m - 1)$, and consider the system of the first m equations. Since the variable $\tilde{f}(m)$ occurs only in the mth equation, the first m equations are satisfied if and only if $\tilde{f}(1), \ldots, \tilde{f}(m - 1)$ are the unique values obtained from the first $m - 1$ equations (according to the induction hypothesis) and

$$(6.5.2) \qquad \tilde{f}(m) = f(m) - \sum_{\substack{d|m \\ d<m}} \tilde{f}(d).$$

This proves the existence and uniqueness of the function $\tilde{f}$. (Formula (6.5.2) serves as a recursion for determining the values of $\tilde{f}$.) □

Examples. Reading the examples after Definition 6.5.1 backwards (and keeping the notation used there), we have

$$\tilde{d}(n) = 1 \qquad \tilde{\sigma}(n) = n \qquad \tilde{g}(n) = \varphi(n) \qquad \tilde{e}(n) = \mu(n).$$

Now we establish a formula for the inversion function:

Theorem 6.5.3 (Möbius Inversion Formula).

$$(6.5.3) \qquad \tilde{f}(n) = \sum_{d|n} \mu(d) f\left(\frac{n}{d}\right).$$

♣

Proof. Since $\tilde{f}$ is unique by Theorem 6.5.2, it is sufficient to verify that the summation function $h^+(n)$ of

$$h(n) = \sum_{d|n} \mu(d) f\left(\frac{n}{d}\right) = \sum_{cd=n} \mu(d) f(c)$$

on the right-hand side of (6.5.3) is $f(n)$. We can do this by rearranging the sums and applying (6.5.1):

$$h^+(n) = \sum_{k|n} h(k) = \sum_{k|n} \sum_{cd=k} \mu(d) f(c) = \sum_{cd|n} \mu(d) f(c)$$

$$= \sum_{c|n} f(c) \left(\sum_{d|\frac{n}{c}} \mu(d) \right) = \sum_{c|n} f(c) \mu^+\left(\frac{n}{c}\right) = \sum_{c|n} f(c) e\left(\frac{n}{c}\right) = f(n). \qquad □$$

Finally, we present the *Smith determinant* as an interesting application of the inversion function:

Theorem 6.5.4. *Let f be an arithmetic function and construct the $n \times n$ matrix*

$$A = \begin{pmatrix} f((1,1)) & f((1,2)) & \cdots & f((1,n)) \\ f((2,1)) & f((2,2)) & \cdots & f((2,n)) \\ \vdots & \vdots & \ddots & \vdots \\ f((n,1)) & f((n,2)) & \cdots & f((n,n)) \end{pmatrix}$$

where (i,j) denotes the gcd of i and j. Then the determinant of A is

$$\det A = \tilde{f}(1)\tilde{f}(2)\ldots\tilde{f}(n). \qquad \clubsuit$$

Proof. Consider the $n \times n$ matrices B and C where the jth element in row i is b_{ij} and c_{ij}, defined as

$$b_{ij} = \begin{cases} 1, & \text{if } j \mid i \\ 0, & \text{if } j \nmid i, \end{cases}$$

and

$$c_{ij} = b_{ij}\tilde{f}(j), \quad \text{i.e.} \quad c_{ij} = \begin{cases} \tilde{f}(j), & \text{if } j \mid i; \\ 0, & \text{if } j \nmid i. \end{cases}$$

Both matrices have only 0s above the main diagonal, hence each determinant is the product of the elements on the main diagonal. The main diagonal of B consists of 1s, whereas the elements on the main diagonal of C are $\tilde{f}(1), \ldots, \tilde{f}(n)$, hence

(6.5.4) $\det B = 1 \quad \text{and} \quad \det C = \tilde{f}(1)\tilde{f}(2)\ldots\tilde{f}(n).$

Now we examine the product $D = BC^T$ where C^T means the transpose of C. The jth element in row i in D is

(6.5.5)
$$d_{ij} = b_{i1}c_{j1} + b_{i2}c_{j2} + \cdots + b_{in}c_{jn} =$$
$$= b_{i1}b_{j1}\tilde{f}(1) + b_{i2}b_{j2}\tilde{f}(2) + \cdots + b_{in}b_{jn}\tilde{f}(n).$$

Here

$$b_{ik}b_{jk}\tilde{f}(k) = \begin{cases} \tilde{f}(k), & \text{if } k \mid i \text{ and } k \mid j \\ 0, & \text{otherwise,} \end{cases}$$

so

(6.5.6) $b_{ik}b_{jk}\tilde{f}(k) = \begin{cases} \tilde{f}(k), & \text{if } k \mid (i,j) \\ 0, & \text{if } k \nmid (i,j). \end{cases}$

Substituting (6.5.6) into (6.5.5) and applying the definition of $\tilde{f}$, we obtain

$$d_{ij} = \sum_{k \mid (i,j)} \tilde{f}(k) = f((i,j)),$$

thus $D = A$.

Finally, (6.5.4) and the product rule of determinants imply

$$\det A = \det D = (\det B)(\det C) = \tilde{f}(1)\tilde{f}(2)\ldots\tilde{f}(n). \qquad \square$$

Exercises 6.5

1. Demonstrate $d_k^+(n) = d_{k+1}(n)$.

2. Prove the assertions:

   (a) f is multiplicative $\iff$ f^+ is multiplicative.

   (b) f is multiplicative $\iff$ $\tilde{f}$ is multiplicative.

   Remark: Exercise 6.5.2 immediately implies that $d(n)$, $\sigma(n)$, and $\varphi(n)$ are multiplicative.

3. (a) Determine all completely multiplicative functions with a completely multiplicative summation function.

   (b) Find all additive functions with an additive summation function.

4. Let $n = p_1^{\alpha_1} \dots p_r^{\alpha_r}$ be the standard form of n. Verify the following statements.

   (a) If f is multiplicative and $f \neq 0$, then

   $$f^+(n) = \prod_{i=1}^{r}(1 + f(p_i) + f(p_i^2) + \cdots + f(p_i^{\alpha_i}))$$

   and

   $$\tilde{f}(n) = \prod_{i=1}^{r}(f(p_i^{\alpha_i}) - f(p_i^{\alpha_i - 1})).$$

   (b) If f is completely multiplicative and its values at primes are all different from 0 and 1, then

   $$f^+(n) = \prod_{i=1}^{r} \frac{f(p_i)^{\alpha_i + 1} - 1}{f(p_i) - 1} \quad \text{and} \quad \tilde{f}(n) = f(n) \prod_{i=1}^{r}\left(1 - \frac{1}{f(p_i)}\right).$$

   Which formulas do we obtain in the special case $f(n) = n$?

5. Determine the inversion function of

   (a) $f(n) = c$ (a constant function)

   (b) $g(n) = \frac{(-1)^n + 1}{2}$

   (c) $\Omega(n)$

   (d) $\omega(n)$.

6. Let f be additive and $\omega(n) \geq 2$. Prove $\tilde{f}(n) = 0$.

7. Find a simpler form for the sum

   $$\sum_{ab=n} \sigma(a)\mu(b).$$

8. Prove the identity

   $$\sum_{d|n} \frac{\mu(d)}{d} = \frac{\varphi(n)}{n}.$$

9. Verify.

   (a) The sum of all primitive complex nth roots of unity is $\mu(n)$.

   * (b) The sum of the kth powers of all primitive complex nth roots of unity is

   $$\frac{\mu(n')\varphi(n)}{\varphi(n')} \qquad \text{where} \qquad n' = \frac{n}{(n,k)}.$$

   (c) For any prime p, the sum of all pairwise incongruent primitive roots modulo p is congruent to $\mu(p-1)$ modulo p.

10. Evaluate the determinants of $n \times n$ matrices whose jth element in row i is

    (a) (i, j)

    (b) $\sigma((i,j))$

    (c) $d((i,j))$

    (d) $\omega((i,j))$.

11. Let $s_1, \ldots, s_n$ be arbitrary distinct integers such that every divisor of each s_i occurs among the numbers s_j. Show that the analog of Theorem 6.5.4 remains valid if the numbers $1, 2, \ldots, n$ are replaced by $s_1, \ldots, s_n$.

6.6. Convolution

Definition 6.6.1. The *convolution* of arithmetic functions f and g is

$$(f * g)(n) = \sum_{d|n} f(d)g(\frac{n}{d}) = \sum_{cd=n} f(d)g(c). \qquad \clubsuit$$

The summation and inversion functions are special cases of convolution: by definition, f^+ is the convolution of f and the constant function 1, and by the Möbius inversion formula, $\tilde{f}$ is the convolution of f and μ, i.e.

$$f^+ = f * 1 \quad \text{and} \quad \tilde{f} = f * \mu.$$

Now we examine the properties of convolution as an operation.

Theorem 6.6.2. *Convolution is associative and commutative, the identity element is*

$$e(n) = \begin{cases} 1, & \text{if } n = 1 \\ 0, & \text{if } n > 1, \end{cases}$$

and f has an inverse if and only if $f(1) \neq 0$. $\qquad \clubsuit$

Proof. The commutative law follows directly from the definition.

Associative law:

$$\big(f * (g * h)\big)(n) = \sum_{bk=n} f(b)\big(\sum_{cd=k} g(c)h(d)\big) = \sum_{bcd=n} f(b)g(c)h(d),$$

and $((f * g) * h)(n)$ can be transformed into the same final form.

Identity element:

$$(e * f)(n) = \sum_{d|n} e(d)f(\frac{n}{d}) = 1 \cdot f(n) + \sum_{1<d|n} 0 \cdot f(\frac{n}{d}) = f(n).$$

Inverse: We can argue similarly as in the proof of Theorem 6.5.2. The inverse g of f has to satisfy $e = f * g$ so

$$1 = e(1) = f(1)g(1)$$
$$0 = e(2) = f(1)g(2) + f(2)g(1)$$
$$0 = e(3) = f(1)g(3) + f(3)g(1)$$
$$0 = e(4) = f(1)g(4) + f(2)g(2) + f(4)g(1)$$
$$0 = e(5) = f(1)g(5) + f(5)g(1)$$
$$0 = e(6) = f(1)g(6) + f(2)g(3) + f(3)g(2) + f(6)g(1)$$
$$\vdots$$

In this system of infinitely many equations, $g(1)$, $g(2)$, ... are the unknowns to be determined. The first m equations contain only the variables $g(1), \ldots, g(m)$, and $g(m)$ occurs first in the mth equation.

If $f(1) = 0$, then the first equation has no solution, hence $f(1) \neq 0$ is a necessary condition for the existence of the inverse. To prove its sufficiency, we have to show that for $f(1) \neq 0$ the system of equations has a (unique) solution.

The first equation holds if and only if

$$g(1) = \frac{1}{f(1)}.$$

The first two equations hold simultaneously if and only if $g(1)$ is the uniquely determined value obtained from the first equation and

$$g(2) = \frac{-f(2)g(1)}{f(1)}.$$

We can proceed similarly by induction. Assume that the system consisting of the first $m - 1$ equations has a unique solution $g(1), \ldots, g(m-1)$, and consider now the system of the first m equations. As $g(m)$ occurs first in the mth equation, the first m equations are satisfied if and only if $g(1), \ldots, g(m-1)$ are the uniquely determined values obtained from the first $m - 1$ equations and

$$g(m) = \frac{-1}{f(1)} \sum_{\substack{d \mid m \\ d < m}} g(d) f\left(\frac{m}{d}\right).$$

This recursion defines the unique inverse g of function f. $\qquad \square$

Convolution gives a simple proof for the Möbius inversion formula and it will also clarify why the function μ plays such a special role.

Using convolution, the inversion function can be written as

$$(6.6.1) \qquad\qquad \tilde{f} * 1 = f,$$

and we have to express $\tilde{f}$. Let g be the inverse of the constant function 1, and multiply (6.6.1) by g, i.e. apply the convolution g to both sides. Then, using also the properties of convolution, we obtain

$$(6.6.2) \qquad\qquad \tilde{f} = f * g.$$

Here g is the inverse of 1, so $1*g = e$, i.e. $g^+ = e$, or equivalently, $g = \tilde{e} = \mu$. Substituting this into (6.5.2), we get

$$\tilde{f} = f * \mu$$

which is precisely the Möbius inversion formula.

In studying arithmetic functions, *Dirichlet series* play a very important role:

Definition 6.6.3. Let f be an arithmetic function and S the set of those real numbers s for which the infinite series

$$(6.6.3) \qquad\qquad \sum_{n=1}^{\infty} \frac{f(n)}{n^s}$$

converges. Then the *Dirichlet series* belonging to f is the function $F : S \to \mathbf{C}$ defined by

$$F(s) = \sum_{n=1}^{\infty} \frac{f(n)}{n^s}. \qquad\qquad \clubsuit$$

Thus the domain of F is the set of those real numbers for which the infinite series (6.5.3) converges.

It is easy to check (see Exercise 6.6.6) that if (6.5.3) converges for some s_0, then it is absolutely convergent for every $s > s_0 + 1$. In the sequel, we shall consider function $F(s)$ only at places s where the series (6.5.3) is *absolutely convergent*. This will have the advantage that we can use theorems on absolutely convergent series that can be roughly summarized as stating that the same rules of computation apply to absolutely convergent series as to the sums with finitely many terms. This means, among other things, that rearranging and grouping the terms of an absolutely convergent series arbitrarily gives an absolutely convergent series again having the same sum as the original one, and multiplying two absolutely convergent series using the every term by every term law (and rearranging and grouping the result in any fashion) yields an absolutely convergent series whose sum is the product of the sums of the two original series.

We note that a Dirichlet series can be investigated as a function of a complex variable, and also as formal series when convergence is not considered, but we do not deal with these variants.

The most famous Dirichlet series is *Riemann's zeta function* belonging to $f = 1$:

$$(6.6.4) \qquad\qquad \zeta(s) = \sum_{n=1}^{\infty} \frac{1}{n^s},$$

defined already in Exercise 5.6.6. The series (6.6.4) is absolutely convergent for $s > 1$, and by Exercise 5.6.6, it can be represented as the infinite product

$$(6.6.5) \qquad\qquad \zeta(s) = \prod_p \frac{1}{1 - \frac{1}{p^s}} = \lim_{n \to \infty} \prod_{p \le n} \frac{1}{1 - \frac{1}{p^s}}.$$

Formula (6.6.5) is due to Euler, and it reveals why the distribution of primes is closely connected to the behavior of the ζ function. Extremely important theorems concerning the primes would follow from the *Riemann Hypothesis* which claims that all non-real roots of the extended version of the zeta function to complex variables have real part 1/2.

The next theorem reveals the connection between Dirichlet series and convolution:

Theorem 6.6.4. *Assume that the Dirichlet series $F(s)$, $G(s)$, and $H(s)$ belonging to the arithmetic functions f, g, and h, are absolutely convergent, and $h = f * g$. Then $H(s) = F(s)G(s)$.* ♣

Proof. Using the properties of multiplication of absolutely convergent series, we get

$$F(s)G(s) = \left(\sum_{k=1}^{\infty} \frac{f(k)}{k^s}\right)\left(\sum_{m=1}^{\infty} \frac{g(m)}{m^s}\right)$$

$$= \sum_{k=1}^{\infty} \sum_{m=1}^{\infty} \frac{f(k)g(m)}{(km)^s}$$

$$= \sum_{n=1}^{\infty} \frac{\sum_{km=n} f(k)g(m)}{n^s}$$

$$= \sum_{n=1}^{\infty} \frac{h(n)}{n^s} = H(s). \qquad \square$$

Theorem 6.6.4 can be used to determine the Dirichlet series

$$M(s) = \sum_{n=1}^{\infty} \frac{\mu(n)}{n^s}$$

belonging to the Möbius function. By $|\mu(n)| \leq 1$, this series is absolutely convergent for $s > 1$. Since $\mu * 1 = e$,

$$M(s)\zeta(s) = \sum_{n=1}^{\infty} \frac{e(n)}{n^s} = \frac{1}{1^s} + \sum_{n=2}^{\infty} \frac{0}{n^s} = 1,$$

hence

(6.6.6) $$M(s) = \frac{1}{\zeta(s)}, \quad \text{i.e.} \quad \sum_{n=1}^{\infty} \frac{\mu(n)}{n^s} = \frac{1}{\sum_{n=1}^{\infty} \frac{1}{n^s}}.$$

Substituting $s = 2$, we get the formula

(6.6.7) $$\sum_{n=1}^{\infty} \frac{\mu(n)}{n^2} = \frac{6}{\pi^2}.$$

Exercises 6.6

1. Which (well known) function will be the kth power by convolution of the function $f = 1$ (i.e. the convolution $1 * 1 * \cdots * 1$ of k factors)?

2. Prove that the arithmetic functions form a commutative ring with identity element and without zero divisors with respect to the operations of addition and convolution.

3. Let f be a (complex-valued) arithmetic function satisfying $f(1) \neq 0$. How many kth roots does f possess with respect to convolution?

4. (a) Verify that the convolution of two multiplicative functions is multiplicative.

   (b) Let f and g be completely multiplicative. Show that $f * g$ is completely multiplicative if and only if $(fg)(n) = 0$ for every $n > 1$.

5. Prove

$$\sum_{d|n} \sigma(d)\varphi(\frac{n}{d}) = nd(n).$$

6. Demonstrate that if the infinite series

$$\sum_{n=1}^{\infty} \frac{f(n)}{n^s}$$

is convergent for $s = s_0$, then it is absolutely convergent for every $s > s_0 + 1$.

7. Let $F(s), F^+(s)$, and $\tilde{F}(s)$ be the Dirichlet series belonging to the functions f, f^+, and $\tilde{f}$. Prove that in the case of absolute convergence,

$$F^+(s) = f(s)\zeta(s) \quad \text{and} \quad \tilde{F}(s) = \frac{F(s)}{\zeta(s)}$$

for every $s > 1$.

8. Show that for $s > 1$

   (a) $\displaystyle\sum_{n=1}^{\infty} \frac{d(n)}{n^s} = \zeta^2(s)$

   (b) $\displaystyle\sum_{n=1}^{\infty} \frac{d_k(n)}{n^s} = \zeta^k(s)$.

9. Prove that if $s > 2$, then

   (a) $\displaystyle\sum_{n=1}^{\infty} \frac{\sigma(n)}{n^s} = \zeta(s)\zeta(s-1)$

   (b) $\displaystyle\sum_{n=1}^{\infty} \frac{\varphi(n)}{n^s} = \frac{\zeta(s-1)}{\zeta(s)}$.

10. In this exercise we generalize the product form of ζ for multiplicative and completely multiplicative functions. The infinite product taken for all primes is defined as in Exercise 5.6.6 (and as in (6.6.5) of this section), and absolute convergence is assumed for all infinite series.

    (a) For a multiplicative f, show

$$\sum_{n=1}^{\infty} \frac{f(n)}{n^s} = \prod_p (\sum_{k=0}^{\infty} \frac{f(p^k)}{p^{ks}}).$$

(b) Let $f \neq 0$, f be completely multiplicative, and $|f(p)| < p^s$ for every prime p. Prove

$$\sum_{n=1}^{\infty} \frac{f(n)}{n^s} = \prod_{p} \frac{1}{1 - \frac{f(p)}{p^s}}.$$

11. Demonstrate

$$\sum_{n=1}^{\infty} \frac{\mu(n)}{n^s} = \prod_{p} \left(1 - \frac{1}{p^s}\right)$$

for $s > 1$.

S 12. Compute the sums

(a) $\displaystyle\sum_{n=1}^{\infty} \frac{d(n)}{n^2}$

* (b) $\displaystyle\sum_{n=1}^{\infty} \left(\frac{d(n)}{n}\right)^2$.

* 13. Determine the sum of squares of reciprocals of all squarefree numbers.

14. (a) Prove that if $|x| < 1$ and both infinite series occurring in

$$\sum_{n=1}^{\infty} \frac{f(n)x^n}{1 - x^n} = \sum_{k=1}^{\infty} f^+(k)x^k$$

are convergent, then equality holds.

(b) Compute the sums

(b1) $\displaystyle\sum_{n=1}^{\infty} \frac{\mu(n)}{2^n - 1}$

(b2) $\displaystyle\sum_{n=1}^{\infty} \frac{\varphi(n)}{2^n - 1}$.

6.7. Mean Value

We proved in Section 6.4 that though the values of $d(n)$ oscillate, the average of the values at the first n integers behaves smoothly. In this section, we investigate the mean value functions of σ, φ, and ω.

Definition 6.7.1. Let f be an arithmetic function and $F(n) = f(1) + f(2) + \cdots + f(n)$. The *mean value* (or *average value*) function of f is defined to be

$$\frac{F(n)}{n} = \frac{(1) + f(2) + \cdots + f(n)}{n}. \qquad \clubsuit$$

We shall often need the following theorem when computing mean value functions.

Theorem 6.7.2. *If $f = g * h$, then*

(6.7.1) $$F(n) = \sum_{i=1}^{n} f(i) = \sum_{j=1}^{n} g(j)\left(\sum_{k=1}^{\lfloor n/j \rfloor} h(k)\right). \qquad \clubsuit$$

Proof. By the definition of convolution,

$$\sum_{i=1}^{n} f(i) = \sum_{i=1}^{n} \sum_{jk=i} g(j)h(k) = \sum_{j=1}^{n} g(j)\left(\sum_{k=1}^{\lfloor n/j \rfloor} h(k)\right).\qquad\square$$

The simplest special case of Theorem 6.7.2 is $f = g^+ = g * 1$. Then

(6.7.2)
$$\sum_{i=1}^{n} f(i) = \sum_{j=1}^{n} g(j)\left(\sum_{k=1}^{\lfloor n/j \rfloor} 1\right) = \sum_{j=1}^{n} g(j)\left\lfloor \frac{n}{j} \right\rfloor.$$

For $f(n) = d(n)$ we have $g = 1$, thus (6.7.2) gives

$$D(n) = \sum_{j=1}^{n} \left\lfloor \frac{n}{j} \right\rfloor$$

which is just equality (6.4.11) in the proof of Theorem 6.4.3.

We determine first the mean value of σ.

Theorem 6.7.3. *Let $\Sigma(n) = \sigma(1) + \sigma(2) + \cdots + \sigma(n)$. Then*

(6.7.3)
$$\Sigma(n) \sim \frac{\pi^2}{12} n^2$$

where $\sim$ stands for asymptotic equality.

Two equivalent forms of (6.7.3) *are*

(6.7.4)
$$\frac{\Sigma(n)}{n} \sim \frac{\pi^2}{12} n$$

and

(6.7.5)
$$\sigma(1) + \sigma(2) + \cdots + \sigma(n) \sim \frac{\pi^2}{6} \cdot 1 + \frac{\pi^2}{6} \cdot 2 + \cdots + \frac{\pi^2}{6} n.\qquad\clubsuit$$

Thus (6.7.4) states that the mean value of σ can be well approximated by $\pi^2 n/12$, and (6.7.5) expresses that the *average order of magnitude* of σ is $\pi^2 n/6$.

Proof. We try first a suitable modification of the method used for $d(n)$, applying (6.7.2). Let $v(n) = n$, then $\sigma = v^+ = v * 1$, so

(6.7.6)
$$\Sigma(n) = \sum_{i=1}^{n} \sigma(i) = \sum_{j=1}^{n} j\left\lfloor \frac{n}{j} \right\rfloor.$$

Estimating the right-hand side of (6.7.6) by the usual inequalities $a - 1 < \lfloor a \rfloor \le a$, we get

$$n^2 - \frac{n(n+1)}{2} < \Sigma(n) \le n^2$$

which does not yield an asymptotic value for $\Sigma(n)$.

Therefore we interchange the roles of 1 and $v(n) = n$, and apply Theorem 6.7.2 with $g = 1$ and $h = v$ for the convolution $\sigma = 1 * v$:

(6.7.7)
$$\Sigma(n) = \sum_{j=1}^{n} \sum_{k=1}^{\lfloor n/j \rfloor} k = \sum_{j=1}^{n} \frac{\left\lfloor \frac{n}{j} \right\rfloor \left(\left\lfloor \frac{n}{j} \right\rfloor + 1\right)}{2}.$$

We estimate the right-hand side of (6.7.7) using

$$a^2 - a = (a-1)a < \lfloor a \rfloor (\lfloor a \rfloor + 1) \le a(a+1) = a^2 + a$$

for $a > 0$ which gives

(6.7.8) $$\left| \lfloor a \rfloor (\lfloor a \rfloor + 1) - a^2 \right| \le a.$$

Applying (6.7.8) with $a = n/j$ to (6.7.7), we obtain

$$\left| \Sigma(n) - \sum_{j=1}^{n} \frac{n^2}{2j^2} \right| \le \sum_{j=1}^{n} \frac{n}{2j} \le \frac{n(1 + \log n)}{2},$$

so

(6.7.9) $$\Sigma(n) = \frac{n^2}{2} \sum_{j=1}^{n} \frac{1}{j^2} + U(n), \qquad \text{where } |U(n)| < n \log n \text{ for } n \ge 3.$$

Dividing (6.7.9) by n^2, we get

(6.7.10) $$\frac{\Sigma(n)}{n^2} = \frac{1}{2} \sum_{j=1}^{n} \frac{1}{j^2} + \frac{U(n)}{n^2}.$$

If $n \to \infty$, then the limit of the first term on the right-hand side of (6.7.10) is

$$\frac{1}{2} \sum_{j=1}^{\infty} \frac{1}{j^2} = \frac{\pi^2}{12},$$

whereas the second term tends to 0, thus

$$\lim_{n \to \infty} \frac{\Sigma(n)}{n^2} = \frac{\pi^2}{12}.$$

This is equivalent to (6.7.3). □

We can treat the mean value of φ with similar methods:

Theorem 6.7.4. *Let* $\Phi(n) = \varphi(1) + \varphi(2) + \cdots + \varphi(n)$. *Then*

(6.7.11) $$\Phi(n) \sim \frac{3}{\pi^2} n^2$$

where $\sim$ *stands for asymptotic equality.*

Two equivalent forms of (6.7.11) *are*

(6.7.12) $$\frac{\Phi(n)}{n} \sim \frac{3}{\pi^2} n$$

and

(6.7.13) $$\varphi(1) + \varphi(2) + \cdots + \varphi(n) \sim \frac{6}{\pi^2} \cdot 1 + \frac{6}{\pi^2} \cdot 2 + \cdots + \frac{6}{\pi^2} n. \qquad \clubsuit$$

Thus (6.7.12) states that the mean value of φ can be well approximated by $3n/\pi^2$, and (6.7.13) expresses that the *average order of magnitude* of φ is $6n/\pi^2$.

Proof. We apply Theorem 6.7.2 now for the convolution $\varphi = \mu * \upsilon$, i.e. with $g = \mu$ and $h = \upsilon$ (where $\upsilon(n) = n$):

$$(6.7.14) \qquad \Phi(n) = \sum_{j=1}^{n} \mu(j) \sum_{k=1}^{\lfloor n/j \rfloor} k = \sum_{j=1}^{n} \mu(j) \frac{\lfloor \frac{n}{j} \rfloor (\lfloor \frac{n}{j} \rfloor + 1)}{2}.$$

We can continue analogously to the proof of Theorem 6.7.3 (for estimating the error term, we use $|\mu(j)| \leq 1$). Finally we arrive at

$$(6.7.15) \qquad \frac{\Phi(n)}{n^2} = \frac{1}{2} \sum_{j=1}^{n} \frac{\mu(j)}{j^2} + \frac{U(n)}{n^2},$$

which corresponds to (6.7.10). If $n \to \infty$, then the second term on the right-hand side of (6.7.15) tends to 0, and the limit of the first term is

$$\frac{1}{2} \sum_{j=1}^{\infty} \frac{\mu(j)}{j^2}.$$

According to formula (6.6.7) after Theorem 6.6.4,

$$\sum_{j=1}^{\infty} \frac{\mu(j)}{j^2} = \frac{6}{\pi^2},$$

therefore

$$\lim_{n \to \infty} \frac{\Phi(n)}{n^2} = \frac{3}{\pi^2}. \qquad \qquad \square$$

As a corollary of Theorem 6.7.4, we can determine the probability of two numbers being coprime. In a more picturesque formulation, what is the probability that a lattice point P can be seen from the origin (since there are no further lattice points on the segment connecting P and the origin if and only if the coordinates of P are coprime)?

We need first an exact definition of this probability. Let Q_n be the square of side length n with the origin as a vertex and two sides lying on the positive halves of the axes. We consider the lattice points in Q_n (apart from the points on the axes), determine the ratio of the ones visible from the origin (i.e. having coprime coordinates), and take the limit of this ratio as the side length of Q_n tends to infinity:

$$(6.7.16) \qquad \lim_{n \to \infty} \frac{H(n)}{n^2}, \quad \text{where} \quad H(n) = \sum_{\substack{1 \leq a \leq n, 1 \leq b \leq n \\ (a,b)=1}} 1.$$

We show that this limit exists and will call it the probability in question.

Theorem 6.7.5. *The probability of two numbers being relatively prime (in the sense of* (6.7.16)*) is* $6/\pi^2$. ♣

It is part of the theorem, of course, that this probability, the limit in (6.7.16), exists.

As indicated earlier, this probability is closely related to the mean value of φ, so Theorem 6.7.5 will follow immediately from Theorem 6.7.4. We shall present also a second proof of Theorem 6.7.5 based on the Inclusion and Exclusion Principle (actually, herewith we obtain another proof also of Theorem 6.7.4).

First proof. We verify that

$$\Phi(n) = \sum_{i=1}^{n} \varphi(i) \quad \text{and} \quad H(n) = \sum_{\substack{1 \le a \le n, 1 \le b \le n \\ (a,b)=1}} 1$$

satisfy

(6.7.17) $$H(n) = 2\Phi(n) - 1.$$

To prove (6.7.17), consider the square Q_n and cut it into two triangles along its diagonal, starting from the origin. $H(n)$ is just the number of lattice points in Q_n with coprime coordinates (disregarding the lattice points on the axes). These lattice points are symmetric about the diagonal starting from the origin. In the lower triangle, a lattice point with first coordinate i counts if and only if its second coordinate t satisfies $1 \le t \le i$ and $(i, t) = 1$. There are $\varphi(i)$ such lattice points, hence there are altogether

$$\sum_{i=1}^{n} \varphi(i) = \Phi(n)$$

suitable lattice points in the lower triangle. By symmetry, the same holds for the upper triangle. We counted twice the lattice points on the diagonal, but $(1, 1)$ is the only relevant point here. Accordingly, the number of lattice points visible from the origin is $2\Phi(n) - 1$.

By Theorem 6.7.4, (6.7.17) implies

$$\lim_{n \to \infty} \frac{H(n)}{n^2} = 2 \lim_{n \to \infty} \frac{\Phi(n)}{n^2} = \frac{6}{\pi^2}. \qquad \square$$

Second proof. We compute $H(n)$ with the Inclusion and Exclusion Principle.

We have to find the number of ordered pairs $\{ (a, b) \mid 1 \le a \le n, 1 \le b \le n \}$ where a and b are coprime.

We exclude the wrong ones, i.e. those for which a and b share one or more prime divisors.

Consider first those pairs where both coordinates are divisible by a prime p (not examining whether or not they have some other common prime divisors too). There are $\lfloor n/p \rfloor^2$ such pairs.

Consider now those pairs where both coordinates are divisible by more than one of the primes p_j (again not caring whether or not they share further common prime divisors). An integer is a multiple of each of them if and only if it is a multiple of their product. Thus there are

$$\left\lfloor \frac{n}{p_1 p_2} \right\rfloor^2$$

pairs where both coordinates are divisible both by p_1 and p_2 where $p_1 < p_2$ are distinct primes, etc.

Hence, by the Inclusion and Exclusion Principle,

(6.7.18) $$H(n) = n^2 - \sum_{p \le n} \left\lfloor \frac{n}{p} \right\rfloor^2 + \sum_{p_1 p_2 \le n} \left\lfloor \frac{n}{p_1 p_2} \right\rfloor^2 \mp \cdots.$$

The right-hand side of (6.7.18) is just the sum of terms

$$\mu(j)\left\lfloor\frac{n}{j}\right\rfloor^2, \qquad j = 1, 2, \ldots, n$$

so

(6.7.19) $$H(n) = \sum_{j=1}^{n} \mu(j)\left\lfloor\frac{n}{j}\right\rfloor^2.$$

To estimate the right-hand side of (6.7.19) we use

$$0 \le a^2 - \lfloor a \rfloor^2 = (a - \lfloor a \rfloor)(a + \lfloor a \rfloor) < 2a$$

for $a > 0$, so

(6.7.20) $$\left|\lfloor a \rfloor^2 - a^2\right| < 2a.$$

Applying $|\mu(j)| \le 1$ and (6.7.20) with $a = n/j$ to (6.7.19), we obtain

$$\left|H(n) - \sum_{j=1}^{n} \mu(j)\left(\frac{n}{j}\right)^2\right| < 2\sum_{j=1}^{n}\frac{n}{j} < 2n(1 + \log n),$$

i.e.

(6.7.21) $$H(n) = n^2 \sum_{j=1}^{n} \frac{\mu(j)}{j^2} + V(n) \quad \text{where} \quad |V(n)| < 4n\log n$$

if $n \ge 3$. Dividing (6.7.21) by n^2 yields

$$\frac{H(n)}{n^2} = \sum_{j=1}^{n} \frac{\mu(j)}{j^2} + \frac{V(n)}{n^2}$$

and we get

$$\lim_{n\to\infty} \frac{H(n)}{n^2} = \sum_{j=1}^{\infty} \frac{\mu(j)}{j^2} = \frac{6}{\pi^2}$$

similar to the end of the proof of Theorem 6.7.4.   □

Now we determine the mean value of ω.

Theorem 6.7.6. *The difference between the mean value function of ω and $\log\log n$ is bounded. In other words, if $z(n) = \omega(1) + \omega(2) + \cdots + \omega(n)$, then there is a constant C such that every integer $n \ge 3$ satisfies*

$$\left|\frac{z(n)}{n} - \log\log n\right| < C.$$ ♣

Proof. We apply Theorem 6.7.2 for the convolution $\omega = \tilde{\omega} * 1$ (then $g = \tilde{\omega}$ and $h = 1$):

(6.7.22) $$z(n) = \sum_{i=1}^{n} \omega(i) = \sum_{j=1}^{n} \tilde{\omega}(j)\left\lfloor\frac{n}{j}\right\rfloor.$$

It is easy to check (see e.g. Exercise 6.5.5d) that

(6.7.23) $$\tilde{\omega}(j) = \begin{cases} 1, & \text{if } j \text{ is a prime} \\ 0, & \text{otherwise.} \end{cases}$$

Substituting (6.7.23) into (6.7.22), we get

(6.7.24)
$$z(n) = \sum_{p \leq n} \left\lfloor \frac{n}{p} \right\rfloor.$$

Applying the inequality

$$a - 1 < \lfloor a \rfloor \leq a$$

with $a = n/p$, we can rewrite (6.7.24) as

$$z(n) = n \sum_{p \leq n} \frac{1}{p} + W(n) \quad \text{where} \quad |W(n)| \leq \pi(n) < n,$$

i.e.

(6.7.25)
$$\left| \frac{z(n)}{n} - \sum_{p \leq n} \frac{1}{p} \right| < 1.$$

Since

$$\sum_{p \leq n} \frac{1}{p} - \log \log n$$

is bounded (for $n \geq 3$) by Theorem 5.6.2, the desired assertion follows from (6.7.25). $\square$

It is easy to see that

$$\sum_{i=2}^{n} \log \log i \sim n \log \log n,$$

therefore Theorem 6.7.6 implies

(6.7.26) $\omega(2) + \cdots + \omega(n) \sim \log \log 2 + \log \log 3 + \cdots + \log \log n.$

Relation (6.7.26) expresses that the average order of magnitude of ω is $\log \log n$.

It is not true in general that an arithmetic function assumes mostly values close to its mean value or average order of magnitude. For example, consider

$$f(n) = \begin{cases} n, & \text{if } n \text{ is a square} \\ 0, & \text{otherwise.} \end{cases}$$

Then

$$F(n) = \sum_{i=1}^{n} f(i) = \sum_{k \leq \sqrt{n}} k^2 \sim \frac{n^{3/2}}{3},$$

which means that the mean value of f is

$$\frac{F(n)}{n} \sim \frac{\sqrt{n}}{3},$$

and we can easily deduce that the average order of magnitude of $f(n)$ is $\sqrt{n}/2$. However, $f(n) = 0$ for almost all n.

A famous theorem of Hardy and Ramanujan states that ω assumes mostly values close to its mean value, i.e. most numbers n have about $\log \log n$ distinct prime divisors. We present the proof of Paul Turán which became the starting-point of applications of probability theory in number theory.

Theorem 6.7.7 (Hardy–Ramanujan Theorem). *Let $\delta > 1/2$ be a fixed real number, $n \geq 3$, and $k(n)$ the number of integers i satisfying $3 \leq i \leq n$ and*

(6.7.27) $$|\omega(i) - \log\log i| < (\log\log i)^\delta.$$

Then

$$\lim_{n\to\infty} \frac{k(n)}{n} = 1. \qquad \clubsuit$$

Since

$$\lim_{i\to\infty} \frac{(\log\log i)^\delta}{\log\log i} = 0$$

(for $\delta < 1$), Theorem 6.7.7 implies that apart from a rare subsequence

$$\omega(i) \sim \log\log i.$$

We shall deduce Theorem 6.7.7 from its finite variant.

Theorem 6.7.7A. *For any $\varepsilon > 0$ there exists a T (depending on ε) such that for any $n \geq 3$ at least $(1 - \varepsilon)n$ integers i among the integers $1, 2, \ldots, n$ satisfy*

(6.7.28) $$|\omega(i) - \log\log n| < T\sqrt{\log\log n}.$$

We call the attention to the difference that the argument of log log is i in (6.7.27) and n in (6.7.28). But as the function log log increases very slowly, this means only a negligible difference for most values of i (as shown in (6.7.41) later).

We prove Theorem 6.7.7A first, and then show how this implies Theorem 6.7.7.

Proof. The basic idea is to verify that the sum of squares

(6.7.29) $$U = \sum_{i=1}^{n}\left(\omega(i) - \log\log n\right)^2$$

is relatively small, hence the non-negative terms $|\omega(i) - \log\log n|$ can be large only for a few values of i.

Let us see the details. We show

(6.7.30) $$U = \sum_{i=1}^{n}\left(\omega(i) - \log\log n\right)^2 < cn\log\log n$$

with a suitable constant c for every $n \geq 3$. We use Theorems 6.7.6 and 5.6.2 stating (for $n \geq 3$)

(6.7.31) $$z(n) = \sum_{i=1}^{n}\omega(i) = n\log\log n + nA(n) \qquad \text{where } A(n) \text{ is bounded,}$$

and

(6.7.32) $$\sum_{p\leq n}\frac{1}{p} = \log\log n + B(n) \qquad \text{where } B(n) \text{ is bounded.}$$

We expand the square in (6.7.29):

$$U = \sum_{i=1}^{n}\omega^2(i) - 2\log\log n \sum_{i=1}^{n}\omega(i) + n(\log\log n)^2.$$

By (6.7.31), we obtain

$$U = \sum_{i=1}^{n} \omega^2(i) - 2\log\log n \big(n\log\log n + nA(n)\big) + n(\log\log n)^2 =$$

(6.7.33)

$$= \sum_{i=1}^{n} \omega^2(i) - n(\log\log n)^2 - 2nA(n)\log\log n.$$

To estimate U from above, we will estimate

(6.7.34)
$$V = \sum_{i=1}^{n} \omega^2(i)$$

from above.

Substituting (partly) the definition of $\omega(i)$ and rearranging the sum, we get

(6.7.35)
$$V = \sum_{i=1}^{n} \omega^2(i) = \sum_{i=1}^{n} \omega(i) \sum_{p|i} 1 = \sum_{p \le n} \sum_{k=1}^{\lfloor n/p \rfloor} \omega(pk).$$

Since

$$\omega(pk) = \begin{cases} \omega(k), & \text{if } p \mid k \\ 1 + \omega(k), & \text{if } p \nmid k, \end{cases}$$

(6.7.35) implies

(6.7.36)
$$V \le \sum_{p \le n} \sum_{k=1}^{\lfloor n/p \rfloor} (1 + \omega(k)) = \sum_{p \le n} \left\lfloor \frac{n}{p} \right\rfloor + \sum_{p \le n} \sum_{k=1}^{\lfloor n/p \rfloor} \omega(k).$$

Let K denote the first sum on the right-hand side of (6.7.36) and L denote the second double sum there.

By (6.7.32), we get an upper estimate for K:

(6.7.37)
$$K = \sum_{p \le n} \left\lfloor \frac{n}{p} \right\rfloor \le n \sum_{p \le n} \frac{1}{p} = n\big(\log\log n + B(n)\big).$$

To estimate L from above, we substitute the definition of $\omega(k)$, rearrange the sum as usual (here p' indicates that the summation is performed for primes), and apply

(6.7.32):

$$L = \sum_{p \le n} \sum_{k=1}^{\lfloor n/p \rfloor} \omega(k)$$

$$= \sum_{p \le n} \sum_{k=1}^{\lfloor n/p \rfloor} \sum_{p'|k} 1$$

(6.7.38)
$$= \sum_{p \le n} \sum_{p' \le n/p} \left\lfloor \frac{n}{pp'} \right\rfloor$$

$$\le n \sum_{pp' \le n} \frac{1}{pp'}$$

$$\le n \Big(\sum_{p \le n} \frac{1}{p} \Big)\Big(\sum_{p' \le n} \frac{1}{p'} \Big)$$

$$= n\big(\log \log n + B(n)\big)^2.$$

Substituting (6.7.37) and (6.7.38) into (6.7.36), we obtain

(6.7.39) $V \le n\big(\log \log n + B(n)\big) + n\big(\log \log n + B(n)\big)^2.$

Combining (6.7.39), (6.7.34), and (6.7.33), the terms $n(\log \log n)^2$ get cancelled and we have

$$U \le \big(1 + 2B(n) - 2A(n)\big)n \log \log n + \big(B(n) + B^2(n)\big)n < cn \log \log n$$

thus proving (6.7.30).

Now we will elaborate the argument indicated at the beginning of the proof that if the sum of squares (6.7.29) is small, then it can contain only few large terms.

Let s denote the number of wrong integers $1 \le i \le n$, those that do not satisfy (6.7.28). Then an equivalent formulation of the theorem is that for any $\varepsilon > 0$ there exists a T such that $s < \varepsilon n$.

We reduce the left-hand side of (6.7.30) by replacing $\big(\omega(i) - \log \log n\big)^2$ with $T^2 \log \log n$ at the s wrong values of i and with 0 at the other values of i. Then (6.7.30) implies

$$sT^2 \log \log n < cn \log \log n \quad \text{or} \quad s < \frac{c}{T^2} n.$$

We get the required estimate $s < \varepsilon n$ by choosing T to satisfy

(6.7.40) $\dfrac{c}{T^2} < \varepsilon.$ □

Proof of Theorem 6.7.7. We will verify that for any $\varepsilon > 0$ there exists an n_0 (depending on ε) such that for every $n > n_0$ there are at most εn numbers i among the integers $3, 4, \ldots, n$ that do not satisfy (6.7.27).

As noted earlier, Theorem 6.7.7A refers to $\log \log n$ in (6.7.28), whereas Theorem 6.7.7 has $\log \log i$ in (6.7.27). The proof basically overcomes this discrepancy.

The main idea is the following observation: $\log\log$ grows so slowly that it can be considered as almost constant between $\sqrt{n}$ and n, and there are so few values i less than $\sqrt{n}$ that they can be included in the set of exceptions.

Let us see the details. We apply Theorem 6.7.7A with $\varepsilon/2$ instead of ε. Then there are at most $\varepsilon n/2$ values i among the integers between $\sqrt{n}$ and n that do not satisfy (6.7.28). As $\sqrt{n} \le i \le n$ implies

$$(6.7.41) \qquad \log\log n - \log 2 = \log\log\sqrt{n} \le \log\log i \le \log\log n,$$

the previous sentence remains valid if we replace both occurrences of $\log\log n$ in (6.7.28) by $\log\log i$; we just have to make T larger than prescribed in (6.7.40). If n is large enough, then the number of values i smaller than $\sqrt{n}$ is less than $\varepsilon n/2$. Summarizing, we infer that with a suitable T and for n large enough, there are at least $(1-\varepsilon)n$ values i among the integers $3, 4, \ldots, n$ satisfying

$$(6.7.42) \qquad |\omega(i) - \log\log i| < T\sqrt{\log\log i}.$$

As $\delta > 1/2$, therefore

$$T\sqrt{\log\log i} < (\log\log i)^\delta$$

if i is sufficiently large depending on T and δ. Thus (6.7.42) implies the statement of Theorem 6.7.7. $\qquad\qquad\qquad\square$

Remark: The probabilistic background in the proof of Theorem 6.7.7A is the following. Let n be fixed, and consider ω as a random variable assuming each of the values $\omega(1)$, $\omega(2), \ldots, \omega(n)$ with the same probability $1/n$. The expectation E of this random variable is, by definition, the mean value of ω at n, which is about $\log\log n$. The expression U in (6.7.29) is around nD^2 where D is the standard deviation of ω. Theorem 6.7.7A then follows from the upper estimation of D (see (6.7.30)) and Chebyshev's inequality about the small probability of the variable being far from its expectation:

$$(6.7.43) \qquad P(|\omega - E| > rD) < \frac{1}{r^2}.$$

Theorems 6.7.6, 6.7.7, and 6.7.7A remain valid also for Ω instead of ω, see Exercise 6.7.5b. Combining these with the inequality

$$2^{\omega(n)} \le d(n) \le 2^{\Omega(n)},$$

we can verify the surprising fact mentioned in Section 6.4 that most n have about

$$(\log n)^{\log 2} = (\log n)^{0.69\ldots}$$

divisors, which is much less than the number $\log n$ corresponding to the mean value of $d(n)$ (see Exercise 6.7.6).

Exercises 6.7

1. Compute
$$\sum_{j=1}^{n} \mu(j)\left\lfloor\frac{n}{j}\right\rfloor.$$

2. What is the probability that a positive integer is squarefree?

* 3. Prove the following asymptotic equalities for $d_3(n)$ and $\sigma_\nu(n)$ (with fixed $\nu > 0$) defined in Exercise 6.2.22:

   (a) $\dfrac{D_3(n)}{n} = \dfrac{d_3(1) + d_3(2) + \cdots + d_3(n)}{n} \sim \dfrac{\log^2(n)}{2}$

   (b) $\dfrac{\Sigma_\nu(n)}{n} = \dfrac{\sigma_\nu(1) + \sigma_\nu(2) + \cdots + \sigma_\nu(n)}{n} \sim \dfrac{n^\nu \zeta(\nu + 1)}{\nu + 1}$

S* 4. Prove that for any k there exist distinct integers $n_1, \ldots, n_k$ satisfying $\sigma(n_1) = \cdots = \sigma(n_k)$.

5. (a) Verify
$$0 \le \sum_{i=1}^{n}\bigl(\Omega(i) - \omega(i)\bigr) < n.$$

   (b) Prove that Theorems 6.7.6, 6.7.7, and 6.7.7A remain valid if ω is replaced by Ω.

6. Show that most integers n have about
$$(\log n)^{\log 2}$$
divisors in the following sense. Let $\varepsilon > 0$ be arbitrary and $k(n)$ denote the number of integers $1 \le i \le n$ satisfying
$$(\log n)^{\log 2 - \varepsilon} < d(i) < (\log n)^{\log 2 + \varepsilon}.$$
Then
$$\lim_{n\to\infty} \frac{k(n)}{n} = 1.$$

* 7. Let $h(n)$ denote the number of integers $1 \le i \le n$ that can be written as the product of two factors each less than $\sqrt{n}$. Compute the limit
$$\lim_{n\to\infty} \frac{h(n)}{n}.$$

8. Formulate precisely and prove the following generalization of the Hardy–Ramanujan Theorem:

   Assume that the real-valued additive function f meets the following requirements.

   (i) There is a K such that $0 \le f(p) \le K$ for all primes p.

   (ii) $f(p^\alpha) = f(p)$ for every prime p and $\alpha > 0$.

   (iii) The infinite series $\sum_p f(p)/p$ is divergent.

Then

$$f(n) \sim \sum_{p \leq n} \frac{f(p)}{p}$$

for almost every n.

6.8. Characterization of Additive Functions

We saw that the oscillation of values is typical for most arithmetic functions. The next theorem of Erdős shows that the only true exception among the additive functions is the logarithm:

Theorem 6.8.1. *Let f be a real-valued additive function and assume that*

  (i) $f(n)$ *is monotone, or*

  (ii) $f(n + 1) - f(n) \to 0$ *if $n \to \infty$.*

Then $f(n) = c \log n$ with a suitable constant c. ♣

Proof. We shall prove a slightly stronger result: If a real-valued additive function f satisfies

(6.8.1) $$\liminf_{n \to \infty}\left(f(n + 1) - f(n)\right) \geq 0,$$

then $f(n) = c \log n$.

    This implies Theorem 6.8.1: if f satisfies (ii) or is monotone increasing, then (6.8.1) holds, and if f is monotone decreasing, then $-f$ satisfies (6.8.1), so $(-f)(n) = c \log n$, i.e. $f(n) = -c \log n$.

    The basic idea of the proof is the following. Let $k > 1$ be a fixed integer, and write an arbitrary n in number system of base k:

(6.8.2) $$n = a_s k^s + \cdots + a_2 k^2 + a_1 k + a_0, \qquad s = \lfloor \log_k n \rfloor.$$

Deleting the last digit of n and modifying suitably the second-to-last digit, we find an integer

(6.8.3) $$n' = a_s k^s + \cdots + a_2 k^2 + a_1' k$$

fairly close to n where $(a_1', k) = 1$. By the condition, $f(n)$ is not too far from

(6.8.4) $$f(n') = f(k) + f(a_s k^{s-1} + \cdots + a_2 k + a_1').$$

We repeat the process for the second term on the right-hand side of (6.8.4), etc., and finally we arrive at

$$f(n) \sim sf(k) \sim \frac{f(k) \log n}{\log k}, \quad \text{so} \quad \lim_{n \to \infty} \frac{f(n)}{\log n} = \frac{f(k)}{\log k}.$$

Hence $f(k)/\log k$ is equal to this limit independent of k, so $f(k)/\log k$ is a constant.

Let us see the detailed and precise elaboration. Let $\varepsilon > 0$ be arbitrary. By (6.8.1), there exists an n_0 (depending on ε) such that every $n > n_0$ satisfies

(6.8.5) $f(n + 1) - f(n) \geq -\varepsilon,$ i.e. $f(n) \leq f(n + 1) + \varepsilon.$

(For technical convenience we assume $n_0 > k^2$.)

Replacing n by $n + 1, n + 2, \ldots, n + t - 1$ in (6.8.5), we obtain

$$f(n + 1) \leq f(n + 2) + \varepsilon, f(n + 2) \leq f(n + 3) + \varepsilon, \ldots, f(n + t - 1) \leq f(n + t) + \varepsilon,$$

thus

(6.8.6) $f(n) \leq f(n + 1) + \varepsilon \leq f(n + 2) + 2\varepsilon \leq \ldots \leq f(n + t) + t\varepsilon.$

Let now n be much bigger than n_0, and consider the representation (6.8.2) (with a fixed $k > 1$). We select the smallest n' according to (6.8.3) satisfying $n' > n$ and $(a'_1, k) = 1$. This means that we delete the last digit a_0 of n, and replace the last but one digit a_1 by a bigger number a'_1 ($a'_1 = k + 1$ is possible). We consider the difference t of n' and n:

(6.8.7) $t = n' - n = (a'_1 - a_1)k - a_0.$

If $a_1 = 0$, then $a'_1 = 1$, and if $a_1 \geq 1$, then $1 \leq a_1 < a'_1 \leq k + 1$, therefore (6.8.7) implies

(6.8.8) $0 < t \leq k^2.$

Applying (6.8.6), (6.8.7), (6.8.8), and (6.8.4) in this order for $n > n_0$, we obtain

(6.8.9) $f(n) \leq f(n + t) + t\varepsilon \leq f(n') + k^2\varepsilon = f(k) + f(a_s k^{s-1} + \cdots + a_2 k + a'_1) + k^2\varepsilon.$

Consider now the number

$$n_1 = a_s k^{s-1} + \cdots + a_2 k + a'_1$$

in the middle term of the right-hand side of (6.8.9). If here $a'_1 = k + 1$, then transform n_1 into the usual representation in the number system (where the coefficient of each power of k is less than k; the last digit will be 1, the last but one digit increases by 1, or if it was $k - 1$, then further changes are possible, too).

Now we repeat the process for n_1 instead of n. We obtain

$$f(a_s k^{s-1} + \cdots + a_2 k + a'_1) = f(n_1) \leq f(k) + f(a_s k^{s-2} + \cdots + a'_2) + k^2\varepsilon.$$

Substituting into (6.8.9), we get

$$f(n) \leq 2f(k) + f(a_s k^{s-2} + \cdots + a'_2) + 2k^2\varepsilon.$$

We proceed similarly as long as the values of the function are greater than n_0. Finally, we have

(6.8.10) $f(n) \leq (s - s_0)f(k) + (s - s_0)k^2\varepsilon + M_0,$

where $s - s_0$ is the number of steps and M_0 is the maximum value of f assumed at integers up to n_0. Here M_0 depends only on ε, and s_0 depends on ε and on (the fixed) k, thus (6.8.10) can be rewritten as

(6.8.11) $f(n) \leq sf(k) + sk^2\varepsilon + M_1$

where M_1 is a constant depending on ε and k.

We can estimate $f(n)$ from below using a similar method. We choose n' close to n with $(a_1', k) = 1$, but instead of the minimal $n' > n$ we take the maximal $n' < n$ (now $a_1' = -1$ can happen). We have to modify the steps of the upper estimate by defining t as $n - n'$ and applying

$$f(n) \geq f(n - t) - t\varepsilon$$

instead of (6.8.6). We get finally

(6.8.12) $$f(n) \geq sf(k) - sk^2\varepsilon - M_2$$

where M_2 is a suitable constant.

Dividing (6.8.11) and (6.8.12) by $s = \lfloor \log_k n \rfloor$, we obtain

(6.8.13) $$\left| \frac{f(n)}{\lfloor \log_k n \rfloor} - f(k) \right| \leq k^2\varepsilon + \frac{M}{\lfloor \log_k n \rfloor}.$$

If $n \to \infty$, then the right-hand side of (6.8.13) tends to $k^2\varepsilon$. But ε was arbitrary, hence

(6.8.14) $$\lim_{n \to \infty} \frac{f(n)}{\lfloor \log_k n \rfloor} = f(k).$$

This clearly implies

$$\lim_{n \to \infty} \frac{f(n)}{\log_k n} = f(k),$$

so

(6.8.15) $$\lim_{n \to \infty} \frac{f(n)}{\log n} = \frac{f(k)}{\log k}.$$

Denote the limit in (6.8.15) by c; as c is independent of k

$$\frac{f(k)}{\log k} = c,$$

i.e.

(6.8.16) $$f(k) = c \log k$$

for any $k > 1$. Finally, $f(1) = \log 1 = 0$, thus (6.8.16) holds for $k = 1$. $\qquad \square$

Exercises 6.8

1. Prove that if a complex valued completely additive function f is bounded, then $f = 0$.

2. Show that if the sequence of values $f(n)$ of a complex-valued completely additive function f is convergent, then $f = 0$.

3. Which are the real-valued monotone multiplicative functions?

4. Verify that if a real-valued additive function f satisfies

$$\limsup_{n \to \infty} (f(n) - f(n-1)) \leq 0,$$

then $f(n) = c \log n$.

5. Prove that if a *complex*-valued additive function f satisfies
$$\lim_{n\to\infty}\left(f(n)-f(n-1)\right)=0,$$
then $f(n)=c\log n$ where c is a suitable complex constant.

6. Verify the following assertions.

   (a) There exists an arbitrary rare subsequence a_n of the natural numbers such that if $f(a_n)$ is monotone for a real-valued additive function f, then $f(n)=c\log n$.

   (b) There exists an arbitrary rare subsequence a_n of the natural numbers such that if
   $$\lim_{n\to\infty}\left(f(a_n)-f(a_{n-1})\right)=0$$
   for a real-valued additive function f, then $f=0$.

(Arbitrary rare means that to any sequence b_n there is a sequence a_n with the prescribed property and $a_n>b_n$.)

Chapter 7

Diophantine Equations

By a Diophantine equation, we generally mean an algebraic equation with integer coefficients where we are looking for integer (or sometimes for rational) solutions. The
Greek mathematician Diophantus lived in Alexandria in the 3rd century CE and investigated many types of such problems. (At that time, it was perfectly natural to search
only for integer or rational solutions, since irrational numbers were not really accepted
though their existence was proved by the Greeks.) The history of Diophantine equations is even older; clay tables show that nearly 4000 years ago the Babylonians were
familiar with Pythagorean triples.

The handling of Diophantine equations requires a large variety of methods, and
there exists no general procedure (as mentioned in Section 5.1, there is no universal
algorithm for answering the simpler question of whether or not an arbitrary Diophantine equation has a solution at all). Also, it is often very hard to decide if an equation
is solvable, not to mention finding the number of solutions or determining them. The
topic is rich in unsolved problems.

After a detailed discussion of linear Diophantine equations, we deal with Pythagorean triples, and then present some useful general elementary methods. The equations of the later sections require seemingly remote mathematical tools: the Gaussian
integers give the key to the representation of integers as the sum of two squares, the
Eulerian integers help to settle the cubic case of Fermat's Last Theorem, and Diophantine approximation serves as a basis to handle Pell's equation. The development of
these aids into independent branches was mainly due to their role played in Diophantine equations. We discuss these areas in more detail in Chapters 8–11. The last section
of this chapter is devoted to partitions, where both the questions and the methods are
significantly different.

7.1. Linear Diophantine Equation

We discuss first the linear Diophantine equation $ax + by = c$ in two variables. Here a, b, and c are fixed integers. We exclude the case $a = b = 0$, and a solution means a pair of integers x and y.

We proved the necessary and sufficient condition of solvability in Theorem 1.3.6, and clarified the relation of the equation to linear congruences in the proof of Theorem 2.5.3. We saw from the proof of Theorem 1.3.6 that the Euclidean algorithm provides a solution. By Theorem 5.7.1, this implied that we can obtain a solution quickly even for large coefficients; we made use of this fact in the RSA scheme (Theorem 5.8.1).

Now we get the number of solutions and the description of all solutions. In the next theorem, for the sake of completeness, we summarize also the statements proved earlier concerning the condition of solvability and the method for solving the equation.

Theorem 7.1.1. *Let a, b, and c be fixed integers where at least one of a and b is not zero, and consider the Diophantine equation $ax + by = c$.*

(i) *There is a solution if and only if $(a, b) \mid c$.*

(ii) *If solvable, there are infinitely many solutions. Let x_0, y_0 be a solution; then all solutions x', y' are given by*

(7.1.1) $$x' = x_0 + t\frac{b}{(a,b)}, \quad y' = y_0 - t\frac{a}{(a,b)}, \quad \text{where} \quad t = 0, \pm 1, \pm 2, \ldots$$

(iii) *We can get a solution using the Euclidean algorithm.* ♣

Proof. As mentioned previously, (i) and (iii) were proved in Theorem 1.3.6.

Turning to (ii), we show first that the integers x', y' given in (7.1.1) give a solution of the equation. Since x_0, y_0 is a solution, $ax_0 + by_0 = c$, so

$$ax' + by' = a\left(x_0 + t\frac{b}{(a,b)}\right) + b\left(y_0 - t\frac{a}{(a,b)}\right) = ax_0 + by_0 = c.$$

To prove the converse, we assume that x', y' is an arbitrary solution, and show that x' and y' are in the prescribed form.

We know that

$$ax_0 + by_0 = c \quad \text{and} \quad ax' + by' = c.$$

Subtracting, we get

$$a(x' - x_0) + b(y' - y_0) = 0.$$

After rearranging the terms and dividing by (a, b), we obtain

(7.1.2) $$\frac{a}{(a,b)}(x' - x_0) = \frac{b}{(a,b)}(y_0 - y').$$

Since

$$\left(\frac{b}{(a,b)}, \frac{a}{(a,b)}\right) = 1,$$

(7.1.2) implies

$$\frac{b}{(a,b)} \mid x' - x_0,$$

so

(7.1.3) $$x' = x_0 + t\frac{b}{(a,b)}$$

with a suitable integer t. Substituting (7.1.3) into (7.1.2), we arrive at

$$y' = y_0 - t\frac{a}{(a,b)}.$$

Thus we have shown that x' and y' are of the form in (7.1.1).   □

To solve a Diophantine equation, it is effective to apply a variant of the Euclidean algorithm that yields all solutions immediately in a parametric form. We present this procedure in an example:

Example. Solve the Diophantine equation $43x + 25y = 98$.

We solve for the variable with coefficient of smaller absolute value and separate the integer parts from the fraction so that the numbers in the numerator of the remaining fraction have minimal absolute value:

(A1) $$y = \frac{98 - 43x}{25} = 4 - 2x + \frac{7x - 2}{25}.$$

The fraction $(7x - 2)/25$ on the right-hand side of (A1) is an integer, we denote it by u. This gives $7x - 2 = 25u$ which is a similar Diophantine equation but the coefficient of x has smaller absolute value than the coefficient of y had in the original equation.

We repeat the process for the equation $7x - 2 = 25u$: we solve for x

(A2) $$x = \frac{25u + 2}{7} = 4u + \frac{2 - 3u}{7}.$$

The fraction $(2 - 3u)/7$ on the right-hand side of (A2) is an integer, we denote it by v, thus $2 - 3u = 7v$. Continuing similarly, we get

(A3) $$u = \frac{2 - 7v}{3} = -2v + \frac{2 - v}{3}.$$

Denoting the integer $(2 - v)/3$ by w, we have $2 - v = 3w$, i.e.

(A4) $$v = 2 - 3w.$$

Since (A4) contains no fractions, we turn and go backwards: we find u, x, and y one after the other from (A3), (A2), and (A1), using w as a parameter:

(B3) $u = -2v + w = -2(2 - 3w) + w = -4 + 7w$

(B2) $x = 4u + v = 4(-4 + 7w) + (2 - 3w) = -14 + 25w$

(B1) $y = 4 - 2x + u = 4 - 2(-14 + 25w) + (-4 + 7w) = 28 - 43w.$

It is clear from the procedure that formulas (B2)–(B1) provide all solutions of the Diophantine equation $43x + 25y = 98$ where the parameter w is an arbitrary integer. If a pair of integers x and y is a solution, then steps (A1)–(A3) lead to w, and then this yields formulas (B2)–(B1) for x and y and, taking an arbitrary integer w, the numbers x and y expressed with it are integers and satisfy the equation.

Remarks: (1) The following pairs of coordinates occur during the procedure:

$$\{43, 25\}; \quad \{25, 7\}; \quad \{7, 3\}; \quad \{3, 1\}.$$

How did we get them? In the first step, the remainder (of least absolute value) on division of 43 by 25 was -7, in the next step the remainder on division of 25 by 7 was -3, etc. Thus we used a variant of the Euclidean algorithm. This implies that we can find the solutions of the equation quickly with this procedure.

(2) The essential point of the method is reducing the absolute values of the coefficients of the variables to eliminate the fractions completely. It is irrelevant from this point of view whether or not we reduce the absolute values also of the constant term; it does not influence the number of steps in the procedure, though it may be slightly easier to work with smaller numbers.

(3) We do not have to check in advance whether the equation is solvable because the procedure decides automatically if there is no solution: we arrive at a fraction that contains no variables but its value is not an integer.

(4) Formulas (B2)–(B1) correspond to (7.1.1) describing all solutions in Theorem 7.1.1; now $x_0 = -14$, $y_0 = 28$, and w plays the role of t. This is a useful tool to detect calculation errors.

We have similar results for linear Diophantine equations with more than two variables. We summarize them in the next theorem, and ask for the proofs in Exercise 7.1.8.

Theorem 7.1.2. *Let $k \geq 2$, $a_1, \ldots, a_k$ integers not all 0, c any integer, and consider the Diophantine equation*

$$a_1 x_1 + \cdots + a_k x_k = c$$

where a solution is a k-tuple of integers $x_1, \ldots, x_k$.

  (i) *The equation is solvable if and only if $(a_1, \ldots, a_k) \mid c$.*

  (ii) *If it is solvable, there are infinitely many solutions. We can describe all solutions with $k - 1$ integer parameters. We can find the solutions with a suitable generalization of the method used for two variables.*   ♣

Exercises 7.1

1. In Crazyland there exist banknotes only of 47 and 79 dollars. How many ways can we pay exactly 10000 dollars?

2. An island is inhabited by dragons with 7 or 11 heads. How many dragons live on the island if they have 118 heads altogether?

3. A shop sells three types of chocolate bars costing 70 cents, 1 dollar and 30 cents, and a dollar and a half. How many ways can we buy (exactly) 50 bars for (exactly) 50 dollars?

S 4. In a certain year of the twentieth century, Alice notes that her age in years equals the sum of digits in the year of her birth date. Bob, who was born in a later year, notes that his age has the same property. How much older is Alice than Bob, if neither of them is older than 99 years?

5. Demonstrate that statement (ii) in Theorem 7.1.1 follows from the proof of Theorem 2.5.4.

6. How many lattice points in the plane can lie on a line if its slope is (a) rational (b) irrational?

7. Find all solutions of the Diophantine equation $6x + 10y + 15z = 7$.

8. Verify the statements in Theorem 7.1.2.

9. Prove that the Diophantine equation $a_1 x_1 + \cdots + a_k x_k = c$ is solvable if and only if the congruence $a_1 x_1 + \cdots + a_k x_k \equiv c \pmod{m}$ is solvable for every positive integer m.

* 10. Characterize the integers $a_1, \ldots, a_k$ for which the Diophantine equation $a_1 x_1 + \cdots + a_k x_k = c$ is solvable in positive integers for every c large enough.

* 11. Let a and b be fixed coprime integers greater than 1. We say that a positive integer c is *assemblable* (from a and b) if c can be represented as $c = ax + by$ with non-negative integers x and y.

  (a) Show that every $c > ab - a - b$ is assemblable, but $c = ab - a - b$ is not assemblable.

  (b) How many positive integers are not assemblable?

  Remark: We can generalize part (a) for more variables. Let $a_1, \ldots, a_k$ be fixed coprime integers greater than 1. Find the maximal integer $F = F(a_1, \ldots, a_k)$ for which the Diophantine equation $a_1 x_1 + \cdots + a_k x_k = F$ has no solutions in non-negative integers. Intensive research has been done to answer this question, called the *problem of Frobenius* for $k > 2$, but we have no completely satisfactory answer even in the case $k = 3$.

* 12. (a) Show that for every sufficiently large n, there exist n (not necessarily congruent) cubes in space such that we can assemble a cube from them (using each exactly once).

  (b) Verify this for every $n \geq 48$.

  (c) Find all n for which there exist n (not necessarily congruent) squares in the plane such that we can assemble a square from them (using each exactly once).

  Remark: It is unknown whether (b) is true for $n = 47$.

7.2. Pythagorean Triples

Pythagorean triples are the positive integer solutions of equation $x^2 + y^2 = z^2$. Geometrically, Pythagorean triples give the lengths of the three sides of a right triangle if these lengths are integers.

We immediately see that the equation is solvable (the triple 3, 4, 5 is a solution), and, multiplying a solution x, y, z by any positive integer d, the new triple dx, dy, dz is a solution. Therefore it is worthwhile to investigate the solutions satisfying $(x, y, z) = 1$ separately. These are called *primitive* Pythagorean triples.

We show that there are infinitely many primitive triples, we describe all of them, and characterize all (primitive and non-primitive) triples with suitable parameters:

Theorem 7.2.1. (i) *All primitive Pythagorean triples, i.e. all positive integer solutions of equation*

$$(7.2.1) \qquad\qquad x^2 + y^2 = z^2$$

satisfying

$$(7.2.2) \qquad\qquad (x, y, z) = 1$$

are

$$(7.2.3) \qquad x = 2mn, \qquad y = m^2 - n^2, \qquad z = m^2 + n^2$$

where the positive integer parameters

$$(7.2.4) \qquad m \text{ and } n \text{ are of opposite parity}, \qquad m > n, \quad and \quad (m, n) = 1.$$

We can interchange the roles of x and y, of course.

(ii) *All Pythagorean triples are multiples of the primitive triples, so*

$$(7.2.5) \qquad x = 2mnd, \qquad y = (m^2 - n^2)d, \qquad z = (m^2 + n^2)d$$

where d is any positive integer and positive integers m and n satisfy (7.2.4).   ♣

Proof. All variables will be positive integers throughout the proof.

(i) We show first that if x, y, and z form a primitive solution (they satisfy (7.2.1) and (7.2.2)), then they are necessarily of the form described in (7.2.3) and (7.2.4).

We start by verifying that x, y, and z are pairwise coprime. We show that $(x, z) = 1$, the other two pairs can be handled similarly. For a proof by contradiction, we assume $p \mid x$ and $p \mid z$ for some prime p. Then $p \mid z^2 - x^2 = y^2$, so $p \mid y$ since p is a prime. But then p is a common divisor of x, y, and z, which contradicts (7.2.2).

Now we show that x and y are of opposite parity. Both cannot be even since $(x, y) = 1$. If both are odd, then their squares' residues are 1 mod 4. Thus the left-hand side of $x^2 + y^2 = z^2$ is 2 mod 4, whereas the right-hand side is 0 or 1, which is a contradiction.

We may assume that x is even and y is odd. Rearranging (7.2.1), dividing by 4, and factoring, we get

$$(7.2.6) \qquad \left(\frac{x}{2}\right)^2 = \frac{z+y}{2} \cdot \frac{z-y}{2}.$$

We prove that the two factors on the right-hand side of (7.2.6) are coprime. Assume that k divides both $(z + y)/2$ and $(z - y)/2$. Then

$$k \mid \frac{z+y}{2} + \frac{z-y}{2} = z \quad \text{and} \quad k \mid \frac{z+y}{2} - \frac{z-y}{2} = y.$$

But $(y, z) = 1$, so $k \mid 1$, thus

(7.2.7)
$$\left(\frac{z+y}{2}, \frac{z-y}{2}\right) = 1.$$

By Exercise 1.6.2a, (7.2.6) and (7.2.7) imply that each of the two (positive) factors on the right-hand side of (7.2.6) is a square, so

(7.2.8)
$$\frac{z+y}{2} = m^2 \quad \text{and} \quad \frac{z-y}{2} = n^2$$

with suitable positive integers m and n. Adding and subtracting the equalities in (7.2.8) and substituting into (7.2.6), we get the required forms (7.2.3) for z, y, and x.

The conditions in (7.2.4) hold, as well; these follow from z (or y) being odd, $y > 0$, and (7.2.7).

Turning to the converse, we show that formulas (7.2.3)–(7.2.4) always define a primitive Pythagorean triple.

The numbers x, y, and z are positive integers due to $m > n > 0$, and a simple substitution verifies that (7.2.1) is true.

We need to prove $(x, y, z) = 1$. This follows if we check that (e.g.) y and z are coprime.

For a proof by contradiction, we assume $p \mid y$ and $p \mid z$ for some prime p. Then $p \mid z + y$ and $p \mid z - y$, so

(7.2.9) $\quad p \mid (m^2 + n^2) + (m^2 - n^2) = 2m^2 \quad \text{and} \quad p \mid (m^2 + n^2) - (m^2 - n^2) = 2n^2.$

As p is a prime, (7.2.9) implies that $p = 2$ or $p \mid m^2$ and $p \mid n^2$.

The case $p = 2$ is impossible since $z = m^2 + n^2$ is odd due to the opposite parity of m and n.

In the other case (using again that p is a prime), we have $p \mid m$ and $p \mid n$, which contradicts the condition $(m, n) = 1$.

(ii) As mentioned before, multiplying a primitive (or any) solution by d, gives a solution again. Conversely, any solution x, y, z can be obtained by multiplying the primitive solution x/d, y/d, z/d by $d = (x, y, z)$. $\quad\square$

Exercises 7.2

1. Show that if the side lengths of a right triangle are integers, then their product is a multiple of 60.

2. Compute the side lengths of a right triangle of area 60 if these lengths are integers.

3. Find all right triangles with integer side lengths whose area and perimeter are equal.

4. For which integers k does there exist a right triangle with integer side lengths one of them being k?

5. Prove that there exist infinitely many three-term arithmetic progressions of co-prime squares.

7.3. Some Elementary Methods

In this section we present a few typical methods for handling Diophantine equations.

I. *A product is a constant*

In each of the four hints to Exercises 7.2.2 and 7.2.3, the key was a Diophantine equation with an integer $c \neq 0$ on one side, and a product on the other side:

$$d^2 mn(m-n)(m+n) = 60, \qquad (x-4)(y-4) = 8, \quad \text{etc.}$$

Using a similar type of factoring, we now determine which integers can be written as the difference of two squares, and in how many ways.

Theorem 7.3.1. *We consider the Diophantine equation* $x^2 - y^2 = n$.

(i) *The equation is solvable if and only if* $n \not\equiv 2 \pmod 4$.

(ii) *The number of solutions is* $2d(n)$ *for n odd and* $2d\left(\frac{n}{4}\right)$ *for $4 \mid n$ (where $d(k)$ means the number of positive divisors of k).* ♣

We count as distinct solutions that differ only in signs. From the theorem, we can easily obtain the number of essentially distinct solutions, see Exercise 7.3.1.

Proof. Equality $(x+y)(x-y) = n$ holds if and only if $x+y$ and $x-y$ are two complementary divisors of n, or

(7.3.1) $$x + y = d_1, \quad x - y = d_2, \quad \text{where} \quad d_1 d_2 = n.$$

Solving system (7.3.1), we get

$$x = \frac{d_1 + d_2}{2}, \qquad y = \frac{d_1 - d_2}{2}.$$

Here x and y are integers if and only if d_1 and d_2 have the same parity.

Accordingly, the Diophantine equation $x^2 - y^2 = n$ is solvable if and only if n is the product of two of its divisors of the same parity, and the number of solutions is the number of such pairs of divisors (where also the signs and the order of the two divisors count).

If n is odd, then its divisors are odd. Therefore the equation is solvable and the number of solutions is the number of all positive and negative divisors of n, i.e. $2d(n)$.

If n is even but not a multiple of 4, then n cannot be written as the product of two divisors of the same parity, since the product of two odd numbers is odd, and the product of two even numbers is divisible by 4. Thus the equation has no solutions for such n.

If $4 \mid n$, then suitable pairs are $2k_1, 2k_2$ where: $n = (2k_1)(2k_2)$. This is equivalent to $n/4 = k_1 k_2$, so the equation is solvable and the number of solutions is the number of all positive and negative divisors of $n/4$, i.e. $2d(n/4)$. □

II. *A product is a power*

The Fundamental Theorem of Arithmetic implies that if a kth power is the product of two coprime factors, then each factor is a kth power, apart from units (see Exercise 1.6.2). This fact played an important role in the proof of Theorem 7.2.1 (see formulas (7.2.6), (7.2.7), and (7.2.8) there), and also in solving Exercise 1.6.3. The next example illustrates that similar arguments can be applied if the factors are not necessarily coprime.

Example. Solve the Diophantine equation $x^3 + 7x = y^3$.

Clearly, $x = y = 0$ is a solution, and if x, y is a solution, then so is $-x, -y$. Therefore we may assume x (and thus y) is positive.

We factor the left-hand side of the equation:

$$(7.3.2) \qquad x(x^2 + 7) = y^3$$

and check the possible values of the gcd of the two factors. Let $d = (x, x^2 + 7)$, then

$$d \mid (x^2 + 7) - x \cdot x = 7,$$

thus $d = 1$ and $d = 7$ are the only potential values.

If $d = 1$, then both x and $x^2 + 7$ are cubes, so

$$(7.3.3) \qquad x = u^3 \quad \text{and} \quad x^2 + 7 = v^3$$

for suitable (positive) integers u and v. Replacing x by u^3 in the second equality, we get

$$(7.3.4) \qquad v^3 - u^6 = 7.$$

The difference of two positive cubes can be 7 only for the pair $(8, 1)$: if $a > b > 0$, then

$$a^3 - b^3 \geq (b + 1)^3 - b^3 = 3b^2 + 3b + 1 \geq 7,$$

and equality holds only for $b = 1$ and $a = b + 1 = 2$. (Another justification: in the product

$$7 = a^3 - b^3 = (a - b)(a^2 + ab + b^2),$$

the factors can only be ± 1 and ± 7 in suitable pairings.)

By (7.3.3) and (7.3.4), we get the solution $x = 1, y = 2$.

Now we consider the case $d = 7$. Then $7 \mid x$ and $x \mid y^3$ imply $7 \mid y^3$, so $7 \mid y$ since 7 is a prime. We check the exponent of 7 on the right-hand side and in the two factors on the left-hand side of (7.3.2). The right-hand side, y^3, is divisible by at least 7^3, whereas the second factor of the left-hand side, $x^2 + 7$ is not divisible by 7^2 since $7^2 \mid x^2$. Therefore the exponent of 7 in the first factor of the left-hand side is at least $3 - 1 = 2$, so $7^2 \mid x$.

Substituting $x = 7^2 r$ and $y = 7s$ into (7.3.2) and cancelling 7^3, we obtain

$$(7.3.5) \qquad r(7^3 r^2 + 1) = s^3.$$

The two factors on the left-hand side of (7.3.5) are coprime, hence each is a cube:

$$r = w^3 \quad \text{and} \quad 7^3 r^2 + 1 = 7^3 w^6 + 1 = z^3.$$

The second equality says $z^3 - (7w^2)^3 = 1$, but this is impossible for non-zero cubes.

Hence, the case $d = 7$ cannot occur.

Thus the equation has three solutions altogether:

$$x = y = 0 \qquad x = 1, y = 2 \qquad x = -1, y = -2.$$

There is another way to treat this equation, see IV below.

III. *Proving insolvability via congruences*

If the two sides of a Diophantine equation are never congruent modulo a suitable integer, then equality cannot hold. (The converse is false!)

Example. Solve the Diophantine equation $x^4 + 5y^4 = 4z^4$.

Clearly, $x = y = z = 0$ is a solution. We claim that there are no other solutions.

For a proof by contradiction, we assume the existence of a solution where x, y, and z are not all 0. We can assume also that x, y, and z are coprime: If $(x, y, z) = d > 1$, then dividing the equation by d^4, we see that x/d, y/d, z/d is a solution and these three numbers are coprime.

If $x^4 + 5y^4 = 4z^4$, then

$$(7.3.6) \qquad\qquad\qquad x^4 + 5y^4 \equiv 4z^4 \pmod{5}.$$

By Fermat's Little Theorem,

$$(7.3.7) \qquad\qquad\qquad a^4 \equiv \begin{cases} 1 \pmod{5}, & \text{if } 5 \nmid a \\ 0 \pmod{5}, & \text{if } 5 \mid a, \end{cases}$$

for any integer a. If $5 \nmid x$, then the left-hand side of (7.3.6) is congruent to 1 and the right-hand side is congruent to 0 or 4 modulo 5, by (7.3.7), which is impossible. The case $5 \nmid z$ leads to a contradiction similarly. Therefore $5 \mid x$ and $5 \mid z$.

Substituting $x = 5x_1$ and $z = 5z_1$ into the original equation, we get

$$5^4 x_1^4 + 5y^4 = 4 \cdot 5^4 z_1^4, \quad \text{i.e.} \quad 5^3 x_1^4 + y^4 = 4 \cdot 5^3 z_1^4.$$

Thus $5 \mid y^4$, and so $5 \mid y$, as 5 is a prime. This, however, contradicts the condition $(x, y, z) = 1$.

Remarks: (1) We can arrive at a contradiction similarly modulo 16.

(2) In general, it is helpful to choose a modulus that divides some coefficient in the equation, or one for which the powers in the equation fall into few residue classes. For example, a square can be congruent only to 0, 1, or 4 modulo 8 and the possible remainders of a fourth power modulo 16 are 0 and 1, thus it is often good to try 8 or 16 as a modulus.

(3) If we get no contradiction for a modulus, this means only the solvability of the corresponding congruence but does not imply that the equation is solvable (and does not imply, of course, that the equation has no solutions). For example, moduli $m = 3$ or $m = 7$ would have not helped with the equation above, as the congruence $x^4 + 5y^4 \equiv 4z^4$ has a non-trivial solution both mod 3 and mod 7:

$$(\pm 1)^4 + 5(\pm 1)^4 \equiv 4 \cdot 3^4 \pmod{3} \quad \text{and} \quad (\pm 1)^4 + 5(\pm 2)^4 \equiv 4(\pm 1)^4 \pmod{7}.$$

(4) We emphasize repeatedly that this method (in itself) can be successful only if the Diophantine equation has no solutions except perhaps a trivial one (as $x = y = z = 0$ at the equation above). If the equation has a non-trivial solution, then it satisfies also the corresponding congruence for every modulus m, so we cannot get a contradiction for any modulus. (Of course, such arguments with congruences can help to exclude solutions of certain types for any Diophantine equation.)

(5) This method is often not effective even if a Diophantine equation has no solutions. It may be that we are not clever or lucky enough to find a suitable modulus leading to a contradiction, but it is possible that no such modulus exists. We saw an equation in Exercise 4.2.8 that had no integer or rational solutions, but the corresponding congruence was solvable for every modulus m.

IV. *Application of inequalities*

Consider a Diophantine equation $f(x) = y^k$. Assume that for some c, every integer x of absolute value greater than c has the property that $f(x)$ is between two consecutive kth powers (not allowing equality). Then only solutions $|x| \leq c$ are possible. Checking these finitely many values, we can obtain all solutions of the equation.

We illustrate the procedure for the Diophantine equation $x^3 + 7x = y^3$ (discussed in II).

As observed previously, we can restrict ourselves to $x > 0$, and we see that $x = 1$, $y = 2$ is a solution.

A simple calculation shows

$$x^3 < x^3 + 7x < (x + 1)^3$$

for $x > 1$. Therefore $x^3 + 7x$ cannot be a cube for $x > 1$.

Thus the three pairs given in II provide all solutions of the equation.

Exercises 7.3

1. Let n be a fixed positive integer. In how many essentially distinct ways can n be represented as the difference of two squares, i.e. what is the number of solutions of the equation $x^2 - y^2 = n$ in non-negative integers?

2. A housewife wants to slice up a rectangular cake (into uniform rectangular pieces) so that she should get as many crispy pieces (that touched the tin's wall) as soft ones (that were away from the tin's wall). How should she do the slicing?

3. Géza Ottlik was a famous Hungarian writer in the twentieth century who also studied mathematics. In his memoirs, he gives a vivid description how he succeeded in defeating the problem:

    Let $p > 2$ be a prime. Verify that $2/p$ has exactly one representation as a sum of reciprocals of two distinct positive integers. (The order of the terms is irrelevant.)

Remark: The reciprocals of positive integers, i.e. the rational numbers having positive denominators and 1 as numerator, are called *unit fractions* or *Egyptian fractions* since the ancient Egyptians generally expressed the rational numbers as the sum of such fractions.

* 4. Which fractions with numerator 4 can be written as the sum of reciprocals of two natural numbers?

5. Show that if n is a positive integer not of the form $24k + 1$, then $4/n$ can be written as a sum of reciprocals of three natural numbers.

   Remark: A long-standing unsolved conjecture of Erdős and Straus claims that every positive integer n has this property.

6. Prove that every positive rational number has infinitely many representations as a sum of reciprocals of finitely many distinct positive integers.

7. Can a fourth power exceed a fifth power by 4?

S 8. Find all solutions of the system of equations

$$t^2 + (s + x)^2 = s^2 + y^2 = (y + t)^2 + x^2$$

in rational numbers x, y, s, and t.

9. Prove that the sum of 99 consecutive squares cannot be a power.

S 10. Determine all integers whose cubes are the sum of eight consecutive cubes.

* 11. Show that 6 consecutive natural numbers cannot be partitioned into two (disjoint) groups so that the product of the elements in the two groups is equal. Demonstrate that this is true also if 6 is replaced by 106.

12. For a given positive integer m, find all positive integers n, x, and y satisfying

$$(n, m) = 1 \quad \text{and} \quad (x^2 + y^2)^m = (xy)^n.$$

13. Solve the Diophantine equations

   (a) $xy + 3x + 5y = 7$
   (b) $x^2 - 2y^2 + 363z^2 = 77$
   (c) $2x^2 + 3y^2 = z^2$
   (d) $x^2 - 230y^2 = 7z^2$
   * (e) $x^5 + 3y^5 = 5z^5$
   (f) $(x^2 - 2)(x^2 + 7) = z^3$
   S* (g) $x^2 - 2y^4 = 1$
   S (h) $x^y = y^x$ (where x and y are positive integers)
   S* (i) $2^x - y^5 = 31$.

14. In which number systems are the following numbers squares?

   (a) 111
   * (b) 11111

(c) 111111.

(See Exercise 7.7.7 for the missing 1111.)

7.4. Gaussian Integers

Theorem 7.3.1 described completely which positive integers can be written as the difference of two squares and in how many ways. Now we raise the analogous question for sums instead of differences, i.e. which positive integers can be represented as the sum of two squares and in how many ways.

In solving $x^2 - y^2 = n$, the key step was factoring the left-hand side. For $x^2 + y^2 = n$, we have no such factorization among the integers (or even among the real numbers), but we can factor over the complex numbers: $(x + yi)(x - yi) = n$. Therefore it is promising to develop number theory for complex numbers $a + bi$ where a and b are integers. These complex numbers are called *Gaussian integers*.

Analogously to the integers, we define here the relevant notions (divisibility, unit, greatest common divisor, irreducible, and prime), show that the Fundamental Theorem of Arithmetic is true for the Gaussian integers and determine all Gaussian primes. This makes it possible to handle our original problem, the Diophantine equation $x^2 + y^2 = n$, in the next section.

Definition 7.4.1. *Gaussian integers* are those complex numbers $\alpha = a + bi$ where both a and b are integers. ♣

To make a clear distinction, Roman letters will denote integers, and Greek letters will denote Gaussian integers.

The Gaussian integers form a commutative ring without zero divisors (i.e. an integral domain) with an identity element under the addition and multiplication of complex numbers.

The norm plays a central role in the number theory of Gaussian integers:

Definition 7.4.2. The *norm* $N(\alpha)$ of a Gaussian integer $\alpha = a + bi$ is the square of the absolute value of α:

$$N(\alpha) = |\alpha|^2 = \alpha\overline{\alpha} = a^2 + b^2.$$ ♣

A few simple but important properties of the norm follow immediately from the definition of Gaussian integers and from the properties of the absolute values of complex numbers:

Theorem 7.4.3. (i) $N(\alpha)$ *is a non-negative integer.*

(ii) $N(\alpha) = 0 \iff \alpha = 0$.

(iii) $N(\alpha\beta) = N(\alpha)N(\beta)$, *for any Gaussian integers α and β.* ♣

To develop number theory for Gaussian integers, we follow the path for integers in Chapter 1; we define the notions and prove the Fundamental Theorem of Arithmetic according to that model. There is some difference in the form of the division algorithm (Theorem 7.4.8), otherwise we just copy the structure for the integers.

Definition 7.4.4. The Gaussian integer β is a *divisor* of the Gaussian integer α if there exists a Gaussian integer γ satisfying $\alpha = \beta\gamma$.   ♣

Similar to the integers, the expressions "α is *divisible* by β" and "α is a *multiple* of β" have the same meaning. We use the notation $\beta \mid \alpha$ for Gaussian integers, too.

If $\beta \neq 0$, then $\beta \mid \alpha$ holds if and only if the complex number $\dfrac{\alpha}{\beta}$ is a Gaussian integer.

Examples.

$$2 + i \mid 7 + i, \quad \text{as} \quad \frac{7+i}{2+i} = 3 - i$$

$$4 + i \nmid 4 - i, \quad \text{since} \quad \frac{4-i}{4+i} = \frac{15}{17} - \frac{8}{17}i.$$

The following (one-way) bridge is an important connection between integers and Gaussian integers:

Theorem 7.4.5. *If $\beta \mid \alpha$ (in the Gaussian integers), then $N(\beta) \mid N(\alpha)$ (in the integers).*
                                                                                                ♣

Proof. The implication follows from Definition 7.4.4 and Theorem 7.4.3(iii).   □

The converse of Theorem 7.4.5 is false, see e.g. the second example above the theorem.

Definition 7.4.6. A Gaussian integer dividing every Gaussian integer is called a *unit*. Multiplying a Gaussian integer γ by a unit, we get an associate of γ.   ♣

Units have several characterizations.

Theorem 7.4.7. *The following statements are equivalent:*

  (i) ε *is a unit.*

 (ii) $\varepsilon \mid 1$.

(iii) $N(\varepsilon) = 1$.

(iv) $\varepsilon = 1, -1, i, \text{ or } -i$.   ♣

Proof. (i)$\Longrightarrow$(ii): If ε divides every Gaussian integer, then it divides 1 in particular.

(ii)$\Longrightarrow$(iii): This follows from Theorem 7.4.5.

(iii)$\Longrightarrow$(iv): $N(a + bi) = a^2 + b^2 = 1$ holds with integers a and b only in the cases $a = \pm 1, b = 0$, or $a = 0, b = \pm 1$.

(iv)$\Longrightarrow$(i): For any Gaussian integer α,

$$\alpha = 1\alpha = (-1)(-\alpha) = i(-i\alpha) = (-i)(i\alpha).$$ □

Now we turn to the division algorithm for Gaussian integers.

Theorem 7.4.8. *For any Gaussian integers α and $\beta \neq 0$, there exist Gaussian integers γ and ϱ satisfying*

(7.4.1) $\alpha = \beta\gamma + \varrho \quad \text{and} \quad N(\varrho) < N(\beta).$ ♣

Proof. Condition (7.4.1) is equivalent to

$$\frac{\alpha}{\beta} - \gamma = \frac{\varrho}{\beta} \quad \text{and} \quad |\varrho| < |\beta|, \quad \text{i.e.} \quad \left|\frac{\varrho}{\beta}\right| < 1.$$

Thus we have to find a Gaussian integer γ satisfying

(7.4.2) $$\left|\frac{\alpha}{\beta} - \gamma\right| < 1.$$

The Gaussian integers form the usual unit square lattice in the complex plane. Hence, (7.4.2) means that the point (with rational coordinates) in the plane corresponding to α/β is closer to lattice point γ than 1, i.e. it falls inside the unit circle around γ.

Consider a unit square in the lattice that contains α/β (inside or on its border; there is more than one such unit square if and only if at least one of the coordinates of α/β is an integer). If we draw unit circles around two opposite vertices, the interiors of these circles cover this unit square entirely except the two other vertices. Thus for any point in the plane, there is a lattice point whose distance from it is less than 1. So to any α/β, there is a suitable γ.

The value of ϱ is determined then by $\varrho = \alpha - \beta\gamma$. $\qquad\square$

Remarks: (1) We see from the proof that the quotient γ and the remainder ϱ are not unique in general; uniqueness holds if and only if α/β itself is a lattice point, i.e. $\beta \mid \alpha$ (and the remainder is 0). Otherwise there are two, three, or four suitable pairs γ, ϱ, depending on the position of α/β.

(2) The proof yields an algorithm to find γ and ϱ: we can choose γ as the closest lattice point to α/β. (Choose one if there exists more than one.) Algebraically, if $\alpha/\beta = r + si$, then choose $\gamma = u + vi$ where u and v are the closest integers to the rational numbers r and s. (Again, just choose in the event of a tie.) Then

$$\left|\frac{\alpha}{\beta} - \gamma\right|^2 = (r - u)^2 + (s - v)^2 \leq \left(\frac{1}{2}\right)^2 + \left(\frac{1}{2}\right)^2 = \frac{1}{2}.$$

For Gaussian integers, we define the greatest common divisor immediately with the special common divisor property seen in Definition 1.3.2 at the integers: it is a common divisor that is a multiple of all common divisors.

Definition 7.4.9. The *greatest common divisor* (or *gcd*) of Gaussian integers α and β is δ if

(i) $\delta \mid \alpha, \delta \mid \beta$

(ii) if γ satisfies $\gamma \mid \alpha$ and $\gamma \mid \beta$, then $\gamma \mid \delta$. ♣

We assume now that at least one of α and β is not zero, and denote the greatest common divisor by (α, β) or $\gcd(\alpha, \beta)$.

The existence of a greatest common divisor follows from the Euclidean algorithm as in the proof of Theorem 1.3.3 (the procedure terminates in finitely many steps also for Gaussian integers since the norms of the remainders form a strictly decreasing sequence of non-negative integers). The Euclidean algorithm is suitable for the practical computation of the greatest common divisor.

The greatest common divisor is unique apart from a unit factor, i.e. if δ is a gcd of the Gaussian integers α and β, then all greatest common divisors are the associates of δ. (This follows from the definition of gcd.)

There are four units, so any two Gaussian integers (not both zero) have exactly four greatest common divisors. Since they are associates, they behave identically concerning divisibility. Also, there is no natural principle to distinguish one of them, as we chose the positive value among the integers. Therefore the notation (α, β) can mean any of the four values.

The relevant further theorems and definitions in Section 1.3 are equally valid for Gaussian integers.

Now we define the notions of Gaussian irreducibles and Gaussian primes on the model of Definitions 1.4.1 and 1.4.2.

Definition 7.4.10. A Gaussian integer π different from units (and zero) is called a *Gaussian irreducible* if it can be factored into the product of two integers only so that one of the factors is a unit:

$$\pi = \alpha\beta \implies \alpha \text{ or } \beta \text{ is a unit.} \qquad \clubsuit$$

Definition 7.4.11. A Gaussian integer π different from units and zero is called a *Gaussian prime* if it can divide the product of two integers only if it divides at least one of the factors:

$$\pi \mid \alpha\beta \implies \pi \mid \alpha \text{ or } \pi \mid \beta. \qquad \clubsuit$$

The analog of Theorem 1.4.3 is valid for Gaussian integers, and the proof is literally the same:

Theorem 7.4.12. *A Gaussian integer is a Gaussian prime if and only if it is a Gaussian irreducible.* $\qquad \clubsuit$

We shall generally use the shorter term Gaussian prime also for a Gaussian irreducible.

We are ready now to state and prove the analog of Theorem 1.5.1:

Theorem 7.4.13 (The Fundamental Theorem of Arithmetic). *Every Gaussian integer different from 0 and units is the product of finitely many Gaussian irreducibles and this decomposition is unique apart from associates and the order of factors.* $\qquad \clubsuit$

Proof. The first proof of uniqueness for integers remains valid literally for Gaussian integers (see Exercise 7.4.11 for the analog of the second proof).

We can apply the same arguments as at the integers for the decomposability with two minor modifications: instead of "minimal positive non-trivial divisor" we need "a(ny) non-trivial divisor of minimal norm", and $|a_i|$ should be replaced by $N(\alpha_i)$. We leave the details to the reader. $\qquad \square$

Remark: As a summary, we can state that we arrived at the Fundamental Theorem of Arithmetic almost identically both for integers and Gaussian integers. We proved decomposability in both cases directly (using similar ideas), and deduced uniqueness with the following steps:

Division algorithm $\Rightarrow$ existence of a greatest common divisor (in the sense of a special common divisor) $\Rightarrow$ every irreducible is a prime $\Rightarrow$ the uniqueness part of the Fundamental Theorem of Arithmetic.

We shall show later that the existence of a division algorithm always implies the Fundamental Theorem of Arithmetic but the converse is false (see Section 11.3).

Our next goal is to characterize all Gaussian primes. As a preparation, we establish a relation between Gaussian primes and ordinary prime numbers in **Z**:

Theorem 7.4.14. (i) *For every Gaussian prime π, there exists exactly one positive prime number p satisfying $\pi \mid p$.*

(ii) *Every positive prime number p is either a Gaussian prime, or it is the product of two complex conjugate Gaussian primes having norm p.* ♣

Proof. (i) As π is different from 0 and units, we have $N(\pi) > 1$, so $N(\pi)$ is the product of positive prime numbers: $N(\pi) = p_1 p_2 \ldots p_r$. Then

$$\pi \mid \pi\overline{\pi} = N(\pi) = p_1 p_2 \ldots p_r,$$

thus π must divide some p_i, as well.

To prove uniqueness by contradiction, we assume $\pi \mid p$ and $\pi \mid q$ for some positive prime numbers $p \neq q$. Since p and q are coprime (among the integers), we have $1 = pu + qv$ with suitable integers u and v. Then $\pi \mid p$ and $\pi \mid q$ imply $\pi \mid pu + qv = 1$, which is a contradiction.

(ii) If the prime number $p > 0$ is not a Gaussian prime, then it is the product of at least two Gaussian primes (by the Fundamental Theorem of Arithmetic):

(7.4.3) $$p = \pi_1 \ldots \pi_r, \quad \text{where} \quad r \geq 2.$$

Taking norms in (7.4.3), we obtain

(7.4.4) $$p^2 = N(p) = N(\pi_1) \ldots N(\pi_r).$$

Every $N(\pi_i) > 1$ since π_i is neither 0 nor a unit. The integer p^2 has only one decomposition into the product of two integers greater than 1: $p^2 = p \cdot p$. Therefore, there are only two factors on the right-hand side of (7.4.4), so the same is true for (7.4.3):

$$p = \pi_1 \pi_2, \quad \text{where} \quad N(\pi_1) = N(\pi_2) = p.$$

Finally,

$$p = \pi_1 \pi_2 \quad \text{and} \quad p = N(\pi_1) = \pi_1 \overline{\pi}_1$$

imply $\pi_2 = \overline{\pi}_1$. □

And now, here is the list of Gaussian primes:

Theorem 7.4.15. *The following Gaussian integers constitute all Gaussian primes (where ε denotes a unit):*

(A) $\varepsilon(1 + i)$

(B) *εq where q is a positive prime number of the form $4k - 1$*

(C) π where $N(\pi)$ is a positive prime number of the form $4k + 1$; to each such prime number, there belong two Gaussian primes (apart from unit factors) that are complex conjugates but not associates.   ♣

Examples. $-1 + i = i(1 + i)$ and $-7i$ are Gaussian primes.

Also $2 - 5i$ is a Gaussian prime since $(2 - 5i)(2 + 5i) = 29$ and 29 is a positive prime number of the form $4k + 1$.

Also $2 + 5i$ is a Gaussian prime that is not an associate of $2 - 5i$.

The factors of the decomposition $29 = (5 - 2i)(5 + 2i)$ can only be associates of the previous two Gaussian primes (by the Fundamental Theorem of Arithmetic): $5 - 2i = (-i)(2 + 5i)$ and $5 + 2i = i(2 - 5i)$.

-37 is not a Gaussian prime, as 37 is a prime number, but not of the form $4k - 1$.

Also $9 + 2i$ is not a Gaussian prime because $(9 + 2i)(9 - 2i) = 85$ is not a prime number.

Proof. By Theorem 7.4.14, we obtain all Gaussian primes from the factorization of positive prime numbers into the product of Gaussian primes. We get different decompositions when the positive prime number is (A) 2, has the form (B) $4k - 1$, or (C) $4k + 1$.

(A) Since $2 = (1 + i)(1 - i) = (-i)(1 + i)^2$, the only Gaussian prime divisor of 2 is $1 + i$, apart from associates.

(B) Let q be a positive prime number of the form $4k - 1$. For a proof by contradiction, we assume that q is not a Gaussian prime. Then, by (ii) in Theorem 7.4.14, there exists a Gaussian prime $\pi = a + bi$ satisfying $q = N(\pi) = a^2 + b^2$. This is impossible, however, as the sum of two squares cannot be of the form $4k - 1$.

(C) Let p be a positive prime number of the form $4k + 1$. We show first that p is not a Gaussian prime.

By Theorem 4.1.4, the congruence $x^2 \equiv -1 \pmod{p}$ is solvable, so $p \mid c^2 + 1$ for some integer c. Hence, p divides the product $(c + i)(c - i)$ among the Gaussian integers. But

$$\frac{c \pm i}{p} = \frac{c}{p} \pm \frac{1}{p}i$$

are not Gaussian integers because their imaginary parts are not integers, thus none of the factors $c + i$ and $c - i$ are divisible by p. Therefore, by definition, p is not a Gaussian prime.

This means, according to Theorem 7.4.14, that $p = \pi\bar{\pi}$ where π and $\bar{\pi}$ are Gaussian primes. By the Fundamental Theorem of Arithmetic, this is the only decomposition of p into the product of Gaussian primes, apart from associates.

Finally, we have to show $\pi \neq \varepsilon\bar{\pi}$ for some unit ε. We can verify this by a simple calculation checking all cases $\varepsilon = 1, -1, i$, and $-i$ for $p = a + bi$. It follows also from Exercise 7.4.3.   □

Exercises 7.4

(α, β, and $a + bi$ denote Gaussian integers throughout.)

1. Which Gaussian integers are divisible by $1 + i$?

2. Verify:

   (a) $\gamma \mid \alpha \iff \overline{\gamma} \mid \overline{\alpha}$
   (b) $(\overline{\alpha}, \overline{\gamma}) = \overline{(\alpha, \gamma)}$
   (c) α is a Gaussian prime $\iff \overline{\alpha}$ is a Gaussian prime.

3. Let $\alpha = a + bi$. Prove

   $$\alpha \mid \overline{\alpha} \iff |a| = |b| \text{ or } ab = 0.$$

4. If a and $b \neq 0$ are two integers, then the divisibility $b \mid a$ and $\gcd(a, b)$ could depend on whether a and b are considered as integers or as Gaussian integers. Show that there is no need for such a distinction:

   (a) $b \mid a$ holds among Gaussian integers if and only if it is true in $\mathbf{Z}$

   (b) the greatest common divisor of a and b in $\mathbf{Z}$ is the same as their gcd among the Gaussian integers, apart from associates.

5. True or false?

   (a) $(N(\alpha), N(\beta)) = 1 \implies (\alpha, \beta) = 1$.
   (b) $(\alpha, \beta) = 1 \implies (N(\alpha), N(\beta)) = 1$.
   (c) $(\alpha, \beta) = (\overline{\alpha}, \beta) = 1 \implies (N(\alpha), N(\beta)) = 1$.

6. Compute the gcd of α and β for

   (a) $\alpha = 8 + i$ and $\beta = 11 - 3i$
   (b) $\alpha = 39(1 - i)^3$ and $\beta = 62(2 + i)^3$
   (c) $\alpha = (4 + i)^{10} + (2 + i)^{11}$ and $\beta = (4 + i)^{10} - (2 + i)^{11}$.

7. Let $\alpha = a + bi$.

   (a) True or false?
       (a1) $(\alpha, \overline{\alpha}) = 1 \implies (a, b) = 1$.
       (a2) $(a, b) = 1 \implies (\alpha, \overline{\alpha}) = 1$.

   (b) What is the connection between $(\alpha, \overline{\alpha})$ and (a, b), in general?

8. Let us call two Gaussian integers α and β *friends* if they are coprime and an ordinary integer is a multiple of α if and only if it is a multiple of β.

   (a) Prove that $a + bi$ has a friend if and only if $(a, b) = 1$ and $a \not\equiv b \pmod 2$.

   (b) How many friends belong to $a + bi$ in this case?

9. Decompose $270 + 2610i$ into a product of Gaussian primes.

10. True or false?

   (a) If α is a Gaussian prime, then $N(\alpha)$ is a prime number.

(b) If $N(\alpha)$ is a prime number, then α is a Gaussian prime.

(c) If α is the cube of a Gaussian integer, then $N(\alpha)$ is the cube of a non-negative integer.

(d) If $N(\alpha)$ is the cube of a non-negative integer, then α is the cube of a Gaussian integer.

(e) If $\alpha \mid \bar{\alpha}$, then $N(\alpha)$ is a square or the double of a square.

(f) If $N(\alpha)$ is a square or the double of a square, then $\alpha \mid \bar{\alpha}$.

* 11. Prove the uniqueness part of the Fundamental Theorem of Arithmetic with a suitable modification of the second proof of Theorem 1.5.1.

7.5. Sums of Squares

In this section, we examine which positive integers can be represented as the sum of two, three, or four squares (allowing also 0 as a summand).

Theorem 7.5.1 (Two Squares Theorem). *Let the standard form of the positive integer n be*

(7.5.1) $$n = 2^\alpha p_1^{\beta_1} \dots p_r^{\beta_r} q_1^{\gamma_1} \dots q_s^{\gamma_s}$$

where the primes p_μ are of the form $4k + 1$, the primes q_ν are of the form $4k - 1$, and the exponents α, β_μ, γ_ν are non-negative integers.

The Diophantine equation

(7.5.2) $$x^2 + y^2 = n$$

is solvable if and only if every γ_ν is even, and then the number of solutions is

$$4 \prod_{\mu=1}^{r} (\beta_\mu + 1). \qquad \clubsuit$$

Similar to Theorem 7.3.1, we consider as distinct solutions differing only in signs or in the order of terms. We can easily deduce from our result also the number of essentially different solutions, see Exercise 7.5.1.

Example. Consider $n = 4050$. Its standard form is $2 \cdot 3^4 \cdot 5^2$. The exponent of 3 is even, thus we have a solution, and the number of solutions is $4(2 + 1) = 12$ obtained from the exponent of 5. The solutions are

$$4050 = (\pm 45)^2 + (\pm 45)^2 = (\pm 9)^2 + (\pm 63)^2 = (\pm 63)^2 + (\pm 9)^2.$$

Proof. The equation $x^2 + y^2 = n$ can be rewritten as

(7.5.3) $$(x + yi)(x - yi) = n.$$

Thus we have to determine which integers n can be factored and in how many ways as a product of two conjugate Gaussian integers.

We determine first the standard form of n among the Gaussian integers. By *standard form*, we mean a representation

$$\varepsilon \varrho_1^{\kappa_1} \dots \varrho_t^{\kappa_t}$$

where no two Gaussian primes ϱ_j are associates and ε is a unit. For example, a standard form of 4 is $(-1)(1+i)^4$ or $(-1)(-1+i)^4$, etc. (We need the extra factor of a unit also among the integers if we want to extend the standard form to negative integers: e.g. -9 can be represented only in the form $(-1)3^2$ or $(-1)(-3)^2$.)

By Theorem 7.4.15, a standard form of n among the Gaussian integers is

$$(7.5.4) \qquad n = (-i)^\alpha (1+i)^{2\alpha} \pi_1^{\beta_1} \overline{\pi}_1^{\beta_1} \dots \pi_r^{\beta_r} \overline{\pi}_r^{\beta_r} q_1^{\gamma_1} \dots q_s^{\gamma_s},$$

where $\pi_\mu \overline{\pi}_\mu = p_\mu$. (No two Gaussian primes on the right-hand side of (7.5.4) are associates.)

As $x + yi \mid n$, the standard form of $x + yi$, according to the Fundamental Theorem of Arithmetic, is

$$(7.5.5) \qquad x + yi = \varepsilon(1+i)^{\alpha'} \prod_{\mu=1}^{r} \left(\pi_\mu^{\beta'_\mu} \overline{\pi}_\mu^{\beta''_\mu} \right) \prod_{\nu=1}^{s} q_\nu^{\gamma'_\nu}$$

where ε is a unit and each Gaussian prime occurs with an exponent not greater than in (7.5.4).

We construct a standard form of $x - yi$ by conjugating (7.5.5) and using $1 - i = (-i)(1+i)$:

$$(7.5.6) \qquad x - yi = \left(\overline{\varepsilon}(-i)^{\alpha'} \right)(1+i)^{\alpha'} \prod_{\mu=1}^{r} \left(\pi_\mu^{\beta''_\mu} \overline{\pi}_\mu^{\beta'_\mu} \right) \prod_{\nu=1}^{s} q_\nu^{\gamma'_\nu}.$$

By the Fundamental Theorem of Arithmetic, (7.5.3) holds if and only if the exponent of each Gaussian prime in (7.5.4) is the sum of the corresponding exponents in (7.5.5) and (7.5.6), and the extra unit factor in (7.5.4) equals the product of the unit factors in (7.5.5) and (7.5.6).

This gives the following equalities:

(7.5.7a)	exponent of $1+i$:	$2\alpha = \alpha' + \alpha'$
(7.5.7b)	exponent of π_μ:	$\beta_\mu = \beta'_\mu + \beta''_\mu$
(7.5.7c)	exponent of $\overline{\pi}_\mu$:	$\beta_\mu = \beta''_\mu + \beta'_\mu$
(7.5.7d)	exponent of q_ν:	$\gamma_\nu = \gamma'_\nu + \gamma'_\nu$
(7.5.7e)	unit:	$(-i)^\alpha = \varepsilon \overline{\varepsilon}(-i)^{\alpha'}$.

Equality (7.5.7a) implies $\alpha' = \alpha$, and then (7.5.7e) is true automatically for any ε. (7.5.7b) and (7.5.7c) mean the same condition that holds if and only if

$$\beta'_\mu = 0, 1, \dots, \beta_\mu \quad \text{and} \quad \beta''_\mu = \beta_\mu - \beta'_\mu, \quad \mu = 1, 2, \dots, r.$$

Finally, (7.5.7d) is valid if and only if γ_ν is even and $\gamma'_\nu = \gamma_\nu/2$.

The above imply that (7.5.2) is solvable if and only if every γ_ν is even.

The number of solutions equals the number of possible choices of ε, α', β'_μ, β''_μ, and γ'_μ. We can select these five values independently in 4, 1, $\beta_\mu + 1$, 1, and 1 ways, thus the number of solutions of (7.5.2) is the product of these numbers, $4 \prod_{\mu=1}^{r} (\beta_\mu + 1)$. $\qquad \square$

Theorem 7.5.2 (Three Squares Theorem). *A positive integer n is not representable as the sum of three squares if and only if n is of the form*

(7.5.8) $n = 4^k(8m + 7)$.   ♣

Proof. We verify only the easier direction that an integer of the form (7.5.8) cannot be written as the sum of three squares. The proof of the converse is significantly harder.

We proceed by induction on k.

For $k = 0$, we have to show that integers of the form $8m + 7$ cannot be represented as a sum of three squares. This holds since a square can have a residue of 0, 1, or 4 modulo 8, and the sum of three such remainders can never produce a remainder of 7.

We assume now that the assertion is true for some k and deduce it for $k + 1$. For a proof by contradiction, let a, b, and c be integers satisfying

(7.5.9) $4^{k+1}(8m + 7) = a^2 + b^2 + c^2$.

The left-hand side of (7.5.9) is divisible by 4. The remainder modulo 4 of an even square is 0 and of an odd square is 1. Thus the right-hand side is divisible by 4 only if each of a, b, and c is even, and so $a/2$, $b/2$, and $c/2$ are integers. Dividing (7.5.9) by 4, we obtain

$$4^k(8m + 7) = \left(\frac{a}{2}\right)^2 + \left(\frac{b}{2}\right)^2 + \left(\frac{c}{2}\right)^2,$$

which contradicts the induction hypothesis.   □

Theorem 7.5.3 (Four Squares Theorem). *Every positive integer is the sum of four squares.*   ♣

Proof. We need the following two lemmas.

Lemma 7.5.4. *If each of two integers is a sum of four squares, then so is their product:*
(7.5.10)
$(a_1^2 + a_2^2 + a_3^2 + a_4^2)(b_1^2 + b_2^2 + b_3^2 + b_4^2) =$

$\quad (a_1b_1 + a_2b_2 + a_3b_3 + a_4b_4)^2 + (a_1b_2 - a_2b_1 + a_3b_4 - a_4b_3)^2 +$

$\quad\quad + (a_1b_3 - a_3b_1 - a_2b_4 + a_4b_2)^2 + (a_1b_4 - a_4b_1 + a_2b_3 - a_3b_2)^2.$ ♣

Lemma 7.5.5. *The congruence*

(7.5.11) $1 + x^2 + y^2 \equiv 0 \pmod{p}$

is solvable for any prime p.   ♣

Proof of Lemma 7.5.4. We can justify identity (7.5.10) by a computation.   □

We note that a natural proof of (7.5.10) arises by using *quaternions*: taking

$$\alpha = a_1 + a_2 i + a_3 j + a_4 k \quad \text{and} \quad \beta = b_1 + b_2 i + b_3 j + b_4 k,$$

(7.5.10) is the expanded version of the identity $N(\alpha)N(\beta) = N(\beta\bar{\alpha})$ for the norms of quaternions. (Of course, the law $N(\alpha)N(\beta) = N(\alpha\beta)$ would also prove the first sentence in Lemma 7.5.4 with another identity instead of (7.5.10), but we shall need (7.5.10) explicitly in the proof of Theorem 7.5.3.)

Proof of Lemma 7.5.5. The statement is obvious for $p = 2$.

For a proof by contradiction, we assume that (7.5.11) has no solution for some prime $p > 2$, i.e.

(7.5.12) $$x^2 \not\equiv -1 - y^2 \pmod{p}$$

for any integers x and y.

If x runs over a complete residue system modulo p, then the values of x^2 are 0 and the quadratic residues modulo p. This gives

$$\frac{p-1}{2} + 1 = \frac{p+1}{2}$$

pairwise incongruent values, by Theorem 4.1.2.

The same applies also for y^2, and thus for $-1 - y^2$. By (7.5.12), this would produce altogether $2\frac{p+1}{2} = p + 1$ pairwise incongruent numbers modulo p, which is clearly nonsense. $\square$

We note that Lemma 7.5.5 can also be easily deduced from Chevalley's Theorem 3.6.1 or from Exercise 3.6.2 (see Exercise 7.5.19).

We turn to the proof of Theorem 7.5.3. For the sake of brevity, we call a positive integer "nice" if it is the sum of four squares. Clearly, 1 and 2 are nice integers.

By Lemma 7.5.4, it is sufficient to show that every prime $p > 2$ is nice.

There exists a nice multiple of p, e.g. $4p^2$. We take the smallest positive m for which mp is nice, and let

(7.5.13) $$mp = a_1^2 + a_2^2 + a_3^2 + a_4^2.$$

We have to prove $m = 1$. We shall show that if $m > 1$, then also $m_1 p$ is nice for some $0 < m_1 < m$. This, however, contradicts the minimality of m, so $m = 1$.

We verify first $m < p$, so p has a nice (positive) multiple less than p^2. By Lemma 7.5.5, (7.5.11) is solvable. Taking the system of residues of least absolute value modulo p, we get a solution x and y satisfying $|x| < \frac{p}{2}$ and $|y| < \frac{p}{2}$. Then

$$v = 1^2 + x^2 + y^2 + 0^2 \text{ is nice}, \quad p \mid v \quad \text{and} \quad 0 < v < 2\left(\frac{p}{2}\right)^2 + 1 < p^2.$$

Next we show that m must be odd. Otherwise, we can partition the four values a_v into two pairs whose elements have the same parity; say a_1 and a_2 are both odd or both even, and the same holds for a_3 and a_4. Then

$$\left(\frac{m}{2}\right)p = \left(\frac{a_1 + a_2}{2}\right)^2 + \left(\frac{a_1 - a_2}{2}\right)^2 + \left(\frac{a_3 + a_4}{2}\right)^2 + \left(\frac{a_3 - a_4}{2}\right)^2,$$

which contradicts the minimality of m.

We shall consider now (7.5.13) modulo m. Let b_1, b_2, b_3, and b_4 be the residues of least absolute value modulo m of a_1, a_2, a_3, and a_4, i.e.

(7.5.14) $$b_v \equiv a_v \pmod{m}, \qquad |b_v| \leq \frac{m-1}{2}, \qquad v = 1, 2, 3, 4.$$

Then

$$b_1^2 + b_2^2 + b_3^2 + b_4^2 \equiv a_1^2 + a_2^2 + a_3^2 + a_4^2 \equiv 0 \pmod{m},$$

thus

(7.5.15) $mm_1 = b_1^2 + b_2^2 + b_3^2 + b_4^2$

for some integer m_1. We show $0 < m_1 < m$ in (7.5.15).

If $m_1 = 0$, then $b_\nu = 0$, so every a_ν is divisible by m. This implies

$$m^2 \mid a_1^2 + a_2^2 + a_3^2 + a_4^2 = mp, \qquad \text{thus} \qquad m \mid p,$$

which contradicts $1 < m < p$.

The inequality $m_1 < m$ follows from

$$mm_1 = \sum_{\nu=1}^{4} b_\nu^2 \le 4\left(\frac{m-1}{2}\right)^2 < 4\left(\frac{m}{2}\right)^2 = m^2.$$

Multiplying (7.5.13) and (7.5.15), we obtain

(7.5.16) $m^2 m_1 p = c_1^2 + c_2^2 + c_3^2 + c_4^2$

where the integers c_ν are determined by (7.5.10).

We show that every c_ν is a multiple of m. Since $b_\nu \equiv a_\nu \pmod{m}$, we obtain

$$c_1 = a_1 b_1 + a_2 b_2 + a_3 b_3 + a_4 b_4 \equiv a_1^2 + a_2^2 + a_3^2 + a_4^2 = mp \equiv 0 \pmod{m},$$

and we derive the divisibility by m for the other three values of c_ν similarly.

Dividing (7.5.16) by m^2, we obtain that $m_1 p$ is the sum of four squares. But $0 < m_1 < m$ contradicts the minimality of m. $\square$

Remark: The method used in the proof was a variant of *infinite descent*. The name will be clearer from the following formulation of our argument: If p itself is not nice, then considering a nice (positive) multiple mp of p, we find another nice multiple $m_1 p$ where $0 < m_1 < m$, then we find similarly a nice multiple $m_2 p$ where $0 < m_2 < m_1$, etc. We get a strictly decreasing infinite sequence $m > m_1 > m_2 > \cdots$ of positive integers so we perform an infinite descent among the positive integers, which is impossible.

The infinite descent for positive integers resembles an induction proof by contradiction. For example, the second proof for the uniqueness of prime factorization in Theorem 1.5.1 was basically an infinite descent.

Despite its connection to induction, infinite descent is based on a different principle: it uses the *well-ordering* property, i.e. every subset has a minimal element, and so we cannot form a sequence of infinite descent. Thus if some property gets inherited in an infinite descent, then no element of a well-ordered set can have this property.

Since the *axiom of choice* implies that every set can be well-ordered, infinite descent can be applied more widely than induction.

Exercises 7.5

1. Let n be a fixed positive integer. In how many essentially different ways can a positive integer be represented as the sum of two squares? (For instance, the example of 4050 after Theorem 7.5.1 has two such representations: $4050 = 45^2 + 45^2 = 9^2 + 63^2$.)

2. How many Gaussian integers have norm 98000?

3. Determine the largest r such that there exist infinitely many sequences of r consecutive integers each being the sum or difference of two squares.

4. Give a new proof to Exercise 4.1.5.

5. For which integers n is the Diophantine equation $x^2 + 4y^2 = n$ solvable, and what is the number of solutions?

* 6. Which positive integers can be represented and in how many ways as the sum of squares of two coprime integers?

* 7. (a) How many (pairwise incongruent) right triangles have integer side lengths one of them being k?

   (b) Solve the same problem if we assume that the side lengths are coprime.

8. Verify that the Diophantine equation $x^2 + y^2 = n$ has $4d'(n) - 4d''(n)$ solutions where $d'(n)$ and $d''(n)$ are the numbers of positive divisors of the form $4k + 1$ and $4k - 1$, of the positive integer n.

* 9. How many representations has a positive integer as a sum of two squares, on average? In a precise formulation, we ask about the approximate behavior of the mean value function
$$\frac{r(1) + r(2) + \cdots + r(n)}{n}$$
for large values of n, where $r(n)$ denotes the number of solutions of the Diophantine equation $x^2 + y^2 = n$.

S* 10. Solve the Diophantine equation $x^2 + 4 = y^3$.

S* 11. Which Gaussian integers are the sum of squares of two Gaussian integers?

12. In the proof of Theorem 7.5.1, we defined standard form for Gaussian integers, and observed that a Gaussian integer can have several standard forms. Prove that the number of standard forms of any Gaussian integer, different from 0 and units, is a power of 4. (Two standard forms are considered the same if they differ only in the order of factors, and we exclude the possibility that some Gaussian prime occurs with exponent 0.)

13. True or false?

   (a) If each of two positive integers is the sum of two squares, then so is their product.

   (b) If the product of two positive integers is the sum of two squares, then so is each factor.

(c) If both the product of two positive integers and one of the factors are sums of two squares, then so is the other factor.

(d) If each of two positive integers is the sum of three squares, then so is their product.

* 14. What is the probability that a positive integer is a sum of three squares?

15. Determine the smallest r such that every sufficiently large positive integer is the sum of at most r odd squares?

16. Deduce the Four Squares Theorem from the Three Squares Theorem.

S 17. Which positive integers can be represented as a sum of four squares so that at least two summands are equal?

18. Is the Diophantine equation $x^2 + 9y^2 + z^2 + w^2 = 10^{11} + 23$ solvable?

19. Give two new proofs for Lemma 7.5.5 based on Chevalley's Theorem 3.6.1 and on Exercise 3.6.2.

20. Theorem 7.5.1 implies that every positive prime of the form $4k + 1$ is the sum of two squares. Give a new proof following the lines of the proof of Theorem 7.5.3.

* 21. The goal of this exercise is to present another proof for the Four Squares Theorem. We shall rely on Lemmas 7.5.4 and 7.5.5 but will establish the existence of a small nice multiple of p using part (a) below instead of infinite descent.

(a) *Thue's lemma.* We call two k-dimensional vectors with integer coordinates *congruent* modulo a prime p if the corresponding coordinates are congruent modulo p. Let C be a $k \times k$ matrix with integer elements and $u_1, \ldots, u_k, v_1, \ldots, v_k$ positive integers satisfying

$$u_1 \ldots u_k v_1 \ldots v_k > p^k.$$

Then there exist vectors

$$\mathbf{x} = \begin{pmatrix} x_1 \\ \vdots \\ x_k \end{pmatrix} \neq \mathbf{0} \text{ and } \mathbf{z} = \begin{pmatrix} z_1 \\ \vdots \\ z_k \end{pmatrix}$$

with integer coordinates such that

$$C\mathbf{x} \equiv \mathbf{z} \pmod{p} \quad \text{and} \quad |x_i| < u_i, \quad |z_i| < v_i, \quad i = 1, 2, \ldots, k.$$

(b) Using a special case with $k = 2$ of part (a) and relying on Lemma 7.5.5, prove that any prime p has a nice multiple less than $4p$.

(c) Finally, verify that if $2p$ or $3p$ is nice for a prime $p > 3$, then p itself is nice.

7.6. Waring's Problem

After sums of squares, we turn to representations as a sum of higher powers in general. Throughout this section, k denotes a positive integer greater than 1, and a kth power means the kth power of a non-negative integer.

Waring stated in 1770 that "every natural number is the sum of 4 squares, 9 cubes, 19 fourth powers, etc." This self-confident declaration hides serious deficiencies, especially concerning the last innocent word "etc." First of all, it is hard to observe any rule for the continuation of the three numbers 4, 9, and 19 and it is absolutely not clear that these numbers can be continued to infinity at all. This requires the proof of: To any k, there exists an r, depending only on k, such that every positive integer is the sum of r terms of kth powers. This was first proved by Hilbert in 1909(!).

Today we already know how to continue Waring's numbers apart from a minimal uncertainty to be specified later. It is interesting that the problem of the 19 fourth powers defied the siege longest, it was proved only in 1986, 6^3 years after Waring's proclamation.

Since a sum of kth powers can always be extended by an arbitrary number of terms 0^k, we are interested in the smallest number of kth powers sufficient for the representation of every positive integer:

Definition 7.6.1. Let $k > 1$. Then $g(k)$ is the minimal r such that every positive integer is the sum of r terms of kth powers of non-negative integers. ♣

Example. $g(2) = 4$, since every positive integer is the sum of four squares by the Four Squares Theorem and there exists a number, e.g. 7, that cannot be written as the sum of three squares.

Theorem 7.6.2.

(7.6.1) $$g(k) \geq 2^k + \left\lfloor \left(\frac{3}{2}\right)^k \right\rfloor - 2. \qquad ♣$$

Proof. To get a lower bound for $g(k)$, it is sufficient to find just one positive integer n that requires many kth powers.

Let n be the greatest integer of the form $t2^k - 1$ that is less than 3^k. Then we can only use terms 1^k and 2^k to represent n, and clearly

$$n = t2^k - 1 = \underbrace{2^k + \cdots + 2^k}_{t-1 \text{ terms}} + \underbrace{1^k + \cdots + 1^k}_{2^k - 1 \text{ terms}}$$

is the representation with the least number of summands. Hence,

$$g(k) \geq 2^k + t - 2.$$

We have to verify $t = \left\lfloor \left(\frac{3}{2}\right)^k \right\rfloor$. This follows from

(7.6.2) $$t2^k - 1 < 3^k \iff t2^k \leq 3^k \iff t \leq \left(\frac{3}{2}\right)^k,$$

and t is the largest integer satisfying (7.6.2). $\qquad \square$

The most important result for $g(k)$ is that (7.6.1) holds with equality in general: There may be only finitely many k for which $g(k)$ is greater than the right-hand side of (7.6.1), and then its value is obtained from the worst n less than 4^k (thus also the term 3^k can be used), similar to the proof of Theorem 7.6.2. This might occur only if $(3/2)^k$ is abnormally close to its ceiling, satisfying some special inequality. No integer less than 471000000 meets this requirement and it is almost certain that there are no such exceptions at all, so

$$g(k) = 2^k + \left\lfloor \left(\frac{3}{2}\right)^k \right\rfloor - 2$$

for every k. Accordingly, the right-hand side of (7.6.1) is the continuation of Waring's numbers. When $k = 2, 3,$ and 4 we obtain the values 4, 9, and 19.

Theorem 7.6.2 shows that some small integers n need extremely many kth powers to be represented. Therefore, it is worthwhile to analyze how many kth powers are necessary to represent every sufficiently large n:

Definition 7.6.3. Let $k > 1$. Then $G(k)$ is the minimal s such that every sufficiently large positive integer is a sum of s terms of kth powers of non-negative integers.   ♣

Example. $G(2) = 4$, as obviously $G(2) \le g(2) = 4$, and by the Three Squares Theorem, infinitely many integers are not representable as the sum of three squares.

The next table summarizes the best known results for $g(k)$ and $G(k)$ for some small values of k:

k	2	3	4	5	6	7	8
$g(k)$	4	9	19	37	73	143	279
$G(k)$	4	4–7	16	6–17	9–24	8–31	32–39

The table reflects the great uncertainty about the exact values of $G(k)$ even for small integers k (e.g. 4–7 at $G(3)$ means that we know no better bounds than $4 \le G(3) \le 7$). The exact values of $G(k)$ were determined only for $k = 2$ and 4 so far.

We know, however, that $G(k)$ is much smaller than $g(k)$ if k is large: e.g. $G(k) < 6k \log k$ for every $k > 1$. The best known result is that to any $\varepsilon > 0$, there exists a $k_0 = k_0(\varepsilon)$ such that $G(k) < (1 + \varepsilon)k \log k$ for every $k > k_0$. Thus $G(k)$ is almost linear in contrast with $g(k)$, which has an exponential order.

Now we get some lower bounds for $G(k)$.

Theorem 7.6.4. $G(k) \ge k + 1$ for every $k > 1$.   ♣

Proof. For a proof by contradiction, we assume $G(k) \le k$ for some k. Then there exists n_0 such that every integer $n > n_0$ is a k-term sum of kth powers, i.e.

$$(7.6.3) \qquad\qquad n = x_1^k + x_2^k + \cdots + x_k^k.$$

We fix (temporarily) a large positive integer M, and let $f(M)$ denote the number of integers n that are representable as k-term sums of kth powers and

$$(7.6.4) \qquad\qquad 0 \le n \le M.$$

By our assumption,

$$(7.6.5) \qquad\qquad f(M) \ge M - n_0.$$

We establish now an upper bound for $f(M)$. Considering the representations (7.6.3) of integers n in (7.6.4), the numbers x_i must satisfy

$$0 \le x_i \le \sqrt[k]{n} \le \sqrt[k]{M}, \qquad i = 1, 2, \dots, k.$$

This means that the values of an x_i can only be

(7.6.6) $$0, 1, \ldots, T = \left\lfloor \sqrt[k]{M} \right\rfloor.$$

We count how many sums $x_1^k + \cdots + x_k^k$ can be formed from the integers listed in (7.6.6); some of these sums may coincide, and many of them exceed M, so the number of such formal sums is $\geq f(M)$. Such a formal sum is equivalent to selecting k elements from the list (7.6.6) so that any element can be chosen arbitrarily many times (since there may be equal terms in (7.6.3)), and the order of selection is irrelevant as the sum remains the same if we permute its terms. Such a selection is called a *combination of k elements out of $T + 1$*, allowing repetitions, and there are $\binom{T+k}{k}$ such combinations. To be self-contained, we give a proof of this formula.

To characterize a combination, assume that we choose m_j pieces of j for every $0 \leq j \leq T$. We write m_0 small circles $\circ$ for the m_0 pieces of 0s followed by a delimiter bar $|$, then draw m_1 small circles $\circ$ for the m_1 pieces of 1s followed again by a delimiter bar $|$, etc. Finally we draw the last m_T small circles $\circ$ for the m_T pieces of terms T. For $k = 5$ and $M = 7^5$, the sum $0^5 + 1^5 + 1^5 + 3^5 + 7^5$ corresponds to the sequence $\circ|\circ\circ||\circ||||\circ$.

We established a bijection between the formal sums and the sequences of k circles and $T = \left\lfloor \sqrt[k]{M} \right\rfloor$ bars. Hence, the number of formal sums is equal to the number of these sequences, which is $\binom{\lfloor \sqrt[k]{M} \rfloor + k}{k}$.

Summarizing, we have proved

(7.6.7) $$f(M) \leq \binom{k + \lfloor \sqrt[k]{M} \rfloor}{k}.$$

Inequalities (7.6.5) and (7.6.7) imply

(7.6.8) $$M - n_0 \leq \binom{k + \lfloor \sqrt[k]{M} \rfloor}{k}.$$

In the expanded form

$$\frac{1}{k!}(k + \lfloor \sqrt[k]{M} \rfloor)(k - 1 + \lfloor \sqrt[k]{M} \rfloor) \ldots (1 + \lfloor \sqrt[k]{M} \rfloor)$$

of the right-hand side in (7.6.8), we can omit the floor signs, which does not decrease the right-hand side in (7.6.8). Dividing both sides by M so that each factor $i + \sqrt[k]{M}$ is divided by $\sqrt[k]{M}$, we obtain

(7.6.9) $$1 - \frac{n_0}{M} \leq \frac{1}{k!}\left(1 + \frac{k}{\sqrt[k]{M}}\right)\left(1 + \frac{k-1}{\sqrt[k]{M}}\right) \ldots \left(1 + \frac{1}{\sqrt[k]{M}}\right).$$

For $M \to \infty$, the left-hand side of (7.6.9) tends to 1 and the right-hand side tends to $1/k!$, which is a contradiction, since $k > 1$. $\qquad\square$

Remark: The proof yields that many integers n are not representable as a k-term sum of kth powers (e.g. for $k = 5$, this happens with probability at least $\dfrac{5! - 1}{5!} = \dfrac{119}{120}$ which is more than 99 percent!). At the same time, this was not a constructive proof because it did not exhibit any n which is not representable.

We show now that the upper bound in Theorem 7.6.4 can be improved e.g. for $k = 6$:

Theorem 7.6.5. $G(6) \geq 9$. ♣

Proof. We use the fact that

$$(7.6.10) \qquad a^6 \equiv \begin{cases} 1 \ (\mathrm{mod}\ 9), & \text{if } 3 \nmid a \\ 0 \ (\mathrm{mod}\ 9), & \text{if } 3 \mid a. \end{cases}$$

The case $3 \nmid a$ follows from the Euler–Fermat Theorem, and if $3 \mid a$, then a^6 is divisible not only by 9, but also by 3^6.

To prove Theorem 7.6.5, we will show that infinitely many n cannot be written as the sum of eight sixth powers. We claim that integers of the form $n = 27t + 9$ have no such representation.

For a proof by contradiction, assume

$$(7.6.11) \qquad n = x_1^6 + \cdots + x_8^6.$$

Considering (7.6.11) modulo 9, (7.6.10) implies

$$(7.6.12) \qquad 0 \equiv u_1 + \cdots + u_8 \ (\mathrm{mod}\ 9), \quad \text{where} \quad u_i = 0 \text{ or } 1, \quad i = 1, 2, \ldots, 8,$$

which can hold only if $u_i = 0$ for every i. Thus each x_i is a multiple of 3. But then (7.6.11) yields $3^6 \mid n$, which is a contradiction. $\square$

Further lower bounds for $G(k)$ occur in Exercise 7.6.2.

Exercises 7.6

1. Verify $G(200) \leq G(600)$.

* 2. (a) Establish the lower bounds for $G(k)$:

   (a1) $G(4) \geq 16$

   (a2) $G(8) \geq 32$

   (a3) $G(24) \geq 32$

   (a4) $G(100) \geq 125$

   (a5) $G(250) \geq 312$.

   (b) For which integers k can we generalize the results in part (a)?

3. Let $k > 1$ be arbitrary. Demonstrate the existence of a positive integer n that has at least 1000 essentially different representations as a $(k+1)$-term sum of kth powers.

4. (a) Verify the identity

$$\sum_{1 \leq i < j \leq 4} \left((a_i + a_j)^4 + (a_i - a_j)^4 \right) = 6(a_1^2 + a_2^2 + a_3^2 + a_4^2)^2$$

   where a_1, a_2, a_3, a_4 are arbitrary complex numbers.

   (b) Prove $g(4) \leq 53$.

5. If we represent the integers as signed sums of kth powers, then generally fewer terms are sufficient than $g(k)$ or even $G(k)$. Show that the minimal number of terms is three for $k = 2$ and, moreover, each of the Diophantine equations $x^2 + y^2 - z^2 = n$ and $x^2 - y^2 - z^2 = n$ has infinitely many solutions for every positive integer n.

7.7. Fermat's Last Theorem

In Section 7.2 we proved that the Pythagorean equation $x^2 + y^2 = z^2$ has infinitely many solutions in positive integers and all solutions can be described by three parameters. According to Fermat's famous conjecture verified recently by Andrew Wiles, the situation is completely different for higher powers:

Theorem 7.7.1 (Fermat's Last Theorem). *For integers $k > 2$, the equation $x^k + y^k = z^k$ has no solutions in positive integers.* ♣

The history of the conjecture started in 1637 when, reading a 1621 edition of Diophantus's book, Fermat added a note to the part about Pythagorean triples: "The sum of two cubes is never a cube, the sum of two fourth powers is never a fourth power, etc. I found a wonderful proof for this but the margin is too small to contain it."

These few lines caused great excitement among both mathematicians and outsiders for three and a half centuries. The problem seems innocent and can be understood without any mathematical training; many amateurs tried to solve it, but in vain. Professional mathematicians did not perform much better either.

It is easy to show (see Exercise 7.7.1) that if the conjecture is true for an exponent k, then it is true for every multiple of k. Hence, it is sufficient to settle the problem for prime exponents k and for $k = 4$. Fermat did prove the case $k = 4$, and more than 100 years later Euler succeeded with exponent $k = 3$. This list was extended with a few more values of k in the first half of the nineteenth century.

The middle of the nineteenth century brought the first major breakthrough with the introduction of "ideal numbers." Today we call them *ideals*, and will discuss them in Chapter 11. Several new criteria were developed that guaranteed that Fermat's Last Theorem is true if a prime exponent k satisfies them. In principle, these criteria can be numerically checked for any particular k, and many such computations were performed, using computers in the last decades.

In spite of all these efforts and results, the conjecture was still verified only for finitely many prime exponents as late as 1980. At the same time, many more general conjectures were formulated, since it was expected that the solution for Fermat's equation will follow from a theorem about a more general problem.

In 1983, Gerd Faltings achieved a sensational new result: For any fixed exponent k, Fermat's equation can have only finitely many *primitive* solutions, those with $(x, y, z) = 1$.

The true sensation, however, occurred in 1993 when Andrew Wiles published a decisive solution after working for many years alone, in secret. It turned out that there was an error in the proof but Wiles, with the help of Richard Taylor, corrected it in 1994.

Thus today Fermat's Last Theorem is no longer a famous unsolved problem but is a valid theorem. The several hundred pages of Wiles' proof are understandable only by a very small group of top specialists, but we can hope for somewhat simpler proofs later.

Fermat's "wonderful proof" was probably either just a vague idea, or a wrong argument that assumed the validity of the Fundamental Theorem of Arithmetic in sets of numbers where it is false (see Section 11.2 for more details). We can practically exclude the possibility that somebody will find a genuinely elementary proof.

During the centuries of assiduous and intensive research on Fermat's Last Theorem, mathematicians elaborated many new, effective theories. Though they brought only partial success in handling the original problem, they became indispensable in some other fields of mathematics. This illustrates well that research on a given problem may help indirectly the development of the entirety of mathematics, too.

We shall prove below, following the historical order, the two easiest special cases of Fermat's Last Theorem for exponents $k = 4$ and 3.

In both cases, we shall prove a slightly stronger result, since we can give an answer to the original problem only by proving sharper theorems.

Fermat's Last Theorem for $k = 4$ follows from the following statement.

Theorem 7.7.2. *The equation $x^4 + y^2 = z^4$ has no solutions in positive integers.*   ♣

Proof. We shall use the following lemma of independent interest.

Lemma 7.7.3. *The sum and difference of two non-zero squares cannot both be squares.*   ♣

Proof. We apply infinite descent (see the Remark after Theorem 7.5.3).

We want to show that the system of equations

(7.7.1a) $x^2 + y^2 = z^2$
(7.7.1b) $x^2 - y^2 = w^2$

has no solutions in positive integers. Assume the converse, and let x_0, y_0, z_0, w_0 be a solution where z_0 is minimal. We prove that then there is a solution x_1, y_1, z_1, w_1 where $0 < z_1 < z_0$, which contradicts the minimality of z_0. Hence no solution can exist.

We may assume $(x_0, z_0) = 1$: if some prime p divides x_0 and z_0, then similar to the Pythagorean triples, (7.7.1a) implies $p \mid y_0$, then we infer $p \mid w_0$ from (7.7.1b), and so $x_0/p, y_0/p, z_0/p, w_0/p$ is a solution with $z_0/p < z_0$, which contradicts the minimality of z_0.

Substituting x_0, y_0, z_0, w_0 into (7.7.1a) and (7.7.1b), adding and subtracting the two equalities, we obtain

(7.7.2a) $$2x_0^2 = z_0^2 + w_0^2$$

(7.7.2b) $$2y_0^2 = z_0^2 - w_0^2.$$

By (7.7.2a), z_0 and w_0 have the same parity. Therefore we can rewrite (7.7.2a) as

(7.7.3) $$x_0^2 = \left(\frac{z_0 + w_0}{2}\right)^2 + \left(\frac{z_0 - w_0}{2}\right)^2.$$

Here

(7.7.4) $$\left(x_0, \frac{z_0 + w_0}{2}, \frac{z_0 - w_0}{2}\right) = 1$$

since x_0 is coprime to the sum of the other two numbers, which is z_0.

By (7.7.3) and (7.7.4), $\frac{z_0+w_0}{2}$, $\frac{z_0-w_0}{2}$, and x_0 form a primitive Pythagorean triple. Thus, by Theorem 7.2.1,

(7.7.5) $$\frac{z_0 + w_0}{2} = 2mn \quad \text{and} \quad \frac{z_0 - w_0}{2} = m^2 - n^2,$$

or vice versa, for some coprime integers of opposite parity with $m > n > 0$.

Using (7.7.5), we rewrite the right-hand side of

$$\frac{y_0^2}{2} = \frac{z_0 + w_0}{2} \cdot \frac{z_0 - w_0}{2},$$

equivalent to (7.7.2b), and divide by 2 to get

(7.7.6) $$\left(\frac{y_0}{2}\right)^2 = mn(m + n)(m - n).$$

Since m and n are coprime and have opposite parity, the four positive integers on the right-hand side of (7.7.6) are pairwise coprime. Therefore each of them is a square, so

(7.7.7) $$m = x_1^2, \quad n = y_1^2, \quad m + n = z_1^2, \quad \text{and} \quad m - n = w_1^2.$$

By (7.7.7), x_1, y_1, z_1, w_1 satisfy the system of equations (7.7.1a)–(7.7.1b) and

$$z_1 \leq z_1^2 = m + n \leq (m + n)(m - n) = \frac{z_0 \pm w_0}{2} < z_0,$$

which contradicts the minimality of z_0. □

We turn now to the proof of Theorem 7.7.2. We assume

(7.7.8) $$c^4 - a^4 = b^2$$

for some positive integers a, b, and c, and find a contradiction. If $(a, b, c) = d$, then a/d, b/d^2, c/d is also a solution of the equation, so we may assume $(a, b, c) = 1$. This implies that a, b, and c are pairwise coprime as we have seen in several similar situations.

Factoring the left-hand side of (7.7.8), we get

(7.7.9) $$(c^2 + a^2)(c^2 - a^2) = b^2.$$

Let h denote the greatest common divisor of the two factors on the left-hand side of (7.7.9): $h = (c^2 + a^2, c^2 - a^2)$. Since $(a^2, c^2) = 1$, h can be only 1 or 2. By the Fundamental Theorem of Arithmetic, the factors on the left-hand side of (7.7.9) are squares themselves in the first case, and are the doubles of squares if the second condition holds.

Thus if $h = 1$, then $c^2 + a^2$ and $c^2 - a^2$ are squares, which contradicts Lemma 7.7.3.

If $h = 2$, then

$$(7.7.10) \qquad\qquad c^2 + a^2 = 2u^2 \quad \text{and} \quad c^2 - a^2 = 2v^2$$

for some integers $u > v > 0$. Taking the sum and the difference of the equalities (7.7.10) and cancelling the results by 2, we obtain

$$c^2 = u^2 + v^2 \quad \text{and} \quad a^2 = u^2 - v^2.$$

But this is impossible according to Lemma 7.7.3. $\qquad\qquad\qquad\qquad\qquad\square$

To prove Fermat's Last Theorem for the exponent $k = 3$, we develop number theory in the ring of Eulerian (or Eisenstein) integers, which behave similarly to Gaussian integers.

Definition 7.7.4. By an *Eulerian integer* (or *Eisenstein integer*), we mean a complex number $a + b\omega$ where a, b are integers, and

$$\omega = \cos\frac{2\pi}{3} + i\sin\frac{2\pi}{3} = -\frac{1}{2} + i\frac{\sqrt{3}}{2}. \qquad\qquad \clubsuit$$

The complex numbers ω and $\omega^2 = -1 - \omega$ are the primitive third roots of unity. The factorization

$$(7.7.11) \qquad\qquad x^3 = z^3 - y^3 = (z - y)(z - y\omega)(z - y\omega^2)$$

reveals the connection between Fermat's equation $x^3 + y^3 = z^3$ and the Eulerian integers. Our proof will be based on the investigation of an equation similar to (7.7.11) and we shall rely heavily on the number theory of Eulerian integers.

Definition 7.7.5. The *norm* $N(\alpha)$ of an Eulerian integer $\alpha = a + b\omega$ is the square of the absolute value of α:

$$N(\alpha) = |\alpha|^2 = \alpha\bar{\alpha} = (a + b\omega)(a + b\omega^2) = a^2 - ab + b^2. \qquad\qquad \clubsuit$$

Clearly, $N(\alpha)$ is a non-negative integer, and $N(\alpha) = 0 \iff \alpha = 0$. We note that the form $a^2 - ab + b^2$ of the norm of the Eulerian integer $a + b\omega$ can be rewritten as $c^2 + 3d^2$ with suitable integers c and d, see Exercise 7.7.10a.

The Eulerian integers form a parallelogram lattice in the complex plane consisting of rhombuses with sides of unit length and with angles of 120 and 60 degrees.

We define divisibility, unit, greatest common divisor, irreducible, and prime exactly as we did for Gaussian integers (see Definitions 7.4.4, 7.4.6, 7.4.9, 7.4.10, and 7.4.11, the adjective "Gaussian" is replaced everywhere by "Eulerian," of course).

Apart from the description of Eulerian units and primes, the theorems and proofs seen for Gaussian integers remain valid for Eulerian integers:

- properties of the norm (Theorems 7.4.3, 7.4.5)
- division algorithm (Theorem 7.4.8; in the proof, we have to consider the corresponding lattice rhombus instead of a square)

- equivalence of prime and irreducible (Theorem 7.4.12)
- the Fundamental Theorem of Arithmetic (Theorem 7.4.13)
- connection between Eulerian primes and positive prime numbers in **Z** (Theorem 7.4.14).

Theorem 7.4.7 and its proof can be adapted to Eulerian integers if we modify the explicit description in part (iv) as follows:

Theorem 7.7.6. *There are six units among the Eulerian integers:*

$$\pm 1, \qquad \pm\omega, \qquad \pm\omega^2 = \mp(1+\omega),$$

which are just the complex sixth roots of unity. ♣

Finally, Eulerian primes are characterized by

Theorem 7.7.7. *All Eulerian primes are (ε denotes an arbitrary unit):*

(A) $\varepsilon(i\sqrt{3}) = \varepsilon(1 + 2\omega)$

(B) εq *where q is a positive prime number of the form $3t - 1$*

(C) π *where $N(\pi)$ is a positive prime number of the form $3t + 1$; to every such prime number, there belong exactly two Eulerian primes (apart from unit factors) that are conjugates but not associates.* ♣

Proof. We have to modify the arguments used in Theorem 7.4.15 accordingly, thus we indicate only the differences briefly.

By the analog of Theorem 7.4.14, we obtain all Eulerian primes from the factorization of positive prime numbers. We get different types of decompositions when this positive prime is (A) 3, (B) is of the form $3t - 1$, or (C) is of the form $3t + 1$.

(A) Since $3 = (-1)(i\sqrt{3})^2$, the only Eulerian prime divisor of 3 is $i\sqrt{3}$, apart from associates.

(B) The positive prime numbers of the form $3t - 1$ are Eulerian primes: we have to show that the norm of an Eulerian integer cannot be of this form, which can be justified as seen at the Gaussian integers.

(C) If p is a positive prime number of the form $3t + 1$, then $\left(\frac{-3}{p}\right) = 1$ (see Exercise 4.2.2c), thus $p \mid c^2 + 3$ for some integer c. We consider the factorization

$$c^2 + 3 = (c + i\sqrt{3})(c - i\sqrt{3}) = (c + 1 + 2\omega)(c - 1 - 2\omega)$$

among the Eulerian integers, and follow the argument used for the Gaussian integers. □

In the proof of Fermat's Last Theorem for cubes, some important properties of the Eulerian prime $i\sqrt{3}$ play an important role. For a convenient formulation, we introduce congruences also among Eulerian integers:

Definition 7.7.8. Let $\mu \neq 0$, α, and β Eulerian integers. We say that α is *congruent* to β modulo μ, if $\mu \mid \alpha - \beta$. ♣

We shall use the notation $\alpha \equiv \beta \pmod{\mu}$ or $\alpha \equiv \beta \,(\mu)$ for short. The elementary properties of congruences for integers are equally valid for Eulerian integers.

We summarize some important properties of the Eulerian prime $i\sqrt{3}$:

Theorem 7.7.9. *Let* $\lambda = i\sqrt{3} = 1 + 2\omega$.

 (i) *The associates of* λ *are* $\pm(1 + 2\omega)$, $\pm(2 + \omega)$, $\pm(1 - \omega)$.

 (ii) *Any Eulerian integer is congruent to exactly one of the three values* 0 *and* ± 1 *modulo* λ.

(iii) $\alpha^3 \equiv \alpha \pmod{\lambda}$ *for every Eulerian integer* α.

(iv) $\alpha \equiv \pm 1 \pmod{\lambda} \implies \alpha^3 \equiv \pm 1 \pmod{\lambda^4}$.   ♣

Proof. (i) By Theorem 7.7.6, the associates of λ are $\pm\lambda$, $\mp\omega\lambda$, and $\pm\omega^2\lambda$. Performing the multiplications and applying the relations $\omega^2 = -1 - \omega$ and $\omega^3 = 1$, we obtain the six Eulerian integers stated in the theorem.

(ii) The identity

$$a + b\omega = a + b - b(1 - \omega) = a + b - b\omega^2\lambda$$

implies

$$a + b\omega \equiv a + b \pmod{\lambda}.$$

Since $a + b \equiv 0, 1,$ or $-1 \pmod 3$ and $\lambda \mid 3$,

$$a + b \equiv 0, 1, \text{ or } -1 \pmod{\lambda}.$$

This proves that any Eulerian integer $a + b\omega$ is congruent to 0, 1, or -1 modulo λ.

We have to show that 0, 1, and -1 are pairwise incongruent modulo λ, so λ does not divide the difference ± 1 or ± 2 of any two of these three numbers. If $\lambda \mid \pm 1$ or $\lambda \mid \pm 2$, then $N(\lambda) \mid 1$ or $N(\lambda) \mid 4$, but this is impossible since $N(\lambda) = 3$.

(iii) This follows immediately from (ii) using the identity

$$\alpha^3 - \alpha = \alpha(\alpha - 1)(\alpha + 1).$$

(iv) If $\alpha \equiv 1 \pmod{\lambda}$, then $\alpha = 1 + \beta\lambda$ for a suitable Eulerian integer β. Cubing both sides, we obtain

$$\alpha^3 = 1 + 3\beta\lambda + 3\beta^2\lambda^2 + \beta^3\lambda^3.$$

From $3 = -\lambda^2$, we get

$$(7.7.12) \qquad \alpha^3 = 1 - \beta\lambda^3 - \beta^2\lambda^4 + \beta^3\lambda^3 = 1 - \beta^2\lambda^4 + (\beta^3 - \beta)\lambda^3.$$

Since $\lambda \mid \beta^3 - \beta$ by (iii), (7.7.12) implies $\alpha^3 \equiv 1 \pmod{\lambda^4}$.

We can proceed similarly in the case $\alpha \equiv -1 \pmod{\lambda}$ or deduce it by applying the previous case to $-\alpha \equiv 1 \pmod{\lambda}$.   □

Remark: Several statements of Theorem 7.7.9 are valid for more general moduli instead of λ (see Exercise 7.7.12):

(ii): The number of residue classes modulo any Eulerian integer $\mu \neq 0$ is $N(\mu)$. Moreover, if $N(\mu)$ is a prime number p, then the integers in a complete residue system in **Z** modulo p form a complete residue system among the Eulerian integers modulo μ. (Theorem 7.7.9 contained the special case $\mu = \lambda$ and $N(\lambda) = 3$.)

(iii): Any Eulerian integer α and Eulerian prime π satisfy

$$\alpha^{N(\pi)} \equiv \alpha \pmod{\pi}.$$

(This is the analog of Fermat's Little Theorem.)

After the preparations we are ready to prove Fermat's Last Theorem for the exponent $k = 3$.

Theorem 7.7.10. *The equation $x^3 + y^3 = z^3$ has no solutions where x, y, and z are non-zero integers.* ♣

Proof. We shall verify a more general statement, namely, the equation

$$(7.7.13) \qquad \xi^3 + \eta^3 + \psi^3 = 0$$

has no solutions where ξ, η, and ψ are non-zero Eulerian integers.

For a proof by contradiction, we assume that there exists such a solution. As we have seen several times, we can restrict ourselves to the case when ξ, η, and ψ are coprime, and moreover, pairwise coprime.

The outline of the proof follows. We show first that exactly one of ξ, η, and ψ is a multiple of λ, say it is ξ. Factoring out the maximal power of λ from ξ and replacing $-\eta$ by κ, we transform (7.7.13) into

$$(7.7.14) \qquad \varepsilon\lambda^{3n}\gamma^3 = \kappa^3 - \psi^3$$

where

$$(7.7.15) \qquad n \geq 1, \quad \varepsilon \text{ is a unit}, \quad \text{and}, \lambda, \gamma, \kappa, \text{ and } \psi \text{ are pairwise coprime.}$$

We show that $n \neq 1$, but on the other hand, if (7.7.14) and (7.7.15) hold for some n, then they hold also for $n - 1$ instead of n with some other values of the variables ε, γ, κ, and ψ. This infinite descent yields the contradiction.

The key step for the infinite descent is a factorization of (7.7.14) as in (7.7.11) where the gcd of the three factors on the right-hand side is λ, so after dividing the equation by λ^3, each of the remaining three pairwise coprime numbers is an associate of a cube of an Eulerian integer, by the Fundamental Theorem of Arithmetic.

We turn now to details.

I. In (7.7.13), at most one of the pairwise coprime ξ, η, and ψ can be divisible by λ.

If none of them were a multiple of λ, then

$$0 = \xi^3 + \eta^3 + \psi^3 \equiv \pm 1 \pm 1 \pm 1 = \pm 1 \text{ or } \pm 3 \pmod{\lambda^4}$$

by (iv) in Theorem 7.7.9. This would imply $\lambda^4 \mid 3$, or $9 \mid 3$, which is impossible.

II. We have thus verified that exactly one of ξ, η, and ψ is a multiple of λ, let it be, say, ξ, so

$$(7.7.16) \qquad \xi = \lambda^n\gamma, \quad \text{where} \quad n > 0 \quad \text{and} \quad \lambda \nmid \gamma.$$

Substituting (7.7.16) into (7.7.13), and denoting $-\eta$ by κ, we get just (7.7.14) and (7.7.15) (now $\varepsilon = 1$).

So it suffices to prove that (7.7.14) and (7.7.15) cannot hold for any unit ε.

III. We show $n \neq 1$ in (7.7.14).

Considering (7.7.14) modulo λ^4, we get

(7.7.17) $$\varepsilon\lambda^{3n}\gamma^3 = \kappa^3 - \psi^3 \equiv \pm 1 \pm 1 = 0 \text{ or } \pm 2 \pmod{\lambda^4}$$

from (iv) in Theorem 7.7.9. The case ± 2 is impossible as it would imply $\lambda \mid 2$. Therefore the right-hand side of (7.7.17) is 0, so

$$\lambda^4 \mid \varepsilon\lambda^{3n}\gamma^3.$$

As $(\lambda, \varepsilon\gamma) = 1$, we get $\lambda^4 \mid \lambda^{3n}$, implying $n \geq 2$.

IV. Now comes infinite descent as key step: If (7.7.14) and (7.7.15) hold for some n, then this can be realized also for $n - 1$ instead of n with some other values of the variables ε, γ, κ, and ψ.

Factoring the right-hand side of (7.7.14), we obtain

(7.7.18) $$\varepsilon\lambda^{3n}\gamma^3 = (\kappa - \psi)(\kappa - \psi\omega)(\kappa - \psi\omega^2).$$

Since the Eulerian prime λ divides the left-hand side of (7.7.18), it must divide at least one factor on the right-hand side, too. The pairwise differences of the three factors are $(\omega - 1)\psi$, $(\omega^2 - 1)\psi$, and $(\omega^2 - \omega)\psi$, which are all multiples of $\omega - 1 = \varepsilon\lambda$. This implies that λ divides all three factors on the right-hand side of (7.7.18).

We verify now that the gcd of any two factors on the right-hand side of (7.7.18) is λ. We check it for the first two factors; the other two pairs can be handled similarly.

Put $\delta = (\kappa - \psi, \kappa - \psi\omega)$. Then

$$\delta \mid (\kappa - \psi) - (\kappa - \psi\omega) = \psi(\omega - 1)$$

and

$$\delta \mid \omega(\kappa - \psi) - (\kappa - \psi\omega) = \kappa(\omega - 1),$$

so

$$\delta \mid (\psi(\omega - 1), \kappa(\omega - 1)) = (\omega - 1)(\kappa, \psi) = \omega - 1 = \varepsilon\lambda.$$

Combined with $\lambda \mid \delta$, shown earlier, this gives $\delta = \lambda$.

Thus

$$\frac{\kappa - \psi}{\lambda}, \quad \frac{\kappa - \psi\omega}{\lambda}, \quad \text{and} \quad \frac{\kappa - \psi\omega^2}{\lambda}$$

are pairwise coprime. By the Fundamental Theorem of Arithmetic,

(7.7.19) $$\begin{aligned} \kappa - \psi &= \varepsilon_1\lambda\nu_1^3 \\ \kappa - \psi\omega &= \varepsilon_2\lambda\nu_2^3 \\ \kappa - \psi\omega^2 &= \varepsilon_3\lambda\nu_3^3 \end{aligned}$$

where $\varepsilon_1, \varepsilon_2, \varepsilon_3$ are units and ν_1, ν_2, ν_3 are pairwise coprime Eulerian integers.

We check now the behavior of ν_i concerning divisibility by λ. Since ν_i are pairwise coprime, two of them, say ν_2 and ν_3 are not multiples of λ. Let s be the exponent of λ in the standard form of ν_1. We claim that $s = n - 1$.

To verify it, we compare the exponents of λ on the two sides of (7.7.18). This exponent is $3n$ on the left-hand side. On the right-hand side, we know from (7.7.19) that each factor contains λ on the first power and ν_1^3 contains it with exponent $3s$. Hence, $3n = 3 + 3s$, so $s = n - 1$.

Thus

(7.7.20) $$\nu_1 = \lambda^{n-1}\gamma_1 \qquad \text{where} \qquad (\gamma_1, \lambda) = 1.$$

Here $n - 1 \geq 1$ as $n \geq 2$.

In the next step we show that taking a suitable combination of the equations in (7.7.19), we get an equality like (7.7.14) with $n - 1$ instead n and this completes the proof.

We multiply the second equation in (7.7.19) by ω and the third equation by ω^2, and add to the first equation:

(7.7.21) $$(\kappa - \psi) + \omega(\kappa - \psi\omega) + \omega^2(\kappa - \psi\omega^2) = \varepsilon_1\lambda\nu_1^3 + \varepsilon_4\lambda\nu_2^3 + \varepsilon_5\lambda\nu_3^3,$$

where $\varepsilon_4 = \varepsilon_2\omega$ and $\varepsilon_5 = \varepsilon_3\omega^2$ are units. The left-hand side of (7.7.21) is

(7.7.22) $$(\kappa - \psi) + \omega(\kappa - \psi\omega) + \omega^2(\kappa - \psi\omega^2) = (1 + \omega + \omega^2)(\kappa - \psi) = 0.$$

By (7.7.20), (7.7.21), and (7.7.22) we obtain

$$0 = \varepsilon_1\lambda^{3(n-1)+1}\gamma_1^3 + \varepsilon_4\lambda\nu_2^3 + \varepsilon_5\lambda\nu_3^3.$$

Dividing by $\varepsilon_5\lambda$ and rearranging the terms yields

(7.7.23) $$\varepsilon_6\lambda^{3(n-1)}\gamma_1^3 = \varepsilon_7\nu_2^3 - \nu_3^3$$

where also ε_6 and ε_7 are units.

We claim $\varepsilon_7 = \pm 1$, so we can rewrite the term $\varepsilon_7\nu_2^3$ as $(\pm\nu_2)^3$. We consider (7.7.23) modulo λ^3. Since $n - 1 \geq 1$, $\lambda \nmid \nu_2$, and $\lambda \nmid \nu_3$, part (iv) of Theorem 7.7.9 implies

$$\varepsilon_7(\pm 1) - (\pm 1) \equiv 0 \pmod{\lambda^3},$$

so λ^3 divides $\varepsilon_7 - 1$ or $\varepsilon_7 + 1$. From

$$N(\lambda^3) \mid N(\varepsilon_7 \mp 1), \qquad N(\lambda^3) = 27, \quad \text{and} \quad N(\varepsilon_7 \mp 1) < 27,$$

we get $\varepsilon_7 \mp 1 = 0$, thus $\varepsilon_7 = \pm 1$.

Therefore we can rewrite (7.7.23) as

$$\varepsilon_6\lambda^{3(n-1)}\gamma_1^3 = (\pm\nu_2)^3 - \nu_3^3.$$

This means that (7.7.14) holds with $n - 1$ instead of n, and the conditions in (7.7.15) are satisfied, with ε, γ, κ, and ψ replaced by ε_6, γ_1, $\pm\nu_2$, and ν_3. $\qquad \square$

Exercises 7.7

1. (a) Show that if $k \mid m$, and the sum of two positive kth powers is never a kth power, then the sum of two positive mth powers cannot be an mth power.

   (b) Explain why it is sufficient to prove Fermat's Last Theorem for prime exponents and for $k = 4$.

2. How many solutions do the following equations have in positive integers?

(a) $x^{20} + y^{24} = z^{28}$

(b) $x^3 + y^4 = z^5$.

3. Solve the exponential version $k^x + k^y = k^z$ of Fermat's equation where k, x, y, and z are positive integers.

4. We examine Fermat's equation in some cases where the exponent is not a positive integer. Find all solutions x, y, z in positive integers.

    (a) $k = -4$: $\dfrac{1}{x^4} + \dfrac{1}{y^4} = \dfrac{1}{z^4}$

    (b) $k = -2$: $\dfrac{1}{x^2} + \dfrac{1}{y^2} = \dfrac{1}{z^2}$

    (c) $k = 1/2$: $\sqrt{x} + \sqrt{y} = \sqrt{z}$

    (d) $k = 1/3$: $\sqrt[3]{x} + \sqrt[3]{y} = \sqrt[3]{z}$.

5. Prove the propositions.

    (a) Both $x^4 + y^2 = z^2$ and $x^2 + y^2 = z^4$ have infinitely many solutions in positive integers satisfying $(x, y, z) = 1$.

  S* (b) The equation $x^4 + y^4 = z^2$ has no solutions in positive integers.
      Remark: Part (b) yields another proof for the case $k = 4$ of Fermat's Last Theorem.

* 6. Solve the Diophantine equation $x^4 - 2y^2 = -1$.

S** 7. In which number systems is the integer 1111 a square?

8. Which Eulerian integers divide their conjugates?

9. Verify the identity

$$(a^2 - ab + b^2)(c^2 - cd + d^2) =$$
$$= (ac - bd)^2 - (ac - bd)(ad + bc - bd) + (ad + bc - bd)^2$$

for any real numbers a, b, c, and d.

10. **S** (a) Prove that the Diophantine equations $x^2 - xy + y^2 = n$ and $x^2 + 3y^2 = n$ are solvable for exactly the same positive integers n.

    * (b) For which values of n are the equations in part (a) solvable and what is the number of solutions?

S* 11. Solve the Diophantine equation $x^2 + 243 = y^3$.

12. Let $\mu \neq 0$ be an Eulerian integer. The Eulerian integers $\varrho_1, \ldots, \varrho_r$ form a complete residue system modulo μ if any Eulerian integer α satisfies the congruence $\alpha \equiv \varrho_i$ (mod μ) for exactly one ϱ_i. Show:

    * (a) There are $N(\mu)$ elements in a complete residue system modulo μ.

    (b) If $N(\mu)$ is a prime number p, then $0, 1, 2, \ldots, N(\mu) - 1$ form a complete residue system modulo μ.

(c) The analog of Fermat's Little Theorem is true:

$$\alpha^{N(\pi)} \equiv \alpha \pmod{\pi}$$

for any Eulerian integer α and Eulerian prime π.

13. Solve the Diophantine equation

$$\frac{u}{v} + \frac{v}{w} = \frac{w}{u}.$$

14. (a) Show that if the side lengths of a right triangle are integers, then the area of the triangle cannot be a square.

(b) Prove that if the side lengths of a right triangle are pairwise coprime integers, then the area of the triangle cannot be a cube.

(c) Does there exist a right triangle with integer side lengths whose area is a cube?

(d) Generalize the problem for higher powers.

7.8. Pell's Equation

Pell's equation is a Diophantine equation of the form

$$(7.8.1) \qquad x^2 - my^2 = 1$$

where m is a positive integer that is not a square. Equation (7.8.1) has two trivial solutions $x = \pm 1, y = 0$; the other solutions (with $y \neq 0$) are called non-trivial solutions.

We can factor the left-hand side of (7.8.1):

$$(7.8.2) \qquad (x + y\sqrt{m})(x - y\sqrt{m}) = 1.$$

So if the pair x, y satisfies (7.8.1), then the (real) numbers $x + y\sqrt{m}$ and $x - y\sqrt{m}$ divide 1 in the ring of numbers of the form $a + b\sqrt{m}$ where a and b are integers, i.e. they are units in this ring. We know that any integer power of a unit is a unit, thus if there exists a unit $\varepsilon \neq \pm 1$, then the powers of ε produce infinitely many units. Returning to Pell's equation, this shows that if (7.8.1) has a non-trivial solution, then there are infinitely many solutions. The special cases $m = 2$ and $m = 3$ occurred essentially in Exercise 1.1.22 and in the proof of Theorem 5.2.4.

Now we show that every Pell's equation has infinitely many solutions (it would be sufficient to prove the existence of at least one non-trivial solution, as discussed above). Then we characterize how to obtain all solutions.

We note that (7.8.1) behaves entirely differently if $m \leq 0$ or $m = k^2$, see Exercise 7.8.1.

Theorem 7.8.1. *Let m be a positive integer that is not square. Then the Diophantine equation (7.8.1) has infinitely many solutions.* ♣

In the proof, we shall rely on Theorem 8.1.1 from the next chapter.

Proof. If $y \neq 0$, then the equivalent form (7.8.2) of (7.8.1) can be written as

$$(7.8.3) \qquad \frac{x}{y} - \sqrt{m} = \frac{1}{y(x + y\sqrt{m})}.$$

We see from (7.8.3) that a pair of positive integers x and y can give a solution only if x/y is very close to $\sqrt{m}$: (7.8.3) implies

(7.8.4)
$$\left| \sqrt{m} - \frac{x}{y} \right| < \frac{1}{y^2}.$$

As $\sqrt{m}$ is irrational, it follows from Theorem 8.1.1 that (7.8.4) holds for infinitely many pairs of integers x, y. Based on this fact, we prove that (7.8.2) has infinitely many solutions. (Conditions (7.8.4) and (7.8.2) are not equivalent; only a small number of x and y satisfying (7.8.4) will be solutions of (7.8.2)).

I. Our first step is to show that there exists an integer $t \neq 0$ such that the Diophantine equation

(7.8.5)
$$x^2 - my^2 = t$$

has infinitely many solutions.

Let c_j, d_j $(j = 1, 2, \ldots)$ be infinitely many pairs of positive integers satisfying (7.8.4), so

$$\left| \sqrt{m} - \frac{c_j}{d_j} \right| < \frac{1}{d_j^2}, \qquad j = 1, 2, \ldots.$$

Then

(7.8.6)
$$|c_j^2 - md_j^2| = d_j^2 \left| \frac{c_j}{d_j} - \sqrt{m} \right| \cdot \left| \frac{c_j}{d_j} + \sqrt{m} \right| < \left| \frac{c_j}{d_j} + \sqrt{m} \right| =$$
$$= \left| \frac{c_j}{d_j} - \sqrt{m} + 2\sqrt{m} \right| < \frac{1}{d_j^2} + 2\sqrt{m} \leq 1 + 2\sqrt{m}.$$

(7.8.6) implies that all values $c_j^2 - md_j^2$ are integers in the interval $(-1 - 2\sqrt{m}, 1 + 2\sqrt{m})$, and none of them is 0 as $\sqrt{m}$ is irrational. By the pigeonhole principle, there must be an integer $t \neq 0$ in this interval for which

$$c_j^2 - md_j^2 = t$$

holds for infinitely many pairs c_j, d_j. This means that the Diophantine equation (7.8.5) has infinitely many solutions.

II. Now we prove that the quotients of suitable solutions of (7.8.5) yield solutions for equation (7.8.1).

Let $x = a_1, y = b_1$ and $x = a_2, y = b_2$ be two solutions of (7.8.5), so

(7.8.7a)
$$a_1^2 - mb_1^2 = (a_1 + b_1\sqrt{m})(a_1 - b_1\sqrt{m}) = t$$

(7.8.7b)
$$a_2^2 - mb_2^2 = (a_2 + b_2\sqrt{m})(a_2 - b_2\sqrt{m}) = t$$

and assume

(7.8.8)
$$a_1 \equiv a_2 \pmod{|t|} \quad \text{and} \quad b_1 \equiv b_2 \pmod{|t|}.$$

If (7.8.7a), (7.8.7b), and (7.8.8) hold, then the pairs a_1, b_1 and a_2, b_2 are called modulo $|t|$ congruent solutions of (7.8.5).

Dividing (7.8.7a) by (7.8.7b), we obtain

$$(7.8.9) \qquad \frac{a_1 + b_1\sqrt{m}}{a_2 + b_2\sqrt{m}} \cdot \frac{a_1 - b_1\sqrt{m}}{a_2 - b_2\sqrt{m}} = 1.$$

The first fraction on the left-hand side of (7.8.9) can be written as

$$\frac{a_1 + b_1\sqrt{m}}{a_2 + b_2\sqrt{m}} = u + v\sqrt{m}$$

where u and v are rational numbers, and then the second fraction is necessarily

$$\frac{a_1 - b_1\sqrt{m}}{a_2 - b_2\sqrt{m}} = u - v\sqrt{m}.$$

We show that u and v are integers if (7.8.8) holds, so u and v provide an integer solution of (7.8.1).

By eliminating the square root in the denominator and using (7.8.7b), we obtain

$$\frac{a_1 + b_1\sqrt{m}}{a_2 + b_2\sqrt{m}} = \frac{(a_1 + b_1\sqrt{m})(a_2 - b_2\sqrt{m})}{a_2^2 - mb_2^2} = \frac{(a_1 + b_1\sqrt{m})(a_2 - b_2\sqrt{m})}{t}.$$

Thus we have to verify that in

$$(a_1 + b_1\sqrt{m})(a_2 - b_2\sqrt{m}) = r + s\sqrt{m}$$

r and s are divisible by t. This follows from (7.8.8) and (7.8.7a):

$$r + s\sqrt{m} = (a_1 + b_1\sqrt{m})(a_2 - b_2\sqrt{m}) \equiv$$

$$\equiv (a_1 + b_1\sqrt{m})(a_1 - b_1\sqrt{m}) = t \equiv 0 \pmod{|t|}.$$

(Congruences for numbers $a + b\sqrt{m}$ are in the usual natural sense.)

III. Since there are $|t|$ possible remainders modulo $|t|$ for each of a and b, the number of pairwise incongruent solutions of (7.8.5) is at most t^2. Applying the pigeonhole principle again, we infer that among the infinitely many solutions of (7.8.5), there are infinitely many such that any two of them are congruent modulo $|t|$. Let $x = f_i, y = g_i$, $i = 1, 2, \ldots$ be such solutions.

Then, according to part II, the values r_i, s_i arising as quotients of f_i, g_i and f_1, g_1 will give infinitely many distinct integer solutions of (7.8.1). $\qquad\square$

Theorem 7.8.2. *Let m be a positive integer which is not a square and x_0, y_0 the uniquely determined solution of the Diophantine equation (7.8.1) where $x_0 > 0$, $y_0 > 0$, and $x_0 + y_0\sqrt{m}$ is minimal. Then all solutions are the pairs of integers x, y such that*

$$(7.8.10) \qquad x + y\sqrt{m} = \pm(x_0 + y_0\sqrt{m})^n, \qquad n = 0, \pm1, \pm2, \ldots \qquad \clubsuit$$

By (7.8.2),

$$(7.8.11) \qquad (x_0 + y_0\sqrt{m})^{-n} = (x_0 - y_0\sqrt{m})^n,$$

thus an equivalent form of (7.8.10) is

$$x + y\sqrt{m} = \pm(x_0 \pm y_0\sqrt{m})^n, \qquad n = 0, 1, 2, \ldots.$$

For $n = 0$, we obtain the two trivial solutions.

Proof. We shall apply several times the fact that the product of two solutions gives a solution.

Assume that both x_1, y_1, and x_2, y_2 are solutions of (7.8.1), i.e.

(7.8.12a) $$(x_1 + y_1\sqrt{m})(x_1 - y_1\sqrt{m}) = 1$$

(7.8.12b) $$(x_2 + y_2\sqrt{m})(x_2 - y_2\sqrt{m}) = 1.$$

Multiplying (7.8.12a) and (7.8.12b), we get

$$(x_1 x_2 + m y_1 y_2 + (x_1 y_2 + y_1 x_2)\sqrt{m})(x_1 x_2 + m y_1 y_2 - (x_1 y_2 + y_1 x_2)\sqrt{m}) = 1.$$

This means that

$$x_3 = x_1 x_2 + m y_1 y_2, \qquad y_3 = x_1 y_2 + y_1 x_2$$

is also a solution of (7.8.1). (In the formulation for units indicated in the introduction, this corresponds to the fact that the product of two units is a unit.)

The above and (7.8.11) imply that the pairs of integers x, y defined by (7.8.10) satisfy (7.8.1).

Now we show that there are no other solutions. Assume that x, y is a solution not of this type. Then also $-x, -y$ is a solution not listed in (7.8.10). Hence we may assume $x + y\sqrt{m} > 0$.

Then

(7.8.13) $$(x_0 + y_0\sqrt{m})^k < x + y\sqrt{m} < (x_0 + y_0\sqrt{m})^{k+1}$$

for some integer k. Multiplying (7.8.13) by $(x_0 - y_0\sqrt{m})^k$, we obtain

(7.8.14) $$1 < (x + y\sqrt{m})(x_0 - y_0\sqrt{m})^k < x_0 + y_0\sqrt{m}.$$

As we multiplied two solutions,

$$(x + y\sqrt{m})(x_0 - y_0\sqrt{m})^k = x' + y'\sqrt{m},$$

x', y' is a solution too, so

(7.8.15) $$(x' + y'\sqrt{m})(x' - y'\sqrt{m}) = 1.$$

By the first inequality in (7.8.14),

(7.8.16a) $$x' + y'\sqrt{m} > 1,$$

thus

(7.8.16b) $$0 < x' - y'\sqrt{m} < 1,$$

by (7.8.15). According to (7.8.16b), the cases $y' = 0$; $x' < 0, y' > 0$; and $x' > 0, y' < 0$ are impossible, and (7.8.16a) excludes $x' < 0, y' < 0$. Therefore $x' > 0, y' > 0$, but then (7.8.14) contradicts the minimality of $x_0 + y_0\sqrt{m}$.   □

Exercises 7.8

1. Determine all solutions of the Diophantine equation $x^2 - my^2 = 1$ if $m \leq 0$ or m is a square.

2. In how many cases does it occur that writing a digit 1 after a square (in decimal system), gives another square?

3. Let m be a positive integer that is not a square, and r be any integer different from 0. Prove that if the Diophantine equation $x^2 - my^2 = r$ is solvable, then it has infinitely many solutions.

4. (a) How many squares are (a1) greater by 1 (a2) smaller by 1
     than the double of a square?

   (b) Investigate the question when double is replaced by triple.

5. For how many integers n is $\binom{n}{2}$ a square?

6. What is the number of pairwise incongruent right triangles whose leg lengths are consecutive integers and the length of the hypotenuse is an integer?

7. Determine the number of solutions of the Diophantine equations

   (a) $x^2 - 3y^2 = 2$
   (b) $x^2 - 3y^2 = 7$
   (c) $x^2 - 3y^2 = 13$
   (d) $x^2 - 3y^2 = 39$
   (e) $2x^2 - 3y^2 = 1$
   (f) $3x^2 - 2y^2 = 1$.

* 8. For which primes $p > 0$ is the Diophantine equation $x^2 - py^2 = -1$ solvable?

9. Let a, b, and c pairwise coprime non-zero integers, and assume that the Diophantine equation $ax^2 + by^2 + cz^2 = 0$ has a non-trivial solution, one differing from $x = y = z = 0$. Verify that the signs of a, b, and c cannot be all the same and the congruences

   $$u^2 \equiv -bc \pmod{|a|}, \qquad v^2 \equiv -ac \pmod{|b|}, \qquad w^2 \equiv -ab \pmod{|c|}$$

   are solvable.

   Remark: It can be shown that these conditions are not only necessary but also sufficient for the non-trivial solvability of the Diophantine equation $ax^2 + by^2 + cz^2 = 0$.

10. For how many integers k is $2 + 2\sqrt{28k^2 + 1}$ a square?

* 11. Prove that Pell's equation $x^2 - 2y^2 = 1$ has no non-trivial solutions when x or y is a square.

7.9. Partitions

Definition 7.9.1. A *partition* of a positive integer n is a representation of n as a sum of positive integers, allowing also a one-term sum. We consider only the essentially different representations, where the order of the terms is irrelevant.

The number of partitions of n is denoted by $p(n)$.   ♣

Example. All partitions of 4 are

$$4 = 3 + 1 = 2 + 2 = 2 + 1 + 1 = 1 + 1 + 1 + 1,$$

so $p(4) = 5$.

We state without proof the asymptotic behavior of $p(n)$: as $n \to \infty$,

$$p(n) \sim \frac{ce^{d\sqrt{n}}}{n} \quad \text{where} \quad c = \frac{1}{4\sqrt{3}} \quad \text{and} \quad d = \frac{\pi\sqrt{6}}{3}.$$

We often investigate special partitions where there are some restrictions on the summands or on their number: e.g. we prescribe that each summand should be odd or the terms should be all distinct, etc.

The basic tools for handling partitions are generating functions. As an illustration, we consider the money changing problem: In how many ways can we pay n dollars with banknotes of denomination less than (say) 50 dollars, including the rare two dollar bills? We want partitions of n into summands 1, 2, 5, 10, and 20 only. Let us denote the number of such partitions by $f(n)$.

We rewrite the problem. Let $u_1, \ldots, u_5$ denote the numbers of 1, 2, 5, 10, and 20 dollar bills, when paying n dollars. Then $f(n)$ is the number of non-negative integer solutions of the Diophantine equation

(7.9.1) $1u_1 + 2u_2 + 5u_3 + 10u_4 + 20u_5 = n.$

The generating function of $f(n)$ is the power series

(7.9.2) $$F(x) = 1 + \sum_{n=1}^{\infty} f(n)x^n.$$

We show first that the series is absolutely convergent for $|x| < 1/2$.

Since $0 \le u_i \le n$ for every i in (7.9.1),

$$0 \le f(n) \le (n + 1)^5.$$

It is easy to see that $(n+1)^5 < 2^n$ for n large enough, thus, for $|x| < 1/2$, the convergent infinite geometric series

$$\sum_{n=0}^{\infty} (2|x|)^n$$

is a majorant of the infinite series (7.9.2). Therefore $F(x)$ is absolutely convergent for $|x| < 1/2$. It can also be proved that this holds for $|x| < 1$.

We write $F(x)$ as a product of convergent geometric series, still assuming $|x| < 1/2$:

$$F(x) = \left(1 + x + x^2 + \dots\right)\left(1 + x^2 + (x^2)^2 + \dots\right)\left(1 + x^5 + (x^5)^2 + \dots\right) \cdot$$

(7.9.3)

$$\cdot \left(1 + x^{10} + (x^{10})^2 + \dots\right)\left(1 + x^{20} + (x^{20})^2 + \dots\right).$$

As the power series expansion of a function around 0 is unique, we have to show that multiplying the finitely many absolutely convergent series on the right-hand side of (7.9.3) we get $f(n)$ as the coefficient of x^n. We map a representation (7.9.1) of n into a product of the terms

$$x^{u_1}, \qquad (x^2)^{u_2}, \qquad (x^5)^{u_3}, \qquad (x^{10})^{u_4}, \qquad (x^{20})^{u_5}$$

on the right-hand side of (7.9.3). This product is

$$x^{u_1}(x^2)^{u_2}(x^5)^{u_3}(x^{10})^{u_4}(x^{20})^{u_5} = x^{1u_1 + 2u_2 + 5u_3 + 10u_4 + 20u_5} = x^n.$$

Thus we established a bijection between the representations and the products, so after performing the multiplication, the coefficient of x^n is $f(n)$.

Using the summation formula for geometric series, we can write (7.9.3) into the form

$$F(x) = \frac{1}{(1-x)(1-x^2)(1-x^5)(1-x^{10})(1-x^{20})} \qquad \text{(for } |x| < 1/2\text{)}.$$

We obtain the following more general result exactly the same way.

Theorem 7.9.2. *Let $a_1, a_2, \dots, a_r$ be distinct positive integers, and let $f(n)$ denote the number of partitions of the positive integer n using no summands other than $a_1, a_2, \dots, a_r$. Then the infinite series $1 + \sum_{n=1}^{\infty} f(n)x^n$ is absolutely convergent for $|x| < 1/2$, and*

$$1 + \sum_{n=1}^{\infty} f(n)x^n = \prod_{i=1}^{r} \frac{1}{1 - x^{a_i}}. \qquad \clubsuit$$

We obtain the generating function of $p(n)$ similarly.

Theorem 7.9.3.

(7.9.4) $$P(x) = 1 + \sum_{n=1}^{\infty} p(n)x^n = \prod_{i=1}^{\infty} \frac{1}{1 - x^i} \qquad \text{(for } |x| < 1\text{)}. \qquad \clubsuit$$

The infinite product on the right-hand side of (7.9.4) is the limit (as seen in Exercises 5.6.6 and 5.6.7)

$$\prod_{i=1}^{\infty} \frac{1}{1 - x^i} = \lim_{r \to \infty} \prod_{i=1}^{r} \frac{1}{1 - x^i}.$$

To prove Theorem 7.9.3, we apply Theorem 7.9.2 with $a_i = i$, and then take the limit when $r \to \infty$. We leave the details to the reader.

Besides generating functions, combinatorial arguments can be applied to partitions. We can plot a partition $n = a_1 + a_2 + \dots + a_r$ satisfying $a_1 \geq a_2 \geq \dots \geq a_r$ as a scheme with a_1 points in the first row, a_2 points in the second row, etc., as the scheme

(7.9.5)

$$
\begin{array}{ccccc}
\bullet & \bullet & \bullet & \bullet & \bullet \\
\bullet & \bullet & \bullet \\
\bullet & \bullet & \bullet \\
\bullet
\end{array}
$$

corresponds to the partition $12 = 5 + 3 + 3 + 1$. It is obvious from the definition that no row can be longer than the row above it.

Thus the rows correspond to the terms of the partition. We can look at the scheme also according to its columns. So scheme (7.9.5) gives the partition $12 = 4+3+3+1+1$. The two interpretations of the schemes yield the result:

Theorem 7.9.4. *Let $g_r(n)$ and $h_r(n)$, be the number of partitions of n where the number of terms and the largest term is r, resp. Then $g_r(n) = h_r(n)$.*   ♣

Proof. Consider the schemes of n points with exactly r rows. Counting the points in the scheme by rows, we get a partition of n into r terms. Considering the scheme by columns, we have a partition of n where the largest term is r. Doing this for all schemes, we infer $g_r(n) = h_r(n)$.   □

Below we investigate partitions of n into an even or odd number of pairwise distinct terms. The next theorem of Euler shows that the difference between the numbers of these two types of partitions is at most 1.

Theorem 7.9.5. *Let $e(n)$ and $o(n)$, be the number of partitions of the positive integer n where all terms are distinct and the number of summands is even or odd. Then*

(7.9.6) $$e(n) - o(n) = \begin{cases} (-1)^k, & \text{if } n = \frac{1}{2}(3k^2 \pm k) \\ 0, & \text{otherwise.} \end{cases}$$ ♣

Example. The partitions of $n = 7$ into distinct terms are

$$6 + 1 = 5 + 2 = 4 + 3 \quad \text{where the number of terms is even}$$
$$7 = 4 + 2 + 1 \quad \text{where the number of terms is odd,}$$

so $e(7) = 3$ and $o(7) = 2$. The equality $e(7) - o(7) = 1 = (-1)^2$ agrees with (7.9.6) as $7 = \frac{1}{2}(3 \cdot 2^2 + 2)$.

Proof. We shall establish an almost bijection between the partitions of n into an even or odd number of distinct summands.

Partitions into distinct terms correspond to schemes where the numbers of elements in the rows are strictly decreasing downwards, as the partition $23 = 7+6+5+3+2$ is represented by

(7.9.7)

$$
\begin{array}{ccccccc}
\bullet & \bullet & \bullet & \bullet & \bullet & \bullet & \bullet \\
\bullet & \bullet & \bullet & \bullet & \bullet & \bullet \\
\bullet & \bullet & \bullet & \bullet & \bullet \\
\bullet & \bullet & \bullet \\
\bullet & \bullet
\end{array}
$$

Let us call the edge of such a partition the longest line of points starting from the upper-right corner and running 45 degrees from northeast to southwest. The edge of scheme (7.9.7) consists of three points. The length of the edge depends on how long the terms decrease one by one, and an edge may contain just a single point.

Let U be the transformation that transfers the edge of a scheme under the last row, creating a new last row provided that we again get a partition into distinct terms, so the new scheme consists of rows with strictly decreasing numbers of points. Similarly, let E be the transformation that transfers the last row of a scheme near the edge (obliquely, as a new edge) provided this creates an appropriate new scheme. Applying E to (7.9.7), we obtain

$$
\begin{matrix}
\bullet & \bullet & \bullet & \bullet & \bullet & \bullet & \bullet & \bullet \\
\bullet & \bullet & \bullet & \bullet & \bullet & \bullet & \bullet \\
\bullet & \bullet & \bullet & \bullet & \bullet \\
\bullet & \bullet & \bullet
\end{matrix}
$$

but U cannot be applied.

We show that apart from a few exceptions, any scheme allows exactly one of U and E to be applied.

Let the number of points be u in the last row and e in the edge.

If $u \le e$, then U cannot be applied, but E can, except if $u = e$ and the last row and the edge have a common point; in this case neither U, nor E can be applied:

(∗)
$$
\begin{matrix}
\bullet & \bullet & \bullet & \bullet & \bullet & \bullet & \bullet \\
\bullet & \bullet & \bullet & \bullet & \bullet & \bullet \\
\bullet & \bullet & \bullet & \bullet & \bullet \\
\bullet & \bullet & \bullet & \bullet
\end{matrix}
$$

If $u > e$, then E is impossible, but U works, except if $u = e + 1$ and the last row and the edge share a point; in this case neither U, nor E can be applied:

(∗∗)
$$
\begin{matrix}
\bullet & \bullet & \bullet & \bullet & \bullet & \bullet \\
\bullet & \bullet & \bullet & \bullet & \bullet \\
\bullet & \bullet & \bullet & \bullet
\end{matrix}
$$

Transformation U increases the number of rows by 1, whereas E reduces it by 1, so the numbers of terms in the original and new partitions have opposite parity. Further, U and E are inverses of each other so their composition in any order restores the original scheme. This means that the pair of transformations U and E establishes, apart from partitions of type (∗) and (∗∗), a bijection between the partitions of n into distinct summands with an even and odd number of terms. This proves $e(n) = o(n)$ except if n has a partition of type (∗) or (∗∗) when $e(n) - o(n)$ is 1 or -1, depending on whether the bad partition consists of an even or odd number of terms (we have to verify that a given n cannot have more than one bad partitions).

If the partition (∗) contains k terms, i.e. the scheme has k rows, then

$$(7.9.8) \qquad n = (2k - 1) + (2k - 2) + \cdots + k = \frac{(3k - 1)k}{2}.$$

We obtain similarly that if the partition (**) consists of k terms, then

(7.9.9) $$n = 2k + (2k - 1) + \cdots + (k + 1) = \frac{(3k + 1)k}{2}.$$

For a given n, neither (7.9.8) nor (7.9.9) can be true for more values of k and n cannot be simultaneously of the form (7.9.8) and (7.9.9) as

$$\frac{(3k - 1)k}{2} = \frac{(3j + 1)j}{2} \iff (3k - 3j - 1)(k + j) = 0,$$

which is impossible for positive integers k and j.

Thus (7.9.8) and (7.9.9) determine the exceptional integers n, and every such n has only one bad transformation. This verifies (7.9.6). $\qquad\square$

As indicated before, Theorem 7.9.5 has important consequences concerning $p(n)$. If $v(n) = e(n) - o(n)$, then the generating function of $v(n)$ is

(7.9.10)
$$V(x) = 1 + \sum_{n=1}^{\infty} v(n)x^n = 1 + \sum_{k=1}^{\infty} (-1)^k \left(x^{\frac{1}{2}(3k^2+k)} + x^{\frac{1}{2}(3k^2-k)} \right) =$$
$$= 1 - x - x^2 + x^5 + x^7 - x^{12} - x^{15} + \ldots,$$

by Theorem 7.9.5. This infinite series is absolutely convergent for $|x| < 1/2$.

On the other hand, we can obtain $V(x)$ as an infinite product, convergent for $|x| < 1/2$:

(7.9.11) $$V(x) = \prod_{i=1}^{\infty} (1 - x^i) = \lim_{r \to \infty} \prod_{i=1}^{r} (1 - x^i).$$

To verify (7.9.11), we consider the product

(7.9.12) $$\prod_{i=1}^{r} (1 - x^i).$$

Performing the multiplication, we get terms of the type

(7.9.13) $$(-x^{i_1})(-x^{i_2}) \ldots (-x^{i_j}) = (-1)^j x^{i_1 + i_2 + \cdots + i_j}$$

where $0 \le j \le r$ and $i_1, \ldots, i_j$ are distinct positive integers not greater than r. (For $j = 0$, we obtain 1 corresponding to the empty product.)

We perform the multiplication in (7.9.12). By (7.9.13), every partition of n with an even or odd number of distinct summands not greater than r generates a term x^n with coefficient $+1$ or -1. If $1 \le n \le r$, then any partition of n can contain only terms not greater than r. This means that for $r \ge n$, expanding (7.9.12) into a polynomial the coefficient of x^n is precisely $e(n) - o(n) = v(n)$.

Finally, we can deduce (7.9.11) taking the limit for $r \to \infty$. We leave the details to the reader.

Theorem 7.9.3 and (7.9.11) imply that the generating functions of $p(n)$ and $v(n)$ are reciprocals of each other:

(7.9.14) $$\left(1 + \sum_{n=1}^{\infty} p(n)x^n \right)\left(1 + \sum_{n=1}^{\infty} v(n)x^n \right) = P(x)V(x) = 1.$$

Thus multiplying the two power series on the left-hand side of (7.9.14), the coefficient of x^n is 0 for every $n \geq 1$ so

(7.9.15) $p(n) + p(n-1)v(1) + p(n-2)v(2) + \cdots + p(1)v(n-1) + v(n) = 0.$

Substituting the values $v(j)$ determined in Theorem 7.9.5 into (7.9.15), we obtain the recursion

(7.9.16) $p(n) = p(n-1) + p(n-2) - p(n-5) - p(n-7) + p(n-12) + \ldots$

We can observe from the right-hand side of (7.9.16) that the recursion contains only about $2\sqrt{2n/3}$ terms, so we can use it to compute $p(n)$ effectively even for relatively large values of n, e.g.

$$p(200) = 3972999029388.$$

Exercises 7.9

1. Prove $p(n+1) \leq 2p(n)$. When does equality hold?

2. Compute the limits

    (a) $\lim_{n \to \infty} (p(n+1) - p(n))$

    (b) $\lim_{n \to \infty} (p(n+1) - 2p(n)).$

3. Which integers have an odd number of partitions into (pairwise) distinct terms?

4. What is the number of representations of n as the sum of positive integers if we consider two representations distinct if they differ in the order of terms?

5. Show that the number of partitions of n into exactly r terms is the same as the number of partitions of $n - r$ into at most r terms.

6. Exhibit the generating function of $h_r(n)$ in Theorem 7.9.4.

7. (a) Let $u(n)$ be the number of partitions of n into pairwise distinct positive integers, and $w(n)$ be the number of partitions into odd, but not necessarily distinct, positive integers. Prove $u(n) = w(n)$.

    (b) (Generalization of part (a).) Let $u_k(n)$ be the number of partitions of n where no integer can occur k times among the summands, and let $w_k(n)$ be the number of partitions where none of the summands is a multiple of k. Then $u_k(n) = w_k(n)$.

8. Verify

$$\sum_{n=1}^{\infty} p(n)x^n = \sum_{r=1}^{\infty} \frac{x^r}{(1-x)(1-x^2)\ldots(1-x^r)}$$

    for $|x| < 1/2$.

** 9. Prove the identity

$$\sigma(n) - \sigma(n-1) - \sigma(n-2) + \sigma(n-5) + \sigma(n-7) - \cdots = \begin{cases} (-1)^{k+1}n, & \text{if } n = \frac{1}{2}(3k^2 \pm k) \\ 0, & \text{otherwise.} \end{cases}$$

Diophantine Approximation

In this chapter we investigate how close irrational numbers can be to rational numbers. The closeness is expressed in terms of the denominator s of the approximating fraction. It turns out that a typical irrational number can be best approximated to an order of magnitude $1/s^2$. To handle the problem, we also use continued fractions and Minkowski's basic theorem in the geometry of numbers. Finally, we deal with the distribution of fractional parts of certain sequences. Diophantine approximation is related to Pell's equation (we used Theorem 8.1.1 in the proof of Theorem 7.8.1), and further applications will appear in the next chapter.

8.1. Approximation of Irrational Numbers

The rational numbers are everywhere dense on the number line, so there are infinitely many rational numbers in any arbitrarily small neighborhood of an irrational number. In this chapter we deal with approximation in a stronger sense, when the difference of the irrational number and the approximating fraction is small as a function of the denominator of the fraction. We have the basic result:

Theorem 8.1.1. *For any irrational number α, there exist infinitely many fractions r/s satisfying*

$$(8.1.1) \qquad \left| \alpha - \frac{r}{s} \right| < \frac{1}{s^2}. \qquad \clubsuit$$

Remark: We always assume $s > 0$ for the approximating rational number r/s. It is clear that if (8.1.1) holds for a fraction with $(r, s) > 1$, then this is even more true for the fraction r'/s' in lowest terms obtained after cancellation (since $s' < s$). On the other hand, we can easily verify (see Exercise 8.1.2) that only finitely many forms r/s of the same rational number can satisfy (8.1.1). By the above, Theorem 8.1.1 and later similar theorems remain equally valid whether we speak about infinitely many distinct rational numbers r/s, or infinitely many distinct fractional forms r/s (in the latter case we count the different forms r/s of the same rational number as distinct). Similarly, the assertion remains true also if we require $(r, s) = 1$ for the approximating fractions r/s.

To prove Theorem 8.1.1, we need

Theorem 8.1.2. *Let α be a real number and n a positive integer. Then there exists at least one fraction r/s satisfying*

$$(8.1.2) \qquad\qquad 1 \le s \le n \quad and \quad \left|\alpha - \frac{r}{s}\right| < \frac{1}{ns}. \qquad\qquad \clubsuit$$

Proof. The *fractional part* of a real number c is $\{c\} = c - \lfloor c \rfloor$. For example, $\{3\} = 0$; $\{2.9\} = 0.9$; $\{-2.9\} = 0.1$. Clearly, $0 \le \{c\} < 1$.

We consider the fractional parts

$$\{\alpha\}, \{2\alpha\}, \dots, \{(n+1)\alpha\}.$$

They are in the interval $[0, 1)$.

We partition the interval $[0, 1)$ into n subintervals of length $1/n$, each closed on the left and open on the right. There are $n + 1$ fractional parts $\{j\alpha\}$ and n subintervals. By the pigeonhole principle, there is a subinterval containing at least two fractional parts, so the distance between them is less than $1/n$, so

$$(8.1.3) \qquad\qquad |\{j\alpha\} - \{i\alpha\}| < \frac{1}{n}$$

for some $1 \le i < j \le n + 1$. We can rewrite (8.1.3) as

$$(8.1.4) \qquad |(j\alpha - \lfloor j\alpha \rfloor) - (i\alpha - \lfloor i\alpha \rfloor)| = |(j - i)\alpha - (\lfloor j\alpha \rfloor - \lfloor i\alpha \rfloor)| < \frac{1}{n}.$$

Let

$$s = j - i \quad and \quad r = \lfloor j\alpha \rfloor - \lfloor i\alpha \rfloor.$$

Then dividing (8.1.4) by s, we get the statement of the theorem. $\square$

Proof of Theorem 8.1.1. Observe that $1 \le s \le n$ in (8.1.2) guarantees

$$\left|\alpha - \frac{r}{s}\right| < \frac{1}{ns} \le \frac{1}{s^2}.$$

Applying Theorem 8.1.2 with α and any positive integer $n = n_1$, we get a fraction r_1/s_1 satisfying

$$\left|\alpha - \frac{r_1}{s_1}\right| < \frac{1}{s_1^2}.$$

Now we repeat this step with a suitable n_2 instead of n_1; we obtain an approximating fraction r_2/s_2. We will show that r_2/s_2 is distinct from r_1/s_1.

Since α is irrational, $\alpha - r_1/s_1 \ne 0$, thus we can choose n_2 to satisfy

$$\left|\alpha - \frac{r_1}{s_1}\right| > \frac{1}{n_2}.$$

By Theorem 8.1.2,

$$\left|\alpha - \frac{r_2}{s_2}\right| < \frac{1}{n_2 s_2} \le \frac{1}{n_2} < \left|\alpha - \frac{r_1}{s_1}\right|,$$

hence

$$\frac{r_2}{s_2} \ne \frac{r_1}{s_1}.$$

Continuing the procedure, we get infinitely many distinct suitable fractions r_i/s_i. $\square$

Remark: If α is rational, then α can be best approximated by itself, obviously. Despite this, the question of approximation is not completely uninteresting even for rational α—e.g. we might need a good approximation with fractions having small denominators both for theoretical and practical purposes. In contrast with the order of magnitude $1/s^2$ for irrational numbers, a rational α can be best approximated, excluding α itself from the approximating fractions, only with order of magnitude c/s, where c is a constant depending on α (see Exercise 8.1.1).

The next theorem is about the simultaneous approximation of more irrational numbers with fractions having the same denominator:

Theorem 8.1.3. *For any irrational numbers $\alpha_1, \ldots, \alpha_k$, there exist infinitely many rational k-tuples with a common denominator*

$$\frac{r_{1i}}{s_i}, \quad \frac{r_{2i}}{s_i}, \quad \ldots, \quad \frac{r_{ki}}{s_i}, \qquad i = 1, 2, \ldots$$

satisfying

(8.1.5) $$\left| \alpha_j - \frac{r_{ji}}{s_i} \right| < \frac{1}{s_i^{1 + \frac{1}{k}}}, \qquad j = 1, 2, \ldots, k, \qquad i = 1, 2, \ldots. \qquad \clubsuit$$

Theorem 8.1.3 can be verified similarly to the proof of Theorem 8.1.1; we require now a k-dimensional version of Theorem 8.1.2:

Theorem 8.1.4. *Let $\alpha_1, \ldots, \alpha_k$ be real numbers and n a positive integer. Then there exist integers $r_1, \ldots, r_k$, and s satisfying*

$$1 \le s \le n^k \quad and \quad \left| \alpha_j - \frac{r_j}{s} \right| < \frac{1}{ns}, \qquad j = 1, 2, \ldots, k. \qquad \clubsuit$$

We leave the details of the proofs to the reader.

We state a sharper version of Theorem 8.1.1 without proof:

Theorem 8.1.5. *For any irrational number α, there exist infinitely many fractions r/s satisfying*

$$\left| \alpha - \frac{r}{s} \right| < \frac{1}{\sqrt{5} s^2}. \qquad \clubsuit$$

We shall verify a slightly weaker statement with 2 instead of $\sqrt{5}$ by two different methods in Sections 8.2 and 8.3.

Theorem 8.1.5 cannot be improved:

Theorem 8.1.6. *Let $\varepsilon > 0$ be arbitrary and $\alpha = (1 + \sqrt{5})/2$. Then only finitely many fractions r/s can satisfy*

(8.1.6) $$\left| \alpha - \frac{r}{s} \right| < \frac{1}{(\sqrt{5} + \varepsilon) s^2}. \qquad \clubsuit$$

Proof. To achieve a contradiction, assume that (8.1.6) holds for infinitely many r/s. The distance between fractions with a given denominator s is (at least) $1/s$, and $s \ge 1$

implies

$$\frac{1}{s} > \frac{2}{(\sqrt{5} + \varepsilon)s^2},$$

so for a given s, (8.1.6) can be valid with at most one r.

Hence, there must occur arbitrarily large integers among the denominators of the infinitely many fractions r/s satisfying (8.1.6).

As α is a root of equation $x^2 - x - 1 = 0$, we have $\alpha(\alpha - 1) = 1$. This helps us to eliminate the square root in α on the left-hand side of (8.1.6):

(8.1.7) $\left(\alpha - \frac{r}{s}\right)\left((\alpha - 1) + \frac{r}{s}\right) = \alpha(\alpha - 1) + \frac{r}{s}(\alpha - (\alpha - 1)) - \frac{r^2}{s^2} = 1 + \frac{r}{s} - \frac{r^2}{s^2}.$

The right-hand side of (8.1.7) is a fraction with denominator s^2 which is not zero as α is irrational, so its absolute value is at least $1/s^2$. Then, by (8.1.7),

(8.1.8) $\left|\alpha - \frac{r}{s}\right| \cdot \left|(\alpha - 1) + \frac{r}{s}\right| \geq \frac{1}{s^2}.$

By (8.1.6), r/s is close to α, thus the second factor on the left-hand side of (8.1.8) is about $2\alpha - 1 = \sqrt{5}$, which contradicts (8.1.6). To see this precisely, we start with the upper estimate

(8.1.9) $\left|(\alpha - 1) + \frac{r}{s}\right| \leq (2\alpha - 1) + \left|\frac{r}{s} - \alpha\right| < \sqrt{5} + \frac{1}{\sqrt{5}s^2}.$

For s large enough,

$$\frac{1}{\sqrt{5}s^2} < \varepsilon,$$

thus (8.1.9) implies

(8.1.10) $\left|(\alpha - 1) + \frac{r}{s}\right| < \sqrt{5} + \varepsilon.$

Combining (8.1.8) and (8.1.10), we get

$$\left|\alpha - \frac{r}{s}\right| > \frac{1}{(\sqrt{5} + \varepsilon)s^2}$$

for s large enough, which contradicts (8.1.6). $\square$

Theorem 8.1.6 shows that Theorems 8.1.5 and 8.1.1 express the right order of magnitude of best approximation for irrational numbers, since some irrational numbers α cannot be approximated substantially better than guaranteed by these theorems.

Now we show that Theorems 8.1.5 and 8.1.1 give the right order of magnitude of best approximation for irrational numbers also in the following sense: only few irrational numbers can be approximated much better. To get a precise meaning of "few", we introduce the notion of *measure zero*:

Definition 8.1.7. A subset H of the real numbers has *measure zero* (or is of measure zero), if to any $\varepsilon > 0$ there exist countably many intervals of total length less than ε so that their union covers H.   ♣

It is easy to see that the set of rational numbers and every countable set has measure zero, but there exist also sets of measure zero that have the cardinality of the continuum (see Exercise 8.1.9).

Theorem 8.1.8. *Let $\kappa > 0$ be a real number and H the set of real numbers α to which there are infinitely many r/s satisfying*

$$(8.1.11) \qquad \left| \alpha - \frac{r}{s} \right| < \frac{1}{s^{2+\kappa}}.$$

Then H has measure zero. ♣

Proof. Let

$$H_i = H \cap [i, i+1), \qquad i = 0, \pm 1, \pm 2, \dots.$$

The approximation property (8.1.11) depends only on the fractional part of α, so any two sets H_i are congruent. Thus it is enough to show that H_0 has measure zero, since

$$H = \bigcup_{i=-\infty}^{\infty} H_i,$$

and the union of countably many sets of measure zero has measure zero (see Exercise 8.1.10c).

For a given integer $s > 1$, let A_s be the set of real numbers $0 \le \alpha < 1$ for which (8.1.11) holds with some r. Clearly, A_s consists of the points in $[0, 1)$ belonging to the open intervals of radius $1/s^{2+\kappa}$ around the points

$$\frac{0}{s}, \quad \frac{1}{s}, \quad \dots, \quad \frac{s}{s},$$

so

$$(8.1.12) \qquad A_s = \left(\bigcup_{r=1}^{s-1} \left(\frac{r}{s} - \frac{1}{s^{2+\kappa}}, \frac{r}{s} + \frac{1}{s^{2+\kappa}} \right) \right) \bigcup \left[0, \frac{1}{s^{2+\kappa}} \right) \bigcup \left(1 - \frac{1}{s^{2+\kappa}}, 1 \right).$$

The total length of the intervals in A_s is

$$(8.1.13) \qquad (s-1)\frac{2}{s^{2+\kappa}} + 2\frac{1}{s^{2+\kappa}} = \frac{2s}{s^{2+\kappa}} = \frac{2}{s^{1+\kappa}}.$$

If $\alpha \in H_0$, then, by the condition, $\alpha \in A_s$ for infinitely many s. This implies

$$(8.1.14) \qquad H_0 \subseteq \bigcup_{s=m}^{\infty} A_s$$

for an arbitrary m. By (8.1.12), (8.1.13), and (8.1.14), H_0 can be covered by countably many intervals of total length

$$(8.1.15) \qquad \sum_{s=m}^{\infty} \frac{2}{s^{1+\kappa}}.$$

The infinite series

$$\sum_{s=1}^{\infty} \frac{1}{s^{1+\kappa}}$$

is convergent, thus we can find for any $\varepsilon > 0$ an m for which the sum in (8.1.15) is less than ε. Thus we have proved that H_0, and so H, have measure zero. $\square$

Generalizing Theorem 8.1.8, we can examine the following question. Let f be a function defined on the positive integers assuming positive values so that $f(s)/s$ is strictly increasing. Let $H(f)$ be the set of real numbers α satisfying

$$\left| \alpha - \frac{r}{s} \right| < \frac{1}{sf(s)}$$

for infinitely many r/s. Similar to the proof of Theorem 8.1.8, it can be shown that if

$$\sum_{s=1}^{\infty} \frac{1}{f(s)} < \infty$$

then $H(f)$ has measure zero.

If, however,

$$\sum_{s=1}^{\infty} \frac{1}{f(s)} = \infty$$

then the situation turns upside down: $H(f)$ contains every real number apart from a set of measure zero. The proof of this result is much more difficult.

Exercises 8.1

1. Let α be a rational number $\alpha = a/b$, where $(a, b) = 1$ and $b > 0$.

    (a) Verify

(8.1.16)
$$\frac{r}{s} \neq \frac{a}{b} \implies \left| \alpha - \frac{r}{s} \right| \geq \frac{1}{bs}.$$

    (b) Show that the right-hand side of (8.1.16) holds with equality for infinitely many fractions r/s.

2. Equality (8.1.1) in Theorem 8.1.1 can be satisfied by more than one fractional form r/s of a rational number. Demonstrate that this cannot hold for infinitely many forms r/s of the same rational number.

3. Let α be an irrational number, and consider infinitely many fractions r_i/s_i satisfying

$$\left| \alpha - \frac{r_i}{s_i} \right| < \frac{1}{s_i^2}, \qquad i = 1, 2, \ldots.$$

    Prove

    (a) $\lim_{i \to \infty} s_i = \infty$

    (b) $\lim_{i \to \infty} \frac{r_i}{s_i} = \alpha.$

4. Verify:

    (a) For any real number α there exist infinitely many integers r and non-negative integers k satisfying

$$\left| \alpha - \frac{r}{2^k} \right| \leq \frac{1}{3 \cdot 2^k}.$$

(b) There exits an α such that

$$\left| \alpha - \frac{r}{2^k} \right| \geq \frac{1}{3 \cdot 2^k}$$

for every fraction $r/2^k$.

(c) For any real number α there exist infinitely many integers r and non-negative integers k satisfying

$$\left| \alpha - \frac{r}{3^k} \right| \leq \frac{1}{2 \cdot 3^k}.$$

(d) There exists an α such that

$$\left| \alpha - \frac{r}{3^k} \right| \geq \frac{1}{2 \cdot 3^k}$$

for every fraction $r/3^k$.

(e) For any irrational $\alpha > 0$ there exist infinitely many fractions r^2/s^2 satisfying

$$\left| \alpha - \frac{r^2}{s^2} \right| < \frac{c(\alpha)}{s^2},$$

where $c(\alpha)$ is a constant depending on α.

(f) There exists an irrational $\alpha > 0$ and a constant $c > 0$ such that

$$\left| \alpha - \frac{r^2}{s^2} \right| > \frac{c}{s^2}$$

for every fraction r^2/s^2.

5. Prove that for any irrational number α there exist infinitely many fractions r/s with distinct numerators satisfying

$$\left| \alpha - \frac{r}{s} \right| < \frac{c(\alpha)}{r^2},$$

where $c(\alpha)$ is a constant depending on α.

6. Find a constant $c > 0$ such that

$$\left| \sqrt{2} - \frac{r}{s} \right| > \frac{c}{s^2}$$

for every fraction r/s.

7. Let $t > 1$ be a real number. We say that a real number α can be approximated to exponential order t if

$$\left| \alpha - \frac{r}{s} \right| < \frac{c(\alpha)}{s^t}$$

holds for infinitely many fractions r/s, where $c(\alpha)$ is a constant depending on α. Thus Theorem 8.1.1 implies that every irrational number can be approximated to order 2, whereas real numbers that can be approximated to order greater than 2 form a set of measure zero.

Assume that the real number α can be approximated to order 20.

Prove:

(a) The number $a\alpha + b$ can be approximated to order 20 if $a \neq 0$ and b are rational numbers.

(b) The number α^2 can be approximated to order 10.

8. Determine all possible values of the following expression as α and β assume all real numbers independently:

   (a) $\{\alpha\} + \{\beta\} - \{\alpha + \beta\}$

   (b) $\{\alpha\}\{\beta\} - \{\alpha\beta\}$

S* (c) $\{\alpha\}^2 - \{\alpha^2\}$.

9. (a) Show that every countable subset of the real numbers has measure zero.

   * (b) Consider those real numbers between 0 and 1 that do not contain the digit 1 in their ternary (base 3) representation (this is the so-called Cantor set). Verify that this set has the cardinality of the continuum, but still has measure zero.

10. Prove:

   (a) A subset of a set of measure zero has measure zero.

   (b) The union of finitely many sets of measure zero has measure zero.

   (c) The union of countably many sets of measure zero has measure zero.

   (d) The union of more than countably many sets of measure zero may, but does not necessarily, have measure zero.

8.2. Minkowski's Theorem

We discuss an important theorem in the geometry of numbers and its applications.

Theorem 8.2.1 (Minkowski's Theorem). *Let L be a parallelogram lattice in the plane and let H be a closed, convex region symmetric around a lattice point. Assume that the area of H is at least 4Δ, where Δ is the area of the fundamental parallelogram in the lattice. Then H contains a lattice point different from its center.* ♣

Remarks: (1) It is easy to check that the conditions of the theorem are necessary.

(2) We may assume that H is bounded. It can be shown that any unbounded convex set can have area only zero or infinity. In the latter, intersecting H by a (closed) circle around its center and with a sufficiently large radius, we get a bounded, closed, convex region symmetric around a lattice point with area at least 4Δ.

(3) For a generalization in higher dimensions and for sets of larger area, see Exercises 8.2.1 and 8.2.2.

We present two proofs of Minkowski's theorem. We denote the center of symmetry of H by O, and the area of H by h.

First proof. We consider first the case when $h > 4\Delta$.

We shrink the lattice L by the ratio $2/k$ from point O, where k is a large integer. Let $N(k)$ be the number of lattice points in the resulting lattice L_k that are elements of H. The area of the fundamental parallelogram of L_k is $4\Delta/k^2$, so the area of H is

$$(8.2.1) \qquad\qquad h = \lim_{k \to \infty} N(k)\frac{4\Delta}{k^2}.$$

Since $h > 4\Delta$, we get from (8.2.1) that $N(k) > k^2$ for k large enough.

Consider the generally skew coordinate system with origin O and axes parallel to the sides of the fundamental parallelogram. Then the lattice points in L have coordinates (ia, jb), and the lattice points in L_k have coordinates

$$\left(\frac{2i}{k}a, \frac{2j}{k}b\right),$$

where a and b are the side lengths of the fundamental parallelogram in L and i and j are arbitrary integers.

Since the pairs (i, j) can give k^2 residues on division by k and $N(k) > k^2$, the pigeonhole principle guarantees the existence of two distinct lattice points

$$Q_1 = \left(\frac{2i_1}{k}a, \frac{2j_1}{k}b\right) \quad \text{and} \quad Q_2 = \left(\frac{2i_2}{k}a, \frac{2j_2}{k}b\right),$$

in L_k satisfying

(8.2.2) $$k \mid i_1 - i_2 \quad \text{and} \quad k \mid j_1 - j_2.$$

As H is symmetric about O, the mirror image

$$Q_2' = \left(\frac{-2i_2}{k}a, \frac{-2j_2}{k}b\right)$$

of Q_2 is in H, and by convexity, the midpoint

$$F = \left(\frac{2i_1 - 2i_2}{2k}a, \frac{2j_1 - 2j_2}{2k}b\right)$$

of the segment Q_1Q_2' belongs to H. The divisibilities (8.2.2) imply $F = (ra, sb)$ with some integers r and s, so F is a lattice point in the original L. Since $Q_1 \neq Q_2$, $F \neq O$. Thus we have proved that H contains a lattice point of L different from O.

We still have to verify the case $h = 4\Delta$. For a proof by contradiction, we assume that the center O is the only lattice point of L in H. Let m be the minimum of the distances of lattice points $P \neq O$ from H. Since H is closed, we have $m > 0$. Thus we can magnify H so that even the resulting H' contains no lattice point besides O. But this is impossible because the area of H' is greater than 4Δ. □

Second proof. We verify first a lemma which expresses an intuitively obvious fact: If translating a bounded set in the plane with all lattice vectors we obtain pairwise disjoint copies, then the area of the set cannot be greater than the area of the fundamental parallelogram in the lattice.

Lemma 8.2.2. *Let Δ be the area of the fundamental parallelogram in the lattice L and t the area of a bounded set K in the plane. For a fixed lattice point O and an arbitrary lattice point P, let K_P denote the copy of K translated by the vector OP, so $K_O = K$. Assume that the sets K_P are disjoint. Then $t \leq \Delta$.* ♣

Proof. The essence of the proof is the observation: We enlarge the fundamental parallelogram (in all directions) by a large r, and place the resulting parallelogram M so that O should be approximately in the center of M. Then the translated copies of K by lattice points in M cannot go much beyond M, so the total area of the translated copies, which is about r^2t, cannot be much bigger than the area of M, which is $r^2\Delta$. The statement now follows by taking the limit for $r \to \infty$.

Let us see all this precisely and in detail. Consider the generally skew coordinate system, used already in the first proof of Theorem 8.2.1, with origin O and axes parallel to the sides of the fundamental parallelogram. Then the coordinates of lattice points in L are (ia, jb), where a and b are the side lengths of the fundamental parallelogram and i and j are arbitrary integers.

Let n be an arbitrary integer and consider the $(2n + 1)^2$ lattice points $P_{ij} = (ia, jb)$ with $|i| \le n$ and $|j| \le n$. Let U_n denote the union of the sets K_P belonging to these points P_{ij}. Then the area of U_n is $(2n + 1)^2 t$. As K is bounded, the coordinates of every point in K are less in absolute value than ca and cb, with a suitable constant $c > 0$. Then U_n is contained in a parallelogram G_n where the coordinates of vertices are

$$\big(\pm a(n + c), \pm b(n + c)\big),$$

so the area of G_n is $(2n + 2c)^2 \Delta$. Being a subset, the area of U_n is not greater than the area of G_n, so

$$(2n + 1)^2 t \le (2n + 2c)^2 \Delta.$$

This implies

$$t \le \left(1 + \frac{2c - 1}{2n + 1}\right)^2 \Delta.$$

Taking the limit as $n \to \infty$, we obtain the desired inequality $t \le \Delta$. □

Turning to the proof of Theorem 8.2.1, it is sufficient to consider the case $h > 4\Delta$.

We shrink H from O by half and denote the resulting set by K. By the condition, the area of K is $t = h/4 > \Delta$, so by Lemma 8.2.2, there exist two distinct lattice points Q and R for which K_Q and K_R share a common point. A translation by the vector QO maps this common point into a common point A of $K_O = K$ and K_P with a suitable lattice point P other than O. We show that P is an element of H, thus proving the statement of the theorem.

Let B be the translate of A by the vector PO, C the reflected image of B through O, and D the midpoint of segment AC.

Since $A \in K_P$, $B \in K$. By the symmetry of K, $C \in K$. As both A and C are in K, by convexity, their midpoint D is in K.

By the construction, $PAOC$ is a parallelogram, since sides OC and AP are parallel and equal. Therefore D is the midpoint of the diagonal OP, and the twofold magnification around O maps D into P. Since this magnification takes K into H, and D is in K, P must be in H. Thus we have proved that H contains the lattice point P different from O. □

We apply Minkowski's theorem in Diophantine approximation to improve Theorem 8.1.1.

Theorem 8.2.3. *For any irrational number α, there exist infinitely many fractions r/s satisfying*

(8.2.3) $$\left|\alpha - \frac{r}{s}\right| < \frac{1}{2s^2}.$$ ♣

Proof. For $s \neq 0$, (8.2.3) is equivalent to

(8.2.4)
$$s(s\alpha - r)| < \frac{1}{2}.$$

We introduce the new variables

$$x = s\alpha - r, \qquad y = s.$$

If r and s assume integer values independently, then the points (x, y) form a lattice where the vertices of the fundamental parallelogram are

(8.2.5)
$$(0, 0), \qquad (-1, 0), \qquad (\alpha, 1), \qquad (\alpha - 1, 1).$$

For the new variables, (8.2.4) implies $|xy| < 1/2$. So we are looking for lattice points (x, y) in the region having the hyperbolas $xy = 1/2$ and $xy = -1/2$ as boundaries, and containing the origin. The equality $xy = \pm 1/2$ cannot hold for a lattice point since α is irrational, hence we can replace $<$ in (8.2.3) and (8.2.4) by $\leq$. Thus we can extend the region with its boundary, and we shall deal with the arising closed set Z in the sequel.

The condition $s \neq 0$ means that we do not count the lattice points on the x-axis.

In the fundamental parallelogram (8.2.5) both the horizontal side and the height have unit length, so the area of the fundamental parallelogram is $\Delta = 1$.

As Z is not convex (and is not bounded either), we cannot apply Minkowski's theorem directly to Z. Instead, we shall consider suitable convex subsets of Z, namely rhombuses touching the four branches of the hyperbolas and with vertices on the axes of the hyperbolas. These rhombuses are convex, closed sets symmetric around the origin.

We show that each such rhombus has area 4. If the rhombus touches the branch of hyperbola in the first quadrant at $\left(a, 1/(2a)\right)$, then the equation of the tangent line is

$$y - \frac{1}{2a} = \frac{-1}{2a^2}(x - a).$$

This line intersects the coordinate axes at points $x = 2a$ and $y = 1/a$. Thus the right triangle with the origin and these two points as vertices has area $\frac{1}{2}(2a)(1/a) = 1$. The rhombus consists of four such triangles, so its area is 4.

Since the area is $4 = 4\Delta$, by Minkowski's theorem, every such rhombus contains a lattice point besides the origin.

Choose the rhombuses to be narrower and narrower in the direction of the y-axis. Thus we can require that each subsequent rhombus does not contain any of the non-trivial lattice points in the previous ones, and it does not contain any lattice point on the x-axis apart from the origin. Thus we obtain infinitely many suitable lattice points, and the usual condition $s > 0$ can be granted by central symmetry. $\qquad \square$

As another application of Minkowski's theorem, we present a new proof for the part of Theorem 7.5.1 stating that every prime $p > 0$ of the form $4k + 1$ is the sum of two squares.

Theorem 8.2.4. *Every prime $p > 0$ of the form $4k + 1$ can be represented as a sum of two squares.* ♣

Proof. By Theorem 4.1.4, $c^2 \equiv -1 \pmod{p}$ with a suitable integer c.

Consider the points on the plane with coordinates

(8.2.6) $x = pu + cv, \qquad y = v,$

where u and v assume integer values independently. These points form a lattice where the area of the fundamental parallelogram is $\Delta = p$.

For any lattice point,

$$x^2 + y^2 = (pu + cv)^2 + v^2 = p(pu^2 + 2cuv) + v^2(c^2 + 1) \equiv 0 \pmod{p},$$

so $p \mid x^2 + y^2$. This means that if $x^2 + y^2 < 2p$ for some lattice point different from the origin, then $x^2 + y^2 = p$.

We apply Minkowski's theorem for the closed circle $x^2 + y^2 \leq 4p/\pi$ around the origin having area $4p = 4\Delta$. Thus this circle contains a lattice point (x, y) different from the origin. For this lattice point, we have

$$x^2 + y^2 \leq \frac{4p}{\pi} < 2p. \qquad \square$$

We note that a similar but much more complex application of the three-dimensional version of Minkowski's theorem (see Exercise 8.2.1a) leads to the proof of the hard part of the Three Squares Theorem 7.5.2 stated there without proof (we have to rely also on Dirichlet's theorem about primes in arithmetic progressions).

For some further applications of Minkowski's theorem, see Exercises 8.2.3–8.2.5.

Exercises 8.2

1. (a) Prove Minkowski's theorem in space: Let L be any parallelepiped lattice in the space and H a closed, convex set symmetric around a lattice point. Assume that the volume of H is at least 8Δ, where Δ is the volume of the fundamental parallelepiped in the lattice. Then H contains a lattice point different from its center.

   (b) Generalize the theorem for arbitrary dimensions.

2. Verify the following generalization of Minkowski's theorem: If L and H meet the requirements of Theorem 8.2.1 and the area of H is at least $4r\Delta$ for some integer $r > 0$, then H contains at least $2r$ lattice points besides its center.

3. Prove that every positive prime of the form $3k + 1$ can be written as $x^2 + 3y^2$ with suitable integers x and y.

4. Let a_{11}, a_{12}, a_{21}, and a_{22} be integers satisfying

$$D = \begin{bmatrix} a_{11} & a_{12} \\ a_{21} & a_{22} \end{bmatrix} \neq 0.$$

   Prove that if $b_1 b_2 \geq |D|$ for the positive (real) numbers b_1 and b_2, then the system of inequalities

$$|a_{11}x_1 + a_{12}x_2| \leq b_1, \qquad |a_{21}x_1 + a_{22}x_2| \leq b_2$$

   has a non-trivial, i.e $(x_1, x_2) \neq (0, 0)$, solution in integers.

* 5. Verify that for any irrational numbers α_1 and α_2 there exist infinitely many pairs r_1/s, r_2/s of rational numbers with a common denominator satisfying

$$\left| \alpha_j - \frac{r_j}{s} \right| < \frac{2}{3} \cdot \frac{1}{s^{3/2}}, \qquad j = 1, 2.$$

8.3. Continued Fractions

For any real number α, consider the following algorithm. Let

(8.3.1) $\qquad c_0 = \lfloor \alpha \rfloor \quad \text{and} \quad \alpha_1 = \{\alpha\}, \qquad \text{then} \qquad \alpha = c_0 + \alpha_1.$

If $\alpha_1 \neq 0$, then let

$$c_1 = \left\lfloor \frac{1}{\alpha_1} \right\rfloor \quad \text{and} \quad \alpha_2 = \left\{ \frac{1}{\alpha_1} \right\}, \qquad \text{then} \qquad \alpha = c_0 + \alpha_1 = c_0 + \frac{1}{c_1 + \alpha_2}.$$

If $\alpha_2 \neq 0$, then we form the floor and fractional part of $1/\alpha_2$, etc. In general, if $c_0, c_1, \dots, c_n$ and $\alpha_1, \dots, \alpha_{n+1}$ have already been determined and $\alpha_{n+1} \neq 0$, then let

(8.3.2) $\qquad c_{n+1} = \left\lfloor \frac{1}{\alpha_{n+1}} \right\rfloor \quad \text{and} \quad \alpha_{n+2} = \left\{ \frac{1}{\alpha_{n+1}} \right\},$

so

(8.3.3) $\qquad \alpha = c_0 + \cfrac{1}{c_1 + \cfrac{1}{c_2 + \cfrac{1}{\ddots \; c_n + \cfrac{1}{c_{n+1} + \alpha_{n+2}}}}}.$

We call the multiple-decked fraction on the right-hand side of (8.3.3) a (finite) *continued fraction*, and for convenience, we denote it by $C(c_0, c_1, \dots, c_n, c_{n+1} + \alpha_{n+2})$. We shall sometimes apply this notation for the right-hand side of (8.3.3) even if the numbers c_i are not integers.)

If $\alpha_{n+1} = 0$, the algorithm terminates.

The integers $c_0, c_1, \dots$ are called the *digits* in the continued fraction expansion of α.

Definition 8.3.1. By the *continued fraction digits* of a real number α, we mean the (finite or infinite) sequence $c_0, c_1, \dots$ defined by (8.3.1) and (8.3.2). ♣

It is clear from the definition that the digits are uniquely determined integers and $c_i > 0$ for $i \geq 1$.

Examples.   **E1** Let $\alpha = 111/25$. Then

$$\frac{111}{25} = 4 + \frac{11}{25}, \qquad c_0 = 4$$

$$\frac{25}{11} = 2 + \frac{3}{11}, \qquad c_1 = 2$$

$$\frac{11}{3} = 3 + \frac{2}{3}, \qquad c_2 = 3$$

$$\frac{3}{2} = 1 + \frac{1}{2}, \qquad c_3 = 1$$

$$\frac{2}{1} = 2 + \quad 0, \qquad c_4 = 2.$$

Thus the digits in the continued fraction expansion of $111/25$ are $4, 2, 3, 1, 2$. This means also

$$\frac{111}{25} = C(4, 2, 3, 1, 2) = 4 + \cfrac{1}{2 + \cfrac{1}{3 + \cfrac{1}{1 + \cfrac{1}{2}}}}.$$

E2 Let $\alpha = \sqrt{2}$. Then

$$\sqrt{2} = \qquad = 1 + (\sqrt{2} - 1), \qquad c_0 = 1$$

$$\frac{1}{\sqrt{2} - 1} = \sqrt{2} + 1 = 2 + (\sqrt{2} - 1), \qquad c_1 = 2$$

$$\frac{1}{\sqrt{2} - 1} = \sqrt{2} + 1 = 2 + (\sqrt{2} - 1), \qquad c_2 = 2$$

$$\vdots$$

Hence the continued fraction digits of $\sqrt{2}$ are $1, 2, 2, 2, \ldots$ We introduce the (so far formal) notation $\sqrt{2} = C(1, 2, 2, \ldots)$ and call this an *infinite continued fraction*.

Theorem 8.3.2. *The sequence of continued fraction digits of α is finite if and only if α is a rational number.* ♣

Proof. Let the sequence of continued fraction digits be finite, so $\alpha = C(c_0, c_1, \ldots, c_k)$ for suitable integers c_i. Then condensing the multiple-decked fraction, we get α as the quotient of two integers, so α is rational.

Conversely, let $\alpha = a/b$, where $b > 0$ and a are integers. We show that the steps defining the continued fraction digits correspond to the steps of the Euclidean algorithm for a and b. This means that the algorithm yielding the continued fraction digits terminates in finitely many steps.

The first step in the Euclidean algorithm is a division of a by b:

$$a = bq_1 + r_1, \qquad 0 \leq r_1 < b.$$

We can rewrite it as

$$\frac{a}{b} = q_1 + \frac{r_1}{b} = \left\lfloor \frac{a}{b} \right\rfloor + \left\{ \frac{a}{b} \right\}$$

so, using the notation of the continued fraction algorithm $c_0 = q_1$ and $\alpha_1 = r_1/b$.

If $r_1 \neq 0$, then the next step in the Euclidean algorithm is

$$b = r_1 q_2 + r_2, \qquad 0 \leq r_2 < r_1,$$

i.e.

$$\frac{1}{\alpha_1} = \frac{b}{r_1} = q_2 + \frac{r_2}{r_1} = \left\lfloor \frac{b}{r_1} \right\rfloor + \left\{ \frac{b}{r_1} \right\},$$

thus $c_1 = q_2$ and $\alpha_2 = r_2/r_1$.

We obtain the same way that also the further continued fraction digits are the quotients occurring in the Euclidean algorithm. □

In the sequel we assume that α is irrational and use continued fractions to exhibit rational numbers approximating α well. They will be the initial finite sections of the infinite continued fraction expansion of α, i.e., the finite continued fractions formed from the first $n + 1$ continued fraction digits for $n \geq 0$. We denote these rational numbers by $C_n(\alpha)$, so if $\alpha = C(c_0, c_1, \dots)$, then

$$(8.3.4) \qquad C_n(\alpha) = C(c_0, c_1, \dots, c_n), \qquad n = 0, 1, 2, \dots.$$

Theorem 8.3.3. *Let c_0, c_1, $\dots$ be the continued fraction digits of an irrational number α, and*

$$(8.3.5) \qquad C_n(\alpha) = C(c_0, c_1, \dots, c_n) = \frac{r_n}{s_n}, \qquad where \qquad (r_n, s_n) = 1, s_n > 0.$$

Then

$$(8.3.6) \qquad \left| \alpha - \frac{r_n}{s_n} \right| < \frac{1}{s_n^2}$$

for any n. Moreover, if $n > 0$, then at least one of the inequalities

$$(8.3.7) \qquad \left| \alpha - \frac{r_n}{s_n} \right| < \frac{1}{2s_n^2}, \qquad \left| \alpha - \frac{r_{n+1}}{s_{n+1}} \right| < \frac{1}{2s_{n+1}^2}$$

holds. ♣

Remarks: (1) Clearly, Theorem 8.3.3 contains the statements of Theorems 8.1.1 and 8.2.3. Besides showing the existence of good approximating fractions, it gives a practical algorithm to find them.

(2) It can be shown that all rational numbers providing a really good approximation are among the fractions r_n/s_n in (8.3.5): If

$$\left| \alpha - \frac{r}{s} \right| < \frac{1}{2s^2}$$

(so r/s approximates α to the order of magnitude in Theorem 8.2.3), then r/s must be equal to some r_n/s_n.

(3) We can use continued fractions to prove Theorem 8.1.5, stated there without proof. By similar but more complex arguments than applied below to verify Theorem 8.3.3, it can be shown that at least one of any three consecutive continued fractions $C_n(\alpha)$ satisfies the approximation in Theorem 8.1.5.

(4) Using Exercise 8.1.3a, Theorem 8.3.3 implies

$$\lim_{n \to \infty} C_n(\alpha) = \alpha, \quad \text{or} \quad \lim_{n \to \infty} C(c_0, \dots, c_n) = C(c_0, c_1, \dots).$$

This gives a natural meaning to the (till now formal) expression "infinite continued fraction" $\alpha = C(c_0, c_1, \dots)$.

To prove Theorem 8.3.3, we need

Lemma 8.3.4. *Let $c_0, c_1, c_2, \dots$ be arbitrary real numbers, where $c_i > 0$ for $i \geq 1$, and form the*

(8.3.8a) $r_0 = c_0,$ $r_1 = c_1 c_0 + 1,$ $r_n = c_n r_{n-1} + r_{n-2},$

(8.3.8b) $s_0 = 1,$ $s_1 = c_1,$ $s_n = c_n s_{n-1} + s_{n-2}.$

Then

(8.3.9) $$C(c_0, c_1, \dots, c_n) = \frac{r_n}{s_n}$$

and

(8.3.10) $$\frac{r_n}{s_n} - \frac{r_{n-1}}{s_{n-1}} = \frac{(-1)^{n-1}}{s_{n-1} s_n} \quad (n \geq 1).$$

If the numbers c_n are integers, then so are also r_n and s_n, further $(r_n, s_n) = 1$, and $s_{n+1} > s_n$ for $n > 0$.   ♣

Remark: It follows from Lemma 8.3.4 that the sequences r_n and s_n defined by (8.3.5) in Theorem 8.3.3 satisfy recursion (8.3.8a)–(8.3.8b), so the notations c_n, r_n, and s_n in Lemma 8.3.4 and Theorem 8.3.3 are in harmony.

Proof.   I. We prove (8.3.9) by induction on n.

In the cases $n = 0, 1$, and 2,

$$C(c_0) = \quad c_0 \quad = \quad \frac{c_0}{1} \quad = \frac{r_0}{s_0}$$

$$C(c_0, c_1) = \quad c_0 + \frac{1}{c_1} \quad = \frac{c_1 c_0 + 1}{c_1} = \frac{r_1}{s_1}$$

$$C(c_0, c_1, c_2) = \frac{c_2 c_1 c_0 + c_2 + c_0}{c_2 c_1 + 1} = \frac{c_2 r_1 + r_0}{c_2 s_1 + s_0} = \frac{r_2}{s_2}$$

So (8.3.9) holds.

Assume now that (8.3.9) is true for $n = m \geq 2$, so

$$C(c_0, c_1, \dots, c_m) = \frac{r_m}{s_m} = \frac{c_m r_{m-1} + r_{m-2}}{c_m s_{m-1} + s_{m-2}},$$

where $r_{m-1}, s_{m-1}, r_{m-2}$, and s_{m-2} depend only on $c_0, \ldots, c_{m-1}$. Then

$$C(c_0, \ldots, c_{m-1}, c_m, c_{m+1}) = C\left(c_0, \ldots, c_{m-1}, c_m + \frac{1}{c_{m+1}}\right)$$

$$= \frac{\left(c_m + \frac{1}{c_{m+1}}\right)r_{m-1} + r_{m-2}}{\left(c_m + \frac{1}{c_{m+1}}\right)s_{m-1} + s_{m-2}}$$

$$= \frac{c_{m+1}(c_m r_{m-1} + r_{m-2}) + r_{m-1}}{c_{m+1}(c_m s_{m-1} + s_{m-2}) + s_{m-1}}$$

$$= \frac{c_{m+1} r_m + r_{m-1}}{c_{m+1} s_m + s_{m-1}}$$

$$= \frac{r_{m+1}}{s_{m+1}}$$

so (8.3.9) holds also for $n = m + 1$.

II. We now verify (8.3.10). By (8.3.8a)–(8.3.8b),

$$r_n s_{n-1} - r_{n-1} s_n = (c_n r_{n-1} + r_{n-2})s_{n-1} - r_{n-1}(c_n s_{n-1} + s_{n-2})$$
$$= -(r_{n-1} s_{n-2} - r_{n-2} s_{n-1}).$$

Repeating this step for $n - 1, n - 2, \ldots, 2$ instead of n, we obtain

(8.3.11) $$r_n s_{n-1} - r_{n-1} s_n = (-1)^{n-1}(r_1 s_0 - r_0 s_1) = (-1)^{n-1}.$$

Dividing by $s_n s_{n-1}$, we get (8.3.10).

III. In the case when every c_i is an integer, all but one of the statements are obvious from the conditions, and $(r_n, s_n) = 1$ follows from (8.3.11). $\qquad\square$

Proof of Theorem 8.3.3. As mentioned before, Lemma 8.3.4 implies that sequences r_n and s_n defined by (8.3.5) satisfy (8.3.8a)–(8.3.8b).

In the sequel we shall use that α itself can be written as a finite continued fraction: by (8.3.3),

(8.3.12) $$\alpha = C(c_0, c_1, \ldots, c_n, c_{n+1} + \alpha_{n+2})$$

for any n, where $0 < \alpha_{n+2} < 1$ by (8.3.2) and the irrationality of α.

To estimate the difference $\alpha - r_n/s_n$, we shall apply Lemma 8.3.4 for $c_0, c_1, \ldots, c_n$, and $c'_{n+1} = c_{n+1} + \alpha_{n+2}$ (instead of c_{n+1}), and stop. Then we get

$$r_0, r_1, \ldots, r_n, r'_{n+1} \quad \text{and} \quad s_0, s_1, \ldots, s_n, s'_{n+1},$$

from (8.3.8a)–(8.3.8b), where

$$r'_{n+1} = c'_{n+1} r_n + r_{n-1} = (c_{n+1} + \alpha_{n+2})r_n + r_{n-1},$$
$$s'_{n+1} = c'_{n+1} s_n + s_{n-1} = (c_{n+1} + \alpha_{n+2})s_n + s_{n-1}$$

for $n \geq 1$. By (8.3.12), (8.3.9), and (8.3.4),

$$\alpha = \frac{r'_{n+1}}{s'_{n+1}} \quad \text{and} \quad C_n(\alpha) = \frac{r_n}{s_n}.$$

Applying (8.3.10), we obtain

(8.3.13)
$$\alpha - \frac{r_n}{s_n} = \frac{r'_{n+1}}{s'_{n+1}} - \frac{r_n}{s_n} = \frac{(-1)^n}{s_n s'_{n+1}}.$$

Since $s'_{n+1} > s_n$, (8.3.13) implies (8.3.6).

To verify (8.3.7) by contradiction, assume

(8.3.14)
$$\left| \alpha - \frac{r_n}{s_n} \right| \geq \frac{1}{2s_n^2} \quad \text{and} \quad \left| \alpha - \frac{r_{n+1}}{s_{n+1}} \right| \geq \frac{1}{2s_{n+1}^2}.$$

By (8.3.13), the differences $\alpha - r_n/s_n$ and $\alpha - r_{n+1}/s_{n+1}$ have opposite signs, so α is between the fractions r_n/s_n and r_{n+1}/s_{n+1}. Accordingly,

(8.3.15)
$$\left| \alpha - \frac{r_n}{s_n} \right| + \left| \alpha - \frac{r_{n+1}}{s_{n+1}} \right| = \left| \frac{r_n}{s_n} - \frac{r_{n+1}}{s_{n+1}} \right|.$$

We estimate the left-hand side of (8.3.15) using (8.3.14), and replace the right-hand side by $1/(s_n s_{n+1})$ based on (8.3.10), so

(8.3.16)
$$\frac{1}{2s_n^2} + \frac{1}{2s_{n+1}^2} \leq \frac{1}{s_n s_{n+1}} \quad \text{so} \quad (s_{n+1} - s_n)^2 \leq 0.$$

But (8.3.16) cannot hold since $s_{n+1} > s_n$ for $n > 0$, and we have reached a contradiction.

$\square$

Exercises 8.3

1. Compute the continued fraction digits of the following numbers:

   (a) $53/11$

   (b) $\sqrt{3}$

   (c) $\sqrt{5}$

   (d) $(1 + \sqrt{5})/2$.

2. Which numbers have the continued fraction expansion

   (a) $1, 2, 3, 4$

   (b) $1, 2, 1, 2, 1, 2, \ldots$?

3. Prove that for any irrational α there exist infinitely many fractions r/s with odd denominators s satisfying

   $$\left| \alpha - \frac{r}{s} \right| < \frac{1}{s^2}.$$

4. Prove

   $$\left| \frac{1 + \sqrt{5}}{2} - \frac{\varphi_{n+1}}{\varphi_n} \right| < \frac{1}{\varphi_n^2}$$

   for every $n \geq 1$, where φ_n is the nth Fibonacci number (see Exercise 1.2.5).

S 5. Prove that if the conditions of Lemma 8.3.4 hold, then

$$\frac{r_n}{s_n} - \frac{r_{n-2}}{s_{n-2}} = \frac{(-1)^n c_n}{s_{n-2}s_n}$$

for any $n \geq 2$.

S* 6. Assume that the continued fraction digits of an irrational α form a periodic sequence (so there exist positive integers k and M such that $c_n = c_{n-k}$ for every $n > M$). Prove that α is a root of a quadratic polynomial with integer coefficients.

Remark: The converse of this statement is also true.

8.4. Distribution of Fractional Parts

We deal with the distribution of fractional parts of sequences of real numbers in this section.

Theorem 8.4.1. *The fractional parts of the multiples of any irrational number are everywhere dense in the interval* $[0, 1]$.

Formally, let α be an irrational number and $v \in [0, 1]$. Then for any $\varepsilon > 0$ there exists an integer $n > 0$ satisfying $|\{n\alpha\} - v| < \varepsilon$. ♣

Proof. By Theorem 8.1.1, there are infinitely many fractions r/s satisfying

$$\left|\alpha - \frac{r}{s}\right| < \frac{1}{s^2}, \quad \text{i.e.} \quad |s\alpha - r| < \frac{1}{s}.$$

Choose a fraction with $s > 1/\varepsilon$ from them, so $|s\alpha - r| < \varepsilon$. Let $d = |s\alpha - r|$ (thus $d < \varepsilon$), and consider the fractional parts

(8.4.1) $$\{s\alpha\}, \{2s\alpha\}, \{3s\alpha\}, \ldots, \{ms\alpha\}$$

where $m = \lfloor 1/d \rfloor$ (we can obviously assume $\varepsilon < 1$, so $m \geq 1$).

Consider first the case $s\alpha - r > 0$. Then for every $1 \leq i \leq m$, we have

$$0 < is\alpha - ir = i(s\alpha - r) = id < 1, \quad \text{i.e.} \quad \{is\alpha\} = id.$$

This means that the fractional parts listed in (8.4.1) form a monotone increasing sequence where the distance between consecutive elements is $d < \varepsilon$, and also the distances between the first element and 0, and between the last element and 1 are less than ε. This implies that there is an element in the sequence that is closer to v than ε (in fact, closer than $\varepsilon/2$).

We can handle the case $s\alpha - r < 0$ similarly: then $\{is\alpha\} = 1 - id$ for $1 \leq i \leq m$, so the fractional parts in (8.4.1) form a monotone decreasing sequence where the distance between consecutive elements, between the first element and 1, and between the last element and 0 are all less than ε. □

Now we consider the variant of the problem of Theorem 8.4.1 for higher dimensions. The simplest case is when α_1 and α_2 are irrational numbers and we investigate the distribution of the points $P_n = (\{n\alpha_1\}, \{n\alpha_2\})$ in the unit square.

Similar to the proof of Theorem 8.4.1, we obtain from Theorem 8.1.3 that for any $\varepsilon > 0$ there exist integers r_1, r_2, and $s > 0$ satisfying

$$|s\alpha_1 - r_1| < \varepsilon \quad \text{and} \quad |s\alpha_2 - r_2| < \varepsilon.$$

This means that $P_s = (\{s\alpha_1\}, \{s\alpha_2\})$ is close to a vertex of the unit square. Similar to the proof of Theorem 8.4.1, it follows that $P_s, P_{2s}, P_{3s}, \ldots$ lie densely on the line connecting P_s with this vertex.

It is not true, however, that the points P_n are dense everywhere in the unit square for every α_1 and α_2. Take $\alpha_2 = \alpha_1 + 1$. Then $\{n\alpha_1\} = \{n\alpha_2\}$ for every n, so each point P_n is on the line $y = x$.

The condition for everywhere dense distribution can be formulated with the help of linear independence.

Theorem 8.4.2. *For real numbers $\alpha_1, \ldots, \alpha_k$, the points*

$$P_n = (\{n\alpha_1\}, \{n\alpha_2\}, \ldots, \{n\alpha_k\}), \qquad n = 1, 2, 3, \ldots$$

are everywhere dense in the k-dimensional unit cube if and only if $1, \alpha_1, \ldots, \alpha_k$ are linearly independent over the rational field. ♣

Linear independence means that $c_0 + c_1\alpha_1 + \cdots + c_k\alpha_k = 0$ can hold with rational numbers c_i only in the trivial case $c_0 = c_1 = \cdots = c_k = 0$. This implies that every α_i must be irrational.

In the example $k = 2$, $\alpha_2 = \alpha_1 + 1$, we saw that the points P_n are not everywhere dense in the unit circle, and $1, \alpha_1, \alpha_2$ are not linearly independent, as $1 \cdot 1 + 1\alpha_1 + (-1)\alpha_2 = 0$.

We do not prove the sufficiency of the condition in Theorem 8.4.2, and have the proof of necessity in Exercise 8.4.3.

Returning to the one-dimensional case, we examine now *uniform distribution*, which is a much stronger requirement than being everywhere dense.

Uniform distribution means that the fractional parts of $u_1, u_2, \ldots$ can be found in any subinterval I of $[0, 1]$ proportional to the length of I: for n large, about dn from the first n fractional parts $\{u_i\}$ are in I where d is the length of I. The precise definition is:

Definition 8.4.3. A sequence of real numbers $u_1, u_2, \ldots$ is *uniformly distributed modulo 1* (or has *uniform distribution*), if for any subinterval I in $[0, 1]$,

$$\lim_{n \to \infty} \frac{f_n(I)}{n} = d,$$

where d is the length of interval I and $f_n(I)$ is the number of those fractional parts among $\{u_1\}, \ldots, \{u_n\}$ that fall into I. ♣

We state Weyl's basic result about uniform distribution without proof.

Theorem 8.4.4. *A sequence of real numbers $u_1, u_2, \ldots$ is uniformly distributed if and only if for every $m \neq 0$,*

$$\lim_{n \to \infty} \frac{1}{n} \sum_{t=1}^{n} e^{2\pi i m u_t} = 0. \qquad ♣$$

The notion of uniform distribution and the criterion of Weyl can be extended to higher dimensions.

Relying on Weyl's condition, we show that the multiples of an irrational number are uniformly distributed.

Theorem 8.4.5. *If α is an irrational number, then α, 2α, ..., $n\alpha$, ... are uniformly distributed.*   ♣

Proof. By Theorem 8.4.4, we have to show

$$(8.4.2) \qquad \lim_{n \to \infty} \frac{1}{n} \sum_{t=1}^{n} e^{2\pi i m t \alpha} = 0$$

for any integer $m \geq 0$. The sum on the left-hand side of (8.4.2) is a geometric series of n terms and with quotient $e^{2\pi i m \alpha} \neq 1$, as α is irrational. Hence

$$\left| \frac{1}{n} \sum_{t=1}^{n} e^{2\pi i m t \alpha} \right| = \frac{|e^{2\pi i m \alpha}| \cdot |e^{2\pi i m n \alpha} - 1|}{n|e^{2\pi i m \alpha} - 1|} \leq \frac{1 \cdot 2}{n|e^{2\pi i m \alpha} - 1|} \to 0,$$

as $n \to \infty$.   □

Exercises 8.4

S 1. Examine whether or not the fractional parts of the sequences below are everywhere dense in the interval $[0, 1]$:

(a) $(1 + \sqrt{2})^n$

(b) $\sqrt{n}$

(c) $\sqrt{n^2 + 1}$

(d) $\sqrt{2n^2 + 1}$

(e) $\sin(n\pi/180)$

(f) $\sin n$

(g) $\log_{10} n$.

* 2. Show the existence of a real number α such that the fractional parts of

$$\alpha, \quad \alpha^2, \quad \alpha^3, \quad \ldots, \quad \alpha^n, \ldots$$

are everywhere dense in $[0, 1]$.

S 3. Prove that the condition of linear independence given in Theorem 8.4.2 is necessary for the points P_n to be everywhere dense in the k-dimensional unit cube.

4. Verify the following statements.

(a) If the fractional parts of a sequence are everywhere dense in $[0, 1]$, then we can rearrange them into a uniformly distributed sequence.

(b) Any uniformly distributed sequence can be rearranged into one that is not uniformly distributed.

5. True or false?

   (a) Any subsequence of a uniformly distributed sequence has uniform distribution.

   (b) If we add the same real number to every element of a uniformly distributed sequence, the new sequence will have uniform distribution.

   (c) If we multiply every element of a uniformly distributed sequence by the same non-zero real number, the new sequence will have uniform distribution.

   (d) The sum of two uniformly distributed sequences is uniformly distributed.

   (e) The product of two uniformly distributed sequences is uniformly distributed.

   (f) The square of a uniformly distributed sequence is uniformly distributed.

   (g) The square of a uniformly distributed sequence can never be uniformly distributed.

6. Demonstrate that the following sequences are *not* uniformly distributed:

   (a) $\log_{10} n$

   (b) $\sin n$.

7. Prove that if a natural number t is not a power of 10, then there exists a positive integer n such that the first five digits of t^n are 54321 in the decimal system.

Chapter 9

Algebraic and Transcendental Numbers

A complex number is algebraic if it is a root of some non-zero polynomial with rational (or equivalently, integer) coefficients, otherwise it is transcendental. We show that most complex numbers are transcendental. On the other hand, however, it is generally very hard to determine, whether a given number is algebraic or transcendental. This will be illustrated by the proof of the transcendence of e and by a long list of unsolved problems.

We discuss first the properties of minimal polynomials, degree, and operations with algebraic numbers. Then we prove that algebraic numbers cannot be approximated well. As a consequence, we construct a transcendental number relatively easily, and we infer that some types of Diophantine equations cannot have infinitely many solutions. At the end of the chapter, we introduce algebraic integers as a generalization of ordinary integers.

Algebraic numbers and algebraic integers also play an important role in the next two chapters.

9.1. Algebraic Numbers

The rational numbers can be characterized among all complex numbers as roots of linear polynomials with rational coefficients. Discarding the restriction on the degree of the polynomial, we get the notion of algebraic numbers:

Definition 9.1.1. A complex number α is an *algebraic number* (or, is *algebraic*), if $f(\alpha) = 0$ for some non-zero polynomial f with rational coefficients. ♣

Remarks: (1) We had to exclude the polynomial $f = 0$, since every complex number is a root of it.

(2) If α is a root of a polynomial with rational coefficients, than multiplying this polynomial by the least common multiple of the denominators in the coefficients, we get a polynomial with integer coefficients having α as a root. Thus we arrive at the same notion if in Definition 9.1.1 we replace "rational coefficients" by "integer coefficients".

(3) The situation changes dramatically, however, if we require real or complex coefficients instead of rational or integer ones: Every complex number is a root of a non-zero polynomial with real coefficients (which are thus complex coefficients), see Exercise 9.1.7.

(4) Instead of "algebraic number" we can say also "algebraic number over the rationals" (or "algebraic element over the rational field") as we can generalize the notion to algebraic elements over other fields than the rationals (see Definition 10.1.4).

Examples. As mentioned before, every rational number is algebraic.

Among the irrational numbers, e.g. $\sqrt{2}$ or $\sqrt[5]{13}$ are algebraic, as they are roots of polynomials $x^2 - 2$ and $x^5 - 13$.

Every complex root of unity is algebraic, being a root of some polynomial $x^n - 1$.

Further examples occur in Exercises 9.1.1 and 9.1.2. With the help of theorems in Section 9.3, we will be able to construct many types of algebraic numbers. The non-algebraic numbers are called transcendental.

Definition 9.1.2. A complex number is a *transcendental number* (or shortly, *transcendental*) if it is not a root of any non-zero polynomial with rational coefficients.   ♣

Theorem 9.1.3. *There exist transcendental numbers, moreover almost all complex numbers are transcendental: the algebraic numbers are countable, whereas the cardinality of transcendental numbers is continuum.*   ♣

Proof. Since the cardinality of complex numbers is that of the continuum, all statements follow if we verify that the algebraic numbers are countable, so we can order the algebraic numbers in a sequence.

The algebraic numbers are the roots of non-zero polynomials with integer coefficients, so first we put these polynomials into a sequence. Then we obtain a sequence of all algebraic numbers by taking all (complex) roots of these polynomials that had not yet been listed as roots of previous polynomials. Let $f = a_0 + a_1 x + \cdots + a_n x^n$ be an arbitrary non-zero polynomial with integer coefficients where $a_n \neq 0$, and define $H(f)$ as

$$H(f) = n + |a_0| + |a_1| + \cdots + |a_n|.$$

For example,

(9.1.1)

$$H(f) = 1 \iff f = \pm 1$$
$$H(f) = 2 \iff f = \pm 2, \pm x$$
$$H(f) = 3 \iff f = \pm 3, \pm x \pm 1, \pm 2x, \pm x^2$$
$$H(f) = 4 \iff f = \pm 4, \pm x \pm 2, \pm 2x \pm 1, \pm 3x,$$
$$\pm x^2 \pm 1, \pm x^2 \pm x, \pm 2x^2, \pm x^3.$$

It is clear from the definition of $H(f)$ that for any k there exist only finitely many f satisfying $H(f) = k$. Therefore we get a suitable sequence by taking one after another the polynomials with $H(f) = 1, 2, 3, \ldots$ From this we get a sequence of all algebraic numbers. The first few elements are, using the order of polynomials in (9.1.1),

$$0, 1, -1, 2, -2, \frac{1}{2}, -\frac{1}{2}, i, -i, \ldots$$

The non-zero constant polynomials have no roots, 0 comes from x, ± 1 comes from $x \mp 1$, etc., the constant multiples, products or divisors of previous polynomials provide no new roots. Thus we can restrict ourselves to polynomials f satisfying $n > 0$, $a_n > 0$, $(a_0, a_1, \ldots, a_n) = 1$, and being irreducible over the rational field. $\square$

In Theorem 9.1.3 we verified the existence of many transcendental numbers without exhibiting a single one. In Section 9.4 we shall construct one, and in Section 9.5 we shall prove that e (the base of the natural logarithm) is transcendental.

Some other notable transcendental numbers are

- π
- $\sin n$, where the angle n is an integer measured in radians
- $\log_{10} n$, where n is a positive integer, except the trivial case when n is a power of 10 (with an integer exponent), see Exercise 9.3.7
- $2^{\sqrt{2}}$, see Theorem 9.3.5
- $\zeta(2) = \sum_{k=1}^{\infty} \frac{1}{k^2}$, see Exercise 9.1.3.

In general, it is difficult to determine whether a given number is algebraic or transcendental; for each of the examples listed, the proof of transcendence is much beyond the scope of this book (Theorem 9.3.5 is stated without proof, and the other two references reduce the question to this theorem and to the transcendence of π).

It is unknown whether $e + \pi$ is transcendental, or at least irrational. It is also unknown whether $\zeta(3) = \sum_{k=1}^{\infty} k^{-3}$ is transcendental; its irrationality was proved only in 1975. About $\zeta(5)$ we do not even know whether it is irrational.

Exercises 9.1

1. Prove that the following numbers are algebraic.

   (a) $\sqrt[20]{7}$
   (b) $\sqrt{2} + 3$
   (c) $\sqrt{2} + \sqrt{3}$
   (d) $\sqrt{2} + \sqrt[3]{4}$
   (e) $\sqrt[3]{2} + \sqrt[3]{4}$
   (f) $\sqrt{2} + \sqrt{3} + \sqrt{5}$

2. Prove that if α is algebraic, then also

   (a) $-\alpha$
   (b) $\bar{\alpha}$
   (c) $1/\alpha$ (for $\alpha \neq 0$)
   (d) $r + \alpha$
   (e) $r\alpha$
   (f) $\sqrt[k]{\alpha}$

   are algebraic, where r is an arbitrary rational number and k is a positive integer.

3. Using the transcendence of π, verify that $\zeta(2) = \sum_{k=1}^{\infty} k^{-2}$ is transcendental.

4. Assume that α is transcendental and $f \neq 0$ is a polynomial with integer coefficients. Show that also $f(\alpha)$ is transcendental.

5. Let g be a non-zero polynomial with complex coefficients. Prove that there exists a non-zero polynomial h with integer coefficients satisfying $g \mid h$ if and only if every (complex) root of g is an algebraic number.

6. Verify that a complex number α is algebraic if and only if $1, \alpha, \ldots, \alpha^n$ are linearly dependent over the rational field for some positive integer n.

7. Prove that every complex number is a root of some non-zero polynomial with (a) complex (b) real coefficients.

9.2. Minimal Polynomial and Degree

Any algebraic number is a root of infinitely many polynomials with rational coefficients: if f is such a non-zero polynomial, then f multiplied by an arbitrary polynomial (with rational coefficients) will have this property, too. Thus it is worthwhile to distinguish those polynomials that have minimal degree:

Definition 9.2.1. The *minimal polynomial* of an algebraic number α is a polynomial with rational coefficients of minimal degree having α as one of its roots. We denote the minimal polynomial of α by m_α.   ♣

The minimal polynomial is not completely unique: if f is a minimal polynomial of α, then cf meets this requirement, where $c \neq 0$ is an arbitrary rational number. Apart from this ambiguity, however, the minimal polynomial is unique:

Theorem 9.2.2. *If f and g are minimal polynomials of the same algebraic number α, then $g = cf$ for some rational number $c \neq 0$.* ♣

Proof. Let

$$
\begin{aligned}
f &= a_0 + a_1 x + \cdots + a_n x^n, & a_n \neq 0 \\
g &= b_0 + b_1 x + \cdots + b_n x^n, & b_n \neq 0.
\end{aligned}
$$

Then α is a root of the polynomial $h = b_n f - a_n g$ which is either the zero polynomial or has degree at most $n - 1$. By the definition of minimal polynomial, the second case is impossible, so $h = 0$. Therefore $g = cf$, where $c = b_n / a_n$. □

The notation m_α can refer in the sequel to any minimal polynomial of α. This can cause no problem by Theorem 9.2.2.

We summarize the most important properties of minimal polynomials in

Theorem 9.2.3.   (i) *Let $g \in \mathbf{Q}[x]$. Then $g(\alpha) = 0 \iff m_\alpha \mid g$.*

(ii) *m_α is irreducible over $\mathbf{Q}$.*

(iii) *If f is irreducible over $\mathbf{Q}$ and $f(\alpha) = 0$, then f is a minimal polynomial of α.* ♣

Proof. (i) We first assume $m_\alpha \mid g$, i.e. $g = h m_\alpha$, for some $h \in \mathbf{Q}[x]$. Then

$$
g(\alpha) = h(\alpha) m_\alpha(\alpha) = h(\alpha) \cdot 0 = 0.
$$

Conversely, we assume $g(\alpha) = 0$. Applying the division algorithm for g and m_α, we get

$$
g = m_\alpha h + r, \qquad \text{where} \qquad h, r \in \mathbf{Q}[x], \quad \text{and} \quad \deg r < \deg m_\alpha \quad \text{or} \quad r = 0.
$$

Then

$$
0 = g(\alpha) = m_\alpha(\alpha) h(\alpha) + r(\alpha) = 0 + r(\alpha) = r(\alpha).
$$

The case $\deg r < \deg m_\alpha$ contradicts the definition of minimal polynomial, so only $r = 0$ is possible, so $m_\alpha \mid g$.

(ii) For a proof by contradiction, assume $m_\alpha = gh$, where g and h are polynomials with rational coefficients of smaller degree than m_α. Then as there are no zero divisors in the complex field,

$$
0 = m_\alpha(\alpha) = g(\alpha) h(\alpha) \implies g(\alpha) = 0 \quad \text{or} \quad h(\alpha) = 0,
$$

which contradicts the definition of minimal polynomial.

(iii) By part (i), $m_\alpha \mid f$. This implies $m_\alpha = c$ or $f = c m_\alpha$ for some constant c, since f is irreducible. The first case is impossible, and the second case says that f is a minimal polynomial. □

Definition 9.2.4. The *degree* of an algebraic number α is the degree of its minimal polynomial: $\deg \alpha = \deg m_\alpha$. ♣

Examples.   **E1** A minimal polynomial of 0 is $m_0 = x$, that of 1 is $m_1 = x - 1$, and in general, $m_r = x - r$ is a minimal polynomial of a rational number r. It is also clear that exactly the rational numbers have degree 1, and there are no algebraic numbers of degree 0.

E2 A minimal polynomial of i is $x^2 + 1$, so $\deg i = 2$.

E3 A minimal polynomial of $\sqrt[5]{3}$ is $x^5 - 3$, since this polynomial is irreducible over $\mathbf{Q}$ by the Schönemann–Eisenstein criterion.

E4 For any positive integer k, there exist infinitely many algebraic numbers of degree k, as, using the Schönemann–Eisenstein criterion, there are infinitely many irreducible polynomials over $\mathbf{Q}$ having degree k.

E5 A minimal polynomial of a primitive complex nth root of unity ϱ is the nth cyclotomic polynomial Φ_n, since $\Phi_n(\varrho) = 0$ and Φ_n is irreducible over $\mathbf{Q}$. Hence, $\deg \varrho = \deg \Phi_n = \varphi(n)$. (Example E2 was a special case $n = 4$.)

Exercises 9.2

1. What is the connection between the degrees of the numbers in Exercise 9.1.2 and the degree of α?

2. Determine the degree of the algebraic numbers

    (a) $\sqrt[7]{12}$

    (b) $\cos 20°$

    (c) $\sqrt[3]{3} - \sqrt[3]{9}$

    (d) $\sqrt{7 - 4\sqrt{3}}$

    (e) $\sqrt[4]{2} + \sqrt{2}$

    (f) $\sqrt[4]{2} + \sqrt{2} + \sqrt[4]{8}$.

3. Prove that α is an algebraic number of degree 2 if and only if $\alpha = r + \sqrt{s}$, where r and s are rational numbers and s is not the square of a rational number.

4. Demonstrate that the algebraic numbers of degree n are everywhere dense

    (a) on the real number line for $n \geq 1$

    (b) on the complex plane for $n \geq 2$.

5. Let f be a polynomial with rational coefficients of degree $n \geq 1$ and $\alpha_1, \ldots, \alpha_n$ its (complex) roots, counted with multiplicity.

    (a) Verify $\sum_{i=1}^{n} \deg \alpha_i \leq n^2$.

    (b) When does (a) hold with equality?

    (c) Show that if (a) holds with strict inequality, then $\sum_{i=1}^{n} \deg \alpha_i \leq n^2 - 2n + 2$.

6. We know that $\deg \alpha = 6$ and α is a root of the polynomial

$$f = x^7 + 8x^6 + 15x^5 + 10x^3 + 35x^2 + 5x - 30.$$

Find a minimal polynomial of α.

7. Assume that the complex numbers α and β are roots of a non-zero polynomial f with rational coefficients and $\deg f < \deg \alpha + \deg \beta$. Prove $m_\alpha = m_\beta$.

S 8. Assume that polynomials with rational coefficients $f \neq 0$ and g, and complex numbers α and β satisfy $f(\alpha) = g(\alpha) = f(\beta) = 0$, $g(\beta) = 1$. Show that f is reducible over $\mathbf{Q}$.

9.3. Operations with Algebraic Numbers

In this section we discuss the connection of algebraic numbers to the four basic arithmetic operations and exponentiation.

Theorem 9.3.1. *The algebraic numbers form a subfield of the complex numbers, so the sum, difference, product, and (if the divisor is not zero, then) quotient of two algebraic numbers are algebraic.* ♣

We prove the theorem using symmetric polynomials. (For another proof, see Section 10.2.)

A polynomial of k variables $F(x_1, \ldots, x_k)$ over a ring R is called *symmetric* if permuting the variables x_i gives the same polynomial. Such polynomials are e.g. the sum or product of the variables, or more generally, the sum of all possible products of j (distinct) variables:

(9.3.1)

$$\sigma_j(x_1, \ldots, x_k) = \sum_{1 \le i_1 < \cdots < i_j \le k} x_{i_1} \ldots x_{i_j}$$

$$= x_1 x_2 \ldots x_{j-1} x_j + x_1 x_2 \ldots x_{j-1} x_{j+1} + \cdots + x_{k-j+1} x_{k-j+2} \ldots x_{k-1} x_k,$$

$$j = 1, \ldots, k.$$

The polynomials σ_j are called the *elementary symmetric polynomials* of variables $x_1, \ldots, x_k$.

Since sums and products of symmetric polynomials are symmetric, thus $\sigma_1 + \sigma_2^3$, or in general, any polynomials formed with coefficients from R in variables σ_j are again symmetric polynomials in variables x_i.

The importance of elementary symmetric polynomials lies primarily in the fact that the converse of the previous observation is true: Every symmetric polynomial can be obtained as a polynomial of elementary symmetric polynomials.

Theorem 9.3.2 (The Fundamental Theorem of Symmetric Polynomials). *Let $F(x_1, \ldots, x_k)$ be a symmetric polynomial (of k variables) over a ring R. Then there exists a polynomial G of k variables over R such that*

$$F(x_1, \ldots, x_k) = G(\sigma_1, \ldots, \sigma_k),$$

where $\sigma_j = \sigma_j(x_1, \ldots, x_k)$ denote the elementary symmetric polynomials with variables x_i defined by (9.3.1). ♣

Example. We can express the sum of squares of x_i with elementary symmetric polynomials σ_j as

$$x_1^2 + x_2^2 + \cdots + x_k^2 = (x_1 + \cdots + x_k)^2 - 2(x_1 x_2 + x_1 x_3 + \cdots) = \sigma_1^2 - 2\sigma_2.$$

The proof of the Fundamental Theorem of Symmetric Polynomials can be found in any introductory algebra textbook.

We can add to the theorem that G is unique, and its coefficients are obtained from the coefficients of F using only addition and subtraction.

We shall apply the theorem mostly for the two cases where R is the rational field or the ring of integers. Thus if the coefficients of a symmetric polynomial F are rational numbers or integers, then the corresponding polynomial G has rational or integer coefficients, resp.

Proof of Theorem 9.3.1. We saw in Exercise 9.1.2 that the negative of an algebraic number and the reciprocal of a non-zero algebraic number are algebraic. Thus it is enough to verify that the sum and product of two algebraic numbers are algebraic.

Assume that the algebraic numbers α and β are roots of polynomials with rational coefficients

$$f = \prod_{i=1}^{m}(x - \alpha_i) \quad \text{and} \quad g = \prod_{j=1}^{n}(x - \beta_j),$$

where $\alpha_1 = \alpha$ and $\beta_1 = \beta$. Then $\alpha + \beta$ is a root of the polynomial

$$h = \prod_{i=1}^{m}\prod_{j=1}^{n}(x - \alpha_i - \beta_j).$$

We show that $h = c_0 + c_1 x + \cdots + c_{nm-1}x^{nm-1} + x^{nm}$ has rational coefficients.

Rewrite h as

$$h = \prod_{i=1}^{m} g(x - \alpha_i),$$

and observe that any permutation of the numbers α_i leaves h, and so each coefficient c_r, unchanged. This means that if we consider $\alpha_1, \ldots, \alpha_m$ as variables, then every coefficient c_r is a symmetric polynomial in the α_i:

$$c_r = F_r(\alpha_1, \ldots, \alpha_m), \qquad r = 0, 1, \ldots, nm - 1,$$

where F_r is a symmetric polynomial with rational coefficients (since these were obtained from the coefficients of g). By Theorem 9.3.2, F_r can be represented as a polynomial in the elementary symmetric polynomials σ_j formed from the numbers α_i, i.e.

$$c_r = F_r(\alpha_1, \ldots, \alpha_m) = G_r(\sigma_1, \ldots, \sigma_m)$$

for a suitable polynomial G_r with rational coefficients. By Viète's formulas about the relation between roots and coefficients, the elementary symmetric polynomials of α_is are just the coefficients of f, possibly with a negative sign. Hence $c_r = G_r(\sigma_1, \ldots, \sigma_m)$ is a rational number. We have thus proved that h has rational coefficients, so $\alpha + \beta$ is an algebraic number.

We can verify similarly that $\alpha\beta$ is algebraic: now we have to consider the polynomial

$$\prod_{i=1}^{m}\prod_{j=1}^{n}(x - \alpha_i\beta_j) = \prod_{i=1}^{m} \alpha_i^n g\left(\frac{x}{\alpha_i}\right).$$

If $\alpha \neq 0$, then we can assume that no α_i is zero, and if $\alpha = 0$, then it is obvious that $\alpha\beta = 0$ is algebraic. $\qquad\square$

An important corollary of Theorem 9.3.1 is that the question whether a complex number is algebraic or transcendental can be reduced to the similar question for real numbers:

Theorem 9.3.3. *A complex number is algebraic if and only if both its real and imaginary parts are algebraic.* ♣

Proof. Let $\alpha = a + bi$ where a and b are real numbers.

We assume first that a and b are algebraic. Since i is algebraic as a root of the polynomial x^2+1, and products and sums of algebraic numbers are algebraic, $\alpha = a+bi$ is algebraic.

Conversely, assume that $\alpha = a + bi$ is algebraic. Then $\overline{\alpha} = a - bi$ is algebraic (see Exercise 9.1.2b). Since sums, differences, and quotients of algebraic numbers, including 2 and $2i$, are algebraic, hence

$$a = \frac{\alpha + \overline{\alpha}}{2} \quad \text{and} \quad b = \frac{\alpha - \overline{\alpha}}{2i}$$

are algebraic. $\qquad\square$

Now we consider powers of algebraic numbers. Since we define the powers of 0 only for positive real exponents and they all are 0, in the sequel we can restrict our investigation to powers of non-zero algebraic numbers.

Theorem 9.3.4. *All powers of algebraic numbers with rational exponents are algebraic.* ♣

Proof. Since products and reciprocals of algebraic numbers are algebraic and 1 is algebraic, the statement holds for integer exponents. The statement for fractional exponents follows from the fact that roots of algebraic numbers with integer exponents are algebraic (see Exercise 9.1.2f). $\qquad\square$

For non-rational exponents, the simplest question is whether $2^{\sqrt{2}}$ is transcendental, or at least irrational. This occurs among the famous Hilbert problems from the year 1900, and Hilbert thought it to be more difficult than Fermat's Last Theorem or Riemann's Hypothesis. This, however, did not discourage researchers, and Gelfond and Schneider proved the following general result in 1934, independently and with different methods, which we state without proof:

Theorem 9.3.5 (Gelfond–Schneider Theorem). *If α and β are algebraic numbers, $\alpha \neq 0$ or 1, and β is not rational, then α^β is transcendental.* ♣

This implies that if an integer n is not a power of 10 with an integer exponent, then $\log_{10} n$ is transcendental (see Exercise 9.3.7).

Theorem 9.3.5 is true also for complex exponents β, when the power generally has infinitely many values. This makes possible a simple verification of the transcendence of e^π (see Exercise 9.3.4b), whereas we cannot answer the weaker question of whether $e+\pi, e-\pi, e\pi, e/\pi$, and π^e are irrational, though most of them must be transcendental, see Exercise 9.3.4a.

We saw in Theorem 9.3.4 (and in Exercise 9.1.2f) that the algebraic numbers are closed under taking roots with integer exponents. Another formulation of this fact is that if α is an algebraic number, then the roots of the polynomial $x^k - \alpha$ having algebraic coefficients are algebraic. This holds not only for such polynomials of special form, but for any polynomials with algebraic coefficients.

Theorem 9.3.6. *If the coefficients of a polynomial $f \neq 0$ are algebraic numbers, then all (complex) roots of f are algebraic, as well.*   ♣

Proof. We shall use again the Fundamental Theorem 9.3.2 of Symmetric Polynomials. We shall see another proof in Section 10.2.

Let $f = \alpha + \beta x + \cdots + \xi x^n$, where $\alpha, \beta, \ldots, \xi$ are algebraic numbers, and let $\alpha_i, \beta_j,$ $\ldots, \xi_k$ denote the other roots of the minimal polynomials of $\alpha, \beta, \ldots, \xi, (\alpha_1 = \alpha,$ etc.). Consider the polynomial

$$h = \prod_{i,j,\ldots,k} (\alpha_i + \beta_j x + \cdots + \xi_k x^n).$$

Since f is a factor of h, all roots of f are roots also of h. Thus it is sufficient to verify that h has rational coefficients.

Let c_r be a coefficient of h. Similar to the arguments in the proof of Theorem 9.3.1, c_r is a symmetric polynomial F_r with variables α_i, where the coefficients of F_r are obtained from the numbers $\beta_j, \ldots, \xi_k$ by addition, subtraction, and multiplication. By Theorem 9.3.2, F_r is a polynomial in elementary symmetric polynomials of variables α_i. Using Viète's formulas connecting the roots and coefficients of the minimal polynomial m_α, we get that these elementary symmetric polynomials are rational numbers. Thus we eliminated the numbers α_i from c_r. Repeating the same argument for β_j, etc., we obtain that c_r is a rational number.   □

Summarizing the statements of Theorems 9.3.1 and 9.3.6, the algebraic numbers form an algebraically closed field.

Exercises 9.3

1. (a) Verify that the sum of an algebraic number and a transcendental number is transcendental.

   (b) Give examples of two transcendental numbers whose sum is (a) transcendental (b) algebraic.

   (c) Investigate similar questions for products instead of sums.

2. What can we assert about α and β (from algebraic/transcendental aspect), if

(a) $\alpha + \beta$ and $\alpha - \beta$ are algebraic

(b) $\alpha + \beta$ is algebraic and $\alpha - \beta$ is transcendental

(c) $\alpha + \beta$ and $\alpha - \beta$ are transcendental

(d) $\alpha\beta$ and α/β are algebraic

(e) $\alpha + \beta$ is algebraic and $\alpha\beta$ is transcendental

(f) $\alpha + \beta$ is transcendental and $\alpha\beta$ is algebraic

(g) $\alpha + \beta$ and $\alpha\beta$ are transcendental

(h) $\alpha + \beta$ and $\alpha\beta$ are algebraic?

What are the changes if α and β are real numbers, and the words "algebraic" and "transcendental" are replaced everywhere by "rational" and "irrational", respectively?

3. Assume that $\alpha + \beta$ and $\alpha + \gamma$ are algebraic and $\beta + \gamma$ is transcendental. Determine for each of the following numbers whether it is algebraic or transcendental

(a) α

(b) $2\alpha + (1 - i)\beta + (1 + i)\gamma$

(c) $3\alpha + (2 - i)\beta + (2 + i)\gamma$.

4. As mentioned before, the transcendence, or even the irrationality, of $e + \pi$, $e - \pi$, $e\pi$, e/π, and π^e is a notorious unsolved problem.

(a) At most how many of $e + \pi$, $e - \pi$, $e\pi$, and e/π can possibly be algebraic?

(b) Prove that (b1) $e + i\pi$ and (b2) e^π are transcendental.

5. For each of the following numbers, determine whether it is algebraic or transcendental

(a) $\sin 7°$

(b) $i\pi + \pi/i$

(c) $\pi^7 + i\pi^5 + \sqrt{2\pi}$.

S 6. Let the trigonometric form of a complex number $\alpha \neq 0$ be $\alpha = r(\cos\varphi + i\sin\varphi)$. Verify that α is algebraic if and only if both r and $\cos\varphi$ are algebraic.

7. Assume that the positive integer n is not a power of 10 with an integer exponent. Prove that $\log_{10} n$ is transcendental.

8. For complex numbers α and β, form the sequence

$$H = (\alpha + \beta, \alpha^2 + \beta^2, \dots, \alpha^k + \beta^k, \dots).$$

Show that if at least two elements of H are algebraic and not both of them are 0, then every element of H is algebraic.

9. Consider the powers with real transcendental exponents of a positive algebraic number different from 1. Prove that there are infinitely many algebraic and infinitely many transcendental numbers among them.

9.4. Approximation of Algebraic Numbers

In this section we discuss the approximation of algebraic numbers and its consequences. As a non-real complex number cannot be approximated well by rationals, we investigate only approximation of real algebraic numbers.

We saw in Exercise 8.1.1 that rational numbers can be approximated only very poorly. We showed in Theorem 8.1.6 and Exercise 8.1.6, that the irrational numbers $(1 + \sqrt{5})/2$ and $\sqrt{2}$ satisfy only a pretty poor approximation. Liouville proved that it is true in general that no algebraic number can be approximated well, in the following sense:

Theorem 9.4.1. *Let $n \geq 2$ and α a (real) algebraic number of degree n. Then there exists a real constant $c = c(\alpha) > 0$ such that every rational number r/s satisfies*

$$(9.4.1) \qquad\qquad \left| \alpha - \frac{r}{s} \right| > \frac{c(\alpha)}{s^n}. \qquad\qquad ♣$$

Remarks: (1) Another formulation of Theorem 9.4.1 is: There exists a real constant $c' = c'(\alpha) > 0$ such that

$$\left| \alpha - \frac{r}{s} \right| < \frac{c'(\alpha)}{s^n}$$

is true only for finitely many rational numbers r/s. This means that we allow finitely many exceptions in (9.4.1). They can easily be eliminated by choosing $c(\alpha)$ so small that the finitely many exceptional fractions r/s satisfy (9.4.1). Thus the two forms of the theorem are equivalent, and each implies the other immediately.

(2) Another corollary of Theorem 9.4.1 is that for any real numbers $t > n$ and $c^* > 0$,

$$\left| \alpha - \frac{r}{s} \right| < \frac{c^*}{s^t}$$

can be valid only for finitely many rational number r/s, since if s is large enough (depending on t and c^*), then

$$\frac{c^*}{s^t} < \frac{c(\alpha)}{s^n}.$$

This means that, using the term introduced in Exercise 8.1.7, an algebraic number of degree n cannot be approximated to a greater order than n, i.e. it cannot be approximated to order t for any $t > n$. However, even a much stronger result holds, see Theorem 9.4.4.

(3) By Exercise 8.1.1, Theorem 9.4.1 is valid also for $n = 1$ if we exclude $\alpha = r/s$.

Proof. For a proof by contradiction, we assume that for every $c > 0$ there exists a rational number r/s (with $s > 0$) satisfying

$$\left| \alpha - \frac{r}{s} \right| < \frac{c}{s^n}.$$

This means

$$(9.4.2) \qquad\qquad \lim_{i \to \infty} s_i^n \left(\alpha - \frac{r_i}{s_i} \right) = 0$$

for a suitable sequence of rational numbers r_i/s_i (where $s_i > 0$). A direct consequence is

(9.4.3) $$\lim_{i \to \infty}\left(\alpha - \frac{r_i}{s_i}\right) = 0 \qquad \text{or} \qquad \lim_{i \to \infty} \frac{r_i}{s_i} = \alpha.$$

We consider a copy of m_α with integer coefficients, and denote its complex roots by $\alpha_1 = \alpha, \alpha_2, \ldots, \alpha_n$. Then

(9.4.4) $$m_\alpha = a_0 + a_1 x + \cdots + a_n x^n = a_n \prod_{j=1}^{n} (x - \alpha_j),$$

where $a_0, a_1, \ldots, a_n$ are integers and $a_n \neq 0$. Since m_α is irreducible over $\mathbf{Q}$, it cannot have multiple roots (see Exercise 9.4.4), so the numbers α_j are distinct.

Substituting r_i/s_i into m_α, we obtain by (9.4.4)

(9.4.5) $$a_0 + a_1\left(\frac{r_i}{s_i}\right) + \cdots + a_n\left(\frac{r_i}{s_i}\right)^n = a_n\left(\frac{r_i}{s_i} - \alpha\right) \prod_{j=2}^{n}\left(\frac{r_i}{s_i} - \alpha_j\right).$$

The left-hand side of (9.4.5) is a rational number with denominator s_i^n, and is not 0, as m_α has no rational roots. Thus the absolute value of the left-hand side in (9.4.5) is at least $1/s_i^n$. Multiplying (9.4.5) by s_i^n we get

(9.4.6) $$1 \leq \left| s_i^n a_n\left(\alpha - \frac{r_i}{s_i}\right) \prod_{j=2}^{n}\left(\frac{r_i}{s_i} - \alpha_j\right) \right|.$$

From (9.4.3) we obtain

$$\lim_{i \to \infty} \prod_{j=2}^{n}\left(\frac{r_i}{s_i} - \alpha_j\right) = \prod_{j=2}^{n}(\alpha - \alpha_j).$$

Combining this with (9.4.2), we see that the right-hand side of (9.4.6) tends to 0 for $i \to \infty$, which is an obvious contradiction. $\square$

Liouville used Theorem 9.4.1 to construct transcendental numbers: If a real number α can be approximated extremely well, then it must be transcendental. We can obtain such an α as the sum of an infinite series where the partial sums converge extremely quickly. Below we present Liouville's construction in detail.

Theorem 9.4.2. *The number*

(9.4.7) $$\alpha = \sum_{k=1}^{\infty} \frac{1}{10^{k!}} = 0.110001000000000000000001\ldots$$

is transcendental. The decimal digits at places $k!$ are 1, all other digits are 0. ♣

Equation (9.4.7) defines a real number, as we see from the decimal representation form, it also follows from the convergence of the infinite series since the infinite geometric series $\sum_{k=1}^{\infty} 10^{-k}$ is its majorant.

Proof. We show that the partial sums of the infinite series (9.4.7) approximate α very well.

We write the mth partial sum as r_m/s_m, where $(r_m, s_m) = 1$ and $s_m > 0$. The common denominator is $10^{m!}$, and

$$\sum_{k=1}^{m} \frac{1}{10^{k!}} = \frac{10A + 1}{10^{m!}},$$

thus $s_m = 10^{m!}$. Then

$$0 < \alpha - \frac{r_m}{s_m} = \sum_{k=m+1}^{\infty} \frac{1}{10^{k!}} < \sum_{j=(m+1)!}^{\infty} \frac{1}{10^j} = \frac{10}{9 \cdot 10^{(m+1)!}} = \frac{10}{9 s_m^{m+1}}.$$

This implies

(9.4.8) $$\left| \alpha - \frac{r_m}{s_m} \right| < \frac{10}{9 s_m^{m+1}}.$$

Assume now that α is algebraic and its degree is n. Since α is not a periodic decimal fraction, α is irrational, so $n \geq 2$. By Theorem 9.4.1 there is a constant $c(\alpha) > 0$ such that (9.4.1) holds for every rational number r/s. Then this is true also for r_m/s_m, so

(9.4.9) $$\left| \alpha - \frac{r_m}{s_m} \right| > \frac{c(\alpha)}{s_m^n}.$$

Combining (9.4.8) and (9.4.9), we get

$$\frac{c(\alpha)}{s_m^n} < \frac{10}{9 s_m^{m+1}}, \quad \text{i.e.} \quad s_m^{m-n+1} < \frac{10}{9 c(\alpha)},$$

which is a contradiction if m is large enough. $\square$

Theorem 9.4.1 can be improved significantly, as we mentioned in Remark 2 after the theorem. Thue and Roth proved the following results that we state without proof:

Theorem 9.4.3 (Thue's Theorem). *Let α be a real algebraic number of degree $n \geq 3$ and c an arbitrarily large constant. Then the inequality*

(9.4.10) $$\left| \alpha - \frac{r}{s} \right| < \frac{c}{s^n}$$

is satisfied only by finitely many rational numbers r/s.   ♣

Theorem 9.4.4 (Roth's Theorem). *Let α be an algebraic number and $\kappa > 0$ arbitrary. Then the inequality*

(9.4.11) $$\left| \alpha - \frac{r}{s} \right| < \frac{1}{s^{2+\kappa}}$$

is satisfied only by finitely many rational numbers r/s.   ♣

Remarks: (1) Roth's theorem is clearly much stronger than Thue's, but Thue's theorem already has important consequences for Diophantine equations (see Theorem 9.4.5).

(2) By Roth's theorem, the exceptional set H in Theorem 8.1.8 consists purely of transcendental numbers. But Theorem 8.1.8 also demonstrates that (besides all algebraic numbers) most transcendental numbers can be approximated very badly.

Diophantine approximation is closely related to the behavior of certain Diophantine equations. We saw in Section 7.8 that if a positive integer m is not a square, then Pell's equation $x^2 - my^2 = 1$ has infinitely many solutions (Theorem 7.8.1); this was based on the fact that the irrational number $\sqrt{m}$ can be approximated to order 2. Now we shall rely on the poor approximation of algebraic numbers to show that certain Diophantine equations of higher degree can have at most finitely many solutions.

Theorem 9.4.5. *Let $f = a_0 + a_1 x + \cdots + a_n x^n$ be a polynomial of degree n with integer coefficients, where $n \geq 3$ and f is irreducible over $\mathbf{Q}$. Then for any (fixed) integer b, the Diophantine equation*

$$(9.4.12) \qquad g(y, z) = y^n f\left(\frac{z}{y}\right) = a_0 y^n + a_1 y^{n-1} z + \cdots + a_n z^n = b$$

can have at most finitely many solutions. ♣

Proof. Assume that infinitely many pairs of integers (y_i, z_i) satisfy (9.4.12). Since for a given y there can be at most n values of z,

$$(9.4.13) \qquad \lim_{i \to \infty} |y_i| = \infty,$$

and we may assume that none of the values y_i is 0.

Substituting (y_i, z_i) into (9.4.12) and dividing by y_i^n, we obtain

$$(9.4.14) \qquad f\left(\frac{z_i}{y_i}\right) = \frac{b}{y_i^n}.$$

From (9.4.13) and (9.4.14) we infer

$$(9.4.15) \qquad \lim_{i \to \infty} f\left(\frac{z_i}{y_i}\right) = 0.$$

Let $\alpha_1, \ldots, \alpha_n$ be the roots of f, so

$$(9.4.16) \qquad f = a_n \prod_{j=1}^{n} (x - \alpha_j).$$

Substituting z_i/y_i, we get

$$(9.4.17) \qquad f\left(\frac{z_i}{y_i}\right) = a_n \prod_{j=1}^{n} \left(\frac{z_i}{y_i} - \alpha_j\right).$$

By (9.4.15), the left-hand side of (9.4.17) tends to 0 for $i \to \infty$, thus, taking a suitable subsequence of the indices i, the limit of some factor on the right-hand side has to be 0. Suppose it is the first factor on the right-hand side, and for convenience we use the notation of the original sequence for the subsequence. So

$$(9.4.18) \qquad \lim_{i \to \infty} \left(\frac{z_i}{y_i} - \alpha_1\right) = 0 \quad \text{or} \quad \lim_{i \to \infty} \frac{z_i}{y_i} = \alpha_1.$$

This implies that α_1 is a real number. By (9.4.18),

$$(9.4.19) \qquad \lim_{i \to \infty} a_n \prod_{j=2}^{n} \left(\frac{z_i}{y_i} - \alpha_j \right) = a_n \prod_{j=2}^{n} (\alpha_1 - \alpha_j).$$

Let d denote the limit in (9.4.19). Due to the irreducibility of f, the numbers α_j are distinct, hence $d \neq 0$. Then

$$(9.4.20) \qquad \left| a_n \prod_{j=2}^{n} \left(\frac{z_i}{y_i} - \alpha_j \right) \right| > \left| \frac{d}{2} \right|$$

for i large enough. Finally, from (9.4.14), (9.4.17), and (9.4.20) we obtain

$$\begin{aligned} \left| \frac{b}{y_i^n} \right| &= \left| f \left(\frac{z_i}{y_i} \right) \right| \\ &= \left| a_n \prod_{j=1}^{n} \left(\frac{z_i}{y_i} - \alpha_j \right) \right| \\ &= \left| \alpha_1 - \frac{z_i}{y_i} \right| \cdot \left| a_n \prod_{j=2}^{n} \left(\frac{z_i}{y_i} - \alpha_j \right) \right| \\ &> \left| \alpha_1 - \frac{z_i}{y_i} \right| \cdot \left| \frac{d}{2} \right|, \end{aligned}$$

if i is sufficiently large, so

$$(9.4.21) \qquad \left| \alpha_1 - \frac{z_i}{y_i} \right| < \left| \frac{2b}{d} \cdot \frac{1}{y_i^n} \right|.$$

Since α_1 is an algebraic number of degree n, (9.4.21) contradicts Theorem 9.4.3. $\square$

If instead of Theorem 9.4.3 we rely on Theorem 9.4.4, then we can prove by similar arguments that a much wider class of Diophantine equations cannot have infinitely many solutions (see Exercise 9.4.3).

Exercises 9.4

1. An irrational number α is a *Liouville number* if for every positive integer n there exists a rational number r/s satisfying $s > 1$ and

$$\left| \alpha - \frac{r}{s} \right| < \frac{1}{s^n}.$$

By Theorem 9.4.1, every Liouville number is transcendental.

S (a) Let α be a Liouville number, $h \neq 0$ a rational number, and k a positive integer. Verify that the following numbers are Liouville numbers:

(i) $h + \alpha$

(ii) $h\alpha$

(iii) α^k

(iv) $1/\alpha$.

(b) Prove that there are infinitely many Liouville numbers, moreover, they have the cardinality of the continuum.

2. Demonstrate that the statement of Theorem 9.4.5 remains valid for a polynomial f of degree at least three with integer coefficients if we replace irreducibility over **Q** with one of the weaker conditions:

   (a) f has no divisor of degree 1 or 2 among polynomials with rational coefficients.

   (b) If $b = 0$, then f has no rational roots, and if $b \neq 0$, then f has no multiple (complex) roots.

3. Let $g(y, z)$ be the polynomial in two variables defined in Theorem 9.4.5, and $h(y, z)$ be any polynomial in two variables of degree at most $n-3$ with integer coefficients. Using Theorem 9.4.4, prove that the Diophantine equation $g(y, z) = h(y, z)$ cannot have infinitely many solutions.

S 4. Show that if a polynomial is irreducible over **Q**, then it cannot have multiple (complex) roots.

9.5. Transcendence of e

First we show that e (the base of natural logarithm) and π are irrational numbers, then we prove that e is transcendental. We note that an improvement of the method can yield the transcendence of π. An important consequence of this is that we cannot get by Euclidean constructions a square having the same area as a given circle.

Theorem 9.5.1. *e is an irrational number.* ♣

Proof. We use the representation of e as the sum of an infinite series:

$$(9.5.1) \qquad e = 1 + \frac{1}{1!} + \frac{1}{2!} + \cdots + \frac{1}{n!} + \ldots.$$

For a proof by contradiction, assume $e = a/b$, where a and b are positive integers. Then $b!\, e$ is an integer. Multiplying (9.5.1) by $b!$, we obtain

$$b!\, e = n_b + \frac{1}{b+1} + \frac{1}{(b+1)(b+2)} + \cdots$$

where n_b is an integer depending on b. We have the following lower and upper bounds for the integer $b!\, e - n_b$:

$$0 < b!\, e - n_b$$
$$= \frac{1}{b+1} + \frac{1}{(b+1)(b+2)} + \cdots$$
$$< \frac{1}{b+1} + \frac{1}{(b+1)^2} + \cdots$$
$$= \frac{1}{b+1} \cdot \frac{1}{1 - \frac{1}{b+1}} = \frac{1}{b}.$$

This means that the integer $b!\,e - n_b$ lies between 0 and $1/b$, which is an obvious contradiction.   □

Theorem 9.5.2. π *is an irrational number.*   ♣

Proof. For a proof by contradiction, we assume $\pi = a/b$, where a and b are positive integers.

Let n be a large positive integer and f be the polynomial of degree $2n$

$$f(x) = \frac{x^n(1-x)^n}{n!}.$$

We consider the integral

$$I = a^{2n+1} \int_0^1 \sin(\pi x) f(x) \, dx.$$

We get a contradiction by showing that, on the one hand,

(A) I is an integer,

but, on the other hand,

(B) $0 < I < 1$ if n is sufficiently large.

We verify (B) first. Since for $0 < x < 1$,

$$0 < \sin(\pi x) \le 1 \quad \text{and} \quad 0 < f(x) < \frac{1}{n!},$$

so

$$0 < I < \frac{a^{2n+1}}{n!}.$$

If n is large enough, then $a^{2n+1}/n! < 1$, verifying (B).

Turning to (A), we show first that f and all its derivatives assume integer values at 0 and 1, i.e.

(9.5.2) $f^{(m)}(0)$ and $f^{(m)}(1)$ are integers, $m = 0, 1, 2, \dots.$

Since $f(x) = f(1-x)$, $f^{(m)}(x) = (-1)^m f^{(m)}(1-x)$ for every m, thus $f^{(m)}(0) = (-1)^m f^{(m)}(1)$. Therefore it is sufficient to deal with $x = 0$.

Another form of f is

$$f(x) = \frac{1}{n!}\left(c_n x^n + c_{n+1} x^{n+1} + \cdots + c_{2n} x^{2n}\right)$$

with integer coefficients c_i. Hence

$$f^{(m)}(0) = \begin{cases} 0, & \text{if } 0 \le m < n \text{ or } m > 2n \\ \dfrac{c_m m!}{n!} = c_m(n+1)(n+2)\dots m, & \text{if } n \le m \le 2n \end{cases}$$

which proves (9.5.2).

We shall integrate by parts several times to show that I is an integer. Assuming $\pi = a/b$, the first such integration yields

$$I = a^{2n+1} \int_0^1 \sin(\pi x) f(x)\, dx =$$

(9.5.3)
$$= a^{2n+1} \left[\frac{-\cos(\pi x) f(x)}{\pi} \right]_0^1 - \frac{a^{2n+1}}{\pi} \int_0^1 -\cos(\pi x) f'(x)\, dx =$$

$$= -a^{2n} b (f(1)\cos\pi - f(0)\cos 0) + I_1$$

where

$$I_1 = a^{2n} b \int_0^1 \cos(\pi x) f'(x)\, dx.$$

As a, b, $f(1)$, $f(0)$, $\cos\pi$, and $\cos 0$ are integers, so by (9.5.3), I is an integer if and only if I_1 is an integer.

We integrate I_1 by parts, using $\pi = a/b$ again:

$$I_1 = a^{2n} b \int_0^1 \cos(\pi x) f'(x)\, dx =$$

$$= a^{2n} b \left[\frac{\sin(\pi x) f'(x)}{\pi} \right]_0^1 - \frac{a^{2n} b}{\pi} \int_0^1 \sin(\pi x) f''(x)\, dx$$

$$= a^{2n-1} b^2 (f'(1)\sin\pi - f'(0)\sin 0) - I_2$$

where

$$I_2 = a^{2n-1} b^2 \int_0^1 \sin(\pi x) f''(x)\, dx.$$

Similar to the previous step, I_1 is an integer if and only if I_2 is an integer.

Continuing the process, we arrive at

$$I_{2n+1} = b^{2n+1} \int_0^1 \cos(\pi x) f^{(2n+1)}(x)\, dx$$

and we have to show that it is an integer. Since f is a polynomial of degree $2n$, $f^{(2n+1)}(x) = 0$, thus $I_{2n+1} = 0$. Hence I_{2n+1}, and so I are integers, proving (A). $\square$

Theorem 9.5.3. *e is a transcendental number.* ♣

Proof. Assume that e is algebraic, i.e.

(9.5.4)
$$a_0 + a_1 e + \cdots + a_n e^n = 0.$$

for some integers $n \geq 1$ and $a_0 \neq 0$, a_1, ..., a_n. Similar to the irrationality of π, a suitable integral will provide the contradiction.

Let f be a polynomial to be specified later, $\deg f = k$, and consider the integral

(9.5.5)
$$I(s) = \int_0^s e^{-x} f(x)\, dx$$

for an integer $s \geq 0$. Integrating by parts, we obtain

(9.5.6) $\qquad I(s) = [-e^{-x}f(x)]_0^s + \int_0^s e^{-x}f'(x)\,dx = f(0) - f(s)e^{-s} + I_1(s)$

where

$$I_1(s) = \int_0^s e^{-x}f'(x)\,dx.$$

Similarly, integrating $I_1(s)$ by parts yields

(9.5.7) $\qquad I_1(s) = [-e^{-x}f'(x)]_0^s + \int_0^s e^{-x}f''(x)\,dx = f'(0) - f'(s)e^{-s} + I_2(s)$

where

$$I_2(s) = \int_0^s e^{-x}f''(x)\,dx.$$

Thus, by (9.5.6) and (9.5.7) we have

$$I(s) = [f(0) + f'(0)] - [f(s) + f'(s)]e^{-s} + I_2(s).$$

Continuing the process, and using $I_{k+1} = 0$ due to $f^{(k+1)} = 0$, we get

(9.5.8)
$$\begin{aligned}
I(s) &= \int_0^s e^{-x}f(x)\,dx \\
&= [f(0) + f'(0) + \cdots + f^{(k)}(0)] - [f(s) + f'(s) + \cdots + f^{(k)}(s)]e^{-s}.
\end{aligned}$$

We multiply (9.5.8) by $a_s e^s$, and add the equalities for $s = 0, 1, \ldots, n$:

(9.5.9)
$$\begin{aligned}
\sum_{s=0}^{n} a_s e^s I(s) &= \sum_{s=0}^{n} a_s e^s \int_0^s e^{-x}f(x)\,dx \\
&= \sum_{s=0}^{n} a_s e^s [f(0) + f'(0) + \cdots + f^{(k)}(0)] \\
&\quad - \sum_{s=0}^{n} a_s [f(s) + f'(s) + \cdots + f^{(k)}(s)].
\end{aligned}$$

The sum in the second line of (9.5.9) is 0 by (9.5.4), thus (9.5.9) is equivalent to

(9.5.10) $\qquad \displaystyle\sum_{s=0}^{n} a_s e^s \int_0^s e^{-x}f(x)\,dx = -\sum_{s=0}^{n} a_s [f(s) + f'(s) + \cdots + f^{(k)}(s)].$

We achieve a contradiction by showing for a suitable f that the left-hand side of (9.5.10) has absolute value less than 1, whereas the right-hand side is a non-zero integer.

Let $p > n|a_0|$ a (large) prime and

(9.5.11) $\qquad\qquad\qquad f(x) = \dfrac{x^{p-1}(x-1)^p \ldots (x-n)^p}{(p-1)!}.$

As a generalization of (9.5.2) in the proof of Theorem 9.5.2, we now show: If $t \geq 0$ and j are integers, and $h(x)$ is a polynomial with integer coefficients, then the polynomial

$$g(x) = \dfrac{(x-j)^t h(x)}{t!}$$

and all its derivatives assume integer values at j, so $g^{(m)}(j)$ is an integer for every integer m. Writing $g(x)$ as

$$g(x) = \frac{d_t(x - j)^t + d_{t+1}(x - j)^{t+1} + \cdots + d_r(x - j)^r}{t!},$$

we obtain

(9.5.12) $\qquad g^{(m)}(j) = \begin{cases} 0, & \text{if } 0 \le m < t \text{ or } m > r \\ \dfrac{d_m m!}{t!} = d_m(t + 1)(t + 2)\ldots m, & \text{if } t \le m \le r. \end{cases}$

Since

$$f(x) = p \cdot \frac{(x - 1)^p h_1(x)}{p!},$$

where the polynomial $h_1(x)$ has integer coefficients, applying (9.5.12) for $g(x) = f(x)/p$, $t = p$, $j = 1$, and $h(x) = h_1(x)$, we obtain that $f^{(m)}(1)$ is an integer divisible by p for every m. Similarly,

(9.5.13) $\qquad p \mid f^{(m)}(j), \qquad j = 1, 2, \ldots, n, \qquad m = 0, 1, 2, \ldots.$

Finally, writing $f(x)$ as

$$f(x) = \frac{x^{p-1} h_0(x)}{(p - 1)!},$$

where the polynomial $h_0(x)$ has integer coefficients, and applying (9.5.12) for $g(x) = f(x)$, $t = p - 1$, $j = 0$ and $h(x) = h_0(x)$, we obtain that also $f^{(m)}(0)$ is an integer for every m, and

(9.5.14) $\qquad p \nmid f^{(p-1)}(0) = (-1)^{np}(n!)^p, \quad \text{but} \quad p \mid f^{(m)}(0), \quad \text{if} \quad m \ne p - 1,$

this holds because (9.5.12) implies $f^{(m)}(0) = 0$ for $m < p - 1$, and the product $f^{(m)}(0) = d_m p \ldots m$ contains a factor p for $m \ge p$.

By (9.5.13) and (9.5.14), we see that every term of the sum on the right-hand side of (9.5.10) is an integer, and each is divisible by p except the term $a_0 f^{(p-1)}(0)$. Thus the right-hand side of (9.5.10) is an integer not divisible by p, so it cannot be 0.

Now we show that the left-hand side of (9.5.10) has absolute value less than 1 for p large enough. If $0 < x < n$, then

$$|e^{-x}| < 1 \quad \text{and} \quad |f(x)| = \left| \frac{x^{p-1}(x - 1)^p \ldots (x - n)^p}{(p - 1)!} \right| < \frac{n^{(n+1)p-1}}{(p - 1)!},$$

hence

(9.5.15) $\qquad \left| \sum_{s=0}^{n} a_s e^s \int_0^s e^{-x} f(x) \, dx \right| \le \frac{e^n \left(\sum_{s=0}^{n} |a_s| \right)\left(n^{n+1} \right)^p}{(p - 1)!}.$

The right-hand side of (9.5.15) is of the form $A \cdot B^p/(p-1)!$, where A and B are constants. This expression tends to 0 for $p \to \infty$, so it will be less than 1 if p is large enough.

Thus we have verified that the left-hand side of (9.5.10) has absolute value less than 1, whereas the right-hand side is a nonzero integer. Thus the assumption (9.5.4) led to a contradiction, and so e cannot be an algebraic number. $\qquad \square$

Exercises 9.5

1. Let $a_1 < a_2 < \cdots < a_n < \ldots$ be a sequence of positive integers where $a_n \mid a_{n+1}$ for every n and every positive integer k is a divisor of at least one a_n. Show that the infinite series $\sum_{n=1}^{\infty} 1/a_n$ is convergent and its sum is an irrational number.

2. Let r denote a rational number. Prove:

   (a) $\sin 1$ and $\cos 1$ are irrational.

   * (b) If $0 < r \le \pi$, then at least one of $\sin r$ and $\cos r$ is irrational.

   (c) If $0 < r < \pi/2$, then $\tan r$ is irrational.

   (The angles are given in radians. Do not rely on the fact stated before without proof that $\sin n$ is transcendental if n is an integer. For trigonometric functions of angles being rational measured in degrees, see Exercise 9.6.11.)

* 3. Refining the proof of Theorem 9.5.2, show that π^2 is irrational.

9.6. Algebraic Integers

Algebraic integers are special algebraic numbers that can be considered as generalizations of (ordinary) integers.

In preparation, we give a characterization of integers within the rationals that can be then extended to algebraic numbers.

The rational numbers are exactly the algebraic numbers of degree one and a minimal polynomial of a rational number r is $x - r$. This minimal polynomial with leading coefficient 1 has integer coefficients if and only if r is an integer. Thus we can distinguish the integers among the rational numbers by observing that they have a minimal polynomial with integer coefficients and leading coefficient 1.

Extending this property to algebraic numbers, we get the notion of algebraic integers.

Definition 9.6.1. An algebraic number is an *algebraic integer* if it has a minimal polynomial with integer coefficients and leading coefficient 1.   ♣

For convenience, minimal polynomials will have leading coefficient 1 in this section.

Examples. **E1** A rational number r is an algebraic integer if and only if r is an integer (this was our starting point in creating the definition of algebraic integers).

E2 $\sqrt[3]{2}$ is an algebraic integer, but $\sqrt[3]{1/2}$ is not, since their minimal polynomials are $x^3 - 2$ and $x^3 - (1/2)$.

E3 The Gaussian integers discussed in Section 7.4 are algebraic integers. Moreover, considering the *Gaussian rationals*, those complex numbers $a + bi$ where a and b are rational, exactly the Gaussian integers are algebraic integers among them (see Exercise 9.6.3 a,f). A similar statement is true also for Eulerian integers and the corresponding Eulerian rationals investigated in Section 7.7. In Chapters 10 and 11 we shall develop number theory in detail for similar types of algebraic integers.

The following theorem makes it possible to verify that a number is an algebraic integer without determining its minimal polynomial, but it cannot be applied to prove that the number is not an algebraic integer.

Theorem 9.6.2. *A complex number α is an algebraic integer if and only if there exists a polynomial f with integer coefficients and leading coefficient 1 with $f(\alpha) = 0$.* ♣

Proof. If α is an algebraic integer, then an appropriate f is its minimal polynomial m_α which has leading coefficient 1.

For the converse, assume that $f(\alpha) = 0$ for some polynomial with integer coefficients and leading coefficient 1. By Theorem 9.2.3, $m_\alpha \mid f$, so $f = g m_\alpha$ for some polynomial g with rational coefficients. The leading coefficients of f and m_α are 1, so g has leading coefficient 1. We use now a basic lemma of Gauss stating that if a non-zero polynomial with integer coefficients is a product of two polynomials with rational coefficients, then it is the product of two polynomials with integer coefficients obtained from the original factors by multiplying them with suitable constants. As f has integer coefficients and $f = g m_\alpha$, there exists a rational number c such that both cg and $(1/c)m_\alpha$ have integer coefficients. Then their leading coefficients are c and $1/c$, which both are integers if and only if $c = \pm 1$. This means that m_α and g have integer coefficients, so α is an algebraic integer by definition, indeed. □

Remarks: (1) We can use Theorem 9.6.2 to show that a complex root of unity is an algebraic integer without referring to the cyclotomic polynomials: An nth root of unity is a root of $x^n - 1$ having integer coefficients and leading coefficient 1.

(2) As we mentioned, we cannot use Theorem 9.6.2 to prove that a given number is not an algebraic integer. If α is a root of even infinitely many polynomials with rational coefficients where not all coefficients are integers and the leading coefficients are 1, we have no information about whether or not α is an algebraic integer. For example, 1 is an algebraic integer, but it is a root of polynomials $f_n = (x - 1)(x - 1/2)^n$ ($n = 1, 2, \dots$), each having rational coefficients not all of which are integers and leading coefficient is 1. We can construct similar examples for any algebraic integer. To verify that an algebraic number is not an algebraic integer, we need its minimal polynomial.

Now we discuss the connection of algebraic integers to operations. The next theorem summarizes the analogs of Theorems 9.3.1, 9.3.4, and 9.3.6 for algebraic integers.

Theorem 9.6.3. (i) *The algebraic integers form a subring of the complex numbers, so sums, differences, and products of algebraic integers are algebraic integers, as well, (though this is not true for quotients in general).*

(ii) *Powers of algebraic integers with rational exponents are algebraic integers.*

(iii) *If the coefficients of a polynomial f are algebraic integers and its leading coefficient is 1, then its roots are algebraic integers.* ♣

Proof. We can adapt the proofs seen for algebraic numbers in Theorems 9.3.1, 9.3.4, and 9.3.6: we just replace the phrases "algebraic number" with "algebraic integer", "rational number" with "integer", and "with rational coefficients" with "with integer coefficients and leading coefficient 1." (Disregard, of course, the parts about reciprocals. In adapting the proof of Theorem 9.3.6 note that $\xi = 1$, so we do not need the ξ_k.) We leave to the reader to check each step in detail. $\square$

Exercises 9.6

1. Show that if α is an algebraic integer, then so are $\bar{\alpha}$, $2\mathrm{Re}(\alpha)$, $2\mathrm{Im}(\alpha)$, and $|\alpha|$.

2. Which are algebraic integers?

   (a) $\sqrt[5]{5} + (\sqrt[3]{7}/2)$

   (b) $(1 + \sqrt{3})/2$

   (c) $(1 + i\sqrt{3})/2$

   (d) $\cos 1°$.

3. Let $\alpha = a + bi$ be a complex number, where a and b are real numbers. True or false?

   (a) If a and b are algebraic integers, then so is α.

   (b) If a is an algebraic integer, then so is α.

   (c) If a and $|\alpha|$ are algebraic integers, then so is α.

   (d) If α is an algebraic integer, then so are a and b.

   (e) If α and a are algebraic integers, then so is b.

   (f) If α is an algebraic integer and a and b are rational numbers, then a and b are integers.

   (g) If $\alpha + 3\beta$ and $5\alpha + 7\beta$ are algebraic integers, then so are α and β.

   (h) If $\alpha + \beta$ and $\alpha\beta$ are algebraic integers, then so are α and β.

4. Investigate the variant of Fermat's Last Theorem for algebraic integers: For an exponent $n \geq 3$, is the equation $x^n + y^n = z^n$ solvable in non-zero algebraic integers?

S 5. Let f be a polynomial with rational coefficients where not all coefficients are integers and the leading coefficient is 1, and consider its (complex) roots. True or false?

   (a) At least one root of f is not an algebraic integer.

   (b) No root of f is an algebraic integer.

   (c) If f is irreducible over $\mathbf{Q}$, then no root of f is an algebraic integer.

   (d) If exactly one of the roots of f is not an algebraic integer, then f has a rational root.

6. Prove that every algebraic number is the quotient of two algebraic integers, moreover, we can require that either of them is an (ordinary) integer.

7. How can we see from the minimal polynomial of an algebraic integer α that also $1/\alpha$ is an algebraic integer?

8. Verify.

   (a) For any algebraic integer α there exist infinitely many algebraic integers β such that α/β is an algebraic integer.

   (b) For an algebraic integer $\alpha \neq 0$ there exist infinitely many algebraic integers β where $1/\beta$ is not, but α/β is, an algebraic integer if and only if $1/\alpha$ is not an algebraic integer.

   (c) For any algebraic integer $\alpha \neq 0$ there exist only finitely many integers b for which α/b is an algebraic integer.

9. Is there a complex number of absolute value one that is not a root of unity, but still is (a) an algebraic number *(b) an algebraic integer?

* 10. (a) Verify that if $n \geq 2$, then the real algebraic integers of degree n are everywhere dense in the real number line.

   (b) Are the algebraic integers of degree n everywhere dense on the complex plane if (b1) $n = 2$ (b2) $n = 4$?

11. (a) Let r be a real number. Prove that at least one of r and $\cos r°$ is irrational, except if r is an integer divisible by 60 or 90.

   (b) Formulate and prove similar statements for sine and tangent.

Chapter 10

Algebraic Number Fields

The simple algebraic extensions of the rational field are called algebraic number fields. In this chapter we deal with such extensions and with the arithmetic properties of algebraic integers in them. We discuss algebraic integers of quadratic fields in detail. As special cases, we have already seen Gaussian and Eulerian integers in Chapter 7, and applied them to handle the Diophantine equations $x^2 + y^2 = n$ and $x^3 + y^3 = z^3$. We continue studying algebraic number fields in the next chapter with the help of ideals.

The general introductory section about extensions is valid for any (commutative) field, but we shall apply these notions and facts for subfields of the complex numbers only. In this chapter we shall often rely on some basic notions and theorems from linear algebra, mostly related to the dimension of vector spaces.

10.1. Field Extensions

Field will always mean a commutative field.

Definition 10.1.1. A field M is an *extension* of the field L if L is a *subfield* of M, i.e. $L \subseteq M$, and the operations in L are the restrictions of the operations in M. ♣

The usual notation for this relation is $M \mid L$ or M/L, but as this might be confused with some other notion, we shall use the notation $M : L$.

If M is an extension of L, then M is also a vector space over L under the naturally arising operations. These vector space operations come from the field operations of M: we add two vectors in M as two elements of the field M, and multiply a vector in M by a scalar in L so that we form the product of these two elements in the field M.

We have a special name and notation for the dimension of M as a vector space over the field L:

Definition 10.1.2. If M is an extension of L, then the dimension of M as a vector space over L is called the *degree* of the extension and is denoted by $\deg(M : L)$. If this dimension is finite, we say that the extension is *finite* (or has a *finite degree*). ♣

Examples. $\deg(\mathbf{C} : \mathbf{R}) = 2, \deg(\mathbf{R} : \mathbf{Q}) = \infty$.

An important fact is that the degree of a chain of extensions is the product of the degrees of the links:

Theorem 10.1.3 (Tower Theorem). *If* $\deg(N : M) < \infty$ *and* $\deg(M : L) < \infty$ *in the chain of extensions* $L \subseteq M \subseteq N$, *then*

$$(10.1.1) \qquad \deg(N : L) = \deg(N : M) \cdot \deg(M : L).$$ ♣

We note that the theorem can be extended to infinite degrees: If at least one of $\deg(N : M)$ and $\deg(M : L)$ is infinite, then $\deg(N : L)$ is infinite and (10.1.1) remains valid in the more refined sense when the degrees mean the cardinalities of the bases.

Proof. We denote the elements of L, M, and N with Greek letters, minuscules, and capitals, respectively.

Let $b_1, \ldots, b_n$ be a basis in the vector space $M : L$, and let $C_1, \ldots, C_k$ be a basis in $N : M$. We are done if we verify that the kn vectors

$$(10.1.2) \qquad b_i C_j, \qquad i = 1, 2, \ldots, n, \qquad j = 1, 2, \ldots, k$$

form a basis in $N : L$.

We show first that the vectors in (10.1.2) are linearly independent in $N : L$. Consider a linear combination

$$(10.1.3) \qquad \sum_{i=1}^{n} \sum_{j=1}^{k} \lambda_{ij}(b_i C_j) = 0$$

with scalars $\lambda_{ij} \in L$. Transforming the left-hand side of (10.1.3) using identities in the field N, we obtain

$$(10.1.4) \qquad \sum_{j=1}^{k} \left(\sum_{i=1}^{n} \lambda_{ij} b_i \right) C_j = 0.$$

Since $C_1, \ldots, C_k$ are linearly independent in $N : M$, (10.1.4) implies

$$(10.1.5) \qquad \sum_{i=1}^{n} \lambda_{ij} b_i = 0, \qquad j = 1, \ldots, k.$$

Now we apply the fact that $b_1, \ldots, b_n$ are linearly independent in $M : L$. Then (10.1.5) yields that every $\lambda_{ij} = 0$. Thus we have proved that $b_i C_j$ are linearly independent in $N : L$.

Now we demonstrate that $b_i C_j$ span $N : L$. As $C_1, \ldots, C_k$ span $N : M$, therefore every $U \in N$ has a representation

$$(10.1.6) \qquad U = v_1 C_1 + \cdots + v_k C_k$$

with some $v_j \in M$. Also, $b_1, \ldots, b_n$ span $M : L$, thus every v_j is a linear combination of the vectors b_i:

$$(10.1.7) \qquad v_j = \alpha_{1j} b_1 + \cdots + \alpha_{nj} b_n, \qquad \alpha_{ij} \in L, \qquad 1 \leq i \leq n, \qquad 1 \leq j \leq k.$$

Substituting the representations in (10.1.7) into (10.1.6), we obtain

$$U = \sum_{i=1}^{n} \sum_{j=1}^{k} \alpha_{ij} b_i C_j,$$

which means that $b_i C_j$ span $N : L$.   □

Now we generalize the notion of algebraic numbers.

Definition 10.1.4. Let L be a subfield in M. An element $\vartheta \in M$ is *algebraic* over the field L if $f(\vartheta) = 0$ for some non-zero polynomial $f \in L[x]$.   ♣

Examples. The algebraic numbers are the special case $L = \mathbf{Q}$, $M = \mathbf{C}$.

Over the real or complex field every complex number is algebraic (see Exercise 9.1.7).

The minimal polynomial and degree of an algebraic element are defined analogously to Definitions 9.2.1 and 9.2.4:

Definition 10.1.5. Let L be a subfield in M. The *minimal polynomial* of an algebraic element $\vartheta \in M$ over L is a polynomial in $L[x]$ of minimal degree having ϑ among its roots. The *degree* of ϑ is the degree of its minimal polynomial.   ♣

The minimal polynomial and the degree depend not only on ϑ but also on over which field L we consider ϑ. For example the minimal polynomial of $\sqrt{2}$ over $\mathbf{Q}$ is $x^2 - 2$, but over $\mathbf{R}$ it is $x - \sqrt{2}$. It can be shown that modifying M does not influence the minimal polynomial of ϑ.

Accordingly, in the notation $m_{\vartheta,L}$ and $\deg_L \vartheta$ of the minimal polynomial and degree we have to indicate also the field L (in the case $L = \mathbf{Q}$ of algebraic numbers we keep the previous fieldless notations m_ϑ and $\deg \vartheta$).

The analogues of Theorems 9.2.2 and 9.2.3 remain valid for minimal polynomials of algebraic elements.

Theorem 10.1.6. *Let L be a subfield in M and $\vartheta \in M$ an algebraic element over L. Then*

  (i) *the minimal polynomial $m_{\vartheta,L}$ is unique apart from a constant factor in L*

 (ii) *for polynomials $f \in L[x]$, we have $f(\vartheta) = 0$ if and only if $m_{\vartheta,L} \mid f$*

(iii) *a polynomial $g \in L[x]$ is a minimal polynomial of ϑ if and only if $g(\vartheta) = 0$ and g is irreducible over L.*   ♣

The proofs are exactly the same as for Theorems 9.2.2 and 9.2.3.

The following fact is useful information about the structure of certain extensions.

Theorem 10.1.7. *If $\deg(M : L) < \infty$, then every element of M is algebraic over L.*   ♣

Proof. Let $\deg(M : L) = n$, and let 1 denote the common identity element of the fields L and M. Then for any $v \in M$, the number of elements $1, v, v^2, \ldots, v^n$ is greater than the dimension of the vector space $M : L$, thus they are linearly dependent. This means

$$\alpha_0 + \alpha_1 v + \cdots + \alpha_n v^n = 0$$

for some scalars $\alpha_0, \ldots, \alpha_n \in L$ not all 0. So v is a root of the non-zero polynomial $f = \alpha_0 + \alpha_1 x + \cdots + \alpha_n x^n$, i.e. v is an algebraic element over L.   □

Remarks: (1) We obtained also $\deg_L v \le \deg(M : L)$ from the proof. We shall show a stronger result, $\deg_L v \mid \deg(M : L)$, in Theorem 10.2.5.

(2) The converse of Theorem 10.1.7 is false. For example, let L be the rational field and M the field of all algebraic numbers (over $\mathbf{Q}$). Then every element in M is algebraic over L (by definition), but $\deg(M : L) = \infty$, because $\deg(M : L) = n < \infty$ would imply by the previous remark that every algebraic number has degree at most n, which contradicts the existence of algebraic numbers of arbitrarily large degrees (Section 9.2, Example E4).

Exercises 10.1

1. Verify that if $\deg(M : L)$ is a prime and a subfield F in M contains L, then $F = M$ or $F = L$.

2. Let $G = \{a + bi \mid a, b \in \mathbf{Q}\}$ be the field of Gaussian rationals and A the field of algebraic numbers. Compute

   (a) $\deg(G : \mathbf{Q})$

   (b) $\deg(\mathbf{C} : A)$

   (c) $\deg(A : G)$.

3. Let $K = \{a + b\sqrt{2} \mid a, b \in \mathbf{Q}\}$. Clearly, K is a subfield in $\mathbf{R}$.

   (a) Prove that a complex number α is algebraic over K if and only if it is algebraic over $\mathbf{Q}$ (i.e. it is an algebraic number).

   (b) Determine the degrees of the complex numbers over K:

      (b1) $3 + 7\sqrt{2}$

      (b2) $\sqrt{2} + i$

      (b3) $\sqrt[4]{2}$

      (b4) $\sqrt[3]{2}$.

4. Consider the chain of extensions $L \subseteq M \subseteq N$, and let $\vartheta \in N$.

   (a) True or false?

      (a1) If ϑ is algebraic over L, then it is algebraic over M.

      (a2) If ϑ is algebraic over M, then it is algebraic over L.

   (b) If ϑ is algebraic over both M and L, what is the relation between $m_{\vartheta, M}$ and $m_{\vartheta, L}$, and between $\deg_M \vartheta$ and $\deg_L \vartheta$?

10.2. Simple Algebraic Extensions

The simplest and most important type of extension occurs when the extension is generated by a single element. For convenience, we discuss this notion only for the special case $\mathbf{Q}(\vartheta)$, when we adjoin a complex number ϑ to the rational field, but everything holds in general when $\mathbf{Q}$ is replaced by any field L and instead of $\vartheta \in \mathbf{C}$ we consider $\vartheta \in M$ where the field M is an arbitrary extension of L.

By a *simple extension* of $\mathbf{Q}$ with a complex number ϑ, we shall mean the set of complex numbers obtained from rational numbers and ϑ by the four arithmetic operations of the complex field. We consider all elements $a_0 + a_1\vartheta + \cdots + a_n\vartheta^n$, where n is an arbitrary non-negative integer and the a_i are rational numbers and form the quotients of such expressions. The number $a_0 + a_1\vartheta + \cdots + a_n\vartheta^n$ is just the value $g(\vartheta) \in \mathbf{C}$ of the polynomial $g = a_0 + a_1 x + \cdots + a_n x^n \in \mathbf{Q}[x]$ at ϑ. Thus the quotients are complex numbers $g(\vartheta)/h(\vartheta)$, where g and h are arbitrary polynomials in $\mathbf{Q}[x]$ and $h(\vartheta) \neq 0$. These elements constitute the smallest subfield of $\mathbf{C}$ containing ϑ and $\mathbf{Q}$. We formulate all this precisely in the following definition and theorem.

Definition 10.2.1. For a complex number ϑ, consider the set of complex numbers

$$(10.2.1) \qquad \frac{g(\vartheta)}{h(\vartheta)}, \quad \text{where } g, h \in \mathbf{Q}[x], \quad h(\vartheta) \neq 0,$$

or, written in detail,

$$(10.2.2) \qquad \frac{\sum_{i=0}^{n} a_i\vartheta^i}{\sum_{j=0}^{k} b_j\vartheta^j}, \quad \text{where } a_i, b_j \in \mathbf{Q}, \quad \sum_{j=0}^{k} b_j\vartheta^j \neq 0, \quad n, k = 0, 1, 2, \ldots.$$

This set is called a *simple extension* of the field $\mathbf{Q}$ with ϑ, and is denoted by $\mathbf{Q}(\vartheta)$. If ϑ is an algebraic number, then we speak about a *simple algebraic extension*. ♣

Theorem 10.2.2. $\mathbf{Q}(\vartheta)$ *is the smallest subfield in the complex field containing ϑ and the rational field, so*

(i) $\mathbf{Q}(\vartheta)$ *is a subfield in* $\mathbf{C}$

(ii) $\vartheta \in \mathbf{Q}(\vartheta), \mathbf{Q} \subseteq \mathbf{Q}(\vartheta)$

(iii) *if F is a subfield in $\mathbf{C}$ and $\vartheta \in F$, $\mathbf{Q} \subseteq F$, then $\mathbf{Q}(\vartheta) \subseteq F$.* ♣

Proof. (i) We have to show that sums, differences, products, and, if the divisor is not 0, quotients of elements in (10.2.1) are in (10.2.1). Clearly,

$$\frac{g_1(\vartheta)}{h_1(\vartheta)} + \frac{g_2(\vartheta)}{h_2(\vartheta)} = \frac{g(\vartheta)}{h(\vartheta)},$$

where $g = g_1 h_2 + g_2 h_1$ and $h = h_1 h_2$ are polynomials with rational coefficients and $h(\vartheta) = h_1(\vartheta)h_2(\vartheta) \neq 0$, since there are no zero divisors in the complex field. The statements for differences, products, and quotients can be verified similarly.

(ii) If $g = x$ and $h = 1$, then $g(\vartheta)/h(\vartheta) = \vartheta$, so $\vartheta \in \mathbf{Q}(\vartheta)$. If r is a rational number, then choosing polynomials $g = r$ and $h = 1$, we have $g(\vartheta)/h(\vartheta) = r$, thus $r \in \mathbf{Q}(\vartheta)$.

(iii) If a subfield F of the complex numbers contains ϑ and $\mathbf{Q}$, then the sums of any products formed from ϑ and rational numbers and the quotients of such sums are in F. This means that every complex number in (10.2.2) is an element of F, hence $\mathbf{Q}(\vartheta) \subseteq F$. □

We show that if ϑ is an algebraic number, then the elements of $\mathbf{Q}(\vartheta)$ have a simpler representation.

As an example, consider the extension $\mathbf{Q}(\sqrt{2})$ of the rational field with $\sqrt{2}$. This is the set F of numbers $a_0 + a_1\sqrt{2}$, where $a_i \in \mathbf{Q}$, since F is a field containing $\sqrt{2}$ and the rational numbers, and it is obviously the smallest field having this property. This means that compared to the form of elements in Definition 10.2.1, we need neither division, nor powers of $\sqrt{2}$ with exponents greater than 1.

If instead of $\sqrt{2}$, we consider the extension $\mathbf{Q}(\sqrt[3]{5})$ with $\sqrt[3]{5}$, then we need only powers of $\sqrt[3]{5}$ with exponents at most 2, since the higher powers can be expressed by these and with suitable rational numbers.

In the general case, we have:

Theorem 10.2.3. *If ϑ is an algebraic number of degree n, then the elements of $\mathbf{Q}(\vartheta)$ can be uniquely represented in the form*

$$a_0 + a_1\vartheta + \cdots + a_{n-1}\vartheta^{n-1}$$

with rational numbers a_i. In other words, to every $\alpha \in \mathbf{Q}(\vartheta)$ there exists exactly one polynomial $f \in \mathbf{Q}[x]$ satisfying

$$\alpha = f(\vartheta) \quad and \quad \deg f \leq n - 1 \text{ or } f = 0. \qquad \clubsuit$$

Proof. I. First we show that there is no need for denominators in (10.2.1), i.e. if $g, h \in \mathbf{Q}[x]$ and $h(\vartheta) \neq 0$, then $g(\vartheta)/h(\vartheta) = t(\vartheta)$ for some polynomial $t \in \mathbf{Q}[x]$.

We perform the following equivalent transformations (relying on the condition $h(\vartheta) \neq 0$ and on Theorem 9.2.3(i)):

$$g(\vartheta)/h(\vartheta) = t(\vartheta) \iff g(\vartheta) = h(\vartheta)t(\vartheta) \iff (g - ht)(\vartheta) = 0 \iff$$
$$\iff m_\vartheta \mid g - ht \iff g = ht + m_\vartheta s, \quad \text{where} \quad s \in \mathbf{Q}[x].$$

Thus we have to verify the existence of polynomials t and s with rational coefficients satisfying

(10.2.3) $g = ht + m_\vartheta s.$

Equality (10.2.3) looks like a linear Diophantine equation, where t and s are the variables, with integers replaced here by polynomials with rational coefficients. The necessary and sufficient condition for the solvability of a linear Diophantine equation was discussed in Theorem 1.3.6, and in the proof we relied only on a consequence of the Euclidean algorithm, i.e. we needed only the division algorithm. Since there is a division algorithm for polynomials over a field, therefore the condition of solvability is the same for Diophantine equations with polynomials. Thus we have to show $(h, m_\vartheta) \mid g$ for the solvability of (10.2.3).

The polynomial m_ϑ is irreducible over $\mathbf{Q}$, so $(h, m_\vartheta) = 1$ or m_ϑ. But the latter would imply $h(\vartheta) = 0$, so only $(h, m_\vartheta) = 1$ is possible and $(h, m_\vartheta) \mid g$. This means, as we have seen before, that (10.2.3) is solvable and we obtain a polynomial t satisfying $t(\vartheta) = g(\vartheta)/h(\vartheta)$.

II. We have proved so far that every $\alpha \in \mathbf{Q}(\vartheta)$ can be written as $\alpha = t(\vartheta)$ with a suitable polynomial $t \in \mathbf{Q}[x]$. Now we show that $\alpha = f(\vartheta)$ can be gotten with a polynomial $f \in \mathbf{Q}[x]$ where $\deg f \leq n - 1$ or $f = 0$.

Apply the division algorithm to t and m_ϑ. We claim that we can choose the remainder as f. If

$$t = qm_\vartheta + f, \qquad \text{where} \qquad \deg f \leq n - 1 \quad \text{or} \quad f = 0,$$

then

$$\alpha = t(\vartheta) = q(\vartheta)m_\vartheta(\vartheta) + f(\vartheta) = 0 + f(\vartheta) = f(\vartheta).$$

III. We show that f is unique. Assume that the polynomials f_1 and f_2 with rational coefficients satisfy

$$f_1(\vartheta) = f_2(\vartheta) \quad \text{and} \quad \deg f_i \leq n - 1 \quad \text{or} \quad f_i = 0, \qquad i = 1, 2.$$

Then the polynomial $f_3 = f_1 - f_2$ has rational coefficients, $f_3(\vartheta) = 0$, and $\deg f_3 < n$ or $f_3 = 0$. Since $\deg \vartheta = n$, only $f_3 = 0$ is possible. So $f_1 = f_2$ and the polynomial f in the theorem is unique. $\qquad \square$

Theorem 10.2.3 expresses that the elements $1, \vartheta, \ldots, \vartheta^{n-1}$ form a basis in the vector space $\mathbf{Q}(\vartheta)$ over $\mathbf{Q}$. Thus the dimension of this vector space, i.e. the degree of the extension $\mathbf{Q}(\vartheta) : \mathbf{Q}$ is equal to the degree of the algebraic number ϑ. We restate this important fact as a theorem:

Theorem 10.2.4. *If ϑ is an algebraic number, then* $\deg\big(\mathbf{Q}(\vartheta) : \mathbf{Q}\big) = \deg \vartheta$. $\qquad \clubsuit$

We can add to Theorems 10.2.3 and 10.2.4, that if ϑ is a transcendental number, then there is no simpler form than given in Definition 10.2.1 for representing the elements of $\mathbf{Q}(\vartheta)$, and the degree of the extension $\mathbf{Q}(\vartheta) : \mathbf{Q}$ is infinite. In this case, the field $\mathbf{Q}(\vartheta)$ is isomorphic to the field of formal quotients of polynomials with rational coefficients, called the quotient field of $\mathbf{Q}[x]$ or the field of *algebraic fractions* over $\mathbf{Q}$ (see Exercise 10.2.13).

We also note that if ϑ is an algebraic number, then by Theorem 10.2.3, we can think of the elements in $\mathbf{Q}(\vartheta)$ as remainders on division by the polynomial m_ϑ: in this case the field $\mathbf{Q}(\vartheta)$ is isomorphic to the *factor ring* $\mathbf{Q}[x]/(m_\vartheta)$ (see Theorem 11.1.6 and Exercise 11.1.9a). This interpretation of simple algebraic extensions makes it possible, for an arbitrary field L instead of $\mathbf{Q}$, to construct $L(\vartheta)$ even if no field M containing L and no element ϑ are given, see Exercise 11.1.9b.

We mention without proof that every finite extension of $\mathbf{Q}$ can be obtained as $\mathbf{Q}(\vartheta)$ with a suitable algebraic number ϑ, so the finite extensions of $\mathbf{Q}$ are the same as the simple algebraic extensions of $\mathbf{Q}$. This is true also if $\mathbf{Q}$ is replaced by any field in which a sum $a + a + \cdots + a$ can be 0 only for $a = 0$.

Sharpening Theorem 10.1.7, we show that the degree of an element in a finite extension must divide the degree of the extension. We formulate the statement for extensions of $\mathbf{Q}$, but it is equally valid for arbitrary fields.

Theorem 10.2.5. *If M is a subfield in $\mathbf{C}$ and $\deg(M : \mathbf{Q}) = k < \infty$, then $\deg \alpha \mid k$ for every $\alpha \in M$.* ♣

Proof. The field $\mathbf{Q}(\alpha)$ is contained in M by Theorem 10.2.2, so

$$(10.2.4) \qquad\qquad \mathbf{Q} \subseteq \mathbf{Q}(\alpha) \subseteq M.$$

The condition $\deg(M : \mathbf{Q}) = k < \infty$ implies that both links in the chain of extensions (10.2.4) are of finite degree, so we can apply the Tower Theorem 10.1.3. This yields $\deg(\mathbf{Q}(\alpha) : \mathbf{Q}) \mid k$. By Theorem 10.1.7, α is an algebraic number, so by Theorem 10.2.4, we have $\deg(\mathbf{Q}(\alpha) : \mathbf{Q}) = \deg \alpha$. Thus $\deg \alpha \mid k$. $\qquad\square$

Now we give new proofs of Theorems 9.3.1 and 9.3.6. For convenience, we restate them with new numbers.

Theorem 10.2.6. *The algebraic numbers form a subfield in the complex field.* ♣

Proof. Let α and β be two algebraic numbers. We have to show that $\alpha + \beta, \alpha - \beta, \alpha\beta$, and α/β ($\beta \neq 0$) are algebraic.

We extend $\mathbf{Q}$ with α, and then extend the resulting field $K = \mathbf{Q}(\alpha)$ with β. This field $N = K(\beta)$ contains both α and β, thus it must contain also their sum, difference, product, and quotient.

Consider the chain of extensions $\mathbf{Q} \subseteq K \subseteq N$ where $K = \mathbf{Q}(\alpha)$ and $N = K(\beta)$. Here

$$\deg(K : \mathbf{Q}) = \deg \alpha \quad \text{and} \quad \deg(N : K) = \deg_K \beta \leq \deg \beta,$$

so $\deg(N : K) < \infty$ by the tower theorem. By Theorem 10.1.7, all elements in N, thus $\alpha + \beta, \alpha - \beta, \alpha\beta$, and α/β are algebraic numbers. $\qquad\square$

Theorem 10.2.7. *If the coefficients of a polynomial $f \neq 0$ are algebraic numbers, then all (complex) roots of f are algebraic numbers.* ♣

Proof. Let $f = \alpha_0 + \alpha_1 x + \cdots + \alpha_n x^n$ and let γ be an arbitrary root of f.

We define a sequence of fields K_i by

$$K_0 = \mathbf{Q}(\alpha_0), \qquad K_j = K_{j-1}(\alpha_j), \qquad j = 1, 2, \ldots, n, \qquad K_{n+1} = K_n(\gamma),$$

and consider the chain of extensions

$$\mathbf{Q} \subseteq K_0 \subseteq K_1 \subseteq \cdots \subseteq K_n \subseteq K_{n+1}.$$

Every link is an extension with an algebraic number over the previous field, thus every link has a finite degree. Thus by the tower theorem, the extension $K_{n+1} : \mathbf{Q}$ is finite, so every element in K_{n+1}, including γ, is algebraic over $\mathbf{Q}$. $\qquad\square$

Exercises 10.2

1. Prove that for a complex number ϑ and rational number $r \neq 0$, the extension $\mathbf{Q}(\vartheta)$ is equal to

   (a) $\mathbf{Q}(r + \vartheta)$

   (b) $\mathbf{Q}(r\vartheta)$

   (c) $\mathbf{Q}(1/\vartheta)$ (if $\vartheta \neq 0$).

2. Let $\alpha \in \mathbf{Q}(\vartheta)$. Verify.

   (a) $\mathbf{Q}(\alpha) \subseteq \mathbf{Q}(\vartheta)$.

   (b) If ϑ is algebraic, then $\mathbf{Q}(\alpha) = \mathbf{Q}(\vartheta)$ if and only if $\deg \alpha = \deg \vartheta$.

   * (c) If ϑ is transcendental, $\mathbf{Q}(\alpha) = \mathbf{Q}(\vartheta)$ if and only if

   $$\alpha = \frac{a_0 + a_1 \vartheta}{b_0 + b_1 \vartheta}, \quad \text{where } a_i, b_i \in \mathbf{Q} \quad \text{and} \quad \alpha \notin \mathbf{Q}.$$

3. True or false?

   (a) $\mathbf{Q}(\vartheta) = \mathbf{Q}(\overline{\vartheta})$.

   (b) If $|\vartheta|^2$ is a rational number, then $\mathbf{Q}(\vartheta) = \mathbf{Q}(\overline{\vartheta})$.

   (c) If $\mathbf{Q}(\vartheta) = \mathbf{Q}(\overline{\vartheta})$, then $|\vartheta|^2$ is a rational number.

   (d) If $\mathbf{Q}(\vartheta) \subseteq \mathbf{Q}(\overline{\vartheta})$, then $\mathbf{Q}(\vartheta) = \mathbf{Q}(\overline{\vartheta})$.

   (e) $\mathbf{Q}(\vartheta) = \mathbf{Q}(\vartheta + \vartheta^2)$.

4. Represent the following numbers in the form $a_0 + a_1\sqrt[3]{2} + a_2\sqrt[3]{4}$ with rational numbers a_0, a_1, and a_2:

   (a) $(\sqrt[3]{4} + 3\sqrt[3]{2})^2$

   (b) $\dfrac{1}{\sqrt[3]{2}}$

   (c) $\dfrac{1 + \sqrt[3]{2}}{1 + 2\sqrt[3]{2}}$.

5. Determine the degree of the algebraic numbers

   S (a) $\sqrt{7} + 3i$

   (b) $i\sqrt[5]{3}$

   (c) $\sqrt[3]{3} + \sqrt[3]{1/3}$

   (d) $\sqrt[4]{2} + \sqrt{2}$.

6. Write in a simpler form:

   (a) $\mathbf{Q}(\sqrt[3]{54}) \setminus \mathbf{Q}(\sqrt[3]{16})$

   (b) $\mathbf{Q}(\sqrt[8]{7}) \cap \mathbf{Q}(\sqrt[9]{7})$

   (c) $\mathbf{Q}(\sqrt[4]{5}) \cap \mathbf{Q}(i\sqrt[4]{5})$.

S 7. Determine the real numbers in $\mathbf{Q}(\vartheta)$ if ϑ is

    (a) $\sqrt[5]{3}(\cos 144° + i \sin 144°)$

    (b) $i\sqrt[8]{3}$

    (c) any value of $\sqrt{i}$.

S* 8. Prove that if $|\vartheta| = 1$, then $\mathbf{Q}(\vartheta) \cap \mathbf{R} = \mathbf{Q}(\operatorname{Re}\vartheta)$.

  9. Let $\alpha = 1 + 3\sqrt[3]{25} + 11\sqrt[3]{125} + 999\sqrt[3]{625}$. Prove the existence of a polynomial f with rational coefficients satisfying $f(\alpha) = \sqrt[3]{5}$.

 10. Let k be the degree of an algebraic number β. What are the possible values of $\deg(\beta^2)$?

S 11. Find all algebraic numbers of odd degree on the unit circle.

 12. (a) Prove that if α and β are algebraic numbers, then the degrees of $\alpha + \beta, \alpha - \beta,$ $\alpha\beta, \alpha/\beta$ $(\beta \neq 0)$ are less than or equal to $(\deg \alpha) \cdot (\deg \beta)$.

    (b) If the coefficients of a polynomial $f = \alpha_0 + \alpha_1 x + \cdots + \alpha_n x^n$ are algebraic numbers and $f(\gamma) = 0$, then $\deg \gamma \leq n \prod_{j=0}^{n} \deg \alpha_j$.

 13. Let $g_1, g_2, h_1 \neq 0, h_2 \neq 0$ be polynomials with rational coefficients and ϑ a transcendental number. Verify the following statements.

    (a) $\dfrac{g_1(\vartheta)}{h_1(\vartheta)} = \dfrac{g_2(\vartheta)}{h_2(\vartheta)} \iff g_1 h_2 = g_2 h_1$.

    (b) The field $\mathbf{Q}(\vartheta)$ is isomorphic to the field of formal quotients of polynomials with rational coefficients, i.e. algebraic fractions over $\mathbf{Q}$.

10.3. Quadratic Fields

In this section we investigate the quadratic extensions of $\mathbf{Q}$ within the complex field and the algebraic integers in them.

Theorem 10.3.1. *All extensions of $\mathbf{Q}$ of degree 2 are of the form $\mathbf{Q}(\sqrt{t})$, where t is a positive or negative squarefree integer and $t \neq 1$. Different values of t induce different extensions.* ♣

Remark: We speak about real or imaginary quadratic extensions according to $t > 0$ or $t < 0$. In the imaginary case, we can take either of the two values of $\sqrt{t}$ since these are negatives of each other and $\mathbf{Q}(\vartheta) = \mathbf{Q}(-\vartheta)$ for every ϑ; in the sequel we let $\sqrt{t}$ be the value of the square root in the upper half plane: $\sqrt{t} = i\sqrt{|t|}$.

Proof. Let M be a subfield in $\mathbf{C}$ with $\deg(M : \mathbf{Q}) = 2$. Then any non-rational element α in M satisfies $\deg \alpha = 2$ and $M = \mathbf{Q}(\alpha)$. We verify that $\mathbf{Q}(\alpha)$ can be given in the form $\mathbf{Q}(\sqrt{t})$ with a suitable squarefree $t \neq 1$.

Let the minimal polynomial of α be $m_\alpha = a_0 + a_1 x + a_2 x^2$, where a_0, a_1, a_2 are integers. Then the quadratic formula yields $\alpha = r_0 + r_1\sqrt{s}$, where $r_1 \neq 0$ and r_0 are rational numbers and $s \neq 0$ is an integer. Factoring out the largest possible square from s, we get $s = k^2 t$, where t is squarefree and $t \neq 1$. Thus $\alpha = r_0 + r_1 k \sqrt{t}$, and so $M = \mathbf{Q}(\alpha) = \mathbf{Q}(\sqrt{t})$ by Exercise 10.2.1.

We will show that t is unique, so if $\mathbf{Q}(\sqrt{t_1}) = \mathbf{Q}(\sqrt{t_2})$, where t_j are squarefree and $t_j \neq 1$, then $t_1 = t_2$.

The conditions imply

$$\sqrt{t_2} \in \mathbf{Q}(\sqrt{t_1}) \quad \text{so} \quad \sqrt{t_2} = a + b\sqrt{t_1}$$

with some rational a and b. Squaring yields

$$t_2 = a^2 + t_1 b^2 + 2ab\sqrt{t_1}.$$

Since $\sqrt{t_1}$ is irrational, $b = 0$ or $a = 0$. In the first case, $\sqrt{t_2}$ is rational, which is impossible. In the second case, $\sqrt{t_2}/\sqrt{t_1}$ is rational, and since t_1 and t_2 are squarefree, we get $t_1 = t_2$. □

Examples. **E1** The Gaussian rationals are the elements of the extension $\mathbf{Q}(i)$; here $t = -1$.

E2 The Eulerian rationals are the elements of the extension $\mathbf{Q}(\omega)$, where

$$\omega = \cos\frac{2\pi}{3} + i\sin\frac{2\pi}{3} = \frac{-1 + i\sqrt{3}}{2}.$$

Here $t = -3$.

Now we investigate how to characterize the algebraic integers in a quadratic extension.

We consider first the Gaussian rationals, those complex numbers $a + ib$, where a and b are rational. We mentioned in Section 9.6 (Example E3, Exercise 9.6.3 a, f) that a Gaussian rational is an algebraic integer if and only if it is a Gaussian integer, i.e. a and b are integers.

Now we look at the Eulerian rationals, i.e. numbers

$$(10.3.1) \quad \alpha = c + d\omega = c + d\frac{-1 + i\sqrt{3}}{2} = \frac{2c - d}{2} + \frac{d}{2}\sqrt{-3} = a + b\sqrt{-3} \quad (a, b, c, d \in \mathbf{Q}).$$

We indicated in Example E3 in Section 9.6 that an Eulerian rational is an algebraic integer if and only if it is an Eulerian integer. This means that the Eulerian rational α in (10.3.1) is an algebraic integer if and only if c and d are integers; that is, if either both a and b are integers, or both of them are fractions with an odd numerator and a denominator 2.

Our examples show that the results obtained for $t = -1$ and $t = -3$ are somewhat different. We have these two possibilities in the general case, depending on the remainder modulo 4 of the integer t characterizing the extension:

Theorem 10.3.2. *Let $t \neq 1$ be a squarefree integer. Then the algebraic integers of the extension $\mathbf{Q}(\sqrt{t})$ are exactly those numbers $c + d\vartheta$ where c and d are integers and*

$$\vartheta = \begin{cases} \sqrt{t}, & \text{if } t \not\equiv 1 \pmod 4 \\ (1 + \sqrt{t})/2, & \text{if } t \equiv 1 \pmod 4. \end{cases}$$

In another formulation, a number $a + b\sqrt{t}$ ($a, b \in \mathbf{Q}$) in $\mathbf{Q}(\sqrt{t})$ is an algebraic integer if and only if

(1) *a and b are integers for $t \not\equiv 1$ (mod 4)*

(2) *$a = u/2$, $b = v/2$, where u and v are integers of the same parity for $t \equiv 1$ (mod 4).*

♣

The two formulations of the theorem are obviously equivalent.

Our results about Gaussian and Eulerian integers were special cases of this theorem for $t = -1 \not\equiv 1$ (mod 4) and $t = -3 \equiv 1$ (mod 4).

Proof. Since a rational number is an algebraic integer if and only if it is an (ordinary) integer, the statement of the theorem is straightforward for the rational elements of the extension $\mathbf{Q}(\sqrt{t})$.

Thus in the sequel we can restrict ourselves to the non-rational elements of the extension. Any such $\alpha \in \mathbf{Q}(\sqrt{t})$ has a unique representation $\alpha = r_0 + r_1\sqrt{t}$, where $r_1 \neq 0$ and r_0 are rational numbers. With a common denominator, we obtain

(10.3.2) $$\alpha = \frac{a + b\sqrt{t}}{c}, \quad \text{where } a, b, c \text{ are integers, } (a, b, c) = 1, c > 0, b \neq 0.$$

Squaring the equality

$$\alpha - \frac{a}{c} = \frac{b\sqrt{t}}{c},$$

we get

(10.3.3) $$\alpha^2 - \frac{2a}{c}\alpha + \frac{a^2 - tb^2}{c^2} = 0.$$

Since $\deg \alpha = 2$, (10.3.3) implies that the minimal polynomial of α is

(10.3.4) $$m_\alpha = x^2 - \frac{2a}{c}x + \frac{a^2 - tb^2}{c^2}.$$

Thus α is an algebraic integer if and only if the minimal polynomial in (10.3.4) has integer coefficients, or

(10.3.5) $c \mid 2a \quad \text{and} \quad c^2 \mid a^2 - tb^2.$

We have to verify that (10.3.5) is equivalent to

(10.3.6a) $c = 1$, if $t \not\equiv 1$ (mod 4)

(10.3.6b) $c = 2$ and a and b are odd, or $c = 1$, if $t \equiv 1$ (mod 4).

We assume first that c is odd. Then the first divisibility in (10.3.5) implies $c \mid a$, and thus we get $c^2 \mid a^2 - (a^2 - tb^2) = tb^2$. Since t is squarefree, we infer by the Fundamental Theorem of Arithmetic that $c^2 \mid b^2$, hence $c \mid b$. Therefore $c \mid (a, b, c) = 1$, so $c = 1$. Conversely, it is obvious that $c = 1$ satisfies (10.3.5) for any integers t, a, and b.

Now let c be even, $c = 2k$. Then the first divisibility in (10.3.5) implies $k \mid a$, so $k^2 \mid a^2 - (a^2 - tb^2) = tb^2$. Similar to the odd case, now $k \mid b$, and therefore $k \mid (a, b, c) = 1$, so $k = 1$, and $c = 2$. So the second divisibility in (10.3.5) means (*) $a^2 - tb^2 \equiv 0$ (mod 4), where at least one of a and b is odd due to $(a, b, c) = 1$, and $t \not\equiv 0 \pmod 4$ as t is squarefree. From these conditions and using that modulo 4 a square is 0 or 1 depending on its parity, we see that the congruence (*) holds if and only if both a and b are odd and $t \equiv 1 \pmod 4$.

Thus we have verified that conditions (10.3.5) and (10.3.6a))–(10.3.6b) are equivalent, and have completed the proof of the theorem. $\qquad\square$

We denote the set of algebraic integers in $\mathbf{Q}(\sqrt{t})$ by $I(\sqrt{t})$. Thus Theorem 10.3.2 states

(10.3.7a) $\qquad I(\sqrt{t}) = \left\{ c + d\sqrt{t} \mid c, d \in \mathbf{Z} \right\},$ $\qquad\qquad$ if $t \not\equiv 1 \pmod 4$

and

(10.3.7b) $\qquad I(\sqrt{t}) = \left\{ c + d\dfrac{1+\sqrt{t}}{2} \mid c, d \in \mathbf{Z} \right\},$ $\qquad$ if $t \equiv 1 \pmod 4$.

As $I(\sqrt{t})$ is the intersection of the ring of all algebraic integers and the field $\mathbf{Q}(\sqrt{t})$, $I(\sqrt{t})$ is a subring in the complex field. It is commutative, free of zero divisors, and has an identity element, but is not a field since it contains only the integers among the rational numbers. Thus—similarly to the Gaussian and Eulerian integers —it is worthwhile to investigate some basic number theoretical questions in $I(\sqrt{t})$.

The notions of divisibility, units, greatest common divisor, irreducible and prime elements can be defined in $I(\sqrt{t})$ exactly as we did for Gaussian integers (see Definitions 7.4.4, 7.4.6, 7.4.9, 7.4.10, and 7.4.11, in which the adjective "Gaussian" should be omitted).

The norm plays an important role in the number theory of $I(\sqrt{t})$:

Definition 10.3.3. The *norm* of an element $\alpha = a + b\sqrt{t} \in I(\sqrt{t})$ is

$$N(\alpha) = a^2 - tb^2 = (a - b\sqrt{t})(a + b\sqrt{t}). \qquad \clubsuit$$

Theorem 10.3.2 implies that the norm of every $\alpha \in E(\sqrt{t})$ is an integer.

We see immediately that Theorems 7.4.3 and 7.4.5 remain valid in every $I(\sqrt{t})$ with $t < 0$, and the only difference for $t > 0$ is that $N(\alpha)$ can be a negative integer, too (and, if $t > 0$ and α is non-rational, then $N(\alpha)$ is not the square of the absolute value of α).

In the general case, Theorems 7.4.7 and 7.7.6, about units, are modified as follows:

Theorem 10.3.4. (A) *The following conditions are equivalent for an element $\varepsilon \in I(\sqrt{t})$:*

$\qquad$ (i) *ε is a unit*

$\qquad$ (ii) *$\varepsilon \mid 1$*

$\qquad$ (iii) *$|N(\varepsilon)| = 1$.*

(B) *If $t > 0$, then there are infinitely many units in $I(\sqrt{t})$.*

(C) *If $t < 0$ and $t \neq -1, -3$, then $I(\sqrt{t})$ has just two units, namely ± 1.* $\qquad \clubsuit$

Proof. (A): (i)$\Longrightarrow$(ii): If ε divides every element in $I(\sqrt{t})$, then in particular it must divide 1.

(ii)$\Longrightarrow$(i): If $\varepsilon \mid 1$, so $\varepsilon\beta = 1$ with some $\beta \in I(\sqrt{t})$, then $\varepsilon(\beta\alpha) = \alpha$, so $\varepsilon \mid \alpha$ for any $\alpha \in E(\sqrt{t})$, and ε is a unit.

(ii)$\Longrightarrow$(iii): If $\varepsilon \mid 1$, then $N(\varepsilon) \mid N(1) = 1$, so $N(\varepsilon) = \pm 1$.

(iii)$\Longrightarrow$(ii): If $\varepsilon = a + b\sqrt{t}$ and

$$N(\varepsilon) = (a + b\sqrt{t})(a - b\sqrt{t}) = \pm 1,$$

then $a - b\sqrt{t} \in I(\sqrt{t})$ implies $\varepsilon \mid 1$.

(B) If $t > 0$, then Pell's equation $x^2 - ty^2 = 1$ has infinitely many solutions in integers x, y (Theorem 7.8.1), so the corresponding elements $\alpha = x + y\sqrt{t} \in I(\sqrt{t})$ have $N(\alpha) = 1$, and thus are units.

(C) If $t < 0, t \not\equiv 1$ (mod 4), then the elements of $I(\sqrt{t})$ are of the form $\alpha = a + b\sqrt{t}$, where a, b are integers. For $t \neq -1$,

$$N(\alpha) = a^2 + |t|b^2 = 1$$

can hold only with $b = 0$ and $a = \pm 1$, so $\alpha = \pm 1$.

If $t < 0, t \equiv 1$ (mod 4), then α can have the form $(u/2) + (v/2)\sqrt{t}$, too, where u and v are odd integers. Then we have to check

(10.3.8) $$N(\alpha) = \frac{u^2 + |t|v^2}{4} = 1 \quad \text{or} \quad u^2 + |t|v^2 = 4.$$

If $|t| > 3$ and u, v are odd, then

$$u^2 + |t|v^2 > 1 + 3 \cdot 1 = 4,$$

thus (10.3.8) cannot hold. $\square$

Remarks: (1) For many values of t, condition (A)(iii) in Theorem 10.3.4 means $N(\varepsilon) = 1$, as $N(\varepsilon) = -1$ cannot occur. This is the case for every $t < 0$, because the norm of every element is non-negative. But we have this situation e.g. for all positive $t \equiv 3$ (mod 4), since then every element in $I(\sqrt{t})$ has the form $\alpha = a + b\sqrt{t}$ with integer a, b, and $N(\alpha) = a^2 - tb^2 \not\equiv -1$ (mod 4).

(2) Related to part (B) in Theorem 10.3.4, we can characterize the units of $I(\sqrt{t})$ for $t > 0$ as follows. If $t \not\equiv 1$ (mod 4), then all units are the elements $x + y\sqrt{t}$ obtained from the integer solutions of equations $x^2 - ty^2 = \pm 1$. If $t \equiv 1$ (mod 4), then besides these $(x + y\sqrt{t})/2$ are units, where x, y are odd solutions of $x^2 - ty^2 = \pm 4$. We can describe these solutions relying on Theorem 7.8.2 (see also the hint to Exercise 7.8.3).

Now we turn to the problem of unique prime factorization, i.e. what can be said concerning the Fundamental Theorem of Arithmetic? The statement about decomposability is valid in all $I(\sqrt{t})$: Every element in $I(\sqrt{t})$ not 0 or a unit can be written as a product of irreducible elements of $I(\sqrt{t})$. This can be verified using the absolute value of the norm as we saw in the proof of Theorem 7.4.13 for Gaussian integers.

The situation is completely different for uniqueness of decomposition, which does not hold in general. We shall investigate first a few concrete extensions, and then state the results and unsolved problems in the general case.

Theorem 10.3.5. *The Fundamental Theorem of Arithmetic is true in* $I(\sqrt{2})$, *but is false in* $I(\sqrt{-5})$ *and in* $I(\sqrt{10})$. ♣

Proof. As indicated before, decomposability holds in every $I(\sqrt{t})$, so it suffices to check uniqueness (or the lack of it).

$I(\sqrt{2})$: We show that similar to the Gaussian and Eulerian integers, we have a division algorithm here, too. As seen several times, this implies the uniqueness part of the theorem.

The division algorithm for Gaussian and Eulerian integers used the norm: The norm is a non-negative integer, only the zero element has norm 0, and we get that the norm of the remainder is smaller than the norm of the divisor. (These properties guarantee that the Euclidean algorithm terminates; see Section 11.3 for a generalization.)

Since the norm of an element in $I(\sqrt{2})$ can be negative, we use the absolute value of the norm, i.e. we verify the possibility of a division algorithm with respect to the absolute value of the norm in $I(\sqrt{2})$.

It is clear that the absolute value of the norm in $I(\sqrt{2})$ is non-negative and only 0 has norm 0.

We have to show that to any $\beta \neq 0$ and α in $I(\sqrt{2})$, we can find γ and ϱ satisfying

(10.3.9) $$\alpha = \beta\gamma + \varrho \quad \text{and} \quad |N(\varrho)| < |N(\beta)|.$$

We can extend the notion of norm to the elements of $\mathbf{Q}(\sqrt{2})$: for $a, b \in \mathbf{Q}$ let

$$N(a + b\sqrt{2}) = (a + b\sqrt{2})(a - b\sqrt{2}) = a^2 - 2b^2.$$

Then clearly, $N(\xi)N(\psi) = N(\xi\psi)$ for any $\xi, \psi \in \mathbf{Q}(\sqrt{2})$.

Thus, dividing (10.3.9) by β, we get an equivalent condition:

(10.3.10) $$\frac{\alpha}{\beta} = \gamma + \frac{\varrho}{\beta} \quad \text{and} \quad \left|N\left(\frac{\varrho}{\beta}\right)\right| < 1.$$

We can formulate (10.3.10) as follows: Given α/β, we need a $\gamma \in I(\sqrt{2})$ satisfying

(10.3.11) $$\left|N\left(\frac{\alpha}{\beta} - \gamma\right)\right| < 1.$$

Let $\alpha/\beta = u + v\sqrt{2}$, where $u, v \in \mathbf{Q}$. We choose the number $c + d\sqrt{2} \in I(\sqrt{2})$ as γ where c and d are integers closest to u and v. Then

$$N\left(\frac{\alpha}{\beta} - \gamma\right) = (u - c)^2 - 2(v - d)^2,$$

and $0 \leq |u - c| \leq 1/2$, $0 \leq |v - d| \leq 1/2$ imply

$$\frac{-1}{2} \leq (u - c)^2 - 2(v - d)^2 \leq \frac{1}{4},$$

so (10.3.11) holds.

$I(\sqrt{-5})$: We show that 6 has two essentially distinct decompositions as a product of irreducible elements in $I(\sqrt{-5})$:

$$6 = 2 \cdot 3 = (1 + \sqrt{-5})(1 - \sqrt{-5}).$$

We have to check that $2, 3, 1 + \sqrt{-5}$, and $1 - \sqrt{-5}$ are irreducible in $I(\sqrt{-5})$, and that 3, for example, is not an associate of $1 \pm \sqrt{-5}$.

The latter statement is obvious, since the only units in $I(\sqrt{-5})$ are ± 1 by part (C) of Theorem 10.3.4.

We verify the irreducibility of 2, we can proceed similarly for the other three numbers.

For a proof by contradiction, assume $2 = \alpha\beta$, where neither α nor β is a unit in $I(\sqrt{-5})$. Then $4 = N(2) = N(\alpha)N(\beta)$, and $N(\alpha) \neq 1, N(\beta) \neq 1$, so $N(\alpha) = N(\beta) = 2$ (as the norm is non-negative in $I(\sqrt{-5})$).

Let $\alpha = a + b\sqrt{-5}$. Now a and b are integers as $-5 \not\equiv 1 \pmod 4$. Then clearly $N(\alpha) = a^2 + 5b^2 = 2$ is impossible. This contradiction justifies that 2 is irreducible in $I(\sqrt{-5})$.

$I(\sqrt{10})$: Note that -9 has two essentially distinct decompositions into the product of irreducible elements:

(10.3.12) $$-9 = 3(-3) = (1 + \sqrt{10})(1 - \sqrt{10}).$$

In (10.3.12), ± 3 is not an associate of $1 \pm \sqrt{10}$, since

$$\frac{1 \pm \sqrt{10}}{\pm 3} = \frac{\pm 1}{3} \pm \frac{\pm 1}{3}\sqrt{10} \notin I(\sqrt{10}).$$

We have to show that all factors in (10.3.12) are irreducible. If ± 3 or $1 \pm \sqrt{10}$ were not irreducible, then similar to the argument seen at $I(\sqrt{-5})$, there would be an $\alpha = a + b\sqrt{10}$ with integers a and b having $N(\alpha) = a^2 - 10b^2 = \pm 3$. This is impossible, however, as $a^2 \not\equiv \pm 3 \pmod 5$.   □

The question of the validity of the Fundamental Theorem of Arithmetic is very hard for general quadratic fields, and is unsolved in general.

We start with real quadratic fields:

R1 It is not known whether the Fundamental Theorem holds in infinitely many $I(\sqrt{t})$ with $t > 0$.

R2 All values $t > 0$ are known where we can perform the division algorithm in $I(\sqrt{t})$ using the absolute value of the norm (see part (iii) in Theorem 10.3.6 below). Thus the Fundamental Theorem is true in $I(\sqrt{t})$ for these values of t. There exist, however, other positive integers t, too, e.g. $t = 14, 22, 23$, or 31, when the Fundamental Theorem holds.

We have had the complete answer for imaginary quadratic fields since 1968:

I1 The Fundamental Theorem is true in exactly nine $I(\sqrt{t})$ with $t < 0$, those listed in part (i) of Theorem 10.3.6. Two of the nine cases are the Gaussian and Eulerian integers discussed earlier.

I2 The division algorithm using the norm works in exactly five cases out of the nine (see part (ii) of Theorem 10.3.6). It can be shown for the other four cases, that there is no division algorithm with any conceivable measure instead of the norm. We return to the precise meaning and proof of this statement in Section 11.3.

We summarize the results indicated in I1, I2, and R2 without proof in

Theorem 10.3.6. (i) *If $t < 0$, then the Fundamental Theorem of Arithmetic holds in* $I(\sqrt{t})$ *if and only if*

$$t = -1, -2, -3, -7, -11, -19, -43, -67, -163.$$

(ii) *We can perform the division algorithm in $I(\sqrt{t})$ with respect to the norm for exactly the first five of the nine values $t < 0$ listed in (i).*

(iii) *If $t > 0$, we can perform the division algorithm in $I(\sqrt{t})$ with respect to the absolute value of the norm if and only if*

$$t = 2, 3, 5, 6, 7, 11, 13, 17, 19, 21, 29, 33, 37, 41, 57, 73. \qquad \clubsuit$$

We ask for the proof of statement (ii) of Theorem 10.3.6 in Exercise 10.3.4.

Finally, we present two theorems about irreducible and prime elements in $I(\sqrt{t})$. The first result is valid in any $I(\sqrt{t})$ independent of the validity of the Fundamental Theorem. Accordingly, we must be careful about the distinction between irreducible and prime, since they are not equivalent due to the lack of the Fundamental Theorem. The second result is about quadratic fields where the Fundamental Theorem is true, so here the two types of elements coincide.

Theorem 10.3.7. *Let $p > 2$ be a prime number and $(p, t) = 1$. Then p is a prime in* $I(\sqrt{t})$ *if and only if* $\left(\frac{t}{p}\right) = -1$. $\qquad \clubsuit$

Proof. First we demonstrate that if $\left(\frac{t}{p}\right) = -1$, then p is a prime in $I(\sqrt{t})$.

We assume $p \mid \alpha\beta$, and want to show that at least one of $p \mid \alpha$ and $p \mid \beta$ must hold. Divisibility $p \mid \alpha\beta$ implies

$$p^2 = N(p) \mid N(\alpha)N(\beta).$$

Since p is a prime in $\mathbf{Z}$, p divides at least one of the factors in the product $N(\alpha)N(\beta)$, say $p \mid N(\alpha)$. Using $\left(\frac{t}{p}\right) = -1$, we shall infer $p \mid \alpha$.

Let $\alpha = a + b\sqrt{t}$. We treat first the case $t \not\equiv 1 \pmod 4$. Then a and b are integers. Thus $p \mid N(\alpha) = a^2 - tb^2$ can be written as

(10.3.13) $$a^2 \equiv tb^2 \pmod p.$$

If $(a, p) = (b, p) = 1$, then (10.3.13) implies

$$1 = \left(\frac{a}{p}\right)^2 = \left(\frac{t}{b}\right)\left(\frac{b}{p}\right)^2 = \left(\frac{t}{p}\right),$$

which contradicts $\left(\frac{t}{p}\right) = -1$. If exactly one of a and b is a multiple of p, then exactly one side of (10.3.13) is divisible by p, which is impossible. Thus (10.3.13) can hold only with $a \equiv b \equiv 0 \pmod{p}$. Then $p \mid a + b\sqrt{t} = \alpha$.

In the case $t \equiv 1 \pmod 4$ we have to consider the possibility of $a = u/2$, $b = v/2$, too, with odd u and v. Then we can work with the congruence $u^2 \equiv tv^2 \pmod{p}$ instead of (10.3.13), and arrive at $p \mid \alpha$ similarly.

For the converse, assume $\left(\frac{t}{p}\right) = 1$. Then $c^2 \equiv t \pmod{p}$ for some integer c. Thus

$$p \mid c^2 - t = (c + \sqrt{t})(c - \sqrt{t}), \quad \text{but} \quad p \nmid c \pm \sqrt{t}.$$

This contradicts the prime property of p in $I(\sqrt{t})$.   □

The following theorem is a generalization of Theorems 7.4.12, 7.4.14, 7.4.15, and 7.7.7 for Gaussian and Eulerian integers, if the Fundamental Theorem is true in $I(\sqrt{t})$:

Theorem 10.3.8. *Assume the validity of the Fundamental Theorem of Arithmetic in $I(\sqrt{t})$. Then:*

(i) *An element in $I(\sqrt{t})$ is irreducible if and only if it is prime. (Thus we shall use the shorter word prime instead of irreducible.)*

(ii) *Every prime π in $I(\sqrt{t})$ has exactly one multiple p among the positive prime numbers (of $\mathbf{Z}$).*

(iii) *Every positive prime number p is either a prime in $I(\sqrt{t})$, or is a product of exactly two primes having norm $\pm p$ and being conjugates in the following sense, (cf. Definition 10.4.1): Let $\pi_1 = a + b\sqrt{t}$, then $\pi_2 = \pm(a - b\sqrt{t})$.*

(iv) *If $p > 2$ is a prime number, $(p, t) = 1$, and $\left(\frac{t}{p}\right) = -1$, then p is a prime in $I(\sqrt{t})$.*

(v) *If $p > 2$ is a prime, $(p, t) = 1$, and $\left(\frac{t}{p}\right) = 1$, then p is the product of two non-associate primes in $I(\sqrt{t})$.*

(vi) *If t is odd, then the behavior of 2 is the following:*

    (a) *If $t \equiv 3 \pmod 4$, then 2 is the product of two associate primes (i.e. 2 is an associate of a prime square);*

    (b) *If $t \equiv 1 \pmod 8$, then 2 is the product of two non-associate primes;*

    (c) *If $t \equiv 5 \pmod 8$, then 2 is a prime.*

(vii) *If a prime number p divides t, then p is a product of two associate primes (i.e. p is an associate of a prime square).*

(viii) *The associates of primes listed in parts (iv)–(vii) provide all primes in $I(\sqrt{t})$.*   ♣

Proof. (i) A prime is necessarily irreducible—see the proof of Theorem 1.4.3. The converse follows from the Fundamental Theorem of Arithmetic, see Exercise 1.5.8 (or Theorem 11.3.1).

(ii) and (iii) can be verified in exactly the same way that Theorem 7.4.14 was proved.

(iv) follows from Theorem 10.3.7.

For (v)–(vii), we first determine whether or not p and 2, are primes in $I(\sqrt{t})$.

For (v), this follows from Theorem 10.3.7.

(vi) If $t \equiv 3 \pmod 4$, then

$$2 \mid t^2 - t = (t + \sqrt{t})(t - \sqrt{t}), \quad \text{but} \quad 2 \nmid t \pm \sqrt{t},$$

so 2 is not a prime.

If $t \equiv 1 \pmod 8$, then

$$2 \mid \frac{1-t}{4} = \frac{1+\sqrt{t}}{2} \cdot \frac{1-\sqrt{t}}{2}, \quad \text{but} \quad 2 \nmid \frac{1 \pm \sqrt{t}}{2},$$

so 2 is not a prime.

If $t \equiv 5 \pmod 8$ and 2 were not a prime, then 2 would have a divisor

$$\alpha = \frac{u + v\sqrt{t}}{2} \in I(\sqrt{t}),$$

where u and v are integers of the same parity, satisfying

$$N(\alpha) = \pm 2, \quad \text{so} \quad u^2 - tv^2 = \pm 8.$$

However, $u^2 - tv^2$ cannot be of the form $16k + 8$, a contradiction.

(vii) Since

$$p \mid t = \sqrt{t} \cdot \sqrt{t}, \quad \text{but} \quad p \nmid \sqrt{t}$$

(this holds also for $p = 2$), p cannot be a prime.

The previous observations imply that in cases (v), (vi)(a), (vi)(b), and (vii), p and 2, are not primes. So by (iii), p and 2, can be written as a product of two primes

$$\pi_1 = a + b\sqrt{t} \quad \text{and} \quad \pi_2 = \pm(a - b\sqrt{t}).$$

Here a and b are integers if $t \not\equiv 1 \pmod 4$, and $a = u/2$, $b = v/2$ for some integers u and v of the same parity if $t \equiv 1 \pmod 4$.

Since $|N(\pi_1)| = |N(\pi_2)| = p$ (or 2), $|N(\pi_1/\pi_2)| = 1$. Thus π_1 and π_2 are associates if and only if

$$(10.3.14) \qquad \frac{\pi_1}{\pi_2} = \frac{a + b\sqrt{t}}{\pm(a - b\sqrt{t})} = \frac{a^2 + tb^2}{p} + \frac{2ab}{p}\sqrt{t} \in I(\sqrt{t}).$$

(10.3.14) is impossible for (v), as $2ab/p$ cannot be an integer or a fraction with denominator 2, because $|N(\pi_1)| = p$.

For (vi)(a), we have $p = 2$ in (10.3.14), and a and b are integers because $t \equiv 3 \pmod 4$, and $a^2 - tb^2 = \pm 2$ implies that a and b are odd. Hence

$$\frac{a^2 + tb^2}{2} \quad \text{and} \quad \frac{2ab}{2} = ab$$

are integers, so π_1 and π_2 are associates.

For (vi)(b), we again have $p = 2$ in (10.3.14), but now

$$a^2 - tb^2 = \pm 2 \quad \text{and} \quad t \equiv 1 \pmod 4$$

imply that a and b are not integers. Hence $a = u/2$ and $b = v/2$ for some odd u and v. Then $2ab/2 = uv/4$ in (10.3.14) is neither an integer nor a fraction with denominator 2, so (10.3.14) is false. So π_1 and π_2 are not associates.

At (vii), we investigate first the case when a and b are integers. Then

$$a^2 - tb^2 = \pm p \quad \text{and} \quad p \mid t$$

imply $p \mid a$, thus both

$$\frac{a^2 + tb^2}{p} \quad \text{and} \quad \frac{2ab}{p}$$

are integers in (10.3.14), so π_1 and π_2 are associates.

We can handle similarly the case when $t \equiv 1 \pmod 4$ and $a = u/2$, $b = v/2$ for some odd u and v.

Finally, (viii) follows immediately from (ii) and (iv)–(vii).   □

Exercises 10.3

1. (a) Verify that the Fundamental Theorem of Arithmetic is true in $I(\sqrt{3})$.
    (b) How can the following equalities be reconciled with the Fundamental Theorem:
    (b1) $7 + 3\sqrt{3} = (1 + \sqrt{3})(1 + 2\sqrt{3}) = (-4 + 3\sqrt{3})(5 + 3\sqrt{3})$
    (b2) $19 + 5\sqrt{3} = (5 - \sqrt{3})(5 + 2\sqrt{3}) = (-4 + 3\sqrt{3})(11 + 7\sqrt{3})$?
    (c) Determine all primes in $I(\sqrt{3})$.
  * (d) For which positive integers n is the Diophantine equation $x^2 - 3y^2 = n$ solvable, and what is the number of solutions?

2. (a) Show that the Fundamental Theorem of Arithmetic holds in $I(\sqrt{-2})$.
    (b) Determine all primes in $I(\sqrt{-2})$.
  * (c) Solve the Diophantine equation $x^2 + 2 = y^3$.

3. Demonstrate that the Fundamental Theorem of Arithmetic is false in $I(\sqrt{t})$ if t is

    (a)   15
    (b)   26
    (c)   −6
    (d)   −10.

 * 4. Prove (ii) of Theorem 10.3.6: There is a division algorithm with respect to the norm in an imaginary $I(\sqrt{t})$ if and only if $t = -1, -2, -3, -7$, or -11.

S* 5. Show that if t is a squarefree composite negative integer, then the Fundamental Theorem of Arithmetic is false in $I(\sqrt{t})$.

S* 6. Let $k > 1$ be an integer and $f = x^2 + x + k$. Show that if the Fundamental Theorem of Arithmetic is true in $I(\sqrt{-4k+1})$, then all integers

$$f(0), f(1), \ldots, f(k-2)$$

are prime numbers.

Remark: It can be shown that the converse holds, too. Thus by Theorem 10.3.6(i), the property in the exercise is true only for $k = 2, 3, 5, 11, 17$, and 41. For $k = 41$, we obtain that $n^2 + n + 41$ is a prime number for $0 \le n \le 39$, as mentioned in Section 5.1. By the above, there is no such sequence of primes for $k > 41$.

7. Show that if $|N(\alpha)|$ is a prime number for some $\alpha \in I(\sqrt{t})$, then α

   (a) is irreducible

   * (b) is a prime

   in $I(\sqrt{t})$ (independent of the validity of the Fundamental Theorem of Arithmetic in $I(\sqrt{t})$).

8. Prove that if $\alpha^2 \mid \beta^2$ for some $\alpha, \beta \in I(\sqrt{t})$, then also $\alpha \mid \beta$, whether or not the Fundamental Theorem of Arithmetic holds in $I(\sqrt{t})$.

9. We investigate which prime numbers $p > 0$ are irreducible or prime elements in $I(\sqrt{-5})$.

   (a) 5 is not irreducible and hence is not a prime.

   (b) 2 is irreducible but is not a prime.

   (c) If $p \equiv 11, 13, 17$, or 19 (mod 20), then p is a prime and thus irreducible.

   (d) If $p \equiv 3$ or 7 (mod 20), then p is irreducible but is not a prime.

   S* (e) If $p \equiv 1$ or 9 (mod 20), then p is not irreducible and so cannot be a prime.

10.4. Norm

In this section, we extend the notion of norm to every extension $\mathbf{Q}(\vartheta)$, where ϑ is an algebraic number. First, for every element $\alpha \in \mathbf{Q}(\vartheta)$, we have to introduce the notions of *conjugates* of α *over* $\mathbf{Q}$ and of *relative conjugates* of α with respect to $\mathbf{Q}(\vartheta)$.

Definition 10.4.1. The complex roots of a minimal polynomial of an algebraic number α are called the *conjugates* of α *over* $\mathbf{Q}$. ♣

Since m_α is irreducible over $\mathbf{Q}$, and an irreducible polynomial has no multiple complex roots (see Exercise 9.4.4), an algebraic number of degree n has n distinct conjugates over $\mathbf{Q}$ including the number itself.

Also the complex conjugate $\overline{\alpha}$ of α occurs among the conjugates of α over $\mathbf{Q}$, as α and $\overline{\alpha}$ have the same minimal polynomial.

In the sequel, we shall say just "conjugate" for brevity instead of "conjugate over $\mathbf{Q}$" in general, but we shall never omit the adjective for the complex conjugate.

Examples. **E1** A rational number has a single conjugate, namely itself.

E2 Let $\alpha = a + bi$ be a non-real Gaussian rational, $a, b \in \mathbf{Q}$, $b \neq 0$. Then one of its conjugates is α itself, and the other is its complex conjugate $\overline{\alpha} = a - bi$. The same holds for non-real Eulerian rationals.

E3 Let $\alpha = a + b\sqrt{2}$ be a non-rational element in $\mathbf{Q}(\sqrt{2})$, i.e. $a, b \in \mathbf{Q}$, $b \neq 0$. Then its two conjugates are itself and $a - b\sqrt{2}$.

E4 The conjugates of $\alpha = \sqrt[5]{2}$ are the numbers $\varrho\alpha$, where ϱ is a fifth complex root of unity.

Definition 10.4.2. Let

$$\vartheta_{(1)} = \vartheta, \quad \vartheta_{(2)}, \quad \ldots, \quad \vartheta_{(n)}$$

be the conjugates of an algebraic number ϑ of degree n, and take $\alpha \in \mathbf{Q}(\vartheta)$. By Theorem 10.2.3, there is a unique polynomial $f \in \mathbf{Q}[x]$ satisfying

$$\alpha = f(\vartheta) \quad \text{and} \quad \deg f \leq n - 1 \text{ or } f = 0.$$

Then the numbers

$$f(\vartheta_{(j)}), \qquad j = 1, 2, \ldots, n$$

are called the *relative conjugates* of α with respect to $\mathbf{Q}(\vartheta)$.   ♣

Thus a relative conjugate $f(\vartheta_{(j)})$ is an element in $\mathbf{Q}(\vartheta_{(j)})$. The extension $\mathbf{Q}(\vartheta_{(j)})$ does not coincide with $\mathbf{Q}(\vartheta)$ in general, so the relative conjugates of α are mostly not contained in $\mathbf{Q}(\vartheta)$.

In Definition 10.4.2 the relative conjugates $f(\vartheta_{(j)})$ seem to depend not only on α and the extension $\mathbf{Q}(\vartheta)$, but also on the choice of ϑ, since a given extension can be generated by many different elements. Theorem 10.4.3, however, will guarantee that this is not the case: If $\mathbf{Q}(\vartheta) = \mathbf{Q}(\psi)$, then the relative conjugates of α will be the same whether they were constructed using ϑ or ψ.

Examples. **E5** All relative conjugates of a rational number r are itself for any extension $\mathbf{Q}(\vartheta)$. The constant polynomial $f = r$ meets the requirements $f(\vartheta) = r$, $\deg f < \deg \vartheta$ or $f = 0$, thus $f(\vartheta_{(j)}) = r$ for every j.

E6 Let $\vartheta = i$, then its conjugates are $\vartheta_{(1)} = i$ and $\vartheta_{(2)} = -i$. Thus the relative conjugates of an element $\alpha = a + bi$ $(a, b \in \mathbf{Q})$ of $\mathbf{Q}(i)$ are

$$a + bi = \alpha \quad \text{and} \quad a + b(-i) = a - bi = \overline{\alpha}.$$

This means that if α is not a rational number, then its relative conjugates are the same as its conjugates over $\mathbf{Q}$. We have the same result also for $\mathbf{Q}(\sqrt{-3})$, $\mathbf{Q}(\sqrt{2})$, and for quadratic fields in general.

E7 If $\vartheta = \sqrt[4]{3}$, then its conjugates are $\pm\vartheta$ and $\pm i\vartheta$. The polynomial representing $\alpha = \sqrt{3} \in \mathbf{Q}(\vartheta)$ according to Theorem 10.2.3 is $f = x^2$, since $\sqrt{3} = (\vartheta)^2$. Thus the relative conjugates of $\sqrt{3}$ are

$$(\pm\vartheta)^2 = \sqrt{3} \quad \text{and} \quad (\pm i\vartheta)^2 = -\sqrt{3}.$$

These four numbers are just the conjugates of $\sqrt{3}$ over $\mathbf{Q}$, each taken twice.

The examples indicate that the relative conjugates of $\alpha \in \mathbf{Q}(\vartheta)$ are the same as the conjugates of α over $\mathbf{Q}$, each counted with a suitable multiplicity:

Theorem 10.4.3. *Let α be an element of degree k in the extension $\mathbf{Q}(\vartheta)$ of degree n. Then we get the relative conjugates of α by taking each conjugate of α over $\mathbf{Q}$ with multiplicity n/k.* ♣

The theorem implies that the relative conjugates remain the same if we replace ϑ by another generating element of $\mathbf{Q}(\vartheta)$, so the relative conjugates depend only on α and the extension itself.

Theorem 10.4.3 gives a new proof for $\deg \alpha$ being a divisor of the extension $\mathbf{Q}(\vartheta)$ (cf. Theorem 10.2.5).

Proof. Let

$$m_\vartheta = \prod_{j=1}^{n}(x - \vartheta_{(j)}), \text{ where } \vartheta_{(1)} = \vartheta, \text{ and } m_\alpha = \prod_{s=1}^{k}(x - \alpha_{(s)}), \text{ where } \alpha_{(1)} = \alpha,$$

be the minimal polynomials of ϑ and α, and f the polynomial representing α according to Theorem 10.2.3, so $f(\vartheta) = \alpha$.

I. We verify first that every relative conjugate $f(\vartheta_{(j)})$ of α coincides with a conjugate α_s of α over $\mathbf{Q}$ and disregard multiplicity temporarily.

Consider the polynomial $g(x) = m_\alpha(f(x))$. Clearly, $g \in \mathbf{Q}[x]$, and

$$g(\vartheta) = m_\alpha(f(\vartheta)) = m_\alpha(\alpha) = 0.$$

Hence $m_\vartheta \mid g$, so

$$0 = g(\vartheta_{(j)}) = m_\alpha(f(\vartheta_{(j)}))$$

for every j. This means that $f(\vartheta_{(j)})$ is a root of m_α, so $f(\vartheta_{(j)})$ is equal to some α_s.

II. We still have to show that each α_s occurs with the same multiplicity among the numbers $f(\vartheta_{(j)})$ $(j = 1, 2, \ldots, n)$. Consider the polynomial

$$h = \prod_{j=1}^{n}(x - f(\vartheta_{(j)})).$$

By the Fundamental Theorem 9.3.2 of Symmetric Polynomials, we obtain similar to the proofs of Theorems 9.3.1 and 9.3.6 that h has rational coefficients: Every coefficient c_r of h is a symmetric polynomial of the variables $\vartheta_{(j)}$, so c_r can be written as a polynomial with rational coefficients of the elementary symmetric polynomials σ_j of the variables $\vartheta_{(j)}$. By Viète's formulas connecting roots and coefficients, the values $\pm\sigma_j$ are just the coefficients of m_ϑ, which are rational numbers, so c_r is rational.

Decompose h into a product of polynomials irreducible over $\mathbf{Q}$. Since the roots of h, i.e. the numbers $f(\vartheta_{(j)})$, are roots of the irreducible polynomial m_α, each factor in the decomposition of h is m_α. Further, both h and m_α have leading coefficient 1, so h is a power of m_α: $h = m_\alpha^t$. Comparing the degrees, we have $t = n/k$ so each root α_s of m_α occurs n/k times among the numbers $f(\vartheta_{(j)})$. □

Now we are ready to give a general definition for the norm:

Definition 10.4.4. The *norm* of an element $\alpha \in \mathbf{Q}(\vartheta)$ is the product of its relative conjugates: If the conjugates of ϑ are $\vartheta_{(1)} = \vartheta, \vartheta_{(2)}, \ldots, \vartheta_{(n)}$ and $\alpha = f(\vartheta)$, then

$$N(\alpha) = \prod_{j=1}^{n} f(\vartheta_{(j)}). \qquad \clubsuit$$

The norm in quadratic number fields introduced in Definition 10.3.3 is a special case of Definition 10.4.4.

We summarize the most important properties of the norm in

Theorem 10.4.5. (i) *Let* $\alpha \in \mathbf{Q}(\vartheta)$, $\deg \vartheta = n$, *and* $\deg \alpha = k$. *Then*

$$N(\alpha) = \left(\prod_{s=1}^{k} \alpha_s\right)^{n/k} = (-1)^n a_0^{n/k},$$

where $\alpha_{(1)} = \alpha, \alpha_{(2)}, \ldots, \alpha_{(k)}$ *are the conjugates of* α *over* $\mathbf{Q}$ *and* a_0 *is the constant term in the minimal polynomial of* α *with leading coefficient* 1.

(ii) $\alpha, \beta \in \mathbf{Q}(\vartheta) \Longrightarrow N(\alpha\beta) = N(\alpha)N(\beta)$.

(iii) *If* α *is an algebraic integer, then* $N(\alpha)$ *is an ordinary integer.* $\qquad \clubsuit$

Proof. The first equality in (i) follows immediately from Theorem 10.4.3, and the second equality is a direct consequence of Viète's formula about the product of roots of the polynomial m_α. This form of $N(\alpha)$ in (i) implies (iii).

To verify (ii), let

$$\alpha = f_1(\vartheta), \qquad \beta = f_2(\vartheta), \quad \text{and} \quad \alpha\beta = f_3(\vartheta).$$

Then ϑ is a root of $h = f_3 - f_1 f_2 \in \mathbf{Q}[x]$, so $m_\vartheta \mid h$. This implies that all other roots of m_ϑ, i.e. all conjugates $\vartheta_{(j)}$ of ϑ are roots of h too, so

$$0 = h(\vartheta_{(j)}) = f_3(\vartheta_{(j)}) - f_1(\vartheta_{(j)})f_2(\vartheta_{(j)}), \qquad j = 1, 2, \ldots, n.$$

Multiplying the equalities $f_3(\vartheta_{(j)}) = f_1(\vartheta_{(j)})f_2(\vartheta_{(j)})$, we obtain

$$N(\alpha\beta) = \prod_{j=1}^{n} f_3(\vartheta_{(j)}) = \left(\prod_{j=1}^{n} f_1(\vartheta_{(j)})\right)\left(\prod_{j=1}^{n} f_2(\vartheta_{(j)})\right) = N(\alpha)N(\beta). \qquad \square$$

Exercises 10.4

1. Determine the conjugates of the algebraic numbers over $\mathbf{Q}$

    (a) $\sqrt{2} + \sqrt{3}$

    (b) $\sqrt{2}(1 + i)$

    (c) $\cos 20°$

    (d) $\cos 1° + i \sin 1°$.

2. Let $\vartheta_{(1)} = \vartheta, \vartheta_{(2)}, \ldots, \vartheta_{(n)}$ denote the conjugates of an algebraic number ϑ over $\mathbf{Q}$. Verify.

   (a) If $\deg \vartheta = 2$, then $\mathbf{Q}(\vartheta_{(1)}) = \mathbf{Q}(\vartheta_{(2)})$.
   (b) If ϑ is a non-real complex number and $\deg \vartheta$ is odd, then $\mathbf{Q}(\vartheta_{(j)}) \neq \mathbf{Q}(\vartheta_{(k)})$ for some j and k.
   (c) If ϑ is a non-real complex number and $\deg \vartheta = 3$, then $j \neq k$ implies $\mathbf{Q}(\vartheta_{(j)}) \cap \mathbf{Q}(\vartheta_{(k)}) = \mathbf{Q}$.

3. Find the relative conjugates and norm of the elements in $\mathbf{Q}(\sqrt[4]{2})$

   (a) $1 + \sqrt[4]{2}$
   (b) $1 + \sqrt{2}$
   (c) $1 + \sqrt[4]{2} + \sqrt{2} + \sqrt[4]{8}$.

4. Prove that an element ε is a unit in the ring $I(\vartheta)$ of all algebraic integers of $\mathbf{Q}(\vartheta)$ if and only if $N(\varepsilon) = \pm 1$.

   Remark: There are infinitely many units in $I(\vartheta)$ except when $\mathbf{Q}(\vartheta)$ is an imaginary quadratic field or $\mathbf{Q}(\vartheta) = \mathbf{Q}$.

5. Verify.

   (a) There exists a Gaussian rational which is not a Gaussian integer, but its norm is an integer.
   (b) There exists an element α in every quadratic field $\mathbf{Q}(\vartheta)$ that is not an algebraic integer, but $N(\alpha)$ is an integer.

10.5. Integral Basis

In this section, ϑ always denotes an algebraic number of degree n.

We know from Theorem 10.2.3 that every $\alpha \in \mathbf{Q}(\vartheta)$ has a unique representation

$$(10.5.1) \qquad \alpha = a_0 + a_1\vartheta + \cdots + a_{n-1}\vartheta^{n-1}, \qquad a_j \in \mathbf{Q}, \qquad j = 0, 1, \ldots, n-1,$$

so $1, \vartheta, \ldots, \vartheta^{n-1}$ form a basis in $\mathbf{Q}(\vartheta)$ considered as a vector space over $\mathbf{Q}$.

Representation (10.5.1) gives no information in general about whether or not α is an algebraic integer. For a quadratic field, however, we proved in Theorems 10.3.1 and 10.3.2 the existence of a basis ω_1, ω_2 that does: Every quadratic field can be written as $\mathbf{Q}(\sqrt{t})$, where t is a squarefree integer different from 1, and taking

$$\omega_1 = 1 \quad \text{and} \quad \omega_2 = \begin{cases} \sqrt{t}, & \text{if } t \not\equiv 1 \pmod 4 \\ (1 + \sqrt{t})/2, & \text{if } t \equiv 1 \pmod 4, \end{cases}$$

every $\alpha \in \mathbf{Q}(\sqrt{t})$ has a unique representation as

$$\alpha = r_1\omega_1 + r_2\omega_2, \qquad r_1, r_2 \in \mathbf{Q},$$

and, α is an algebraic integer if and only if both r_1 and r_2 are ordinary integers.

In general, a basis with this property is called an *integral basis* of $\mathbf{Q}(\vartheta)$.

Definition 10.5.1. The elements $\omega_1, \ldots, \omega_n$ of an extension $\mathbf{Q}(\vartheta)$ form an *integral basis* in $\mathbf{Q}(\vartheta)$ if every $\alpha \in \mathbf{Q}(\vartheta)$ has a unique representation

$$(10.5.2) \qquad \alpha = r_1\omega_1 + r_2\omega_2 + \cdots + r_n\omega_n, \qquad r_j \in \mathbf{Q}, \qquad j = 1, 2, \ldots, n,$$

and α is an algebraic integer if and only if every r_j is an ordinary integer. ♣

Our goal is to prove that every extension $\mathbf{Q}(\vartheta)$ possesses an integral basis.

Let ϑ be an algebraic number of degree n. We consider the extension $\mathbf{Q}(\vartheta)$. To make a clear distinction, bases of the vector space $\mathbf{Q}(\vartheta)$ over $\mathbf{Q}$ will be called v-bases, and the integral bases among them will be referred to as i-bases.

We examine first how to determine whether n elements of $\mathbf{Q}(\vartheta)$ form a v-basis. Let $\alpha_1, \ldots, \alpha_n \in \mathbf{Q}(\vartheta)$,

$$(10.5.3a) \quad \alpha_i = f_i(\vartheta), \quad \text{where } f_i \in \mathbf{Q}[x], \deg f_i \leq n - 1 \text{ or } f_i = 0, \quad i = 1, \ldots, n$$

so

$(10.5.3b)$

$$\alpha_i = a_{0i} + a_{1i}\vartheta + \cdots + a_{n-1,i}\vartheta^{n-1}, \quad a_{ki} \in \mathbf{Q}, \quad 0 \leq k \leq n - 1, \quad 1 \leq i \leq n.$$

Consider the linear transformation $\mathcal{A}$ of the vector space $\mathbf{Q}(\vartheta)$ that maps the elements $1, \vartheta, \ldots, \vartheta^{n-1}$ of the v-basis to the vectors $\alpha_1, \ldots, \alpha_n$, in this order. Then the matrix of the transformation $\mathcal{A}$ in the v-basis $1, \vartheta, \ldots, \vartheta^{n-1}$ is

$$(10.5.4) \qquad A = \begin{pmatrix} a_{01} & a_{02} & \cdots & a_{0n} \\ a_{11} & a_{12} & \cdots & a_{1n} \\ \vdots & \vdots & \ddots & \vdots \\ a_{n-1,1} & a_{n-1,2} & \cdots & a_{n-1,n} \end{pmatrix}$$

where a_{ki} are the rational numbers in (10.5.3b).

We know from elementary linear algebra that the vectors $\alpha_1, \ldots, \alpha_n$ form a v-basis if and only if matrix A has an inverse, or $\det A \neq 0$.

Observe that the numbers $\alpha_1, \ldots, \alpha_n$ can be expressed as

$$(10.5.3c) \qquad \begin{pmatrix} \alpha_1 \\ \alpha_2 \\ \vdots \\ \alpha_n \end{pmatrix} = A^T \begin{pmatrix} 1 \\ \vartheta \\ \vdots \\ \vartheta^{n-1} \end{pmatrix}$$

where A^T denotes the transpose of the matrix A.

To verify the existence of an i-basis, we shall use the *discriminant* which is the square of the determinant of a matrix closely related to A.

Let V be the Vandermonde matrix generated by the conjugates of ϑ over $\mathbf{Q}$:

$$(10.5.5) \qquad V = V(\vartheta_{(1)}, \vartheta_{(2)}, \ldots, \vartheta_{(n)}) = \begin{pmatrix} 1 & 1 & 1 & \cdots & 1 \\ \vartheta_{(1)} & \vartheta_{(2)} & \vartheta_{(3)} & \cdots & \vartheta_{(n)} \\ \vartheta_{(1)}^2 & \vartheta_{(2)}^2 & \vartheta_{(3)}^2 & \cdots & \vartheta_{(n)}^2 \\ \vdots & \vdots & \vdots & \ddots & \vdots \\ \vartheta_{(1)}^{n-1} & \vartheta_{(2)}^{n-1} & \vartheta_{(3)}^{n-1} & \cdots & \vartheta_{(n)}^{n-1} \end{pmatrix}$$

and

(10.5.6) $$\tilde{A} = A^T V.$$

Then the jth element in row i of the matrix $\tilde{A}$ is the inner product of row i in A and column j in V, or

(10.5.7) $$a_{0i} + a_{1i}\vartheta_{(j)} + \cdots + a_{n-1,i}\vartheta_{(j)}^{n-1}.$$

By (10.5.3a)–(10.5.3b), the sum in (10.5.7) is just the jth relative conjugate $f_i(\vartheta_{(j)})$ of α_i.

The discriminant $\Delta(\alpha_1, \ldots, \alpha_n)$ of the numbers $\alpha_1, \ldots, \alpha_n$ is the square of the determinant of matrix $\tilde{A}$:

Definition 10.5.2. Consider the extension $\mathbf{Q}(\vartheta)$, where $\deg \vartheta = n$, and let $\vartheta_{(1)} = \vartheta$, $\vartheta_{(2)}, \ldots, \vartheta_{(n)}$ denote the conjugates of ϑ. The *discriminant* $\Delta(\alpha_1, \ldots, \alpha_n)$ of the numbers $\alpha_1, \ldots, \alpha_n$ is the square of the determinant of the matrix $\tilde{A}$, i.e. using (10.5.3a)–(10.5.7),

$$\Delta(\alpha_1, \ldots, \alpha_n) = (\det(A^T V))^2 = \begin{vmatrix} f_1(\vartheta_{(1)}) & f_1(\vartheta_{(2)}) & \cdots & f_1(\vartheta_{(n)}) \\ f_2(\vartheta_{(1)}) & f_2(\vartheta_{(2)}) & \cdots & f_2(\vartheta_{(n)}) \\ \vdots & \vdots & \ddots & \vdots \\ f_n(\vartheta_{(1)}) & f_n(\vartheta_{(2)}) & \cdots & f_n(\vartheta_{(n)}) \end{vmatrix}^2 . \qquad \clubsuit$$

We summarize the most important properties of the discriminant in

Theorem 10.5.3. (i) *The discriminant* $\Delta(\alpha_1, \ldots, \alpha_n)$ *is a rational number, and for algebraic integers* α_i, *it is an integer.*

(ii) $\alpha_1, \ldots, \alpha_n$ *is a v-basis if and only if* $\Delta(\alpha_1, \ldots, \alpha_n) \neq 0$.

(iii) *If C is an $n \times n$ matrix with rational elements and*

$$\begin{pmatrix} \beta_1 \\ \beta_2 \\ \vdots \\ \beta_n \end{pmatrix} = C \begin{pmatrix} \alpha_1 \\ \alpha_2 \\ \vdots \\ \alpha_n \end{pmatrix}$$

then

$$\Delta(\beta_1, \ldots, \beta_n) = (\det C)^2 \Delta(\alpha_1, \ldots, \alpha_n). \qquad \clubsuit$$

Proof. (i) The discriminant is a symmetric polynomial in the variables $\vartheta_{(j)}$. Interchanging two $\vartheta_{(j)}$ means interchanging two columns in the determinant, which gives a sign change for the determinant, so its square remains the same. This implies in the usual way (as in the proofs of Theorems 9.3.1, 9.3.6, or 10.4.3) that the discriminant is a rational number.

If every α_i is an algebraic integer, then their conjugates, and so their relative conjugates, are algebraic integers. The discriminant is computed from the relative conjugates using addition, subtraction, and multiplication. As the algebraic integers form a ring, the discriminant is an algebraic integer, too. Hence the discriminant is both a rational number and an algebraic integer, so it is necessarily an integer.

(ii) By the rule of multiplication of determinants,

$$\Delta(\alpha_1, \ldots, \alpha_n) = (\det A)^2 (\det V)^2.$$

Since the generating elements $\vartheta_{(j)}$ of the Vandermonde determinant V are all distinct, $\det V \neq 0$. Thus

$$\Delta(\alpha_1, \ldots, \alpha_n) \neq 0 \iff \det A \neq 0.$$

And as we showed earlier, $\alpha_1, \ldots, \alpha_n$ is a v-basis if and only if $\det A \neq 0$.

(iii) By (10.5.3c),

$$\begin{pmatrix} \alpha_1 \\ \alpha_2 \\ \vdots \\ \alpha_n \end{pmatrix} = A^T \begin{pmatrix} 1 \\ \vartheta \\ \vdots \\ \vartheta^{n-1} \end{pmatrix} \quad \text{and} \quad \begin{pmatrix} \beta_1 \\ \beta_2 \\ \vdots \\ \beta_n \end{pmatrix} = B^T \begin{pmatrix} 1 \\ \vartheta \\ \vdots \\ \vartheta^{n-1}. \end{pmatrix}$$

Thus

$$\begin{pmatrix} \beta_1 \\ \beta_2 \\ \vdots \\ \beta_n \end{pmatrix} = C \begin{pmatrix} \alpha_1 \\ \alpha_2 \\ \vdots \\ \alpha_n \end{pmatrix} = CA^T \begin{pmatrix} 1 \\ \vartheta \\ \vdots \\ \vartheta^{n-1}, \end{pmatrix}$$

so $B^T = CA^T$ by the uniqueness of matrix B belonging to the numbers β_i. This implies

$$\Delta(\beta_1, \ldots, \beta_n) = (\det(B^T V))^2 = (\det(CA^T V))^2$$
$$= (\det C)^2 (\det(A^T V))^2 = (\det C)^2 \Delta(\alpha_1, \ldots, \alpha_n). \qquad \square$$

Now we are ready to prove the existence of an i-basis.

Theorem 10.5.4. *There exists an integral basis in $\mathbf{Q}(\vartheta)$ for any algebraic number ϑ.* ♣

Proof. We establish first a few properties of i-bases that will help to find an i-basis among the v-bases.

If $\omega_1, \ldots, \omega_n$ is an i-basis, then every ω_i is an algebraic integer, since every coefficient is an integer in the representation

$$\omega_i = 0 \cdot \omega_1 + \cdots + 1 \cdot \omega_i + \cdots + 0 \cdot \omega_n.$$

If $\omega_1, \ldots, \omega_n$ is an i-basis and $\beta_1, \ldots, \beta_n$ is a v-basis of algebraic integers, then every β_i is a linear combination with integer coefficients of the basis vectors ω_j, so

$$\begin{pmatrix} \beta_1 \\ \beta_2 \\ \vdots \\ \beta_n \end{pmatrix} = C \begin{pmatrix} \omega_1 \\ \omega_2 \\ \vdots \\ \omega_n \end{pmatrix}$$

with a suitable invertible matrix C with integer elements. Then Theorem 10.5.3(iii) implies

$$\Delta(\beta_1, \ldots, \beta_n) = \Delta(\omega_1, \ldots, \omega_n)(\det C)^2.$$

Since $\det C$ is a non-zero integer, $(\det C)^2 \geq 1$, so

$$|\Delta(\beta_1, \ldots, \beta_n)| \geq |\Delta(\omega_1, \ldots, \omega_n)|.$$

This says that the absolute value of the discriminant of an i-basis is less than or equal to the absolute value of the discriminant of any v-basis consisting of algebraic integers.

Accordingly, a v-basis can be an i-basis only if its elements are algebraic integers and the absolute value of its discriminant is minimal among all v-bases of this type.

We verify that there exists a v-basis with this property, and it is also an i-basis.

We show first that there are v-bases consisting of algebraic integers. Let $\gamma_1, \ldots, \gamma_n$ be an arbitrary v-basis. By Exercise 9.6.6, every γ_i can be written as $\gamma_i = \alpha_i/c_i$, where α_i is an algebraic integer and $c_i \neq 0$ is an ordinary integer. Then clearly $\alpha_1, \ldots, \alpha_n$ is a v-basis.

Consider all v-bases of algebraic integers. The discriminant of each is a non-zero integer, by Theorem 10.5.3(i)–(ii). Choose a v-basis $\omega_1, \ldots, \omega_n$ that has a discriminant of minimal absolute value. We prove that $\omega_1, \ldots, \omega_n$ is an i-basis. Thus, we have to verify that $\alpha \in \mathbf{Q}(\vartheta)$ is an algebraic integer if and only if every r_j is an integer in representation (10.5.2)

$$\alpha = r_1\omega_1 + r_2\omega_2 + \cdots + r_n\omega_n, \qquad r_j \in \mathbf{Q}, \qquad j = 1, 2, \ldots, n.$$

Assume first that $r_1, \ldots, r_n$ are integers. Since every ω_i is an algebraic integer and the algebraic integers form a ring, $\alpha = \sum_{j=1}^{n} r_j\omega_j$ is an algebraic integer.

Conversely, let $\alpha \in \mathbf{Q}(\vartheta)$ be an algebraic integer. Assume that, say, r_1 is not an integer in representation (10.5.2)

$$\alpha = r_1\omega_1 + r_2\omega_2 + \cdots + r_n\omega_n.$$

Let

$$\beta_1 = \alpha - \lfloor r_1 \rfloor \omega_1 = \{r_1\}\omega_1 + r_2\omega_2 + \cdots + r_n\omega_n \quad \text{and} \quad \beta_j = \omega_j \quad \text{for } 2 \leq j \leq n.$$

Then the numbers $\beta_1, \ldots, \beta_n$ are algebraic integers, and

$$\begin{pmatrix} \beta_1 \\ \beta_2 \\ \vdots \\ \beta_n \end{pmatrix} = C \begin{pmatrix} \omega_1 \\ \omega_2 \\ \vdots \\ \omega_n \end{pmatrix}$$

where

$$C = \begin{pmatrix} \{r_1\} & r_2 & r_3 & \cdots & r_n \\ 0 & 1 & 0 & \cdots & 0 \\ 0 & 0 & 1 & \cdots & 0 \\ \vdots & \vdots & \vdots & \ddots & \vdots \\ 0 & 0 & 0 & \cdots & 1 \end{pmatrix}$$

By Theorem 10.5.3(iii),

$$\Delta(\beta_1, \ldots, \beta_n) = \Delta(\omega_1, \ldots, \omega_n)(\det C)^2 = \Delta(\omega_1, \ldots, \omega_n)\{r_1\}^2,$$

and $0 < \{r_1\} < 1$ implies

$$0 < |\Delta(\beta_1, \ldots, \beta_n)| < |\Delta(\omega_1, \ldots, \omega_n)|,$$

which contradicts the minimality of $|\Delta(\omega_1, \ldots, \omega_n)|$. $\qquad\square$

Remarks: (1) We see from the proof that the absolute values of the discriminants are the same for any two integral bases in $\mathbf{Q}(\vartheta)$. It can be shown that the discriminants themselves are equal, see Exercise 10.5.2b. This common value is called the discriminant of the extension $\mathbf{Q}(\vartheta)$.

(2) The proof above shows only the existence of an integral basis, and is not suitable to construct one explicitly.

(3) We can exhibit an integral basis in a quadratic field by Theorem 10.3.2, but for extensions of higher degree, it is hard to find an integral basis. It can be shown that if ϑ is a pth primitive complex root of unity for a prime $p > 2$, then $1, \vartheta, \ldots,$ ϑ^{p-2} form an integral basis in $\mathbf{Q}(\vartheta)$.

Exercises 10.5

1. Compute the discriminant $\Delta(1, \vartheta, \ldots, \vartheta^{n-1})$ in $\mathbf{Q}(\vartheta)$ if ϑ is

   (a) i

   (b) $\cos(2\pi/3) + i \sin(2\pi/3)$

   (c) $\sqrt[3]{2}$;

   * (d) $\sqrt[n]{2}$.

2. Consider an extension $\mathbf{Q}(\vartheta)$, where $\deg \vartheta = n$. Prove.

   (a) If $\omega_1, \ldots, \omega_n$ is an integral basis and $\beta_1, \ldots, \beta_n \in \mathbf{Q}(\vartheta)$ are algebraic integers, then $\Delta(\omega_1, \ldots, \omega_n) \mid \Delta(\beta_1, \ldots, \beta_n)$.

   (b) Any two integral bases have the same discriminant.

3. Determine the discriminant of an integral basis in a quadratic field.

4. Let $\deg \vartheta = n$ and $\alpha_1, \ldots, \alpha_n$ algebraic integers in $\mathbf{Q}(\vartheta)$ such that $\Delta(\alpha_1, \ldots, \alpha_n)$ is squarefree. Show that $\alpha_1, \ldots, \alpha_n$ is an integral basis in $\mathbf{Q}(\vartheta)$.

5. (a) Find a necessary and sufficient condition for Gaussian rationals $a + bi$ and $c + di$ to form an integral basis in $\mathbf{Q}(i)$.

   (b) Answer the similar question for Eulerian rationals.

S 6. In which quadratic fields does there exist an integral basis ω_1, ω_2, where ω_2 is the conjugate of ω_1 over $\mathbf{Q}$?

7. Let $\deg \vartheta = n$, and assume that the minimal polynomial m_ϑ has only real roots. Then $\Delta(\beta_1, \ldots, \beta_n) \geq 0$ for any elements $\beta_1, \ldots, \beta_n$ in $\mathbf{Q}(\vartheta)$.

8. (a) Exhibit an example showing that the discriminant $\Delta(\alpha_1, \ldots, \alpha_n)$ can be a non-zero integer even if not all the α_i are algebraic integers.

   (b) Prove that if $\mathbf{Q}(\vartheta) \neq \mathbf{Q}$, then there exists a v-basis in $\mathbf{Q}(\vartheta)$ such that none of its elements is an algebraic integer, but its discriminant is an integer.

Ideals

Ideals play a central role in ring theory, but we restrict ourselves to the number theoretic relations. We establish a necessary and sufficient condition for the validity of the Fundamental Theorem of Arithmetic, and show that it always holds in principal ideal domains and Euclidean rings. Then we build number theory for ideals, and prove that unique prime factorization is true among the ideals of algebraic integers in an algebraic number field. As an application, we illustrate through an example that ideals can help to handle Diophantine equations even if the Fundamental Theorem of Arithmetic is false for the algebraic integers of the corresponding extension.

11.1. Ideals and Factor Rings

"Ideal numbers" were introduced by Kummer in the middle of the 19th century for a more efficient approach to Fermat's Last Theorem. We shall discuss this in Section 11.2 in more detail. The notion of ideals, developed from ideal numbers, has become a fundamental tool in ring-theoretical investigations, independent of its impact on number theory.

Definition 11.1.1. A non-empty subset I of a ring R is an *ideal* in R, if

(A) I is closed for addition and taking negatives so

$$i, j \in I \Longrightarrow i + j \in I, -i \in I;$$

(B) Multiplying any element of I by an arbitrary element of R gives a product that is in I:

$$i \in I, r \in R \Longrightarrow ri \in I, ir \in I. \qquad \clubsuit$$

In an equivalent formulation, an ideal I is a special subring where also those products are in I when one of the factors is in I and the other factor is not.

Examples of Ideals. **E1** The set of multiples of m in the ring of integers.

E2 The set of polynomials having a given complex number α among their roots in the ring of polynomials with rational coefficients.

E3 The set of polynomials having an even constant term in the ring of polynomials with integer coefficients.

E4 In any ring, the ring itself and the one-element subset containing the zero alone. These are called *trivial ideals*. A field has just the two trivial ideals (see Exercise 11.1.3).

Since we investigate only number-theoretic connections of ideals, we shall restrict ourselves in this chapter to rings R that are commutative, free of zero divisors, and have an identity element or identity, for short.

The first two properties mean that R is an *integral domain* (see Exercise 1.1.23). As we deal mostly with polynomial rings and subrings of the complex field, we shall denote the identity element by 1.

The simplest, but also most important, ideals are the ones generated by a single element. These are the *principal ideals*.

Definition 11.1.2. Let a be an arbitrary element in an integral domain R with identity. The set $\{\, ra \mid r \in R \,\}$ is called a *principal ideal* generated by a and is denoted by (a).   ♣

Thus the principal ideal (a) consists of the multiples of a formed by elements of R.

The phrases "generated by a" and "ideal" in the definition are justified by

Theorem 11.1.3. *The principal ideal (a) is the smallest ideal containing a, i.e.*

  (i) (a) *is an ideal in R*

 (ii) $a \in (a)$

(iii) *if I is an ideal in R and $a \in I$, then $(a) \subseteq I$.*   ♣

Proof. (i) We verify that the non-empty set $\{\, ra \mid r \in R \,\}$ satisfies Definition 11.1.1. To avoid ambiguity in the formulas, we use square brackets for the usual meaning of parentheses, and keep round parentheses for denoting ideals.

$$r_1 a + r_2 a = [r_1 + r_2]a, \quad -[ra] = [-r]a, \quad \text{and} \quad [r_1 a]r_2 = r_2[r_1 a] = [r_2 r_1]a.$$

(ii) $a = 1a \in \{\, ra \mid r \in R \,\}$.

(iii) If ideal I contains a, then by (B) in Definition 11.1.1, it must also contain ra for every $r \in R$, so $(a) \subseteq I$.   □

We used the identity element and the commutative law in verifying (ii) and (i), and we did not need the lack of zero divisors.

Examples. The two trivial ideals (in Example E4) are principal ideals, generated by the identity and zero: $R = (1)$ and $\{0\} = (0)$.

The ideals in examples E1 and E2 are principal ideals: The multiples of m in $\mathbf{Z}$ constitute the principal ideal (m); the polynomials satisfying $f(\alpha) = 0$ in $\mathbf{Q}[x]$ constitute (0) or (m_α) according to α being transcendental or algebraic (m_α is the minimal polynomial of α).

The ideal in example E3, however, is not a principal ideal. Let I denote this set of polynomials with integer coefficients having an even constant term, and assume $I = (f)$ for some f. Then f is a divisor of every element in I, including 2. Therefore only $f = \pm 1$ or ± 2 are possible. However, (± 1) contains all polynomials with integer coefficients, whereas (± 2) is the polynomials where every coefficient is even. Hence these principal ideals are not equal to I. This contradiction shows that I is not a principal ideal.

As a generalization of principal ideals, we introduce the notion of finitely generated ideals.

Definition 11.1.4. Let $a_1, \ldots, a_k$ be elements of an integral domain R with identity. Then the set $\{\sum_{j=1}^{k} r_j a_j \mid r_j \in R\}$ is called the *principal ideal* generated by $a_1, \ldots, a_k$ and is denoted by $(a_1, \ldots, a_k)$.

An ideal I is *finitely generated* if $I = (a_1, \ldots, a_k)$ for some suitable elements $a_1, \ldots, a_k$.
♣

The analog of Theorem 11.1.3 holds for finitely generated ideals:

Theorem 11.1.5. *The ideal $(a_1, \ldots, a_k)$ is the smallest ideal containing the elements a_j:*

(i) $(a_1, \ldots, a_k)$ *is an ideal in R*

(ii) $a_j \in (a_1, \ldots, a_k)$, $j = 1, 2, \ldots, k$

(iii) *if I is any ideal in R and $a_j \in I$, $j = 1, 2, \ldots, k$, then $(a_1, \ldots, a_k) \subseteq I$.* ♣

The proof of Theorem 11.1.5 is similar to that seen in Theorem 11.1.3, so we leave the details to the reader.

Examples. Clearly, every principal ideal is a finitely generated ideal, generated by a single element.

Also, the ideal I in Example E3 is finitely generated: $I = (2, x)$.

In the ring U of all algebraic integers,

$$K = \left\{ \xi \sqrt[k]{2} \mid \xi \in U, k = 2, 3, 4, \ldots \right\}$$

is an ideal, but cannot be generated by finitely many elements (see Exercise 11.1.4).

If ϑ is an algebraic number, then every ideal in $I(\vartheta)$ is finitely generated (see Exercise 11.1.10). (As earlier, $I(\vartheta)$ denotes the ring of algebraic integers in the extension $\mathbf{Q}(\vartheta)$.)

Finally we present the construction of factor rings with respect to ideals (or, for short, modulo ideals). This is a generalization of the ring of modulo m residue classes (see Section 2.8).

We saw in Example E1 after Definition 11.1.1 that the multiples of m form an ideal I in the ring $\mathbf{Z}$. The residue class modulo m containing the integer a (the one represented by a) has the form

(11.1.1) $$a + I = \{a + i \mid i \in I\}.$$

We defined addition and multiplication for residue classes using their representatives. Using (11.1.1), this means

(11.1.2) $[a + I] + [b + I] = [a + b] + I$ and $[a + I][b + I] = ab + I$.

We had to verify that (11.1.2) defines operations for the classes, i.e. the resulting class is unique, it does not depend on the choice of the representatives taken from the two classes. Analyzing the proof, it turns out that uniqueness is guaranteed by I being an ideal. Thus we arrive at the generalization:

Theorem 11.1.6. *Let I be an ideal in a ring R. Then the* residue classes (11.1.1) *modulo I are disjoint subsets in R and their union equals R. Further, they form a ring with respect to the addition and multiplication defined by* (11.1.2). *This ring is called the* factor ring *of R modulo I, and is denoted by R/I.*   ♣

Accordingly, the ring of residue classes modulo m is the factor ring $\mathbf{Z}/(m)$ of the integers modulo the principal ideal (m).

We leave the proof of Theorem 11.1.6 to the reader. One has to use the ideal properties of I to show that the classes (11.1.1) cover R, any two of them either coincide or are distinct, and the operations in R/I are uniquely defined. The commutative, associative, and distributive laws in R/I follow from the ones in R, the zero element of R/I is the residue class $0 + I$, that is the ideal I itself, and the negative of a residue class $a + I$ is the residue class $[-a] + I$.

Example. We analyze the factor ring $\mathbf{Q}[x]/(x^2 - 2)$ of the ring of polynomials with rational coefficients modulo the principal ideal of polynomials divisible by $x^2 - 2$.

We can apply similar considerations to those we used when we constructed the ring of residue classes modulo m at the integers, which is in fact the factor ring $\mathbf{Z}/(m)$. Polynomials fall into the same residue class modulo the principal ideal $(x^2 - 2)$ if they give the same remainder on division by $x^2 - 2$. Thus every residue class can be characterized uniquely by a remainder, i.e. by a polynomial $a + bx$ (with rational coefficients) of degree at most one (including the 0 polynomial representing the ideal itself).

Computations in the factor ring are actually done with these remainders, so to multiply two residue classes we multiply the corresponding remainders and take the remainder of the product on division by $x^2 - 2$ (just as the product of 7 and 6 modulo 15 is 12). Thus we perform addition as

$$[a + bx] + [c + dx] = [a + c] + [b + d]x,$$

and the rule for multiplication is

$$a + bx][c + dx] = ac + [ad + bc]x + bdx^2 =$$
$$= ac + [ad + bc]x + 2bd + bd[x^2 - 2] = [ac + 2bd] + [ad + bc]x,$$

exactly as in $\mathbf{Q}(\sqrt{2})$ (imagine everywhere $\sqrt{2}$ instead of the letter x).

This means that the factor ring $\mathbf{Q}[x]/(x^2 - 2)$ is isomorphic to (or, in literal translation, "is of the same form as") the field $\mathbf{Q}(\sqrt{2})$.

Similar to this example, we can characterize $\mathbf{Q}(\vartheta)$ as a factor ring for any algebraic number ϑ, the field $\mathbf{Q}(\vartheta)$ is isomorphic to $\mathbf{Q}[x]/(m_\vartheta)$. See Exercise 11.1.9.

Exercises 11.1

1. Consider the sets of Gaussian integers $\alpha = a + bi$ with the properties:

   (a) both a and b are even

   (b) $a \equiv b \pmod 2$

   (c) $a \equiv b \pmod 3$

   (d) $2 \mid N(\alpha)$

   (e) $5 \mid N(\alpha)$

   (f) $7 \mid N(\alpha)$.

   Which of the sets form an ideal in the ring of Gaussian integers? Which of them are principal ideals? Find a generating element for each of them.

2. Consider the sets of polynomials f with integer coefficients having the properties:

   (a) $f(1/2) = 0$

   (b) $f(\sqrt{2}) = f(\sqrt{3}) = 0$

   (c) $f(\sqrt{2}) = f(\sqrt{3})$

   (d) $f(3)$ is even

   (e) the leading coefficient of f is even or $f = 0$.

   Which of the sets are ideals in the ring $\mathbf{Z}[x]$, and what is the minimal number of generators?

3. Prove that a non-zero, commutative ring with identity element and no zero divisors is a field if and only if it has only trivial ideals.

4. Let U be the ring of all algebraic integers and

$$K = \left\{ \xi \sqrt[k]{2} \mid \xi \in U, k = 2, 3, 4, \ldots \right\}.$$

   Show that K is an ideal in U, but cannot be generated by finitely many elements.

5. Let $\alpha_1, \ldots, \alpha_k$ and ξ arbitrary elements of an integral domain R with identity. Verify

$$(\alpha_1, \alpha_2, \ldots, \alpha_k) = (\alpha_1 - \xi \alpha_2, \alpha_2, \ldots, \alpha_k).$$

6. Let G be the ring of Gaussian integers.

   (a) How many elements are there in the factor rings modulo the ideals below, and which of them are fields:

   (a1): (2)

(a2): (3)

(a3): $(2 + i)$?

* (b) Answer these questions in general for an arbitrary principal ideal in G.

7. Consider the ring $I(\sqrt{-5})$.

   (a) Show that the ideal $(2, 1 + \sqrt{-5})$ is not a principal ideal in $I(\sqrt{-5})$.

   (b) How many elements are there in the factor rings modulo the ideals below, and which of them are fields:

   (b1): $(2, 1 + \sqrt{-5})$

   (b2): $(1 + \sqrt{-5})$

   (b3): (11)?

S 8. (a) Which of the factor rings are fields:

   (a1): $\mathbf{R}[x]/(x^2 - 2)$

   (a2): $\mathbf{R}[x]/(x^2 + 1)$

   (a3): $\mathbf{C}[x]/(x^2 + 1)$?

   (b) Let F be an arbitrary commutative field and $g \in F[x]$. Find a necessary and sufficient condition for the factor ring $F[x]/(g)$ to be a field.

   (c) Verify that the factor ring $\mathbf{Z}[x]/(2, x^2 + x + 1)$ is a field.

* 9. (a) Let ϑ be an algebraic number. Prove that the field $\mathbf{Q}(\vartheta)$ is isomorphic to the factor ring $\mathbf{Q}[x]/(m_\vartheta)$.

   (b) Let L be an arbitrary commutative field and f an irreducible polynomial over L. Construct a field M satisfying the properties:

   (i) M has a subfield L^* isomorphic to L

   (ii) If we obtain the coefficients of the polynomial $f^* \in L^*[x]$ from the coefficients of f using the isomorphism $L \to L^*$, then f^* has a root $\vartheta \in M$

   (iii) $M = L^*(\vartheta)$.

   Remark: This construction enables extending L by a—not yet existing(!)—root of an irreducible polynomial even if no field containing L is given.

* 10. (a) Let ϑ be an algebraic number and $K \neq 0$ an ideal in $I(\vartheta)$. Show that the factor ring $I(\vartheta)/K$ has finitely many elements.

   (b) Verify that in $I(\vartheta)$ there is no infinite strictly increasing chain of ideals

   $$A_1 \subset A_2 \subset \cdots \subset A_j \subset \ldots.$$

   (c) Prove that every ideal in $I(\vartheta)$ is finitely generated.
   Remark: We sharpen this result in Theorem 11.5.9 by proving that every ideal in $I(\vartheta)$ can be generated by at most two elements.

11.2. Elementary Connections to Number Theory

In this section, we discuss how ideals are related to divisibility, units, and greatest common divisor.

Divisibility and units can be defined in any integral domain R with identity the usual way (as in Definitions 1.1.1 and 1.1.2), and the elementary properties listed in Theorems 1.1.4 and 1.1.5 are valid in general, too.

We show first that divisibility and the role of units can be described simply using principal ideals.

Theorem 11.2.1. *In any integral domain R with identity,*

(i) $a \mid b \iff b \in (a) \iff (b) \subseteq (a)$

(ii) *a and b are associates if and only if $(a) = (b)$.*   ♣

Proof. Using the definition of principal ideals, we can rewrite the three conditions in (i) as

> a is a divisor of b
> b occurs among the multiples of a
> all multiples of b occur among the multiples of a,

so the three conditions are equivalent.

(ii) $(a) = (b)$ means by part (i) that both $a \mid b$ and $b \mid a$ hold, or equivalently, that a and b are associates (see Theorem 1.1.5/(iii)).   □

Now we turn to the connection of ideals to the greatest common divisor.

The greatest common divisor is a common divisor that is a multiple of all common divisors according to Definitions 1.3.2 or 7.4.9.

In the rings of integers, Gaussian integers, or Eulerian integers, any two elements have a greatest common divisor as guaranteed by the Euclidean algorithm.

Two elements can have a gcd even if there is no division algorithm in the ring. For example, the polynomials with integer coefficients form such a ring. (We shall return to this problem later.)

There are, however, rings where two elements do not necessarily possess a gcd, e.g. $2 + 2\sqrt{-5}$ and 6 have no greatest common divisor in the ring $I(\sqrt{-5})$ (see Exercise 11.2.4).

It is true in any ring R that if two elements have a gcd, then it is unique up to associates; this follows from the definition of the greatest common divisor.

The greatest common divisor of two elements is closely related to the ideal generated by them. To avoid ambiguity in notation, the greatest common divisor of a and b will be denoted in this chapter by $\gcd\{a, b\}$, whereas (a, b) stands for ideal generated by a and b.

Consider first the ring of integers. Here the ideal $(6, 15)$ is the set of numbers $6u + 15v$, where u and v are arbitrary integers. By Theorem 1.3.6 about the solvability of linear Diophantine equations, this set is equal to the set of all multiples of $\gcd\{6, 15\} = 3$

which set is just the principal ideal (3). It is true in general among the integers that if $d = \gcd\{a, b\}$, then $(a, b) = (d)$. In an arbitrary ring, the situation is slightly more complicated.

Theorem 11.2.2. *Let R be an arbitrary integral domain with identity.*

  (i) *If* $(a, b) = (d)$, *then* $d = \gcd\{a, b\}$.

 (ii) $d = \gcd\{a, b\}$ *implies* $(a, b) \subseteq (d)$, *but* $(a, b) \neq (d)$ *in general.*

(iii) $(a, b) = (d)$ *if and only if* $d = \gcd\{a, b\}$ *and* $d = au + bv$ *with suitable* $u, v \in R$.   ♣

Proof. (i) From $(a, b) = (d)$ we have $a \in (a, b) = (d)$, so $d \mid a$, and similarly $d \mid b$, so d is a common divisor of a and b.

Let now c be an arbitrary common divisor, so $c \mid a$ and $c \mid b$. By Theorem 11.2.1, $a \in (c)$ and $b \in (c)$. Since (a, b) is the smallest ideal containing a and b, $(d) = (a, b) \subseteq (c)$, so, using Theorem 11.2.1 again, $c \mid d$.

(ii) If $d = \gcd\{a, b\}$, then $d \mid a$ and $d \mid b$. This means that a and b are in the ideal (d), hence (d) must be at least as large as the smallest ideal containing a and b. So $(a, b) \subseteq (d)$.

The next example shows that equality does not necessarily hold. Among the polynomials with integer coefficients, the greatest common divisor of 2 and x is 1, but $(2, x) \neq (1)$. We saw in the previous section that $(2, x)$ is not even a principal ideal.

We shall give another type of counterexample in Exercise 11.2.4c.

(iii) If $(a, b) = (d)$, then $d = \gcd\{a, b\}$ as was proved in (i) and we have $d \in (a, b)$, so $d = au + bv$ with suitable elements of R, by definition.

For the converse, we assume $d = \gcd\{a, b\}$ and $d = au + bv$. The first condition implies $(a, b) \subseteq (d)$ by (ii), the second condition means $d \in (a, b)$, so $(d) \subseteq (a, b)$, so $(a, b) = (d)$.   □

Remark: There are many rings where Theorem 11.2.2 can be reduced to

(11.2.1) $$d = \gcd\{a, b\} \iff (a, b) = (d).$$

For example, the ring of integers has this property, as we sketched before stating the theorem. Similar considerations show that (11.2.1) holds in every ring with a division algorithm.

As mentioned earlier, ideals appeared first in Kummer's investigations of Fermat's Last Theorem. To understand the situation, consider Fermat's equation

(11.2.2) $$x^p + y^p = z^p$$

for a prime $p > 2$. The factorization

(11.2.3) $$x^p + y^p = \prod_{j=0}^{p-1}(x + y\varrho^j), \qquad \varrho = \cos\left(\frac{2\pi}{p}\right) + i\sin\left(\frac{2\pi}{p}\right)$$

shows that (11.2.2) is closely related to the number theory of the ring $I(\varrho)$.

Combining (11.2.2) and (11.2.3), we get

(11.2.4)
$$\prod_{j=0}^{p-1}(x + y\varrho^j) = z^p.$$

The product on the left-hand side is a pth power. We might try the tactics successful many times earlier to show that each factor is itself a pth power in $I(\varrho)$, and use the resulting p equations $x + y\varrho^j = \alpha_j^p$ to arrive at a contradiction (assuming a non-trivial solution x, y, z).

We know in the integers that if the factors of a product are pairwise coprime and the product is a pth power, then each factor is an associate of a pth power. The same holds for Gaussian or Eulerian integers, and in general, in every ring where the Fundamental Theorem of Arithmetic is valid. However, this is no longer true in the lack of the Fundamental Theorem: $3^2 = (2 + \sqrt{-5})(2 - \sqrt{-5})$ in $I(\sqrt{-5})$, and though the factors on the right-hand side are coprime, they are not associates of squares (in fact, they are irreducible).

Thus, our attempt above to prove Fermat's Last Theorem can be promising only if the Fundamental Theorem of Arithmetic is true in $I(\varrho)$. It can be shown, however, that this is not the case for $p > 19$, and so other approaches have to be applied.

We mention as a historical curiosity that Lamé, a member of the French Academy, gave an erroneous proof in 1847 of Fermat's Last Theorem along the lines of the argument above, taking the Fundamental Theorem for granted for $I(\varrho)$. (It is conceivable that Fermat's "wonderful proof"—if it existed at all—was based on a similar mistake.) It was Liouville, who pointed out the gap in Lamé's argument (at that time Liouville was not yet aware in which cases the Fundamental Theorem holds). Lamé made another mistake by not considering that even if the factors on the left-hand side of (11.2.4) are pairwise coprime and the Fundamental Theorem is true, we cannot infer that they are necessarily pth powers, only that they are associates of pth powers. And since there are infinitely many units in $I(\varrho)$ for $p > 3$, this minor inattentiveness causes another hardly repairable gap in Lamé's argument.

At roughly the same time, the German Kummer followed a similar path, but he realized the importance of the Fundamental Theorem in $I(\varrho)$, and observed that it does not always hold. He knew also that if any two elements have a greatest common divisor, then one can deduce the Fundamental Theorem easily. This gave him the idea to adjoin ideal numbers to the rings $I(\varrho)$ where the Fundamental Theorem was false: these were intended to make up the missing greatest common divisors in $I(\varrho)$. Kummer hoped that any two elements will have a gcd in this enlarged set and also the Fundamental Theorem will hold.

Kummer based the construction of ideal numbers on the following property of the greatest common divisor. We know in the integers that if $\gcd\{a, b\} = d$, then the multiples of d are just the numbers of the form $au + bv$, and as we indicated, the same applies also for every $I(\vartheta)$. Thus, Kummer defined the ideal number belonging to a fixed pair α and β as the set of numbers

$$\{\alpha\xi + \beta\psi \mid \xi, \psi \in I(\vartheta)\}.$$

In modern terminology, this is just the ideal (α, β) generated by α and β. If α and β have a greatest common divisor δ, then this set is the multiples of δ, and so we can identify it with δ. If, however, $\gcd\{\alpha, \beta\}$ does not exist, then this ideal number can compensate the lack of the greatest common divisor. Then Kummer built number theory among the ideal numbers (i.e. ideals), and achieved significant progress concerning Fermat's Last Theorem. (We shall discuss number theory for ideals in Section 11.4.)

Exercises 11.2

1. Verify that the following subsets form a principal ideal in the ring of integers, and exhibit a generating element for each of them.

   (a) $(30, 50, 75)$

   (b) $(20) \cap (30)$.

2. Consider the ring G of Gaussian integers.

   (a) In how many ways can we generate a given non-zero principal ideal with a single element?

   (b) How many principal ideals contain $22 + 6i$?

3. Let R be an integral domain with identity and $a, b \in R$. Demonstrate

$$a + b \in (a) \cap (b) \iff (a) = (b).$$

4. Consider the ring $I(\sqrt{-5})$.

   (a) Show that $2 + 2\sqrt{-5}$ and 6 have no greatest common divisor.

   (b) Find all principal ideals containing the ideal $(2 + 2\sqrt{-5}, 6)$.

   (c) Exhibit an example where α and β have a greatest common divisor δ, but $(\alpha, \beta) \neq (\delta)$.

11.3. Unique Factorization, Principal Ideal Domains, and Euclidean Rings

This section is devoted to the Fundamental Theorem of Arithmetic.

The notions of irreducible and prime elements are defined in an arbitrary integral domain R with identity exactly as in the rings discussed earlier (see Definitions 1.4.1 and 1.4.2 for integers, or Definitions 7.4.10 and 7.4.11 for Gaussian integers).

Saying that the Fundamental Theorem of Arithmetic (or unique prime factorization) holds in R has the usual meaning that every element in R different from 0 and units is the product of finitely many irreducible elements, and this decomposition is unique apart from associates and the order of factors (see e.g. Theorem 1.5.1).

As we have pointed out several times, one of the crucial questions of number theoretic investigations is (also from the point of view of applications) whether the Fundamental Theorem in a ring is true or false. Nearly every part of number theory for integers uses the Fundamental Theorem in $\mathbf{Z}$. In handling the Diophantine equations

$x^2 + y^2 = n$ and $x^3 + y^3 = z^3$, we applied the Fundamental Theorem for Gaussian and Eulerian integers. In Section 11.2 we indicated that the proof of Fermat's Last Theorem would have been much easier if the Fundamental Theorem were true for algebraic integers in certain algebraic number fields. The Fundamental Theorem for polynomials with rational coefficients played an important role in the theory of algebraic numbers.

Below, we establish first a necessary and sufficient condition for the validity of the Fundamental Theorem in a ring (Theorem 11.3.1). Then we demonstrate that the division algorithm always implies the Fundamental Theorem. The proof will be somewhat different from those seen for integers, Gaussian integers, etc.: We deduce from the division algorithm that every ideal is a principal ideal (Theorem 11.3.5) and show that the Fundamental Theorem holds in rings with this property (Theorem 11.3.3).

In the proofs, several parts will be literally identical to the arguments seen for the integers, so we shall just refer to them without repeating them in detail.

Before turning to the general theorems, we say a few words about the relation between irreducible and prime elements. In formulating the Fundamental Theorem, we use only the naturally occurring notion of irreducible and we do not need the notion of prime at all. The validity of the Fundamental Theorem, however, depends strongly on the relation of primes and irreducibles.

A prime must be irreducible in every integral domain R with identity, as we can see from the first part of the proof for Theorem 1.4.3. The converse is false, however, e.g. in $I(\sqrt{-5})$, where the Fundamental Theorem does not hold, the number 2 is irreducible, but is not a prime. On the other hand, every irreducible is a prime in the rings of integers, Gaussian integers, and Eulerian integers, and this was a decisive step in proving uniqueness in the Fundamental Theorem. The result below shows that in the general case, one of the essential conditions for the Fundamental Theorem to be true is that every irreducible is a prime.

Theorem 11.3.1. *The Fundamental Theorem of Arithmetic holds in an integral domain R with identity if and only if*

(i) *a strictly increasing sequence*

$$(a_1) \subset (a_2) \subset \cdots \subset (a_j) \subset \ldots$$

of ideals cannot be infinite,

(ii) *every irreducible element is a prime.* ♣

Proof. We prove first the sufficiency of conditions (i) and (ii).

Uniqueness follows from (ii) exactly as in the first proof of uniqueness in Theorem 1.5.1.

We shall use (i) to establish decomposability. Let a be an arbitrary element in R different from 0 and units. First we show that a has an irreducible divisor.

If a is irreducible, we are done. Otherwise, $a = a_1 b_1$, where neither of a_1 and b_1 is a unit. Then $(a) \subset (a_1)$ by Theorem 11.2.1 with strict containment, as b_1 is not a unit.

If a_1 is irreducible, then it is an irreducible divisor of a. Otherwise, $a_1 = a_2 b_2$, where neither of a_2 and b_2 is a unit. Then $(a_1) \subset (a_2)$ with strict containment.

We show that continuing the procedure, some a_i is necessarily irreducible. If this were not the case, then

$$(a) \subset (a_1) \subset \cdots \subset (a_j) \subset \ldots$$

would be an infinite strictly ascending chain of principal ideals, contradicting thus (i). Thus we have proved that a has an irreducible divisor.

Now we show that a can be written as a product of irreducible elements. If a is irreducible, then we are done. Otherwise, $a = p_1 c_1$, where p_1 is irreducible and c_1 is not a unit. Since p_1 is not a unit either, $(a) \subset (c_1)$ (with strict containment).

If c_1 is irreducible, then both factors in $a = p_1 c_1$ are irreducible and we are done. Otherwise, $c_1 = p_2 c_2$, where p_2 is irreducible and c_2 is not a unit. Thus $(c_1) \subset (c_2)$ (with strict containment).

Continuing the procedure, some c_i is necessarily a unit, since otherwise the infinite strictly ascending chain

$$(a) \subset (c_1) \subset \cdots \subset (c_j) \subset \ldots$$

contradicts (i). This means that we arrived at a decomposition of a into the product of irreducible elements.

Turning to necessity, assume that the Fundamental Theorem holds in R. We can prove (ii) exactly as in the solution of Exercise 1.5.8.

To prove (i) by contradiction, assume the existence of an infinite strictly increasing chain

$$(a_1) \subset (a_2) \subset \cdots \subset (a_j) \subset \ldots$$

of principal ideals. Here $a_2 \neq 0$, and a_3, a_4, ... are infinitely many pairwise non-associate divisors of a_2. But this is impossible, since if $a_2 = p_1 \ldots p_k$ where every p_i is irreducible, then the Fundamental Theorem implies that every divisor of a_2 is either a unit or an associate of the product of some factors p_i (and if a_2 is a unit, then so are all its divisors, too). □

Remark: We saw several examples where the uniqueness part of the Fundamental Theorem was false (see Theorems 10.3.5 and 10.3.6, and the paragraphs about Fermat's Last Theorem in Section 11.2). But we can easily find a ring where there is a problem with decomposability: there are no irreducible elements at all in the ring U of all algebraic integers (see Exercise 11.3.1), so no element can be written as a product of irreducible elements.

Now we show that if every ideal in R is a principal ideal, then the Fundamental Theorem of Arithmetic is valid in R.

Definition 11.3.2. An integral domain R with identity is a *principal ideal domain*, if every ideal in R is a principal ideal.   ♣

Theorem 11.3.3. *The Fundamental Theorem of Arithmetic is true in every principal ideal domain.*   ♣

Proof. We verify that a principal ideal domain satisfies conditions (i) and (ii) of Theorem 11.3.1.

(i) To achieve a contradiction, assume the existence of an infinite strictly increasing chain

$$(a_1) \subset (a_2) \subset \cdots \subset (a_j) \subset \ldots$$

of principal ideals. A calculation shows that $A = \bigcup_{j=1}^{\infty}(a_j)$ is an ideal (see Exercise 11.3.4). As R is a principal ideal domain, A is a principal ideal, $A = (b)$. Then

$$b \in A = \bigcup_{j=1}^{\infty}(a_j),$$

so $b \in (a_k)$, and $(b) \subseteq (a_k)$ for some k. Thus

$$A = (b) \subseteq (a_k) \subset (a_{k+1}) \subset \bigcup_{j=1}^{\infty}(a_j) = A,$$

a contradiction.

(ii) We verify that any two elements a and b have a greatest common divisor. Since (a, b) is a principal ideal, we know $(a, b) = (d)$ and Theorem 11.2.2 implies $d = \gcd\{a, b\}$.

The existence of a greatest common divisor yields (ii): see the proof for Theorem 1.3.4 given in the solution of Exercise 1.3.11, the proof of Theorem 1.3.9, and finally, part II in the proof of Theorem 1.4.3.   □

Remarks: (1) There exist rings that are not principal ideal domains, but for which the Fundamental Theorem still holds, the simplest example being $\mathbf{Z}[x]$. On the one hand, we saw in Section 11.1 that $(2, x)$ is not a principal ideal in $\mathbf{Z}[x]$. On the other hand, the Fundamental Theorems in $\mathbf{Z}$ and in $\mathbf{Q}[x]$ imply its validity also in $\mathbf{Z}[x]$: It follows from a basic lemma of Gauss used in the proof of Theorem 9.6.2 that a polynomial f is irreducible over $\mathbf{Z}$ if and only if f is either a constant that is a prime number, or the coefficients of f are coprime (not necessarily pairwise) and f is irreducible over $\mathbf{Q}$.

(2) Among the algebraic number fields, the principal ideal domains are exactly the same as the ones where the Fundamental Theorem holds: A ring $I(\vartheta)$ is a principle ideal domain if and only if the Fundamental Theorem is true in it (see Exercise 11.3.9b).

We turn to the general formulation of the division algorithm and prove that if there is a division algorithm in R, then R is a principal ideal domain, and so (by Theorem 11.3.3,) the Fundamental Theorem of Arithmetic is true in R.

Definition 11.3.4. An integral domain R with identity is a *Euclidean ring*, if we can assign to every $c \in R$ a non-negative integer $f(c)$ such that $f(c) = 0 \iff c = 0$ and to every $a, b \in R, b \neq 0$ there exist $q, r \in R$ satisfying

(11.3.1) $a = bq + r$ and $f(r) < f(b)$.   ♣

Remarks: (1) An equivalent definition of Euclidean rings is if we assign only to the non-zero elements $c \in R$ a non-negative integer $f(c)$, and in (11.3.1) we allow the possibility that $r = 0$ (besides $f(r) < f(b)$).

(2) We do not have to assume in Definition 11.3.4 that R has an identity because this follows from the division algorithm (see Exercise 11.3.6).

(3) We investigated several rings with a division algorithm earlier; see most examples below. In them, the function f had further useful properties, such as $f(ab) = f(a)f(b)$ or at least $f(a) \leq f(ab)$. However, we do not have to require such properties in the definition of Euclidean rings.

Examples.   **E1** For the integers, we can choose $f(c) = |c|$, i.e. we will have $|r| < |b|$. We note that in this case, the quotient and the remainder are generally not unique, as for $a = 33$ and $b = 5$, we can satisfy (11.3.1) in two ways:

$$33 = 6 \cdot 5 + 3 = 7 \cdot 5 + (-2).$$

In Theorems 1.2.1 and 1.2.1A, we required the stronger conditions $0 \leq r < |b|$ and $-|b|/2 < r \leq |b|/2$, instead of $|r| < |b|$, to guarantee the uniqueness of quotient and remainder. This uniqueness, however, has no impact on the proof of the Fundamental Theorem.

E2 For Gaussian or Eulerian integers, we can take $f(c) = N(c)$. (We saw during the proof of Theorem 7.4.8 that the quotient and the remainder are not unique in general.)

E3 In $I(\sqrt{2})$, we may choose $f(c) = |N(c)|$.

E4 In a polynomial ring over a field, we can perform a division algorithm with respect to the degree. To satisfy Definition 11.3.4 formally, we define $f(0) = 0$ and $f(c) = 1 + \deg c$ for $c \neq 0$.

E5 Finite decimal fractions form a Euclidean ring, see Exercise 1.5.5c.

Theorem 11.3.5. *If R is a Euclidean ring, then R is a principal ideal domain.*   ♣

Proof. We have to verify that every ideal I of R is a principal ideal.

If the only element in I is 0, then $I = (0)$. Otherwise, consider the values $f(c)$ of the non-zero elements of I. They are positive integers, so there must be a smallest among them, let it be $f(b)$ (here b is not unique in general). We prove $I = (b)$.

As $b \in I$, $(b) \subseteq I$. Conversely, let a be an element in I. We have to show $a \in (b)$, i.e. $b \mid a$.

We apply the division algorithm for a and b: there exist $q, r \in R$ satisfying (11.3.1). Since $a, b \in I$ and I is an ideal, so $r = a - bq \in I$. Further, $f(b)$ was minimal and $f(r) < f(b)$, hence only $r = 0$ is possible and $b \mid a$.   □

Remark: The converse of Theorem 11.3.5 is false as there exist principal ideal domains which are not Euclidean rings. Some examples are

(11.3.2) $I(\sqrt{-19}), \quad I(\sqrt{-43}), \quad I(\sqrt{-67}), \quad \text{and} \quad I(\sqrt{-163})$

(see Exercise 11.3.10).

In general, it is difficult to determine whether or not a ring R is Euclidean. Of course, if we find a suitable f, then R is Euclidean, and if the Fundamental Theorem of Arithmetic is false in R or R is not a principal ideal domain, then it follows from Theorems 11.3.3 and 11.3.5 that R cannot be Euclidean either. However, we have basically no tools to show that some principal ideal domain is not a Euclidean ring. It is not enough to verify that some given or naturally arising function f does not meet the requirements in Definition 11.3.4, but we have to do that for every posssible f.

Let us take a closer look at the rings of algebraic integers in algebraic number fields. In Remark 2 after Theorem 11.3.3, we indicated that if the Fundamental Theorem holds in $I(\vartheta)$, then $I(\vartheta)$ is a principal ideal domain (see Exercise 11.3.9b). As to the division algorithm, we performed it in the previously discussed cases (Gaussian or Eulerian integers, $I(\sqrt{2})$, etc.) with respect to the absolute value of the norm. There were some rings $I(\vartheta)$ where this failed though the Fundamental Theorem was true (see Theorem 10.3.6). However, we cannot exclude the possibility that there still is a division algorithm with respect to some other function f. Our present (lack of) knowledge is slightly paradoxical:

(A) Some deep conjectures make probable that apart from the imaginary quadratic fields, every $I(\vartheta)$ where the Fundamental Theorem holds is Euclidean, so it satisfies Definition 11.3.4 for some function f (even if $|N(\alpha)|$ is not suitable for this purpose).

(B) For a long time, however, no ring $I(\vartheta)$ was known where the Fundamental Theorem is true, the ring is not Euclidean with respect to the absolute value of the norm, but there exists some other division algorithm. The first such proven example was $I(\sqrt{69})$ in 1994, which opened new horizons: today we know that apart from at most 2(!) exceptions, all real quadratic fields with unique factorization are Euclidean.

For imaginary quadratic fields, it can be shown that if $I(\vartheta)$ is Euclidean, then the division algorithm can be performed with respect to $|N(\alpha)|$ (see Exercise 11.3.10). Thus Theorem 10.3.6 implies that the four examples listed in (11.3.2) are principal ideal domains but are not Euclidean rings.

Exercises 11.3

1. Let U be the ring of all algebraic integers.

    (a) Characterize the units in U with the help of their minimal polynomials.

    (b) Show that there are no irreducible elements in U, thus the Fundamental Theorem of Arithmetic is false.

2. We can perform number theoretic investigations also in a (commutative) field F (but it does not makes much sense, as we see below).

    (a) Which $a, b \in F$ satisfy $a \mid b$?

    (b) Determine the units, irreducible, and prime elements in F.

    (c) Show that the Fundamental Theorem of Arithmetic is true in F, that F is a principal ideal domain, and is a Euclidean ring.

3. Let W be the set of rational numbers with odd denominators.

    (a) Verify that that there is just one irreducible element in W apart from associates.

    (b) If we try to adapt the argument for proving the existence of infinitely many prime numbers in the integers (see Theorem 5.1.1), why does it not work in W?

    (c) Show that W is a Euclidean ring.

    (d) Determine all ideals in W.

4. Let $I_1 \subseteq I_2 \subseteq \ldots$ be arbitrary ideals in a ring R. Demonstrate that also $\bigcup_{j=1}^{\infty} I_j$ is an ideal in R.

S 5. Let R be an integral domain with identity. Prove that the polynomial ring $R[x]$ is a principal ideal domain if and only if R is a field.

6. Show that there is no need to require in Definition 11.3.4 of Euclidean rings that R has an identity, since this follows from the other conditions.

7. Let R be a Euclidean ring, f a function meeting the requirements in Definition 11.3.4, and k the minimal positive value of f. True or false?

    (a) If $f(c) = k$, then c is a unit.

    (b) If c is a unit, then $f(c) = k$.

8. Verify that in the integers, the division algorithm can be performed not only with respect to the absolute value, but also using the following function f:

$$f(c) = \begin{cases} 1 + \lfloor \log_2 |c| \rfloor, & \text{if } c \neq 0 \\ 0, & \text{if } c = 0, \end{cases}$$

so

$$f(0) = 0, \qquad f(\pm 1) = 1, \qquad f(\pm 2) = f(\pm 3) = 2,$$
$$f(\pm 4) = f(\pm 5) = f(\pm 6) = f(\pm 7) = 3, \ldots .$$

9. **S*** (a) Assume that the Fundamental Theorem of Arithmetic holds in an integral domain R with identity and the factor ring R/I has finitely many elements for every ideal $I \neq 0$. Prove that R is a principal ideal domain.

    (b) Verify that if ϑ is an algebraic number and the Fundamental Theorem of Arithmetic holds in $I(\vartheta)$, then $I(\vartheta)$ is a principal ideal domain.

S* 10. Let t be a negative squarefree integer. Show that the algebraic integers of the imaginary quadratic field $\mathbf{Q}(\sqrt{t})$ form a Euclidean ring if and only if $t = -1, -2, -3, -7,$ or -11.

11.4. Divisibility of Ideals

In this section we define a multiplication for ideals in an integral domain R with identity. This makes it possible to introduce divisibility of ideals and to study the greatest common divisor, irreducible ideals, and prime ideals.

By developing number theory for ideals, our main goal is to get further information about the rings $I(\vartheta)$. Therefore, when introducing the notions, we shall require further restrictions that are satisfied by the ideals in $I(\vartheta)$, but do not hold in every ring R.

Definition 11.4.1. Let A and B be two ideals in an integral domain R with identity. We define the *product* of A and B by

$$(11.4.1) \qquad AB = \left\{ \sum_{i=1}^{n} a_i b_i \mid n = 1, 2, \ldots, a_i \in A, b_i \in B, i = 1, \ldots, n \right\}. \qquad \clubsuit$$

Thus the product of two ideals is the set of all possible sums (of arbitrarily many terms) of products where the factors are taken from A and B.

We summarize some important properties of multiplication of ideals in

Theorem 11.4.2. (i) *The product AB of ideals A and B is the smallest ideal containing all elements ab, where $a \in A$ and $b \in B$.*

(ii) *The product of finitely generated ideals is finitely generated, as well.*

(iii) *The product of principal ideals is a principal ideal.*

(iv) $AB \subseteq A \cap B$.

(v) *The multiplication of ideals in a ring R is a commutative and associative operation, with identity element $(1) = R$:*

$$(11.4.2) \qquad AB = BA, \qquad (AB)C = A(BC), \qquad (1)A = A(1) = A.$$

Only the identity has an inverse and

$$AB = (0) \iff A = (0) \text{ or } B = (0). \qquad \clubsuit$$

Proof. (i) We have to verify that

(a) AB is an ideal

(b) $a \in A, b \in B \implies ab \in AB$

(c) if an ideal I contains all elements ab, where $a \in A$, $b \in B$, then $AB \subseteq I$.

(a) We show that AB satisfies Definition 11.1.1. The elements of AB are of the form $u = \sum_{i=1}^{n} a_i b_i$. The sum of two such elements has the same form. The negative

$$-u = -\sum_{i=1}^{n} a_i b_i = \sum_{i=1}^{n} (-a_i) b_i \in AB,$$

since $-a_i \in A$ as A is an ideal. Similarly, for any $r \in R$,

$$ur = ru = r \sum_{i=1}^{n} a_i b_i = \sum_{i=1}^{n} (r a_i) b_i \in AB,$$

since $r a_i \in A$ because A is an ideal.

(b) For $n = 1$, we get just the elements ab in (11.4.1).

(c) If an ideal I contains the elements $a_i b_i$, then the ideal property implies that I must contain also their sum, so every $\sum_{i=1}^{n} a_i b_i \in AB$ and $AB \subseteq I$.

(ii) We show that if

$$A = (\alpha_1, \ldots, \alpha_k) \quad \text{and} \quad B = (\beta_1, \ldots, \beta_m)$$

then

$$AB = (\alpha_1 \beta_1, \alpha_1 \beta_2, \ldots, \alpha_i \beta_j, \ldots, \alpha_k \beta_m),$$

so the products of the generator elements of A and B form a (possible) generating system of AB.

The products $\alpha_i \beta_j$ are in AB by definition, so the ideal generated by them is a subset of AB.

For the reverse containment, it is sufficient by (i) to verify that every ab with $a \in A$, $b \in B$ lies in the ideal generated by the elements $\alpha_i \beta_j$. This means that ab can be expressed as a combination of the elements $\alpha_i \beta_j$ with coefficients from R. This holds as for suitable $r_i, s_j \in R$, we have

$$ab = \left(\sum_{i=1}^{k} r_i \alpha_i \right) \left(\sum_{j=1}^{m} s_j \beta_j \right) = \sum_{i=1}^{k} \sum_{j=1}^{m} (r_i s_j)(\alpha_i \beta_j).$$

(iii) Applying the proof of (ii) for the special case $k = m = 1$, we get $(\alpha)(\beta) = (\alpha\beta)$.

(iv) Since A is an ideal, $a_i b_i \in A$ for any $a_i \in A$ and $b_i \in B$, and thus $\sum_{i=1}^{n} a_i b_i \in A$, so $AB \subseteq A$. We get $AB \subseteq B$ similarly.

(v) The properties in (11.4.2) follow immediately from the definition of multiplication of ideals and from the ring properties of R.

The inverse of the identity $R = (1)$ is itself. Conversely, if the ideal I has an inverse, so $JI = R$ for some ideal J, then $R \subseteq I$ by (iv) and so $I = R$.

If $A = (0)$ or $B = (0)$, then every sum in the definition of AB is 0, thus $AB = (0)$. If, however, there exist non-zero elements $a \in A$ and $b \in B$, then $ab \neq 0$ since R is free of zero divisors. As $ab \in AB$, $AB \neq (0)$. $\square$

Remarks: (1) The products ab with $a \in A$ and $b \in B$ do not form an ideal in general (see Exercise 11.4.1a). This is the reason why we had to take sums of such products in the definition of AB.

(2) We defined only multiplication for ideals so far. Addition can be defined, see Remark 4 after Theorem 11.4.5. Some of the usual nice properties do not hold for it (only the zero element has a negative), and so the ideals in R do not form a ring for this addition and multiplication.

Examples. **E1** Let $R = \mathbf{Z}[x]$, and let A and B be the sets of polynomials having constant terms divisible by 2 and 3. Then AB is the set of polynomials with constant term divisible by 6:

$$AB = (2, x)(3, x) = (6, 2x, 3x, x^2)$$

$$= (6, 2x, 3x - 2x, x^2) = (6, 2x, x, x^2) = (6, x).$$

E2 Let $R = E(\sqrt{-5})$, $A = (3, 1+\sqrt{-5})$, and $B = (3, 1-\sqrt{-5})$. Then AB is the principal ideal (3):

$$AB = (3, 1 + \sqrt{-5})(3, 1 - \sqrt{-5}) = (9, 3 + 3\sqrt{-5}, 3 - 3\sqrt{-5}, 6)$$
$$= (9 - 6, 3 + 3\sqrt{-5}, 3 - 3\sqrt{-5}, 6) = (3).$$

Using multiplication, we can introduce divisibility among ideals of R:

Definition 11.4.3. An ideal B is a *divisor* of an ideal A if there is an ideal C satisfying $BC = A$. We denote this relation by $B \mid A$, as usual. ♣

Remarks: (1) We can easily see that divisibility of principal ideals is equivalent to the divisibility of their generators (in R):

$$(\beta) \mid (\alpha) \iff \beta \mid \alpha.$$

Moreover, if $\beta \neq 0$ and $(\alpha) = (\beta)C$, then C is a principal ideal, $C = (\gamma)$, where γ can be chosen to satisfy $\alpha = \beta\gamma$ (see Exercise 11.4.3). This means that divisibility of ideals can be considered as a generalization of divisibility in R.

(2) We discuss some elementary properties of divisibility in Exercise 11.4.2. An important one is

(11.4.3) $$B \mid A \implies A \subseteq B.$$

The converse of (11.4.3) is true for principal ideals by the previous remark and Theorem 11.2.1. It is false, however, in general for arbitrary ideals, see Exercise 11.4.6.

In the sequel we deal only with rings R where multiplication of ideals obeys the cancellation law:

(11.4.4) $$AB = AC, A \neq (0) \implies B = C,$$

and the converse of (11.4.3) is true:

(11.4.5) $$B \mid A \iff A \subseteq B.$$

We shall show in Section 11.5 that the rings $I(\vartheta)$ constituting the main direction of our investigation meet requirements (11.4.4) and (11.4.5).

Now we define the greatest common divisor of two ideals in the usual way as a common divisor that is a multiple of every common divisor:

Definition 11.4.4. An ideal D is the greatest common divisor of ideals A and B if

(i) $D \mid A, D \mid B$

(ii) if $C \mid A$ and $C \mid B$ for some ideal C, then $C \mid D$. ♣

Theorem 11.4.5. *Any two ideals A and B have a unique greatest common divisor D and*

(11.4.6) $$D = \{ a + b \mid a \in A, b \in B \}. \qquad ♣$$

Proof. Based on (11.4.5), we can characterize the greatest common divisor by containment: it is the smallest ideal containing A and B. We can verify easily (see Exercise 11.4.4a) that D defined by (11.4.6) is the unique ideal with this property. $\qquad \square$

Remarks: (1) We can consider D as the ideal generated by A and B. Thus the notation $D = (A, B)$ agrees with the usual notation both for greatest common divisor and generated ideal.

(2) If A and B are principal ideals, $A = (\alpha)$ and $B = (\beta)$, then their greatest common divisor by (11.4.6) is $D = \{r\alpha + s\beta \mid r, s \in R\}$, which is just the ideal (α, β). This shows again that an ideal generated by two elements can be considered as a generalization of the notion of greatest common divisor.

(3) If A and B are finitely generated ideals,

$$A = (\alpha_1, \ldots, \alpha_k) \quad \text{and} \quad B = (\beta_1, \ldots, \beta_m),$$

then their greatest common divisor by (11.4.6) is

$$D = (\alpha_1, \alpha_2, \ldots, \alpha_k, \beta_1, \beta_2, \ldots, \beta_m),$$

so the union of generators of A and B form a (possible) generating system of D.

(4) By (11.4.6), we can interpret D also as the sum of ideals A and B. We emphasize once again that the ideals of R do not form a ring for this addition and the multiplication introduced in Definition 11.4.1 (see Exercise 11.4.4b).

Now we turn to the notion, properties, and relation of irreducible and prime ideals.

The definitions of irreducible and prime are analogous to the previous definitions of irreducible and prime. The only unit among the ideals of R is $(1) = R$, since this is the only ideal dividing every ideal (see Exercise 11.4.2e).

Definition 11.4.6. An ideal I of R is *irreducible* if it is non-trivial (differs from (0) and (1)), and it can be written as a product of two ideals only if one of the factors is (1), so

(11.4.7) $\qquad\qquad I = AB \Longrightarrow A = (1) \text{ or } B = (1).$ ♣

By (11.4.4) and (11.4.5), irreducibility of a non-trivial ideal I is equivalent to either of the two conditions (here A denotes an arbitrary ideal):

I has only trivial divisors:

(11.4.8) $\qquad\qquad A \mid I \Longrightarrow A = (1) \text{ or } A = I.$

There is no non-trivial ideal containing I as a proper subset:

(11.4.9) $\qquad\qquad I \subseteq A \subseteq R \Longrightarrow A = R \text{ or } A = I.$

Ideals satisfying (11.4.9) are called *maximal* ideals (also in rings where (11.4.5) is not valid).

Definition 11.4.7. An ideal P of R is a *prime ideal* if it is non-trivial (differs from (0) and (1)), and it can divide a product of two ideals only if it divides at least one of the factors:

(11.4.10) $\qquad\qquad P \mid AB \Longrightarrow P \mid A \text{ or } P \mid B.$ ♣

By (11.4.5), we can rephrase the definition of a prime ideal using containment: an ideal different from (0) and (1) is a prime ideal if and only if

(11.4.11) $$AB \subseteq P \Longrightarrow A \subseteq P \text{ or } B \subseteq P.$$

Another equivalent condition is

(11.4.12) $$ab \in P \Longrightarrow a \in P \text{ or } b \in P.$$

The equivalence of (11.4.11) and (11.4.12) holds also in rings where (11.4.5) is not valid (see Exercise 11.4.7), and in this case, (11.4.11) or (11.4.12) serves as a definition for prime ideals.

If (11.4.4) and (11.4.5) hold, then prime ideals are the same as irreducible ideals:

Theorem 11.4.8. *An ideal P is a prime ideal if and only if it is an irreducible ideal.*   ♣

Proof. We follow the lines of the proof of Theorem 1.4.3. We can assume that P is a non-trivial ideal.

First we assume that P is a prime ideal, and want to show that it is also irreducible. Consider a product representation $P = AB$; we have to verify $A = (1)$ or $B = (1)$.

Since $P = AB$, so also $P \mid AB$. As P is a prime ideal, we infer $P \mid A$ or $P \mid B$.

If $P \mid A$, then $A = PC = ABC$ with a suitable ideal C. Combining it with the equality $A = A(1)$, we obtain $ABC = A(1)$. Cancelling by $A \neq 0$, we get $BC = (1)$. This implies $B = (1)$ (and $C = (1)$).

If $P \mid B$, then we obtain $A = (1)$ similarly.

Now we assume that P is irreducible, and show that it is a prime ideal. Starting from divisibility $P \mid AB$, we have to verify that at least one of $P \mid A$ and $P \mid B$ holds.

If $P \mid A$, we are done. If $P \nmid A$, then $(P, A) = (1)$ since P is irreducible.

Since $P \mid PB$ and $P \mid AB$, we infer $P \mid (PB, AB)$. By Exercise 11.4.4c, we obtain

$$(PB, AB) = (P, A)B = (1)B = B, \quad \text{and so} \quad P \mid B. \qquad \square$$

Exercises 11.4

Throughout, A, B, and C denote ideals in an integral domain R with identity. In the exercises related to greatest common divisor, irreducible ideals and prime ideals, we assume, unless stated otherwise, the validity of (11.4.4) and (11.4.5), so the cancellation law for ideals, and the equivalence of divisibility and (opposite) containment (these hold in rings $I(\vartheta)$, as mentioned earlier).

1. Let H be the set of products ab formed from elements of A and B: $H = \{ ab \mid a \in A, b \in B \}$.

   (a) Give an example where H is not an ideal.

   (b) Prove that if at least one of A and B is a principal ideal, then H is an ideal (and so $H = AB$).

2. Verify the elementary properties of divisibility of ideals:

    (a) $A \mid A$ for every A.

    (b) $C \mid B,\ B \mid A \Longrightarrow C \mid A$.

    (c) $B \mid A \Longrightarrow A \subseteq B$.

    (d) $A \mid B,\ B \mid A \Longrightarrow A = B$.

    (e) $B \mid A$ for every $A \iff B = (1)$.

3. Verify the statements about divisibility of principal ideals:

    (a) $(\beta) \mid (\alpha) \iff \beta \mid \alpha$.

    (b) If $\beta \neq 0$ and $(\alpha) = (\beta)C$, then C is necessarily a principal ideal, as well, $C = (\gamma)$, where γ can be chosen to satisfy $\alpha = \beta\gamma$.

4. Let $D = \{a + b \mid a \in A, b \in B\}$.

    (a) Prove that D is the smallest ideal containing A and B.

    (b) Interpreting D as the sum of A and B, show that this addition of ideals is commutative and associative, (0) is a zero element, but only (0) has a negative. *Remark*: By part (a), we can consider D as the ideal generated by A and B, and so D is also the greatest common divisor of A and B by the connection between containment and divisibility (see Theorem 11.4.5). Accordingly, we use the notation $D = (A, B)$. For part (b), however, it is better to write $D = A + B$.

    (c) Demonstrate the distributive law $A(B, C) = (AB, AC)$ (or, $A(B+C) = AB+AC$, by the other notation).

5. Define least common multiple for ideals, and show that if (11.4.4) and (11.4.5) are valid, then two ideals A and B have a unique least common multiple M, namely $M = A \cap B$.

6. Give an example of ideals A and B, such that $A \subseteq B$, but $B \nmid A$.

7. Prove that (11.4.11) and (11.4.12) about prime ideals (after Definition 11.4.7) are equivalent in any integral domain R with identity (even if (11.4.4) or (11.4.5) is false in R).

8. Consider the ring $I(\sqrt{-5})$.

  S (a) Find all divisors of the ideals:

        (a1): $(2, 1 + \sqrt{-5})$

        (a2): (2)

        (a3): $(1 + \sqrt{-5})$.

    (b) Compute the gcd of the ideals:

        (b1): (2) and $(1 + \sqrt{-5})$

        (b2): $(2, 1 + \sqrt{-5})$ and $(3, 1 - \sqrt{-5})$.

    (c) Determine which of the ideals are irreducible:

        (c1): $(2, 1 + \sqrt{-5})$

        (c2): (2)

        (c3): (11).

S 9. True or false?

    (a) If α is an irreducible element in R, then (α) is an irreducible ideal.

    (b) If (α) is an irreducible ideal, then α is an irreducible element in R.

    (c) If α is a prime element in R, then (α) is a prime ideal.

    (d) If (α) is a prime ideal, then α is a prime element in R.

10. (a) Show by an example that ideals of $\mathbf{Z}[x]$ do not obey the cancellation law (11.4.4): $AB = AC, A \neq (0) \not\Rightarrow B = C$.

    (b) Prove that we can cancel by a non-zero principal ideal in any integral domain R with identity: If $A = (\alpha) \neq (0)$, then $AB = AC \Rightarrow B = C$.

11. Consider the ring R of polynomials with non-negative rational exponents and real coefficients ($3 + 7x^{4/7} + 11x^{5/3}$ is such a polynomial).

    (a) Verify that the elements without term x^0 (i.e. having constant term 0) form an ideal I in R.

    (b) Demonstrate that I can be decomposed into the product of two ideals only as $I = (1)I = I(1) = I \cdot I$.
    Remark: The ideal I meets the requirements (11.4.8) and (11.4.9) concerning irreducible ideals, but it has also a non-trivial factorization $I = I \cdot I$ (which shows that R does not obey the cancellation law (11.4.4)). Due to this and similar singularities, we generally discuss irreducibility (and other number-theoretic notions) only for ideals of rings where (11.4.4) and (11.4.5) are valid.

12. Let R be an integral domain with identity (but we do not require (11.4.4) and (11.4.5) now). Among the non-trivial ideals, we define maximal and prime ideals by (11.4.9) and (11.4.12). We call a non-trivial ideal Q *quasi-irreducible* if $Q = AB$ implies $A = Q$ or $B = Q$ (or both, cf. the previous exercise).

    (a) Prove that every prime ideal is quasi-irreducible.

    (b) Exhibit a quasi-irreducible ideal that is not a prime ideal.

    (c) Verify that every maximal ideal is also a prime ideal and hence quasi-irreducible.

    (d) Give an example of a prime ideal, that is not maximal.

    (e) Demonstrate that I is a maximal ideal if and only if the factor ring R/I is a field, and I is a prime ideal if and only if R/I contains no zero divisors.

    Remark: We introduced quasi-irreducible ideals just for the exercise, but maximal and prime ideals in this interpretation play an important role in arbitrary rings (as is suggested by part (e)).

11.5. Dedekind Rings

In this section, ϑ denotes an algebraic number.

    We show that the Fundamental Theorem of Arithmetic holds for the ideals of $I(\vartheta)$, so every ideal different from (0) and (1) is the product of irreducible ideals, and this decomposition is unique apart from the order of factors. Rings with this property are called *Dedekind rings*.

First we verify a result of independent interest about the product of polynomials with algebraic integer coefficients (Theorem 11.5.1). It is a generalization of a basic lemma of Gauss for polynomials with rational coefficients (see Exercise 11.5.9) that occurred in the proof of Theorem 9.6.2 and was referred to in Remark 1 after the proof of Theorem 11.3.3. We apply Theorem 11.5.1 to show that to every ideal $A \neq (0)$ of $I(\vartheta)$ there exists an ideal $B \neq (0)$ such that AB is a principal ideal (Theorem 11.5.5). As a consequence, we obtain the cancellation law for ideals (Theorem 11.5.6), and the equivalence of divisibility and the (opposite) containment of ideals (Theorem 11.5.7); we required these properties in the previous section when discussing general number-theoretic notions for ideals. Then we prove unique prime factorization for ideals (Theorem 11.5.8). Finally, we establish the surprising result that every ideal of $I(\vartheta)$ can be generated by at most two elements (Theorem 11.5.9).

Theorem 11.5.1. *Let*

$$f(x) = \alpha_0 + \alpha_1 x + \cdots + \alpha_m x^m \quad and \quad g(x) = \beta_0 + \beta_1 x + \cdots + \beta_n x^n$$

be polynomials with algebraic integer coefficients and consider their product

$$f(x)g(x) = \gamma_0 + \gamma_1 x + \cdots + \gamma_{m+n} x^{m+n}.$$

Assume that some algebraic integer δ divides all coefficients of the product:

(11.5.1) $\delta \mid \gamma_k, \qquad k = 0, 1, \ldots, m + n.$

Then

$$\delta \mid \alpha_i \beta_j, \qquad i = 0, 1, \ldots, m, \qquad j = 0, 1, \ldots, n. \qquad \clubsuit$$

Proof. We need a chain of three lemmas.

Lemma 11.5.2. *The product of a root and the leading coefficient of a polynomial with algebraic integer coefficients is an algebraic integer.* $\qquad \clubsuit$

Proof of Lemma 11.5.2. Let

(11.5.2) $h(x) = \lambda_0 + \lambda_1 x + \cdots + \lambda_r x^r = \lambda_r \prod_{i=1}^{r} (x - \xi_i).$

We verify that $\lambda_r \xi_1$ is an algebraic integer. Multiplying

$$0 = f(\xi_1) = \lambda_0 + \lambda_1 \xi_1 + \cdots + \lambda_r \xi_1^r$$

by λ_r^{r-1}, we obtain

$$0 = \lambda_0 \lambda_r^{r-1} + \lambda_1 \lambda_r^{r-2}(\lambda_r \xi_1) + \cdots + \lambda_{r-1}(\lambda_r \xi_1)^{r-1} + (\lambda_r \xi_1)^r.$$

This means that $\lambda_r \xi_1$ is a root of the polynomial

$$\lambda_0 \lambda_r^{r-1} + \lambda_1 \lambda_r^{r-2} x + \cdots + \lambda_{r-1} x^{r-1} + x^r$$

with algebraic integer coefficients and leading coefficient one. Hence, $\lambda_r \xi_1$ is an algebraic integer by Theorem 9.6.3/(iii). $\qquad \square$

Lemma 11.5.3. *Dividing a polynomial with algebraic integer coefficients by any of its root factors, we obtain a polynomial with algebraic integer coefficients again. (A root factor of a polynomial f is a linear polynomial $x - \alpha$ where $f(\alpha) = 0$.)*   ♣

Proof of Lemma 11.5.3. Let h be the polynomial in (11.5.2). We show that the coefficients of

$$h_1(x) = \frac{h(x)}{x - \xi_1}$$

are algebraic integers.

We proceed by induction on the degree r of h.

The statement is true for $r = 1$ because then $h_1(x)$ is the constant polynomial λ_1.

Assume now that the statement holds for every polynomial of degree not greater than $r - 1$. Consider the polynomial

$$s(x) = h(x) - \lambda_r(x - \xi_1)x^{r-1}.$$

Clearly, the degree of $s(x)$ is at most $r - 1$, $s(\xi_1) = 0$, and $\lambda_r \xi_1$ is an algebraic integer by Lemma 11.5.2, so the coefficients of $s(x)$ are algebraic integers.

By the induction hypothesis, the coefficients of

$$s_1(x) = \frac{s(x)}{x - \xi_1} = \frac{h(x)}{x - \xi_1} - \lambda_r x^{r-1} = h_1(x) - \lambda_r x^{r-1}$$

are algebraic integers. Since λ_r is an algebraic integer, we obtain that $h_1(x)$ has algebraic integer coefficients.   □

Lemma 11.5.4. *The product of the leading coefficient and arbitrarily many roots of a polynomial with algebraic integer coefficients is an algebraic integer.*   ♣

Proof of Lemma 11.5.4. Let h be the polynomial in (11.5.2). We verify that $\lambda_r \xi_1 \ldots \xi_k$ is an algebraic integer.

Dividing $h(x)$ by the missing root factors, those with indices greater than k, we obtain the polynomial

$$t(x) = \lambda_r \prod_{j=1}^{k}(x - \xi_j),$$

which has algebraic integer coefficients by repeated applications of Lemma 11.5.3. Thus its constant term

$$(-1)^k \lambda_r \xi_1 \ldots \xi_k$$

is an algebraic integer.   □

Now we turn to the proof of Theorem 11.5.1.

Let the roots of the polynomials f and g be $\xi_1, \ldots, \xi_m$, and $\eta_1, \ldots, \eta_n$. Then

$$(11.5.3) \qquad f(x)g(x) = \sum_{k=0}^{m+n} \gamma_k x^k = \alpha_m \beta_n \prod_{i=1}^{m}(x - \xi_i) \prod_{j=1}^{n}(x - \eta_j).$$

Dividing (11.5.3) by δ (appearing in the statement of Theorem 11.5.1), we get

$$(11.5.4) \qquad \sum_{k=0}^{m+n} \frac{\gamma_k}{\delta} x^k = \frac{\alpha_m \beta_n}{\delta} \prod_{i=1}^{m} (x - \xi_i) \prod_{j=1}^{n} (x - \eta_j).$$

By (11.5.1), the polynomial on the left-hand side of (11.5.4) has algebraic integer coefficients. Thus an arbitrary product

$$(11.5.5) \qquad \frac{\alpha_m \beta_n}{\delta} \xi_{i_1} \dots \xi_{i_r} \eta_{j_1} \dots \eta_{j_s}$$

is an algebraic integer by Lemma 11.5.4.

Using the root factor decomposition, we get any coefficient α_i of f by adding some terms of the form $\pm \alpha_m \xi_{i_1} \dots \xi_{i_r}$, and we have a similar result for g. Therefore every $\alpha_i \beta_j$ can be written as

$$\alpha_i \beta_j = \left(\sum \pm \alpha_m \xi_{i_1} \dots \xi_{i_r} \right) \left(\sum \pm \beta_n \eta_{j_1} \dots \eta_{j_s} \right),$$

so

$$(11.5.6) \qquad \frac{\alpha_i \beta_j}{\delta} = \frac{\alpha_m \beta_n}{\delta} \left(\sum \pm \xi_{i_1} \dots \xi_{i_r} \right) \left(\sum \pm \eta_{j_1} \dots \eta_{j_s} \right).$$

The right-hand side of (11.5.6) is a sum with signs of algebraic integers of the form in (11.5.5), so it is an algebraic integer. This proves that $\alpha_i \beta_j / \delta$ on the left-hand side is an algebraic integer. $\qquad \square$

The following result of Kronecker plays a central role in studying ideals of $I(\vartheta)$: It makes it possible to get at least partial answers for many questions by reducing the problems to principal ideals, as these have a much more transparent structure.

Theorem 11.5.5. *To every ideal $A \neq (0)$ in $I(\vartheta)$ there exists an ideal $B \neq (0)$ such that AB is a principal ideal.* ♣

Remark: It turns out from the proof that we can choose $B \neq (0)$ to yield $AB = (c)$ with an integer c. This can also be easily deduced from the statement of the theorem (see Exercises 11.5.1 and 11.5.2).

Proof. By Exercise 11.1.10c, the ideal A is finitely generated:

$$A = (\alpha_0, \alpha_1, \dots, , \alpha_k).$$

Let $\vartheta_{(1)} = \vartheta, \vartheta_{(2)}, \dots, \vartheta_{(n)}$ be the conjugates of ϑ over $\mathbf{Q}$ (i.e. the roots of its minimal polynomial), and let $f_\nu(\vartheta_{(j)})$ denote the jth relative conjugate of the generator α_ν (see Section 10.4); in particular, $f_\nu(\vartheta_{(1)}) = \alpha_\nu$.

Consider the polynomials

$$F_j(x) = f_0(\vartheta_{(j)}) + f_1(\vartheta_{(j)})x + \dots + f_k(\vartheta_{(j)})x^k, \qquad j = 1, 2, \dots, n.$$

(Thus the coefficient of x^i in $F_j(x)$ is the jth relative conjugate of the generator α_i.) In particular,

$$F_1(x) = \alpha_0 + \alpha_1 x + \dots + \alpha_k x^k.$$

Let $G(x) = \prod_{j=1}^{n} F_j(x)$.

$G(x)$ is a symmetric polynomial in variables ϑ_j, so the same applies for its coefficients. By the fundamental theorem of symmetric polynomials and the Viète formulas for the minimal polynomial of ϑ, we infer (in the usual way we have seen several times) that $G(x)$ has rational coefficients.

As the coefficients of $G(x)$ are obtained from the algebraic integers α_ν and their algebraic integer relative conjugates with the help of addition and multiplication, the coefficients of $G(x)$ are algebraic integers. Since they are also rational, they must be integers,

$$G(x) = a_0 + a_1 x + \cdots + a_{kn} x^{kn}, \qquad a_s \in \mathbf{Z}, \qquad s = 0, 1, \ldots, kn.$$

Let

$$H(x) = \frac{G(x)}{F_1(x)} = \prod_{j=2}^{n} F_j(x).$$

Since the coefficients of every $F_j(x)$ are algebraic integers, $H(x)$ has algebraic integer coefficients. Further, the coefficients of $G(x)$ and $F_1(x)$ are in $\mathbf{Q}(\vartheta)$, and the division algorithm has terms that are in the field containing the coefficients, so the coefficients of $H(x)$ belong to $\mathbf{Q}(\vartheta)$. Combining the two observations, we see that the coefficients of $H(x)$ are in $I(\vartheta)$,

$$H(x) = \beta_0 + \beta_1 x + \cdots + \beta_{kn-k} x^{kn-k}.$$

We show that

$$B = (\beta_0, \beta_1, \ldots, \beta_{kn-k}) \quad \text{and} \quad c = \gcd\{a_0, a_1, \ldots, a_{kn}\}$$

satisfy $AB = (c)$.

As c is the greatest common divisor of the coefficients of $G \neq 0$, $c \neq 0$ (and so clearly $B \neq (0)$).

We verify first $AB \subseteq (c)$. By the definition of c, it divides every coefficient of the polynomial $G(x) = F_1(x)H(x)$. Then, by Theorem 11.5.1, c divides every product $\alpha_i \beta_j$, so $\alpha_i \beta_j \in (c)$, hence $AB \subseteq (c)$.

To prove the reverse containment $(c) \subseteq AB$, observe that $G(x) = F_1(x)H(x)$, so

$$a_0 + a_1 x + \cdots + a_{kn} x^{kn} = (\alpha_0 + \alpha_1 x + \cdots + \alpha_k x^k)(\beta_0 + \beta_1 x + \cdots + \beta_{kn-k} x^{kn-k})$$

implies

$$a_s = \sum_{i+j=s} \alpha_i \beta_j \in AB, \qquad s = 0, 1, \ldots, kn.$$

By Theorem 1.3.5, which refers to a property of the gcd for integers, we have

$$c = \sum_{s=0}^{kn} a_s u_s$$

with suitable integers u_s, thus

$$c \in (a_0, a_1, \ldots, a_{kn}) \subseteq AB \quad \text{so} \quad (c) \subseteq AB.$$

The mutual containment proves $AB = (c)$. $\qquad\qquad\qquad\qquad\qquad\qquad\qquad\qquad\square$

Theorem 11.5.6. *The cancellation law holds for ideals of $I(\vartheta)$:*

$$AB = AC, \ A \neq (0) \implies B = C. \qquad\qquad\qquad\qquad\qquad\qquad\clubsuit$$

Proof. By Theorem 11.5.5, given an ideal $A \neq (0)$ there exists an ideal $D \neq (0)$ such that AD is a principal ideal, so $AD = (\psi)$ for some $(0 \neq)\psi \in E(\vartheta)$ (moreover, ψ can be chosen to be an integer).

Multiplying $AB = AC$ by D, we obtain $(\psi)B = (\psi)C$. Then $B = C$ follows by Exercise 11.4.10b. □

Theorem 11.5.7. *For ideals of $I(\vartheta)$, $B \mid A \iff A \subseteq B$.* ♣

Proof. We saw in Exercise 11.4.2c that the implication $\Rightarrow$ holds in any integral domain with identity.

For the converse, assume $A \subseteq B$. We may clearly restrict ourselves to the case $B \neq (0)$. By Theorem 11.5.5, multiplying $B \neq (0)$ by a suitable ideal $D \neq (0)$, the product is a principal ideal: $BD = (\psi)$. Then $AD \subseteq BD = (\psi)$.

Every ideal in $I(\vartheta)$, including AD, is finitely generated. By the condition $AD \subseteq (\psi)$, every generator is divisible by ψ:

$$AD = (\eta_1\psi, \ldots, \eta_s\psi) = (\psi)(\eta_1, \ldots, \eta_s).$$

Denoting the ideal $(\eta_1, \ldots, \eta_s)$ by K, we obtain

$$AD = (\psi)K = BDK.$$

Cancelling $D \neq (0)$, we get

$$A = BK, \quad \text{so} \quad B \mid K. \qquad □$$

By Theorems 11.5.6 and 11.5.7, the ideals of $I(\vartheta)$ obey the cancellation law, and divisibility is equivalent to containment (in the opposite direction). Accordingly, the results of Section 11.4 depending on these properties are valid for ideals in $I(\vartheta)$. We stress among them the equivalence of irreducible and prime ideals (Theorem 11.4.8). This will have a crucial role in the proof of the next theorem: We show that the Fundamental Theorem of Arithmetic holds for the ideals of $I(\vartheta)$.

Theorem 11.5.8. *Every ideal in $I(\vartheta)$ different from (0) and (1) is the product of finitely many irreducible ideals, and the decomposition is unique apart from the order of factors.* ♣

Proof. We follow closely the argument in the proof of sufficiency in Theorem 11.3.1.

Decomposability. Let A be a non-trivial ideal. We show first that A has a divisor among the irreducible ideals.

If A itself is irreducible, then we are done.

Otherwise, $A = A_1B_1$, where $A_1 \neq (1)$, $B_1 \neq (1)$. Then $A \subset A_1$ with strict containment, since if $A = A_1$, then cancelling A in $A(1) = A = AB_1$ would imply $(1) = B_1$.

If A_1 is irreducible, then it is an irreducible divisor of A. Otherwise, $A_1 = A_2B_2$, where $A_2 \neq (1)$, $B_2 \neq (1)$. Then $A_1 \subset A_2$ (with strict containment).

Continuing the procedure similarly, we get a strictly ascending chain of ideals

$$A \subset A_1 \subset A_2 \subset \cdots \subset A_j \subset \ldots.$$

It cannot be infinite by Exercise 11.1.10b, so some A_i must be irreducible.

Now we verify that A is a product of irreducible ideals. If A is irreducible, then we are done. Otherwise, $A = P_1 C_1$, where P_1 is irreducible and $C_1 \neq (1)$. Since $P_1 \neq (1)$, we have $A \subset C_1$ (with strict containment).

If C_1 is irreducible, then both factors are irreducible in the product $A = P_1 C_1$, and we are done. Otherwise, $C_1 = P_2 C_2$, where P_2 is irreducible and $C_2 \neq (1)$. Therefore $C_1 \subset C_2$ (with strict containment).

Continuing the procedure, we get a strictly increasing chain of ideals

$$A \subset C_1 \subset \cdots \subset C_j \subset \dots.$$

It cannot be infinite by Exercise 11.1.10b, so some $C_i = (1)$, and A is a product of irreducible ideals.

Uniqueness: Assume that some A has at least two essentially distinct decompositions into the product of irreducible ideals:

(11.5.7) $$A = P_1 P_2 \dots P_r = Q_1 Q_2 \dots Q_s.$$

If a P_i is equal to a Q_j, then we can cancel by the common factor. Thus we may assume that $P_i \neq Q_j$ in (11.5.7).

We have $P_1 \mid Q_1 Q_2 \dots Q_s$. Since P_1 is irreducible, it is a prime ideal by Theorem 11.4.8. Therefore P_1 divides at least one factor Q_j.

As Q_j is irreducible, $P_1 \mid Q_j$ implies $P_1 = (1)$ or $P_1 = Q_j$, but both are impossible. $\square$

Example. Factor the principal ideal (6) in $I(\sqrt{-5})$ into a product of irreducible ideals.

We saw earlier that 6 has two essentially distinct representations as a product of irreducible elements in $I(\sqrt{-5})$:

$$6 = 2 \cdot 3 = [1 + \sqrt{-5}][1 - \sqrt{-5}].$$

Accordingly, the principal ideal (6) has two decompositions into the product of principal ideals:

$$(6) = (2)(3) = (1 + \sqrt{-5})(1 - \sqrt{-5}).$$

Each factor can be written as a product of two irreducible ideals:

$$(2) = (2, 1 + \sqrt{-5})(2, 1 - \sqrt{-5}) = (2, 1 + \sqrt{-5})^2$$

$$(3) = (3, 1 + \sqrt{-5})(3, 1 - \sqrt{-5})$$

$$(1 + \sqrt{-5}) = (2, 1 + \sqrt{-5})(3, 1 + \sqrt{-5})$$

$$(1 - \sqrt{-5}) = (2, 1 - \sqrt{-5})(3, 1 - \sqrt{-5}).$$

Thus the principal ideal (6) has the following factorization into the product of irreducible ideals:

$$(6) = (2, 1 + \sqrt{-5})^2 (3, 1 + \sqrt{-5})(3, 1 - \sqrt{-5}).$$

The irreducible ideals arise from the two decompositions of 6 into irreducible factors: we can interpret the ideal $(3, 1 + \sqrt{-5})$ as a "hiding common divisor ideal number" in the factors 3 and $1 + \sqrt{-5}$, and in fact, we refined the two distinct decompositions of 6 into a common decomposition of the principal ideal (6) with the help of these hidden factors.

Because of the equivalence of irreducible and prime ideals, we shall use the name prime ideal for both notions in the sequel.

We can introduce the standard form of ideals by Theorem 11.5.8: If $A \neq (0)$ and $A \neq (1)$, then

$$A = P_1^{\alpha_1} \dots P_r^{\alpha_r} = \prod_{i=1}^{r} P_i^{\alpha_i},$$

where $P_1, \dots, P_r$ are distinct prime ideals and $\alpha_1, \dots, \alpha_r$ are positive integers.

The standard forms for the greatest common divisor (see Definition 11.4.4. and Theorem 11.4.5) and the least common multiple (see Exercise 11.4.5) of ideals have the same well-known formulas as for integers: Every prime ideal occurring in the ideals has to be taken with the minimal or maximal exponent, respectively, and $P^0 = (1)$. The proof is the same as for integers.

As an application of Theorems 11.5.5 and 11.5.8, we prove that every ideal in $I(\vartheta)$ is almost a principal ideal:

Theorem 11.5.9. *Every ideal in $I(\vartheta)$ can be generated by at most two elements.*   ♣ ♣

Proof. We can assume $A \neq (0)$ and $A \neq (1)$.

By Theorem 11.5.5, $AB = (\psi)$ for some ideal $B \neq (0)$. We want to find a principal ideal (γ) such that the greatest common divisor of (γ) and $AB = (\psi)$ is A, since then

(11.5.8) $A = (\psi, \gamma)$

by Remark 2 after Theorem 11.4.5.

Let $P_1, \dots, P_r$ be all prime ideals that divide at least one of A and B and let

$$A = P_1^{\alpha_1} \dots P_r^{\alpha_r} = \prod_{i=1}^{r} P_i^{\alpha_i}$$

be the standard form of A where $\alpha_i = 0$, so $P_i^0 = (1)$ may occur, if P_i divides only B.

Consider the ideals:

$$C = \prod_{i=1}^{r} P_i^{1+\alpha_i}, \quad \text{and} \quad C_j = P_j^{\alpha_j} \prod_{i \neq j} P_i^{1+\alpha_i}, \quad j = 1, 2, \dots, r.$$

Then

$$C_j \mid C, \quad \text{so} \quad C \subset C_j, \quad \text{but} \quad C \neq C_j.$$

Choose $\gamma_1, \dots, \gamma_r$ satisfying

$$\gamma_j \in C_j, \quad \text{but} \quad \gamma_j \notin C, \quad j = 1, 2, \dots, r.$$

We prove

(11.5.9a) $\gamma_j \in P_i^{1+\alpha_i} \quad \text{if } j \neq i$

(11.5.9b) $\gamma_i \in P_i^{\alpha_i}$

(11.5.9c) $\gamma_i \notin P_i^{1+\alpha_i}.$

If $j \neq i$, then $P_i^{1+\alpha_i} \mid C_j$ implies $C_j \subseteq P_i^{1+\alpha_i}$, and so $\gamma_j \in P_i^{1+\alpha_i}$, since $\gamma_j \in C_j$. This verifies (11.5.9a), and (11.5.9b) can be shown similarly.

We prove (11.5.9c) by contradiction. If $\gamma_i \in P_i^{1+\alpha_i}$, then combining it with (11.5.9a), we obtain

$$(11.5.10) \qquad \gamma_i \in \bigcap_{t=1}^{r} P_t^{1+\alpha_t}.$$

By Exercise 11.4.5, the intersection of ideals is their least common multiple, so

$$(11.5.11) \qquad \bigcap_{t=1}^{r} P_t^{1+\alpha_t} = \mathrm{lcm}\{P_1^{1+\alpha_1}, \ldots, P_r^{1+\alpha_r}\} = \prod_{t=1}^{r} P_t^{1+\alpha_t} = C.$$

From (11.5.10) and (11.5.11) we get $\gamma_i \in C$, which contradicts the selection of γ_i.

We claim that for

$$\gamma = \gamma_1 + \cdots + \gamma_r$$

the greatest common divisor of AB and (γ) is A, and so (11.5.8) holds.

We have to verify that in the standard form of the gcd of AB and (γ)

(i) only the prime ideals P_i can occur

(ii) the exponent of P_i is α_i ($i = 1, 2, \ldots, r$).

Condition (i) holds, as only the prime ideals P_i appear also in the standard form of AB.

Since $A \mid AB$, the exponent of P_i in the standard form of AB is at least α_i. Thus to show (ii), we have to verify that the exponent of P_i in the standard form of (γ) is exactly α_i, or

(iii) $P_i^{\alpha_i} \mid (\gamma)$, but

(iv) $P_i^{1+\alpha_i} \nmid (\gamma)$.

(iii): By (11.5.9a) and (11.5.9b),

$$\gamma_t \in P_i^{\alpha_i}, \qquad t = 1, 2, \ldots, r,$$

so every term in the sum defining γ is an element of $P_i^{\alpha_i}$. As $P_i^{\alpha_i}$ is an ideal, $\gamma \in P_i^{\alpha_i}$, and thus

$$(\gamma) \subseteq P_i^{\alpha_i}, \quad \text{so} \quad P_i^{\alpha_i} \mid (\gamma).$$

(iv): By (11.5.9a) and (11.5.9c),

$$\gamma_j \in P_i^{1+\alpha_i} \quad \text{if} \quad j \neq i, \quad \text{but} \quad \gamma_i \notin P_i^{1+\alpha_i},$$

so every term in the sum defining γ is an element of $P_i^{1+\alpha_i}$ with the exception of exactly one term. As $P_i^{1+\alpha_i}$ is an ideal, $\gamma \notin P_i^{1+\alpha_i}$, and so

$$(\gamma) \not\subseteq P_i^{1+\alpha_i}, \quad \text{i.e.} \quad P_i^{1+\alpha_i} \nmid (\gamma). \qquad \square$$

Exercises 11.5

All exercises refer to ideals of $I(\vartheta)$.

1. Prove that to an element $\alpha \in I(\vartheta)$ and an ideal A there exists an ideal B satisfying $AB = (\alpha)$ if and only if $\alpha \in A$.

2. (a) Demonstrate $\alpha \mid N(\alpha)$ for any $\alpha \in I(\vartheta)$.

   (b) Show that every ideal $A \neq (0)$ in $I(\vartheta)$ contains infinitely many integers that form a principal ideal in **Z**.

3. Verify that an ideal $A \neq (0)$ has only finitely many divisors.

4. Consider the prime ideals in a given $I(\vartheta)$.

   (a) Demonstrate that every prime ideal contains exactly one positive prime number.

   (b) Prove that there are infinitely many prime ideals.

   (c) Can a prime number be an element of two different prime ideals?

   (d) Can a prime number be an element of infinitely many different prime ideals?

5. Prove that the product of any two ideals in $I(\vartheta)$ equals the product of their sum and intersection.

6. Show $\alpha\beta \in (\alpha^2, \beta^2)$ for any $\alpha, \beta \in I(\vartheta)$.

7. Consider the ring $I(\sqrt{-5})$.

   (a) Factor the principal ideal (21) into a product of prime ideals.

   (b) For which primes $p > 0$ is $(p, 1 + \sqrt{-5})$ a prime ideal?

   S* (c) For which primes $p > 0$ is $(p, a + \sqrt{-5})$ a prime ideal for a suitable a?

8. Prove that the Fundamental Theorem of Arithmetic holds for the elements of $I(\vartheta)$ if and only if every prime ideal is a principal ideal.

S 9. A non-constant polynomial with integer coefficients is called *primitive* if its coefficients are coprime. Deduce from Theorem 11.5.1:

   (a) (*First form of Gauss's Lemma.*) The product of two primitive polynomials is primitive, too.

   (b) (*Second form of Gauss's Lemma.*) If a polynomial H with integer coefficients is the product of polynomials F and G with rational coefficients, $H = FG$, then $H = F_1 G_1$, where F_1 and G_1 are polynomials with integer coefficients and are constant multiples of F and G.

   Remark: Statements (a) and (b) can be deduced from each other easily, therefore both are referred to as Gauss's lemma. Some books, however, call only statement (a) by this name.

11.6. Class Number

We assume also in this section that ϑ is an algebraic number and introduce an equivalence relation among the non-zero ideals of $I(\vartheta)$. The number of resulting equivalence classes plays an important role in the number theory of $I(\vartheta)$. As an application of ideals, we show that the Diophantine equation $x^2 + 17 = y^3$ has no solution.

Definition 11.6.1. The ideals $A \neq (0)$ and $B \neq (0)$ are *equivalent* if there exist principal ideals $(\alpha) \neq (0)$ and $(\beta) \neq (0)$ such that

$$(\alpha)A = (\beta)B. \qquad \clubsuit$$

Notation: $A \sim B$.

In the sequel we always assume that ideals (including the principal ideals) are not zero.

We summarize some simple but important properties of equivalence in

Theorem 11.6.2. (i) *The relation $\sim$ in Definition* 11.6.1 *is reflexive, symmetric, and transitive, so it is an equivalence relation.*

(ii) $A \sim B$, $C \sim D \Longrightarrow AC \sim BD$.

(iii) $A \sim B \iff AC \sim BC$.

(iv) $A \sim (1) \iff A$ *is a principal ideal.* $\qquad \clubsuit$

Proof. (i) $A \sim A$, since $(1)A = A$. Symmetry is obvious from the definition. If $A \sim B$ and $B \sim C$, or

$$(\alpha)A = (\beta)B \quad \text{and} \quad (\gamma)B = (\delta)C$$

with suitable non-zero principal ideals $(\alpha), (\beta), (\gamma),$ and (δ), then

$$(\alpha\gamma)A = (\beta\gamma)B = (\beta\delta)C.$$

(ii) If $A \sim B$ and $C \sim D$, or

$$(\alpha)A = (\beta)B \quad \text{and} \quad (\varrho)C = (\xi)D,$$

then

$$(\alpha\varrho)AC = (\beta\xi)BD.$$

(iii) As $C \neq (0)$, so $(\alpha)A = (\beta)B \iff (\alpha)AC = (\beta)BC$.

(iv) If $A = (\varrho)$, then $(1)A = (1)(\varrho) = (\varrho)(1)$ guarantees $A \sim (1)$. Conversely, if $A \sim (1)$, so $A(\alpha) = (1)(\beta) = (\beta)$, then A is a principal ideal by Exercise 11.4.3b. $\qquad \square$

The equivalence relation $\sim$ partitions the non-zero ideals of $I(\vartheta)$ into disjoint classes. We state the following fundamental result without proof:

Theorem 11.6.3. *There are finitely many ideal classes of $I(\vartheta)$.* $\qquad \clubsuit$

We denote the number of ideal classes of $I(\vartheta)$ by $h(\vartheta)$.

The following table contains the class number of $I(\sqrt{t})$ for some negative integers t:

t	-1	-3	-5	-17	-31	-35	-74
$h(\sqrt{t})$	1	1	2	4	3	2	10

It is easy to verify that the Fundamental Theorem of Arithmetic holds for the elements of $I(\vartheta)$ if and only if $h(\vartheta) = 1$ (see Exercise 11.6.2).

We show now that any (non-zero) ideal in $I(\vartheta)$ raised to the power $h(\vartheta)$ is always a principal ideal:

Theorem 11.6.4. *Let $h(\vartheta)$ be the number of ideal classes of $E(\vartheta)$ and let $A \neq (0)$ be an ideal. Then $A^{h(\vartheta)}$ is a principal ideal.* ♣

Proof. We follow the proof of the Euler–Fermat Theorem 2.4.1. Let $h(\vartheta) = h$ and

(11.6.1) $$A_1, A_2, \ldots, A_h$$

be any representatives of the (distinct) ideal classes. We show that

(11.6.2) $$AA_1, AA_2, \ldots, AA_h$$

fall into different classes. If $AA_i \sim AA_j$, or

$$(\varrho)AA_i = (\tau)AA_j,$$

then cancellation by $A \neq (0)$ yields $A_i \sim A_j$, so $i = j$.

This means that the ideals listed in (11.6.2) are equivalent to the ideals in (11.6.1) in some order. Thus to every $1 \leq i \leq h$ there exists exactly one j, $1 \leq j \leq h$, such that $AA_i \sim A_j$. We denote this A_j by B_i:

(11.6.3)
$$AA_1 \sim B_1$$
$$AA_2 \sim B_2$$
$$\vdots$$
$$AA_h \sim B_h.$$

The ideals $B_1, \ldots, B_h$ are a permutation of the ideals $A_1, \ldots, A_h$.

Multiplying the equivalences in (11.6.3), we obtain

(11.6.4) $$A^h A_1 A_2 \ldots A_h \sim B_1 B_2 \ldots B_h = A_1 A_2 \ldots A_h$$

by Theorem 11.6.3(ii). By Theorem 11.6.3(iii), we can cancel by every ideal $A_i \neq (0)$ yielding $A^h \sim (1)$. It follows from Theorem 11.6.3/(iv) that A^h is a principal ideal. □

We close the chapter with an illustration that shows that ideals may be suitable to handle Diophantine equations even if the Fundamental Theorem of Arithmetic is false for the algebraic integers of the corresponding number field.

Theorem 11.6.5. *The Diophantine equation $x^2 + 17 = y^3$ has no solution.* ♣

Proof. We discussed similar Diophantine equations earlier: $x^2 + 4 = y^3$ (Exercise 7.5.10) and $x^2 + 243 = y^3$ (Exercise 7.7.11). We factored the left-hand side in the Gaussian and Eulerian integers and showed by the Fundamental Theorem of Arithmetic that each factor is an associate of a cube, which made it possible to determine the solutions.

Now we have to overcome the difficulty that after the factorization

$$(11.6.5) \qquad [x + \sqrt{-17}][x - \sqrt{-17}] = y^3$$

we cannot proceed similarly, since the Fundamental Theorem of Arithmetic is false in $I(\sqrt{-17})$. Therefore we have to switch from (11.6.5) involving numbers to the corresponding equation with principal ideals:

$$(11.6.6) \qquad (x + \sqrt{-17})(x - \sqrt{-17}) = (y)^3.$$

We show that the ideals $(x + \sqrt{-17})$ and $(x - \sqrt{-17})$ are coprime. Assume that a prime ideal P is their common divisor. Then P divides $(y)^3$, and as P is a prime ideal, it divides (y), as well. Switching to the corresponding inclusions,

$$x + \sqrt{-17} \in P, \qquad x - \sqrt{-17} \in P, \quad \text{and} \quad y \in P.$$

Then

$$\sqrt{-17}\big[[x - \sqrt{-17}] - [x + \sqrt{-17}]\big] = 2 \cdot 17 = 34 \in P$$

holds.

We show that y and 34 are coprime in the integers.

If $17 \mid y$, then we see from the original equation that 17 divides x. Then $x^2 + 17$ and y^3 are divisible by exactly the first and at least the third powers of 17, which is impossible.

If $2 \mid y$, then x is odd, and the residues of the two sides modulo 8 of the equation are 2 and 0, which cannot hold.

Thus we have proved that y and 34 are coprime. Then $1 = yu + 34v$ for some integers u and v. Since 34 and y are elements of P, 1 lies in P, so $P = (1)$, which contradicts the definition of a prime ideal.

Thus the two (principal) ideals on the left-hand side of (11.6.6) are coprime. It follows from the unique prime factorization for ideals (Theorem 11.5.8) that both ideals are cubes of ideals, and so

$$(11.6.7) \qquad (x + \sqrt{-17}) = A^3.$$

Since the number of ideal classes in $I(\sqrt{-17})$ is $h(\sqrt{-17}) = 4$, therefore by Theorem 11.6.4, A^4 is a principal ideal, $A^4 = (\gamma)$. Multiplying (11.6.7) by A, we obtain

$$A(x + \sqrt{-17}) = (\gamma).$$

By Exercise 11.4.3b, A is a principal ideal, so $A = (\alpha)$. We can rewrite (11.6.7) as

$$(11.6.8) \qquad (x + \sqrt{-17}) = (\alpha^3) \quad \text{or} \quad x + \sqrt{-17} = \varepsilon\alpha^3,$$

where ε is a unit in $I(\sqrt{-17})$. The only units in $I(\sqrt{-17})$ are ± 1, which are cubes themselves and the elements of $I(\sqrt{-17})$ are of the form $a + b\sqrt{-17}$ with integer a and b, since $-17 \equiv -1 \pmod 4$. Therefore (11.6.8) is equivalent to

$$x + \sqrt{-17} = \beta^3 = [a + b\sqrt{-17}]^3.$$

Cubing and comparing the imaginary parts gives

$$1 = 3a^2 b - 17 b^3 = b[3a^2 - 17b^2].$$

This implies $b = \pm 1$, but we get no integer values for a. Thus the Diophantine equation $x^2 + 17 = y^3$ has no solution. $\qquad\qquad\square$

Exercises 11.6

1. Verify that the ideals $(2, \sqrt{-6})$ and $(3, \sqrt{-6})$ are equivalent in $E(\sqrt{-6})$.

2. Prove that the Fundamental Theorem of Arithmetic holds for the elements of $I(\vartheta)$ if and only if $h(\vartheta) = 1$.

S 3. Assume that the integers $k > 0$ and $h = h(\vartheta)$ are coprime. Prove:

    (a) $A^k \sim B^k \implies A \sim B$.

    (b) If A^k is a principal ideal, then so is A.

4. Solve the Diophantine equations:

    (a) $x^2 + 5 = y^3$

    (b) $17x^2 + 1 = y^3$

    (c) $x^2 + 74 = y^3$

  S (d) $x^2 + 35 = y^3$.

Chapter 12

Combinatorial Number Theory

The intersection of number theory and combinatorics is a relatively young area (at least compared to other branches of number theory) as its classical results, the theorems of Schur and Van der Waerden, are barely a century old. The field is extremely rich both in content and methods and its far-reaching questions can be attacked by ingenious elementary ideas combined with delicate arguments of analysis, algebra, and probability theory. Its continuous dynamic progress has been motivated greatly by the work of Paul Erdős, and nearly all problems discussed in this chapter are connected to him.

12.1. All Sums Are Distinct

In 1993 University Eötvös Loránd (ELTE, Budapest) awarded an honorary doctoral degree to the eighty-year-old Paul Erdős and asked him to give a talk on "The Actual Problems of Mathematics." The university's largest auditorium was completely full on this occasion. The lecture was recorded and quoting its start evokes the fascinating personality of Erdős.

"Can you hear me well? Also in the back rows? If you cannot hear me, please protest.

"Well, the title of the talk is a bit arrogant, but it was not my formulation; it cannot be said that the questions I will talk about are the actual problems of mathematics. The last such lecture was held by Hilbert at the Paris mathematical congress in 1900, and it is not sure that now we could find a human being capable to perform such a talk. Anyway, it would need years of preparations, and a mathematical congress would be the suitable scene for it. I cannot undertake this task, partly because of my high age, but also because I know nothing about many areas, for example I am not an expert in algebraic topology, algebraic geometry, or logic. Thus a more suitable title of my talk is

"My favorite problems", and since some people in the audience are not mathematicians, I will speak about elementary geometry and number theory.

"Let us start with elementary number theory. I will tell you now two problems. I raised the first one in 1931, so long ago, that I am not certain whether it was before or after Christ. By the way, an old joke of mine is that I am two and a half billion years old. To prove it, the age of Earth was two billion years when I was a child, and now it is well known to be 4.6 billion years. Obviously, the difference is my age, and once I gave a talk in Los Angeles with the title "My First Two Billion Years in Mathematics", and the students made a figure with a diagram "Earth born, Erdős born, dinosaur born", and drew a picture where I was riding a dinosaur.

"But putting the joke aside, the problem is the following, I pay 500 dollars for a proof or disproof, maybe there is some chalk around, can I get some chalk please, because I am captured by the wire [of the microphone], thank you very much, thus here is the problem:

"Let be given a sequence of integers: $a_1 < a_2 < \cdots < a_k \le n$, and assume that all subset sums

$$\sum_{j=1}^{k} \varepsilon_j a_j, \quad \varepsilon_j = 0 \text{ or } 1,$$

are distinct. Such numbers are for example the powers of two: 1, 2, 4, 8, 16, ..., since every baby knows that each number has a unique representation as the sum of [distinct] powers of two. Now the 500 dollar problem is to determine max k, i.e. maximally how many numbers can be given up to n so that all these sums should be distinct."

For powers of two (including $2^0 = 1$), we have $k = 1 + \lfloor \log_2 n \rfloor$, and at first glance one could think that this gives the maximum. This is false, however: for $n = 2^{21}$ Conway and Guy found a sequence that was denser by one element, for which $k = 2 + \lfloor \log_2 n \rfloor$. This implies that such a sequence exists for every $n \ge 2^{21}$, see Exercise 12.1.12. It is unknown whether further improvements are possible or not.

On the other hand, Erdős proved that the maximum cannot be much bigger than $\log_2 n$:

Theorem 12.1.1. *Assume that all sums formed from distinct integers $1 \le a_1 < a_2 < \cdots < a_k \le n$ are distinct. Then*

(12.1.1) $k \le \log_2 n + \log_2 \log_2 n + 1,$

and (for $n > 8$)

(12.1.2) $k \le \log_2 n + \dfrac{\log_2 \log_2 n}{2} + 2.$ ♣

Combined with the lower bound max$k > \log_2 n$, the estimates guarantee the asymptotic equality max$k \sim \log_2 n$ with fairly good error terms. The sharper estimate (12.1.2) is a joint result of Erdős and Leo Moser, and this is the best upper bound currently known (apart from the fact that the number 2 at the end of the formula can be replaced by a slightly smaller constant, see Exercise 12.1.13).

Thus the maximum wanted by Erdős falls between the two bounds

(12.1.3) $$\lfloor \log_2 n \rfloor + 2 \leq \max k \leq \log_2 n + \frac{\log_2 \log_2 n}{2} + 2.$$

The 500 dollar prize was offered by Erdős to clarify whether or not the difference $\max k - \log_2 n$ remains bounded as n grows to infinity. This problem is still unsolved.

Proof. We can form 2^k sums u_j from the numbers a_i (also $Z = \sum_{i=1}^{k} a_i$ and the empty sum 0 appear among the integers u_j). Each u_j falls into the interval $[0, nk - 1]$ (if $k > 1$). According to the assumption, the values u_j are distinct, hence the number of these sums must be less than or equal to the number of integers in the interval, i.e.

(12.1.4) $$2^k \leq nk.$$

Taking logarithms, we obtain

(12.1.5) $$k \leq \log_2 n + \log_2 k.$$

Now we will establish an upper bound to the second term on the right-hand side of (12.1.5) in terms of n. Since clearly $k \leq n$, therefore $\log_2 k \leq \log_2 n$, so (12.1.5) implies

(12.1.6) $$k \leq 2 \log_2 n.$$

Taking logarithms again, we have

(12.1.7) $$\log_2 k \leq 1 + \log_2 \log_2 n,$$

and substituting this into (12.1.5) we arrive at (12.1.1).

To prove the stronger result, we shall make use of the fact that the sums u_j are not evenly distributed in the interval $[0, nk - 1]$, but their major part clusters around the mean. We shall get the precise formulation using elementary probability theory (though everything could be discussed even without this, but the essential point will be seen much better with a probabilistic view).

Consider the random variable η that assumes each of the 2^k sums u_j with probability 2^{-k}. Denoting expectation by E, standard deviation by D, and probability by P, *Chebyshev's inequality*

(12.1.8) $$P\big(|\eta - E(\eta)| < cD(\eta)\big) > 1 - c^{-2}$$

says that the number of sums u_j in the interval with center $E(\eta)$ and length $2cD(\eta)$ is at least $1 - c^{-2}$ times the number of all values u_j. We shall repeat the argument used to verify (12.1.1) for this interval (with a suitable c).

Turning to the details, the expectation is $E(\eta) = Z/2$, since pairing the complementary sums u_j, the sum of every pair is Z. To compute the variance, we introduce the random variables ξ_i, $i = 1, 2, \ldots, k$, where ξ_i assumes each of the values a_i and 0 with probability $1/2$. Then the variables ξ_i are independent and their sum is η, so we get

$$D^2(\eta) = \sum_{i=1}^{k} D^2(\xi_i) = \frac{1}{4} \sum_{i=1}^{k} a_i^2 < \frac{kn^2}{4}.$$

We now apply Chebyshev's inequality (12.1.8) with $c = 2$ for $E(\eta) = Z/2$ and $D(\eta) < n\sqrt{k}/2$. We obtain that at least 75% of the 2^k (distinct) sums u_j are in the interval with center $Z/2$ and of length $2n\sqrt{k}$. Therefore

$$(12.1.9) \qquad \frac{3 \cdot 2^k}{4} \leq 2n\sqrt{k} \quad \text{or} \quad 2^k \leq \frac{8n\sqrt{k}}{3}$$

(compared to the similar estimate in (12.1.4), the factor k on the right-hand side has changed to $\sqrt{k}$).

Taking logarithms in (12.1.9), we obtain

$$(12.1.10) \qquad k < \log_2 n + \frac{\log_2 k}{2} + \log_2 \left(\frac{8}{3}\right).$$

Inequality (12.1.10) clearly implies (12.1.6) (for $n > 8$), thus also (12.1.7) is valid, which substituted into (12.1.10) gives us (12.1.2). $\square$

Sets with distinct subset sums give rise to another interesting problem of Erdős:

Theorem 12.1.2. *If all sums formed from distinct integers* $1 \leq a_1 < a_2 < \cdots < a_k$ *are distinct, then*

$$(12.1.11) \qquad \sum_{i=1}^{k} \frac{1}{a_j} < 2. \qquad \clubsuit$$

The example of powers of two shows that 2 cannot be replaced by a smaller number on the right-hand side of (12.1.11) (if there is no bound on k). For k fixed, the maximal sum of reciprocals is attained exactly if we take the first k powers of 2 $(1, 2, 4, \ldots 2^{k-1})$; this will be clear from the second and third proofs. If we allow also infinite sets then the theorem remains valid in the form that the sum of reciprocals is less than or equal to 2, and equality holds only in case we take all powers of two. This result can be verified by a suitable modification of any of the proofs below.

The statement of Theorem 12.1.2 was conjectured by Erdős, and was first proved by Ryavec using a series of ingenious tricks (see the first proof). This proof, however, relies quite strongly on analysis, and it is hard to see why it works. Many years later two further proofs were given that use only high school mathematics and their ideas (differing also from each other) are very natural (see the second and third proofs created by Bruen and Borwein, and Peter Frenkel; the third proof was found by Frenkel when he was still a high school student). This example shows that in combinatorial number theory it is sometimes possible to achieve new results using completely elementary methods.

The first proof is the most difficult, but besides keeping the chronological order, it is worth to wade through this argument first to enjoy the natural beauty of the second and third proofs even better.

First proof. Consider the product

$$(12.1.12) \qquad (1 + x^{a_1})(1 + x^{a_2}) \ldots (1 + x^{a_k}).$$

Performing the multiplication, we obtain terms x^m where m is the sum of some distinct exponents a_i (here $1 = x^0$ represents the empty sum). According to the assumption, all terms x^m are distinct, hence for $0 < x < 1$ the product (12.1.12) is less than the sum of the infinite geometric series

$$1 + x + x^2 + \cdots + x^n + \cdots = \frac{1}{1-x},$$

so

$$(12.1.13) \qquad (1 + x^{a_1})(1 + x^{a_2}) \ldots (1 + x^{a_k}) < \frac{1}{1-x}, \quad \text{if} \quad 0 < x < 1.$$

Now we apply the following trick: Take the (natural) log of both sides, divide by x, and integrate from 0 to 1:

$$(12.1.14) \qquad \sum_{i=1}^{k} \int_0^1 \frac{\log(1 + x^{a_i})}{x} \, dx < - \int_0^1 \frac{\log(1 - x)}{x} \, dx.$$

We make a substitution in the integrals on the left-hand side:

$$x^{a_i} = y, \quad \text{then} \quad dy = a_i x^{a_i - 1} dx, \quad \text{hence} \quad dx = \frac{dy}{a_i x^{a_i - 1}},$$

and thus

$$(12.1.15) \qquad \int_0^1 \frac{\log(1 + x^{a_i})}{x} \, dx = \int_0^1 \frac{\log(1 + y)}{x a_i x^{a_i - 1}} \, dy = \frac{1}{a_i} \int_0^1 \frac{\log(1 + y)}{y} \, dy.$$

Using (12.1.15), we can rewrite (12.1.14) as

$$(12.1.16) \qquad \left(\sum_{i=1}^{k} \frac{1}{a_i} \right) \int_0^1 \frac{\log(1 + y)}{y} \, dy < - \int_0^1 \frac{\log(1 - x)}{x} \, dx.$$

To complete the proof, we will show that the integral $A = \int_0^1 \frac{\log(1+x)}{x} \, dx$ on the left-hand side is half of $B = - \int_0^1 \frac{\log(1-x)}{x} \, dx$ on the right-hand side. Taking

$$A - B = \int_0^1 \left(\frac{\log(1 + x)}{x} + \frac{\log(1 - x)}{x} \right) dx = \int_0^1 \frac{\log(1 - x^2)}{x} \, dx,$$

and substituting $t = x^2$, $dt = 2x dx$, we obtain

$$A - B = \int_0^1 \frac{\log(1 - t)}{x \cdot 2x} \, dt = \frac{1}{2} \int_0^1 \frac{\log(1 - t)}{t} \, dt = -\frac{1}{2} B, \quad \text{so} \quad A = \frac{B}{2}. \qquad \square$$

Remark: We can also finish the proof by computing the integrals in (12.1.16); we expand the integrands into power series and integrate term by term (which is allowed given our present conditions):

$$\frac{-\log(1 - x)}{x} = 1 + \frac{x}{2} + \frac{x^2}{3} + \cdots + \frac{x^{j-1}}{j} + \ldots,$$

thus

$$(12.1.17) \quad B = - \int_0^1 \frac{\log(1 - x)}{x} \, dx = \left[x + \frac{x^2}{4} + \frac{x^3}{9} + \cdots + \frac{x^j}{j^2} + \ldots \right]_0^1 = \sum_{j=1}^{\infty} \frac{1}{j^2} = \frac{\pi^2}{6}.$$

Similarly,

$$A = \int_0^1 \frac{\log(1+x)}{x}\, dx = \left[x - \frac{x^2}{4} + \frac{x^3}{9} - \cdots + (-1)^{j+1}\frac{x^j}{j^2} + \cdots \right]_0^1 =$$

(12.1.18)

$$= \sum_{j=1}^{\infty} (-1)^{j+1}\frac{1}{j^2} = \sum_{j=1}^{\infty} \frac{1}{j^2} - 2\sum_{t=1}^{\infty} \frac{1}{(2t)^2} = \left(1 - \frac{2}{4}\right)\sum_{j=1}^{\infty} \frac{1}{j^2} = \frac{\pi^2}{12}.$$

So $A = B/2$.

Second proof. According to the condition, the $2^i - 1$ non-empty sums formed from the numbers $a_1, a_2, \ldots, a_i$ give distinct positive integers for every i, $1 \le i \le k$, hence the largest of these sums is at least $2^i - 1$, so

(12.1.19) $a_1 + a_2 + \cdots + a_i \ge 2^i - 1, \qquad i = 1, 2, \ldots, k.$

Introducing the notation

$$b_i = 2^{i-1}, \qquad i = 1, 2, \ldots, k,$$

we can rewrite (12.1.19) as

(12.1.20) $a_1 + a_2 + \cdots + a_i \ge b_1 + b_2 + \cdots + b_i, \qquad i = 1, 2, \ldots, k.$

To prove our theorem, it is sufficient to show that

(12.1.21) $$\frac{1}{a_1} + \cdots + \frac{1}{a_k} \le \frac{1}{b_1} + \cdots + \frac{1}{b_k}$$

since the right-hand side of (12.1.21) is

$$1 + \frac{1}{2} + \frac{1}{4} + \cdots + \frac{1}{2^{k-1}} = 2 - \frac{1}{2^{k-1}} < 2.$$

We shall prove the (stronger) statement that (12.1.21) holds with equality only if $a_i = b_i$, $i = 1, 2, \ldots, k$, hence the maximal sum of reciprocals is obtained for $a_i = 2^{i-1}$.

We show that (12.1.20) implies (12.1.21) for any real numbers

(12.1.22) $0 < a_1 < a_2 < \cdots < a_k, \qquad 0 < b_1 < b_2 < \cdots < b_k.$

Rearranging (12.1.21) and (12.1.20), we have to verify the inequality

(12.1.21a) $$\frac{1}{b_1} - \frac{1}{a_1} + \frac{1}{b_2} - \frac{1}{a_2} + \cdots + \frac{1}{b_k} - \frac{1}{a_k} \ge 0$$

assuming (12.1.22) and

(12.1.20a) $c_i = a_1 - b_1 + a_2 - b_2 + \cdots + a_i - b_i \ge 0, \quad i = 1, 2, \ldots, k.$

We can transform the left-hand side of (12.1.21a) to get (in Steps 2 and 3 we apply the so-called Abelian summation)

(12.1.23)

$$\frac{1}{b_1} - \frac{1}{a_1} + \frac{1}{b_2} - \frac{1}{a_2} + \cdots + \frac{1}{b_k} - \frac{1}{a_k} = \frac{a_1 - b_1}{a_1 b_1} + \frac{a_2 - b_2}{a_2 b_2} + \cdots + \frac{a_k - b_k}{a_k b_k}$$

$$= \frac{c_1}{a_1 b_1} + \frac{c_2 - c_1}{a_2 b_2} + \cdots + \frac{c_k - c_{k-1}}{a_k b_k}$$

$$= c_1 \left(\frac{1}{a_1 b_1} - \frac{1}{a_2 b_2} \right) + c_2 \left(\frac{1}{a_2 b_2} - \frac{1}{a_3 b_3} \right) + \cdots$$

$$+ c_{k-1} \left(\frac{1}{a_{k-1} b_{k-1}} - \frac{1}{a_k b_k} \right) + \frac{c_k}{a_k b_k}.$$

In the sum obtained at the end of (12.1.23), the numbers $c_i \geq 0$ are multiplied by positive numbers according to (12.1.20a) and (12.1.22), so this sum is non-negative, as claimed.

We obtained also that we have equality in (12.1.21) if and only if every $c_i = 0$, which implies by (12.1.20a) that $a_i = b_i$ for every i. This means that if all sums are distinct then the maximal sum of reciprocals is attained for $a_i = 2^{i-1}$, as indicated. $\square$

Third proof. We shall use only (12.1.19), established in the beginning of the second proof, and will show that if it holds for positive integers $a_1 < a_2 < \cdots < a_k$, then the sum of reciprocals is less than 2.

If for every i we have equality in (12.1.19), then $a_i = 2^{i-1}$, and the sum of reciprocals is $2 - 1/2^{k-1} < 2$.

If we do not have equality in (12.1.19) for every i, then modifying one or two values of a_i we shall increase the sum of reciprocals whereas (12.1.19) remains valid. It will be clear from the process that in finitely many steps we shall have equality in (12.1.19) for every i. This completes the proof that the sum of reciprocals is maximal for $a_i = 2^{i-1}$ (which is somewhat stronger than the original assertion of the theorem).

Let r be the smallest number for which we have strict inequality in (12.1.19) ($r = 1$ is possible), so

(12.1.24)
$$a_1 + a_2 + \cdots + a_i = 2^i - 1, \qquad i = 1, 2, \ldots, r - 1, \quad \text{and}$$
$$a_1 + a_2 + \cdots + a_r > 2^r - 1.$$

We distinguish two cases: (A) We have strict inequality in (12.1.19) for every $i > r$ and (B) There exists an $i > r$, for which (12.1.19) holds with equality.

(A) We put $a'_r = a_r - 1$ and the other integers a_i remain unchanged. The sum of reciprocals is clearly larger (since $1/a'_r > 1/a_r$), but (12.1.19) remained valid, as the left-hand side of (12.1.19) decreased by 1 for $i \geq r$, so the inequality is preserved (possibly with $\geq$ instead of $>$).

We have to show that our new numbers form a positive increasing sequence. For $r = 1$, we have $a_1 > 1$ from (12.1.24), hence $a'_1 > 0$. For $r > 1$, we have to exhibit $a'_r = a_r - 1 > a_{r-1}$, so $a_r \geq a_{r-1} + 2$. Using (12.1.24) again,

$$a_r = (a_1 + \cdots + a_r) - (a_1 + \cdots + a_{r-1}) \geq 2^r - (2^{r-1} - 1) = (2^{r-1} - 1) + 2 = a_{r-1} + 2.$$

(B) Let s be the smallest number greater than r for which we have equality in (12.1.19) ($s = r + 1$ is possible), so

$$
(12.1.25) \quad
\begin{aligned}
a_1 + a_2 + \cdots + a_i &> 2^i - 1, \qquad i = r, r+1, \ldots, s-1, \quad \text{and} \\
a_1 + a_2 + \cdots + a_s &= 2^s - 1.
\end{aligned}
$$

Put $a'_r = a_r - 1$, $a'_s = a_s + 1$, and let the other integers a_i be unchanged. Then (12.1.19) is still valid, because for $r \le i \le s - 1$ the left-hand side of (12.1.19) became smaller by 1, hence the inequality still holds (some $>$ may be replaced by $\ge$), and for $i \ge s$ (and also for $i < r$) the left-hand side of (12.1.19) was not affected.

We show that the sum of reciprocals has increased, so

$$
\frac{1}{a_r} + \frac{1}{a_s} < \frac{1}{a'_r} + \frac{1}{a'_s},
$$

or

$$
\frac{a_r + a_s}{a_r a_s} < \frac{(a_r - 1) + (a_s + 1)}{(a_r - 1)(a_s + 1)}.
$$

As the numerators are equal, this is equivalent to the converse inequality for the denominators (all occurring numbers are positive), and a calculation gives $a_r - 1 < a_s$ which clearly holds.

Finally, we can show that our numbers form a positive, strictly increasing sequence in a similar way as seen in case (A).

It is clear from the algorithm that applying the steps a finite number of times we get that there should be equality in (12.1.19) also for $i = r$. Then we repeat the whole process with the first value $i > r$ for which there is a strict inequality in (12.1.19) till we get equality for this value i. This proves that in finitely many steps we arrive at the state when we have equality everywhere in (12.1.19), as stated.   □

Exercises 12.1

In the exercises $1 \le a_1 < a_2 < \cdots < a_k \le n$ denote integers satisfying various conditions.

1. (a) Find the maximum of k (in terms of n), if no a_i is the sum of (more than one) distinct integers a_j.

    * (b) Let $a_1 < a_2 < \ldots$ be an infinite sequence of positive integers such that no a_i is the sum of (more than one) distinct integers a_j. Let $A(n)$ denote the number of elements in the sequence not exceeding n. Prove $\lim_{n \to \infty} A(n)/n = 0$.

2. Assume that no a_i can be written as $a_j + a_{j+1}$. Let $f(n)$ be the maximum of k with this condition. Show that $\lim_{n \to \infty} f(n)/n = 2/3$.

S 3. We examine the number of representations of an integer t as a sum of consecutive elements a_i, i.e in the form $t = a_i + a_{i+1} + \cdots + a_j$ (there is no restriction on the number of terms and we allow $i = j$). Let $L(k)$ be the maximal number of solutions of the equation $t = a_i + a_{i+1} + \cdots + a_j$ taken for all possible systems a_i and t (also n can be arbitrary). Verify $L(k) = \lceil k/2 \rceil$.

4. Assume $[a_i, a_j] > n$ for every $i \neq j$ (where [ ] stands for the least common multiple). Prove that the sum of reciprocals of the numbers a_i is less than (a) 2 (b) 3/2.

   Remark: Schinzel and Szekeres showed that the maximal sum of reciprocals is 31/30, and this occurs only for the numbers 2, 3, 5, and $n = 5$.

5. Show $\displaystyle\sum_{i=1}^{k-1} \frac{1}{[a_i, a_{i+1}]} < 1$ (for any integers a_i).

6. Assume that $a_i + a_j$ is never a square. Let $g(n)$ be the maximum of k with this condition.

   (a) Verify
   $$\frac{1}{3} \leq \liminf_{n\to\infty} \frac{g(n)}{n} \quad \text{and} \quad \limsup_{n\to\infty} \frac{g(n)}{n} \leq \frac{1}{2}.$$

   * (b) Improve the lower bound to 11/32 in the previous inequality.

   Remark: In 2002 Endre Szemerédi proved that $\lim_{n\to\infty} g(n)/n = 11/32$.

* 7. Assume that $a_i - a_j$ is never a square (for $i \neq j$). Let $h(n)$ be the maximum of k with this condition. Verify $h(n) \geq n^{0.7}$ if n is large enough.

   Remark: This result is due to Ruzsa. Sárközy and Fürstenberg proved that (in contrast to the sum problem in the previous exercise) $\lim_{n\to\infty} h(n)/n = 0$, but the exact order of magnitude of $h(n)$ in not known.

* 8. Assume that the products formed from arbitrarily many (distinct) numbers a_i are distinct. Let $s(n)$ be the maximum of k with this condition. Prove
   $$|s(n) - \pi(n)| < 2n^{2/3},$$
   where $\pi(n)$ is the number of primes not exceeding n.

   Remark: Erdős proved that there exist positive constants c_1 and c_2 such that for every n large enough
   $$\pi(n) + c_1 \frac{\sqrt{n}}{\log n} < s(n) < \pi(n) + c_2 \frac{\sqrt{n}}{\log n}.$$

   A related question is a multiplicative variant of the (additive) Sidon problem (to be investigated in the next section), when we require only products $a_i a_j$ composed of two factors ($i < j$) to be distinct. Erdős showed that (for suitable positive constants c_3 and c_4, and for every n large enough)
   $$\pi(n) + c_3 \frac{n^{3/4}}{(\log n)^{3/2}} < \max k < \pi(n) + c_4 \frac{n^{3/4}}{(\log n)^{3/2}}.$$

9. Determine the maximum of k (in terms of n) if no a_i divides the product of some other integers a_j.

10. Assume $6 \mid n$. Find the maximum of k (in terms of n) if among any three a_i there are two numbers that are not coprime.

11. Show that if k is a prime, then $\dfrac{a_i}{(a_i, a_j)} \geq k$ for some i and j.

    Remark: The result is true also for every k. This long-standing unsolved conjecture of R. L. Graham was proved (for k large enough) by Mario Szegedy in 1985, when he was still a university student.

12. Assume that for $n = 2^j$ there exist $k = 2 + \lfloor \log_2 n \rfloor$ integers a_i between 1 and n so that all subset sums are distinct. Show that the same holds also for every $n \geq 2^j$.

13. How far can be improve the upper bound (12.1.2) in Theorem 12.1.1 if we use an optimal c in Chebyshev's inequality in the proof?

12.2. Sidon Sets

In this section we deal with Sidon sets, another favorite of Erdős. Sidon sets are finite or infinite sequences $a_1 < a_2 < \cdots$ of positive integers where the sums $a_i + a_j$, $i \leq j$ (or equivalently, the differences $a_i - a_j$, $i \neq j$) are all distinct. These occurred first in Simon Sidon's investigations of Fourier series around 1930.

We consider finite Sidon sets first.

At most how many elements can a Sidon set have in the interval $[1, n]$? We shall prove that the maximum is about $\sqrt{n}$. This contains two statements: On the one hand, there exists a Sidon set between 1 and n that has about $\sqrt{n}$ elements (a lower estimate for the maximum), and on the other hand, no Sidon set can have substantially more elements within these limits (an upper estimate for the maximum).

Erdős and Turán showed in 1941 that the maximum is at most $n^{1/2} + 2n^{1/4}$. Later Lindström improved this to $n^{1/2} + n^{1/4} + 1$ by a different method, but it also follows from a more precise execution of the Erdős–Turán proof (Theorem 12.2.4). Relying on a result of J. Singer, Erdős and S. Chowla proved independently in 1944 that there exists a Sidon set of size $n^{1/2} - n^\varrho$ for a suitable positive constant $\varrho < 1/2$ (Theorem 12.2.3). These two theorems together mean that the maximal size of a Sidon set in the interval $[1, n]$ is asymptotically $\sqrt{n}$ with very good (upper and lower) error terms. It is a much stronger conjecture that the difference of the maximum and $\sqrt{n}$ is bounded (independently of n). For a proof or disproof Erdős offered 1000 US dollars, but there has been no improvement in the last 70 years.

Let $s = s(n)$ denote the maximal possible number of a Sidon set up to n. We first establish some simple upper bound for s. There are $\binom{s+1}{2}$ sums $a_i + a_j$ that are distinct and fall between 2 and $2n$, hence $\binom{s+1}{2} < 2n$, and so $s < 2\sqrt{n}$. We get a better upper bound by considering the differences $a_i - a_j > 0$; these $\binom{s}{2}$ numbers are distinct and they are less than n, hence $\binom{s}{2} < n$, and so $s < \sqrt{2n} + 1$. Thus we attained immediately the order of magnitude $\sqrt{n}$, and only the coefficient $\sqrt{2}$ of $\sqrt{n}$ has to be reduced to 1.

In the opposite direction, it is much less clear how we can get the order of magnitude $\sqrt{n}$. The example of powers of two yields $\log_2 n$, and the greedy algorithm guarantees only $\sqrt[3]{n}$ (see Exercise 12.2.1). But a nice elementary construction of Erdős provides $\sqrt{n/2}$ elements (see Exercise 12.2.2), and as mentioned, we can lift the coefficient of $\sqrt{n}$ to 1.

Let us start constructing really big Sidon sets. We do this for some special n first, and then use the result to handle general n.

Theorem 12.2.1. *Let p be an arbitrary positive prime and $n = p^2 + p + 1$. There exists a Sidon set in the interval $[1, n]$ that has $\lceil \sqrt{n} \rceil = p + 1$ elements.* ♣

Theorem 12.2.1 is a consequence of a surprising and much sharper statement of independent interest.

Theorem 12.2.2. *Let p be a positive prime. Then there exist $p + 1$ integers a_i such that the differences $a_i - a_j$, $i \neq j$ are pairwise incongruent modulo $p^2 + p + 1$.* ♣

Remark: The number of differences in Theorem 12.2.2 is $p^2 + p$, and there are just that many non-zero residues modulo $p^2 + p + 1$. This means that each non-zero residue has exactly one representation as a difference $a_i - a_j$.

It is clear that the integers a_i in Theorem 12.2.2 must be pairwise incongruent themselves, so they can be chosen between 1 and $n = p^2 + p + 1$. Thus Theorem 12.2.1 follows.

Proof. We use some basic facts about finite fields and a bit of linear algebra.

Consider the finite field F_3 of p^3 elements and its subfield F_1 of p elements. Let Δ be a generator of the cyclic multiplicative group of F_3, so

$$(12.2.1) \qquad F_3 = \{0, \Delta, \Delta^2, \ldots, \Delta^{p^3-1} = 1\}.$$

The non-zero elements of F_1 form a subgroup of the multiplicative group of F_3. This cyclic subgroup is generated by Δ^n, where $n = (p^3 - 1)/(p - 1) = p^2 + p + 1$. Thus

$$F_1 = \{0, \Delta^n, \Delta^{2n}, \ldots, \Delta^{(p-1)n} = \Delta^{p^3-1} = 1\}.$$

Consider F_3 as a vector space over F_1. By the above, Δ^i and Δ^j in F_3 are linearly dependent over F_1 if and only if

$$(12.2.2) \qquad i \equiv j \pmod{n}.$$

We now construct the integers a_i. We fix $\Theta \in F_3 \setminus F_1$, and form the elements $\Theta + \gamma_i$, where $\gamma_1, \ldots, \gamma_p$ are the elements of F_1. By (12.2.1), we can write

$$(12.2.3) \qquad \Theta + \gamma_i = \Delta^{a_i},$$

and so we obtained p integers a_i, $1 \leq i \leq p$. Let $a_{p+1} = 0$.

We verify that these numbers meet the requirements, so the differences $a_i - a_j$, or equivalently the sums $a_i + a_j$ are pairwise incongruent modulo $p^2 + p + 1$.

Assume $a_i + a_j \equiv a_k + a_m \pmod{p^2 + p + 1}$. If none of the four numbers is $a_{p+1} = 0$, then by (12.2.2) and (12.2.3), this gives

$$\Delta^{a_i} \Delta^{a_j} = \Delta^{a_i + a_j} = \gamma \Delta^{a_k + a_m} = \gamma \Delta^{a_k} \Delta^{a_m},$$

so

$$(\Theta + \gamma_i)(\Theta + \gamma_j) - \gamma(\Theta + \gamma_k)(\Theta + \gamma_m) = 0$$

for some $\gamma \in F_1$. Since the degree of Θ is 3 over F_1, it cannot be a root of a polynomial of degree at most 2. Therefore only $\gamma = 1$ and $\{\gamma_i, \gamma_j\} = \{\gamma_k, \gamma_m\}$ are possible which means that the corresponding pairs of integers a_i are the same, as stated.

The proof runs the same way if $a_{p+1} = 0$ occurs among the four integers a_i. $\square$

Remark: Theorem 12.2.2 and its proof remain valid if p is a prime power. All this is closely related to finite projective planes.

Theorem 12.2.3. *If n is large enough, then the interval $[1, n]$ contains a Sidon set having at least $n^{1/2} - n^{0.27}$ elements.* ♣

Proof. Consider the biggest prime p less than or equal to $p^2 + p + 1 \leq n$ and perform the previous construction of $p + 1$ elements for $p^2 + p + 1$. By Theorem 5.5.4(A), there is a prime between $n^{1/2} - n^{0.27}$ and $n^{1/2}$ if n is large enough, so $p > n^{1/2} - n^{0.27}$, thus verifying the theorem for a general n. $\square$

Remark: In the transition to an arbitrary n, we used that the primes occur densely. If we know that there is a prime between N and $N + N^c$ for N large enough, then the error term in our theorem can be reduced to $n^{c/2}$. As we experienced in Section 5.5, the question of the size of the gaps between consecutive primes is a very hard problem.

For other proofs of Theorem 12.2.3, see Exercises 12.2.3 and 12.2.4.

Now we turn to the sharp upper bound of the size of Sidon sets.

Theorem 12.2.4. *A Sidon set in the interval $[1, n]$ has at most $n^{1/2} + n^{1/4} + 1$ elements.* ♣

First proof. Let t be an integer to be specified later. We push a segment of length $t - 1$ through the interval $[0, n]$, i.e. we consider the intervals $[-t + 1, 0]$, $[-t + 2, 1]$, ..., $[n, n + t - 1]$. We take a Sidon set of size s and denote the number of its elements in these intervals by $A_1, A_2, \ldots, A_{n+t}$. Then

$$(12.2.4) \qquad \sum_{i=1}^{n+t} A_i = ts.$$

We now count the pairs $\{a_i, a_j\}$, $i > j$ that fall into such an interval. We count them with the suitable multiplicity, so each pair is counted as many times as the number of intervals that contain it. Let D be the total number of such pairs. Then

$$(12.2.5) \qquad D = \sum_{i=1}^{n+t} \binom{A_i}{2} = \sum_{i=1}^{n+t} \frac{A_i^2}{2} - \sum_{i=1}^{n+t} \frac{A_i}{2}.$$

On the other hand, if the difference $a_i - a_j$ in a pair is d, then it falls into exactly $t - d$ intervals. By the Sidon property, every such d can occur at most once, so

$$(12.2.6) \qquad D \leq \sum_{d=1}^{t-1} (t - d) = \frac{t(t-1)}{2}.$$

Combining (12.2.5) and (12.2.6), we obtain

$$\sum_{i=1}^{n+t} A_i^2 - \sum_{i=1}^{n+t} A_i \le t(t-1). \tag{12.2.7}$$

Using (12.2.4) and the inequality for arithmetic and quadratic means, we can estimate the left-hand side of (12.2.7) from below:

$$\sum_{i=1}^{n+t} A_i^2 - \sum_{i=1}^{n+t} A_i \ge \frac{\left(\sum_{i=1}^{n+t} A_i\right)^2}{n+t} - ts = \frac{t^2 s^2}{n+t} - ts. \tag{12.2.8}$$

From (12.2.7) and (12.2.8), we infer

$$s^2 - s\left(\frac{n}{t} + 1\right) - \left(\frac{n}{t} + 1\right)(t-1) \le 0.$$

Solving this quadratic inequality, we get

$$s \le \frac{n}{2t} + \frac{1}{2} + \sqrt{n + t + \frac{n^2}{4t^2} - \frac{n}{2t} - \frac{3}{4}}.$$

Choosing $t = \lfloor n^{3/4} \rfloor + 1$, we arrive at the statement of the theorem. $\square$

Second proof. We shall estimate the sum of certain differences $a_i - a_j$ from both sides. Let

$$K = \sum_{0 < i-j \le r} (a_i - a_j), \tag{12.2.9}$$

where r will be specified later. The sum in (12.2.9) contains

$$(s-1) + (s-2) + \cdots + (s-r) = rs - \frac{r(r+1)}{2} = rw$$

terms, where

$$w = s - \frac{r+1}{2}. \tag{12.2.10}$$

Each term is a difference $a_i - a_j$, and these are distinct by the Sidon property. Thus K is not less than the sum of the first rw positive integers, so

$$K \ge \frac{rw(rw+1)}{2} > \frac{r^2 w^2}{2}. \tag{12.2.11}$$

On the other hand, the sum in (12.2.9) contains e.g.

$$(a_s - a_{s-1}) + (a_{s-1} - a_{s-2}) + \cdots + (a_2 - a_1) < a_s \le n,$$

and many other telescoping sums, which can be estimated from above similarly. Their general form is

$$(a_{s-\nu} - a_{s-\nu-\mu}) + (a_{s-\nu-\mu} - a_{s-\nu-2\mu}) + \cdots < a_{s-\nu} \le n, \quad 0 \le \nu < \mu \le r.$$

Moreover, the entire K consists of such telescoping sums; the indices cover all maximal arithmetic progressions between 1 and s where the difference is at most r. There are μ arithmetic progressions with difference μ, so there are $1 + 2 + \cdots + r = r(r+1)/2$ telescoping sums. Each such telescoping sum is not greater than n, hence

$$K \le \frac{nr(r+1)}{2}. \tag{12.2.12}$$

Combining (12.2.11) and (12.2.12) and multiplying by $2/r^2$, we obtain the inequality $w^2 < n + n/r$. Taking a square root and substituting (12.2.10), we get

$$s < \frac{r+1}{2} + \sqrt{n + \frac{n}{r}}.$$

Choosing $r = \lfloor n^{1/4} \rfloor + 1$, we arrive at the statement of the theorem. $\qquad\square$

Now we turn to infinite Sidon sets. Erdős showed in 1955 that an infinite Sidon set is necessarily less dense; it cannot be of about the maximal finite size, i.e. $\sqrt{n}$ in interval $[1, n]$ for every n:

Theorem 12.2.5. *Let $A(n)$ denote the number of elements in an infinite sequence A up to n. If A is an infinite Sidon set, then*

$$\liminf_{n\to\infty} \frac{A(n)}{\sqrt{n}} = 0, \quad moreover \quad \liminf_{n\to\infty} \frac{A(n)}{\sqrt{n/\log n}} < \infty. \qquad\clubsuit$$

Proof. Let N be a large integer and A_i the number of elements of A in the interval $[(i-1)N + 1, iN]$, so

$$A_i = A(iN) - A((i-1)N), \quad i = 1, 2, \ldots, N.$$

There are altogether $\sum_{i=1}^{N} \binom{A_i}{2}$ pairs of points in the intervals and their differences are all distinct and do not exceed the length $N - 1$ of an interval, therefore

$$\sum_{i=1}^{N} \binom{A_i}{2} < N.$$

Hence

$$2N > \sum_{i=1}^{N} A_i(A_i - 1) \geq \frac{1}{2} \sum_{i=1}^{N} (A_i^2 - 1),$$

so

(12.2.13)
$$\sum_{i=1}^{N} A_i^2 < 5N.$$

We shall estimate

$$S = \sum_{i=1}^{N} \frac{A_i}{\sqrt{i}}$$

from both directions. On the one hand, applying Cauchy's inequality and (12.2.13), we obtain

(12.2.14)
$$S \leq \sqrt{\left(\sum_{i=1}^{N} A_i^2\right)\left(\sum_{i=1}^{N} \frac{1}{i}\right)} \approx \sqrt{5N \log N}.$$

On the other hand, using Abelian summation, we can estimate S as

(12.2.15)
$$S = \sum_{i=1}^{N} \frac{A(iN) - A((i-1)N)}{i} >$$
$$> \sum_{i=1}^{N-1} A(iN)\left(\frac{1}{\sqrt{i}} - \frac{1}{\sqrt{i+1}}\right) > \sum_{i=1}^{N-1} \frac{A(iN)}{2(i+1)\sqrt{i}}.$$

Assuming now

(12.2.16)
$$A(iN) > c\sqrt{\frac{iN}{\log(iN)}}, \quad i = 1, 2, \dots, N,$$

for some $c > 0$, then (12.1.15) implies

(12.2.17)
$$S > c\sum_{i=1}^{N-1} \frac{\sqrt{iN}}{2(i+1)\sqrt{i\log N^2}} = \frac{c\sqrt{N}}{\sqrt{8\log N}} \sum_{i=1}^{N-1} \frac{1}{i+1} \approx \frac{c}{\sqrt{8}}\sqrt{N\log N}.$$

Since (12.1.17) contradicts (12.1.14) for $c > \sqrt{40}$, (12.1.16) cannot hold for $c > \sqrt{40}$, which proves the statement of the theorem. $\qquad\square$

Theorem 12.2.5 does not assert that an infinite Sidon set could not be now and then as dense as a finite one. In fact, Erdős and later Krückeberg constructed an infinite Sidon set that has nearly $\sqrt{n}$ elements in the interval $[1, n]$ for infinitely many integers n (see Exercise 12.2.5).

If we want to construct an infinite Sidon set that is sufficiently dense in every finite initial segment, then the greedy algorithm provides one having always at least $\sqrt[3]{n}$ elements up to n (see Exercise 12.2.1). It is surprising that it took a long time to surpass this order of magnitude: Ajtai, Komlós, and Szemerédi proved in 1981 the existence of an infinite Sidon set that has at least $c\sqrt[3]{n\log n}$ elements up to every (sufficiently large) n with a suitable positive constant c. Even this was just slightly better than $\sqrt[3]{n}$ obtained by the greedy algorithm. In 1997 Ruzsa improved the bound significantly to $cn^{\sqrt{2}-1-\varepsilon}$, though even this is far from the order of magnitude $n^{1/2-\varepsilon}$ conjectured by Erdős (where ε is an arbitrarily small positive number).

Finally, we consider infinite sequences where the Sidon property is replaced by a weaker condition: the number of representations of positive integers as $a_i + a_j$ is bounded (this bound is 1 for Sidon sets). We show that we can achieve the order of magnitude $n^{1/2-\varepsilon}$ for such sequences:

Theorem 12.2.6. *For every $\varepsilon > 0$ there exist an integer m and an infinite sequence $A = \{1 \le a_1 < a_2 < \cdots\}$ such that*
$$\liminf_{n\to\infty} \frac{A(n)}{n^{1/2-\varepsilon}} > 0,$$
and every positive integer has at most m representations in the form $a_i + a_j$. $\qquad\clubsuit$

Theorem 12.2.6 is due to Erdős and Rényi. Their proof was among the first probabilistic constructions in number theory: Introducing a suitable probability space on the set of sequences of positive integers, they verified that (with respect to this probability) almost all sequences meet the requirements. We shall use this type of argument to prove Theorem 12.6.3.

The elementary proof below to Theorem 12.2.6 is due to Ruzsa.

Proof. We shall use a number system with varying bases, so we write integers in the form

$$c_0 + c_1 k_1 + c_2 k_1 k_2 + \cdots + c_i k_1 \ldots k_i + \cdots,$$

where $k_1, k_2, \ldots$ are integers greater than 1 (these are the varying bases) and the digits are $0 \leq c_i < k_{i+1}$. We choose the bases as a slowly increasing sequence to satisfy

$$(12.2.18) \qquad\qquad k_{i+1} \approx k_i^{1+\delta}$$

for some small positive δ. We also fix a finite Sidon set S_i of maximal size between 0 and $k_i/2$ for every i (we may clearly assume that the smallest element is 0 in S_i).

We can construct the required infinite set as follows. We take integers with digits from the corresponding Sidon sets, i.e. $c_i \in S_{i+1}$, and at most t digits can differ from 0.

Adding two such numbers, there occurs no carrying, so every integer can be written as the sum of two such numbers in at most 2^t ways (at every place the digits can be interchanged between the two numbers), so $m = 2^t$. We shall guarantee the required density by a suitable choice of δ and t.

For any n

$$(12.2.19) \qquad\qquad k_1 k_2 \ldots k_j \leq n < k_1 k_2 \ldots k_j k_{j+1}$$

for some j. Our sequence definitely contains all the integers with digits

$$(12.2.20) \qquad c_0 = c_1 = \cdots = c_{j-t-1} = 0 \quad \text{and} \quad c_i \in S_{i+1}, i = j-t, \ldots, j-1.$$

We shall show that the number of these integers alone is greater than $n^{1/2-\varepsilon}$, if we choose a sufficiently small δ and a sufficiently large t.

Let us see the details. Let $k_1 = r$ and

$$(12.2.21) \qquad\qquad k_i = \left\lfloor r^{(1+\delta)^{i-1}} \right\rfloor$$

according to (12.2.18). Then

$$(12.2.22) \qquad |S_i| > \sqrt{\frac{k_i}{3}} > r^{h_i}, \quad \text{where} \quad h_i = \frac{(1+\delta)^{i-1} - \log_r 4}{2}.$$

Let K denote the number of integers satisfying (12.2.20). Then it is enough to show

$$(12.2.23) \qquad\qquad K > n^{1/2-\varepsilon}, \quad \text{or} \quad \log_r K > \left(\frac{1}{2} - \varepsilon\right) \log_r n$$

for every sufficiently large n. We estimate $\log_r n$ from above with the help of (12.2.19) and (12.2.21):

$$(12.2.24) \quad \log_r n < \log_r(k_1 k_2 \ldots k_{j+1}) \leq 1 + (1+\delta) + \cdots + (1+\delta)^j < \frac{(1+\delta)^{j+1}}{\delta}.$$

Now we estimate $\log_r K$ from below. Since $K = |S_{j-t+1}| \cdot \cdots \cdot |S_j|$, (12.2.22) yields

$$\log_r K > \frac{(1+\delta)^{j-t} + \cdots + (1+\delta)^{j-1} - t\log_r 4}{2}$$

(12.2.25)
$$= \frac{(1+\delta)^{j-t}((1+\delta)^t - 1)}{2\delta} - \frac{t\log_r 4}{2}$$

$$= \frac{(1+\delta)^j}{2\delta}(1 - (1+\delta)^{-t}) - \frac{t\log_r 4}{2}.$$

By (12.2.24) and (12.2.25),

$$\frac{2\log_r K}{\log_r n} > \frac{(1+\delta)^j(1 - (1+\delta)^{-t}) - t\delta\log_r 4}{(1+\delta)^{j+1}}$$

(12.2.26)
$$= \frac{1 - (1+\delta)^{-t}}{1+\delta} - \frac{t\delta\log_r 4}{(1+\delta)^{j+1}}.$$

Now we choose a sufficiently small δ, and then a sufficiently large t so that the first term in the last row of (12.2.26) is greater than $1-\varepsilon$. As δ and t are fixed, the numerator of the second term is a constant whereas the denominator tends to infinity with $j \to \infty$, hence the second term is less than ε if j (i.e. if n) is large enough. So the entire expression is greater than $1 - 2\varepsilon$, which proves (12.2.23). $\qquad\square$

Exercises 12.2

1. Show that the greedy algorithm yields a Sidon set between 1 and n of at least $\sqrt[3]{n}$ elements.

2. Let $p > 0$ be a prime and $a_i = 1 + 2ip + \langle i^2 \bmod p \rangle$, $i = 0, 1, \ldots, p - 1$, where $\langle i^2 \bmod p \rangle$ denotes the least non-negative residue of i^2 modulo p. Verify that this is a Sidon set in $[1, n]$ of size $\sqrt{n/2}$ for $n = 2p^2$.

S* 3. Let $p > 0$ be a prime. There exist p integers a_i such that the sums $a_i + a_j$, $i \le j$, are (not just distinct, but are) pairwise incongruent modulo $p^2 - 1$.

   Remark: An equivalent formulation is that the differences $a_i - a_j$, $i \ne j$, are (not just distinct, but are) pairwise incongruent modulo $p^2 - 1$. There are $p^2 - p$ such differences and $p^2 - 2$ non-zero residues modulo $p^2 - 1$. This means that nearly all residues can be represented as a difference $a_i - a_j$. We can see from the proof that the missing residues are the multiples of $p+1$. We can deduce Theorem 12.2.3 also from this exercise just as we did it from Theorem 12.2.2. (The same holds also for the next exercise.)

S* 4. Let $p > 0$ be prime. There exist $p - 1$ integers a_i such that the differences $a_i - a_j$, $i \ne j$, are (not just distinct, but are) pairwise incongruent modulo $p^2 - p$.

5. Construct a Sidon set satisfying $A(n) > (1/\sqrt{2}-\varepsilon)\sqrt{n}$ for every $\varepsilon > 0$ with infinitely many n (i.e. $\limsup_{n\to\infty} A(n)/\sqrt{n} \ge 1/\sqrt{2}$).

   Remark: It is unknown whether the same holds with 1 instead of $1/\sqrt{2}$.

6. *Sums of more terms.* Let $h \geq 2$ be a fixed integer, and consider sequences in the interval $[1, n]$ such that the h-fold sums are all distinct. (The Sidon sets are the special case $h = 2$.)

    * (a) Prove the existence of a sequence having about $n^{1/h}$ elements.

    (b) Show that there is a constant $c = c(h)$ depending only on h such that every sequence has at most $c(h)n^{1/h}$ elements.

Remark: It is an unsolved problem whether, similar to the Sidon sets, we can reduce $c(h)$ to $1 + \varepsilon$, i.e. the maximal size is asymptotically $n^{1/h}$ for $h > 2$. The proof of Theorem 12.2.4 does not work since we cannot switch sums to differences if $h \neq 2$.

7. Show that there exists an infinite sequence of integers $a_1 < a_2 < \cdots$ such that every non-zero integer has a unique representation as $a_i - a_j$.

8. Two infinite sequences of positive integers A and B form a good pair if the sums $a + b\ (a \in A, b \in B)$ are distinct. We get a good pair if we cut a Sidon set into two parts. Show that there exist denser good pairs, too: Construct a good pair such that both $A(n) > c\sqrt{n}$ and $B(n) > c\sqrt{n}$ for every n with a suitable constant $c > 0$.

12.3. Sumsets

In this section we deal with sets of the type $A + A = \{a_i + a_j \mid a_i, a_j \in A\}$, where the elements of A are either integers in the interval $[0, n - 1]$ or residue classes modulo p for some prime p. Let the number of elements in A be $|A| = k$.

The size of $A+A$ is maximal if A is a Sidon set when $|A+A| = \binom{k+1}{2}$. We examine now first the opposite extreme: What is the minimal value of $|A + A|$? If the elements of A are integers, then in conformity with expectations, the minimum occurs when A consists of consecutive terms of an arithmetic progression, so $\min |A + A| = 2k - 1$ (see Exercise 12.3.1). We get a similar result also if $A \subseteq \mathbf{Z}_p$ (so the elements of A are modulo p residue classes), we verify this (by no means obvious) fact in Theorem 12.3.1. This result was found by Cauchy, but was rediscovered 120 years later independently by Davenport and Chowla. We give two proofs and present several interesting applications of the theorem and method in Exercises 12.3.3–12.3.8.

Our second topic concerning sumsets is a dual of the Sidon property in a certain sense. For finite Sidon sets, the main goal was to find large sets A such that every integer has at most one representation in the form $a_i + a_j$. Now we are looking for small sets A such that every integer in the interval $[0, n - 1]$ has at least one representation in the form $a_i + a_j$. Sets A with this property are called (additive) *bases* (of second order). Theorem 12.3.3 provides lower and upper bounds for the minimal number of elements in a basis.

Let us turn to determine the minimum of $|A + A|$ if $A \subseteq \mathbf{Z}_p$. More generally, we shall find the minimal number of elements in sets $A + B = \{a + b \mid a \in A, b \in B\}$ as a function of $|A|$ and $|B|$. This is not just in order to have a more general result, but—as so often in mathematics—this generalization gives the key to the proof of the original statement.

Theorem 12.3.1 (Cauchy–Davenport–Chowla). *If p is a prime, $A, B \subseteq \mathbf{Z}_p$, $|A| = k(> 0)$, $|B| = r(> 0)$, then*

$$(12.3.1) \qquad\qquad |A + B| \geq \min(p, k + r - 1). \qquad\qquad \spadesuit$$

The upper bound p in (12.3.1) cannot be omitted, as $A + B \subseteq \mathbf{Z}_p$, and so $|A+B| \leq p$.

The inequality is sharp: If $A = \{0, 1, \ldots, k-1\}$ and $B = \{0, 1, \ldots, r-1\}$, then (assuming $k + r \leq p + 1$) $A + B = \{0, 1, \ldots, k + r - 2\}$, so $|A + B| = k + r - 1$ yielding equality in (12.3.1).

In the special case $A = B$, we obtain $|A + A| \geq \min(p, 2k - 1)$, and equality holds if, for example, $A = \{0, 1, \ldots, k-1\}$.

First proof. To get a contradiction, we assume that for some p there exist A and B for which (12.3.1) is false. Let us call such a pair of sets *ugly*.

We consider an ugly pair A, B with $|A| = k$, $|B| = r$, where r is minimal. We shall construct an ugly pair A', B' with $|A'| = k'$, $|B'| = r'$, and $r' < r$, which contradicts the minimality of r. This means that there cannot be ugly pairs.

If $k + r - 1 > p$, then delete $k + r - 1 - p(< r)$ elements from B, denote the remaining set by B', and let $A' = A$. Clearly,

$$|A' + B'| \leq |A + B| < \min(p, k + r - 1) = p = \min(p, k' + r' - 1),$$

so A', B' is an ugly pair and $(0 <)r' < r$, which is impossible. Therefore $k + r - 1 \leq p$.

Clearly $k \geq r \geq 2$, since if $k < r$, then interchanging the roles of A and B contradicts the minimality of r, and if $r = 1$, then (12.3.1) holds with equality, so A, B is not an ugly pair. As $r \geq 2$ and $k + r - 1 \leq p, k < p$.

We may assume $0 \in B$, since adding the same value to every element in B causes no changes in $|A|$, $|B|$, and $|A + B|$.

We show that if $b \neq 0$ is any fixed element in B, then $A + b = \{a + b \mid a \in A\} \not\subseteq A$. Otherwise, we have $A + b = A$, and so the sums of elements on the two sides are the same:

$$\sum_{a \in A} a = \sum_{a \in A} (a + b) = kb + \sum_{a \in A} a, \quad \text{so} \quad kb = 0,$$

which is impossible as $k < p$ and $b \neq 0$.

Thus there exist $a_1 \in A$ and $b_1 \in B$ such that $a_1 + b_1 \notin A$. Let

$$A' = A \cup \{a_1 + b \mid b \in B, a_1 + b \notin A\} \quad \text{and} \quad B' = \{b \mid a_1 + b \in A\}.$$

Then clearly $k' + r' = k + r$ and $0 < r' < r$ (since $0 \in B'$, but $b_1 \notin B'$). We show $A' + B' \subseteq A + B$. Let $a' + b' \in A' + B'$. If $a' \in A$, then $a' + b' \in A + B$. If $a' = a_1 + b$, then

$$a' + b' = (a_1 + b) + b' = (a_1 + b') + b \in A + B,$$

since $a_1 + b' \in A$ by the definition of B'. Therefore

$$|A' + B'| \leq |A + B| < \min(p, k + r - 1) = k + r - 1 = k' + r' - 1 = \min(p, k' + r' - 1),$$

so the pair A', B' is ugly, and $r' < r$, providing the contradiction. $\qquad\square$

For a second proof of Theorem 12.3.1, we need a lemma.

Lemma 12.3.2. *Let F be a commutative field, $A, B \subseteq F$, $|A| = k$, $|B| = r$, and $f(x, y)$ a polynomial in two variables over F such that its degrees with respect to x and y are less than k and r, (so $f(x, y) = \sum_{i<k, j<r} \alpha_{ij} x^i y^j$). Assume $f(a, b) = 0$ for every $a \in A$ and $b \in B$. Then f is the zero polynomial with every coefficient 0.* ♣

Proof. We write $f(x, y)$ as a polynomial in y, so the coefficients are polynomials in x:

$$(12.3.2) \qquad f(x, y) = h_0(x) + h_1(x)y + \cdots + h_{r-1}(x)y^{r-1}, \quad \deg h_i \le k - 1.$$

For $a \in A$, let

$$g_a(y) = f(a, y) = h_0(a) + h_1(a)y + \cdots + h_{r-1}(a)y^{r-1}.$$

Then $\deg g_a \le r - 1$, but $g_a(b) = f(a, b) = 0$ for every $b \in B$, so g_a has at least r roots. This is possible only if every coefficient of g_a is 0. This means that every $a \in A$ is a root of each polynomial h_i of degree at most $k - 1$, so each h_i has at least k roots, which implies $h_i = 0$ (i.e. every coefficient is 0). Therefore $f = 0$ by (12.3.2). □

Second proof of Theorem 12.3.1. We assume that (12.3.1) is false for some A and B. As in the first proof, we may restrict ourselves to $k + r - 1 \le p$ (where $|A| = k$ and $|B| = r$). Let $C = A + B$, so $|C| \le k + r - 2 < p$. Let

$$(12.3.3) \qquad f_1(x, y) = (x + y)^m \prod_{c \in C}(x + y - c), \quad \text{where } m = k + r - 2 - |C|.$$

Then $f_1(a, b) = 0$ for every $a \in A$ and $b \in B$.

We cannot apply Lemma 12.3.2 to $f_1(x, y)$ directly because it contains terms $x^i y^j$ with $i \ge k$ or $j \ge r$. Consider an x^i where $i \ge k$, and replace it with a polynomial $u_i(x)$ of degree at most $k - 1$ that has the same values as x^i in A, i.e. $u_i(a) = a^i$ for every $a \in A$. It is well known that such an interpolation polynomial $u_i(x)$ always exists and is unique. We proceed similarly if $j \ge r$. Then y^j is replaced by a polynomial $v_j(y)$ of degree at most $r - 1$ such that $v_j(b) = b^j$ for every $b \in B$.

Thus we obtain a polynomial $f(x, y)$ satisfying $f(a, b) = f_1(a, b) = 0$ for every $a \in A$ and $b \in B$, that contains only terms $x^i y^j$ with $i \le k - 1$ and $j \le r - 1$. By Lemma 12.3.2, every coefficient of f is 0.

We now compute the coefficient of $x^{k-1} y^{r-1}$ in f directly, and show that it is not 0, which gives the desired contradiction.

By (12.3.3), terms $x^i y^j$ in f_1 with $i + j = k + r - 2$ arise only from $(x + y)^{k+r-2}$ as we have $i + j < k + r - 2$ for every other term. In reducing f_1 to f, the terms x^i and y^j with $i \ge k$ and $j \ge r$, are replaced by polynomials of smaller degree. Therefore f has exactly one term $x^{k-1} y^{r-1}$ that is obtained from the expansion of $(x + y)^{k+r-2}$; its coefficient is $\binom{k+r-2}{k-1}$. Since $k + r - 2 < p$, this coefficient is not 0 (in $\mathbf{Z}_p$), as claimed. □

Now we turn to the second topic of the section. Repeating the definition, an additive basis of order 2 in $[0, n - 1]$ is a set A of non-negative integers such that every integer $0 \le r \le n - 1$ is the sum of two elements in A, i.e. $r = a_i + a_j$ ($a_i, a_j \in A$).

If $|A| = k$, then there are $\binom{k+1}{2}$ sums $a_i + a_j$, and if A is a basis, then there are at least n distinct integers among them, so

$$\binom{k+1}{2} \geq n, \quad \text{hence} \quad k > \sqrt{2n} - \frac{1}{2}.$$

On the other hand, if n is a square, $n = s^2$, then the integers less than n have (at most) two digits in the number system with base s, so they can be written in the form $i + sj$, where $0 \leq i, j \leq s - 1$. This means that

$$A = \{0, 1, \ldots, s - 1, s, 2s, \ldots, (s - 1)s\}$$

is a basis of second order having $2s = 2\sqrt{n}$ elements. If n is not a square, then we do the same for the smallest square greater than n, and so $s = \lceil \sqrt{n} \rceil$.

These observations yield estimates for the minimal size of a basis:

(12.3.4) $$\sqrt{2n} - \frac{1}{2} < \min k < 2\sqrt{n} + 2.$$

We show in the next theorem that the coefficients of $\sqrt{n}$ can be slightly improved in both bounds:

Theorem 12.3.3. *Let $f(n)$ denote the minimal number of additive bases of second order in $[0, n - 1]$. Then*

(12.3.5) $$\sqrt{\frac{289}{144}}\sqrt{n} - 2 < f(n) < (\sqrt{3.5} + \varepsilon)\sqrt{n}$$

if n is large enough, depending on $\varepsilon > 0$. ♣

This is the currently known best upper bound due to Katalin Fried. The lower estimate comes from a simplified version of Moser's method. Moser's original bound is somewhat better.

Proof. For the upper estimate, we observe that the construction using number systems establishes the basis as the union of two arithmetic progressions. As a variant of this idea, our basis will be the union of five arithmetic progressions.

Let t be a positive integer, and consider the following five disjoint arithmetic progressions:

$$\begin{aligned}
B &= \{b_0, \ldots, b_t\} &&= \{j \mid 0 \leq j \leq t\} \\
C &= \{c_0, \ldots, c_{3t-1}\} &&= \{2t + 1 + j(t + 1) \mid 0 \leq j \leq 3t - 1\} \\
D &= \{d_0, \ldots, d_t\} &&= \{3t^2 + 5t + 1 + j \mid 0 \leq j \leq t\} \\
E &= \{e_0, \ldots, e_t\} &&= \{6t^2 + 12t + 3 + jt \mid 0 \leq j \leq t\} \\
F &= \{f_0, \ldots, f_t\} &&= \{10t^2 + 18t + 5 + jt \mid 0 \leq j \leq t\}
\end{aligned}$$

The differences of the progressions in order are $1, t + 1, 1, t$, and t, and they have $t + 1$, $3t, t + 1, t + 1$, and $t + 1$ elements.

Let A_t be the union of the five sets, so $|A_t| = 7t + 4$. We verify that A_t is a basis of second order for $n = 14t^2 + 24t + 7$, so every integer up to $14t^2 + 24t + 6$ is the sum of two elements in A_t.

For an arbitrary n, we take the smallest t satisfying $n \le 14t^2 + 24t + 7$. Then A_t is a suitable basis for n, and $|A_t| = 7t + 4 \sim \sqrt{3.5n}$, as $n \to \infty$ (since $t \sim \sqrt{n/14}$).

So we have to prove that every integer $0 \le r \le 14t^2 + 24t + 6$ is the sum of two elements in A_t. Let $[[x, y]]$ denote the set of integers in the interval $[x, y]$. Clearly,

$$B + B = [[0, 2t]] \quad \text{and} \quad B + C = [[2t + 1, 3t^2 + 5t]].$$

We obtain similarly

$$B + D = [[3t^2 + 5t + 1, 3t^2 + 7t + 1]]$$
$$C + D = [[3t^2 + 7t + 2, 6t^2 + 10t + 1]]$$
$$D + D = [[6t^2 + 10t + 2, 6t^2 + 12t + 2]]$$
$$B + E = [[6t^2 + 12t + 3, 7t^2 + 13t + 3]].$$

So far $A_t + A_t \supseteq [[0, 7t^2 + 13t + 3]]$.

Now we show $C + E \supseteq [[7t^2 + 13t + 4, 9t^2 + 17t + 3]]$. We start with

$$c_0 + e_{t-1} = (2t + 1) + (7t^2 + 11t + 3) = 7t^2 + 13t + 4.$$

As the differences in C and E are $t + 1$ and t, therefore it is worthwhile to combine the consecutive elements of C with the corresponding earlier elements of E:

$$c_1 + e_{t-2} = c_0 + e_{t-1} + 1$$
$$c_2 + e_{t-3} = c_0 + e_{t-1} + 2$$
$$\vdots$$
$$c_{t-1} + e_0 = c_0 + e_{t-1} + (t - 1).$$

The next integer is obtained as the sum $c_0 + e_t = c_0 + e_{t-1} + t$, and we proceed forward in C and backward in E to represent every integer in the form $c_i + e_{t-i}$ up to $c_t + e_0 = c_0 + e_t + t$. We jump now to $c_1 + e_t = c_0 + e_t + (t + 1)$, and the sums $c_{1+i} + e_{t-i}$ give the next $t + 1$ integers. Continuing the procedure similarly, we arrive at the sum $c_{3t-1} + e_1 = 9t^2 + 17t + 3$, so $C + E \supseteq [[7t^2 + 13t + 4, 9t^2 + 17t + 3]]$.

The next observation is $D + E = [[9t^2 + 17t + 4, 10t^2 + 18t + 4]]$.

Similar to the previous considerations, we can show

$$B + F = [[10t^2 + 18t + 5, 11t^2 + 19t + 5]]$$
$$C + F = [[11t^2 + 19t + 6, 13t^2 + 23t + 5]]$$
$$D + F = [[13t^2 + 23t + 6, 11t^2 + 24t + 6]].$$

Thus we have verified $r \in A_t + A_t$ for every integer $0 \le r \le 14t^2 + 24t + 6$, which completes the proof of the upper bound.

Turning to the lower estimate, we consider an arbitrary basis $A = \{0 \le a_1 < \cdots < a_k \le n - 1\}$ of second order in the interval $[0, n - 1]$. Let

(12.3.6)
$$h(x) = \sum_{i=1}^{k} x^{a_i}$$

be the generating function belonging to A, so

$$h^2(x) = \left(\sum_{i=1}^{k} x^{a_i}\right)\left(\sum_{j=1}^{k} x^{a_j}\right) = \sum_{i,j=1}^{k} x^{a_i+a_j} = 2\sum_{1\leq i<j\leq k} x^{a_i+a_j} + \sum_{i=1}^{k} x^{2a_i} =$$

$$= 2\sum_{1\leq i\leq j\leq k} x^{a_i+a_j} - \sum_{i=1}^{k} x^{2a_i} = 2\sum_{1\leq i\leq j\leq k} x^{a_i+a_j} - h(x^2).$$

Thus

$$(12.3.7) \qquad g(x) = \sum_{1\leq i\leq j\leq k} x^{a_i+a_j} = \frac{h^2(x) + h(x^2)}{2}.$$

The coefficient of x^r in $g(x)$ is the number of representations of r in the form $a_i + a_j$ where $i \leq j$. Since every $0 \leq r \leq n-1$ can be written as $a_i + a_j$, the coefficient of x^r is at least 1, so

$$(12.3.8) \qquad g(x) = 1 + x + \cdots + x^{n-1} + \sum_{m=0}^{2n-2} u_m x^m, \quad \text{where } u_m \geq 0.$$

By (12.3.7) and (12.3.8),

$$(12.3.9) \qquad g(1) = \frac{h^2(1) + h(1)}{2} = \frac{k^2 + k}{2} = n + \sum_{m=0}^{2n-2} u_m.$$

Since $u_m \geq 0$, (12.3.9) implies $(k^2 + k)/2 \geq n$, which leads to the estimate $k \geq \sqrt{2n} - (1/2)$ obtained before stating our theorem. To improve this bound, we replace $\sum_{m=0}^{2n-2} u_m \geq 0$ by a significantly better lower estimate.

We show

$$(12.3.10) \qquad S = \sum_{m=0}^{2n-2} u_m > \nu k^2,$$

where we shall determine the constant $\nu > 0$ explicitly, which substituted back into (12.3.9) will give the lower bound claimed in the theorem.

Let $B = \tau k$ and $L = (1 - \tau)k$ be the number of those elements in A for which $a_i > (n-1)/2$ and $a_i \leq (n-1)/2$, (so $B + L = k$ and τ is the ratio of the big elements a_i in this basis).

Observe that $S' = \sum_{m=n}^{2n-2} u_m$ is just the number of sums $a_i + a_j$, $i \leq j$, that are greater than $n-1$. If both a_i and a_j are larger than $(n-1)/2$, then $a_i + a_j > n-1$, so

$$(12.3.11) \qquad S \geq S' \geq \frac{(B+1)B}{2} = \frac{(\tau k + 1)(\tau k)}{2} \geq \frac{\tau^2}{2} \cdot k^2.$$

(This means informally that if there are many elements a_i greater than $(n-1)/2$, then many sums get wasted, and thus we need a larger basis to represent the integers up to $n-1$. To elaborate this idea precisely, we do not even need generating functions. But we cannot get along without them if the small elements dominate A, see below.)

We now substitute a complex nth root of unity $\varrho \neq 1$ into x in (12.3.8). Then the sum $1 + \varrho + \cdots + \varrho^{n-1}$ in (12.3.8) is 0, so

$$g(\varrho) = \sum_{m=0}^{2n-2} u_m \varrho^m.$$

Taking the absolute value of both sides,

$$|g(\varrho)| = \left| \sum_{m=0}^{2n-2} u_m \varrho^m \right| \leq \sum_{m=0}^{2n-2} |u_m| \cdot |\varrho|^m = \sum_{m=0}^{2n-2} u_m = S,$$

since $u_m \geq 0$ and $|\varrho| = 1$. By (12.3.7),

$$(12.3.12) \qquad S \geq |g(\varrho)| = \frac{|h^2(\varrho) + h(\varrho^2)|}{2} \geq \frac{|h^2(\varrho)|}{2} - \frac{|h(\varrho^2)|}{2}.$$

To continue this chain of inequalities, we need a lower bound for the difference at the right of (12.3.12), so we estimate the subtrahend from above and the minuend from below.

By definition (12.3.6) of the generating function $h(x)$,

$$|h(\varrho^2)| = \left| \sum_{i=1}^{k} \varrho^{2a_i} \right| \leq \sum_{i=1}^{k} |\varrho|^{2a_i} = k,$$

since $|\varrho| = 1$, so

$$(12.3.13) \qquad \frac{|h(\varrho^2)|}{2} \leq \frac{k}{2},$$

which will be negligible compared to the minuend $|h^2(\varrho)|/2$ having an order of magnitude k^2.

Thus we seek a lower bound for

$$(12.3.14) \qquad |h(\varrho)| = \left| \sum_{i=1}^{k} \varrho^{a_i} \right|.$$

Recall that we have to cope basically with the case when the small elements are dominant, i.e. $L = (1 - \tau)k$ is big. Accordingly, in (12.3.14) we separate the parts belonging to the small and large elements a_i:

$$(12.3.15)$$
$$|h(\varrho)| = \left| \sum_{i=1}^{L} \varrho^{a_i} + \sum_{i=L+1}^{k} \varrho^{a_i} \right| \geq \left| \sum_{i=1}^{L} \varrho^{a_i} \right| - \left| \sum_{i=L+1}^{k} \varrho^{a_i} \right| \geq$$
$$\geq \left| \sum_{i=1}^{L} \varrho^{a_i} \right| - \sum_{i=L+1}^{k} |\varrho^{a_i}| = \left| \sum_{i=1}^{L} \varrho^{a_i} \right| - B.$$

Therefore, we have to find a good lower bound for

$$(12.3.16) \qquad T(\varrho) = \left| \sum_{i=1}^{L} \varrho^{a_i} \right|.$$

Let $\omega = \cos(2\pi/n) + i\sin(2\pi/n)$ and $z_j = \omega^{a_j}$, $j = 1, \ldots, L$. Since $0 \leq a_j \leq (n-1)/2$, every complex number z_j has a non-negative imaginary part, and thus they all are in the upper half-plane.

Let α be an acute angle to be specified later, and let U denote how many numbers z_j have an angle β_j satisfying $\alpha \leq \beta_j \leq \pi - \alpha$. Let us call them *upper numbers*. Thus the other $S - U$ *lower numbers* z_j have angles $0 \leq \beta_j < \alpha$ or $\pi - \alpha < \beta_j < \pi$.

Concerning the imaginary parts, $\text{Im}(z_j) \geq \sin\alpha$ for upper numbers, and $\text{Im}(z_j) \geq 0$ for lower ones, so

$$(12.3.17) \qquad \Big| \sum_{j=1}^{L} z_j \Big| \geq \text{Im}\Big(\sum_{j=1}^{L} z_j \Big) \geq U \cdot \sin\alpha.$$

If we choose $\varrho = \omega$, then $\varrho^{a_j} = z_j$, so

$$(12.3.18) \qquad T(\omega) \geq U \cdot \sin\alpha$$

by (12.3.16) and (12.3.17).

We next choose $\varrho = \omega^2$, so $\varrho^{a_j} = z_j^2$. For lower numbers, the angle of z^2 lies between -2α and 2α, so $\text{Re}(z_j^2) > \cos(2\alpha)$, and $\text{Re}(z_j^2) \geq -1$ for upper numbers, trivially. Therefore

$$(12.3.19) \qquad T(\omega^2) = \Big| \sum_{j=1}^{L} z_j^2 \Big| \geq \text{Re}\Big(\sum_{j=1}^{L} z_j^2 \Big) \geq (L - U)\cos(2\alpha) - U.$$

Choosing $\alpha = \pi/6$, (12.3.18) and (12.3.19) yield

$$(12.3.20) \qquad T(\omega) \geq U/2 \quad \text{and} \quad T(\omega^2) \geq (L - 3U)/2.$$

Let

$$M = \max\big(T(\omega), T(\omega^2)\big).$$

Then (12.3.20) implies

$$M \geq \frac{3T(\omega) + T(\omega^2)}{4} \geq \frac{L}{8}.$$

Thus we obtained that $T(\varrho)$ in (12.3.16) satisfies $T(\varrho) \geq L/8$ for a suitable ϱ (where $\varrho = \omega$ or $\varrho = \omega^2$). Substituting into (12.3.15), we get

$$(12.3.21) \qquad |h(\varrho)| \geq \frac{L}{8} - B = \frac{1 - 9\tau}{8}k.$$

So

$$(12.3.22) \qquad S \geq \frac{(1 - 9\tau)^2}{128}k^2 - \frac{k}{2}$$

by (12.3.12), (12.3.13), and (12.3.21). Taking (12.3.11) into consideration, we obtain

$$(12.3.23) \qquad S \geq \max\Big(\frac{\tau^2}{2}k^2, \frac{(1 - 9\tau)^2}{128}k^2 - \frac{k}{2} \Big).$$

The worst case is if the coefficients of k^2 are the same in the two expressions, i.e. $\tau = 1/17$, and then

$$(12.3.24) \qquad S \geq \frac{k^2}{578} - \frac{k}{2},$$

so (12.3.10) is satisfied with the constant $\nu = 1/578$ (disregarding the error term $k/2$).

Substituting (12.3.24) into (12.3.9), we get

$$\frac{k^2 + k}{2} \geq n + \frac{k^2}{578} - \frac{k}{2},$$

so

$$\frac{144}{289}k^2 + k \geq n, \quad \text{and so} \quad (k+2)^2 > \frac{289}{144}n,$$

which is just the lower bound claimed in the theorem.   □

Exercises 12.3

1. Verify the following statements for sets of real numbers.

   (a) If $|A| = k$, then $|A + A| \geq 2k - 1$, and equality holds if and only if the elements of A form an arithmetic progression.

   (b) If $|A| = k$, $|B| = r$, then $|A + B| \geq k + r - 1$, and equality holds if and only if either $k = 1$, or $r = 1$, or A and B are arithmetic progressions with a common difference.

   (c) If $|A_i| = k_i$, $i = 1, 2, \ldots, t$, then $|A_1 + \cdots + A_t| \geq k_1 + \cdots + k_t + 1 - t$, and if $k_i > 1$, $i = 1, 2, \ldots, t$, then equality holds if and only if every A_i is an arithmetic progression with the same difference.

2. Prove the following generalization of Theorem 12.3.1 for an arbitrary modulus m: Let $A, B \subseteq \mathbf{Z}_m$, $0 \in B$. Then $|A + B| \geq \min(m, |A| + s)$, where s is the number of elements in B coprime to m. Show an example for a composite m and $s < |B| - 1$ when we have equality.

3. Prove the statements.

   (a) Let $A, B \subseteq \mathbf{Z}_m$, $0 \in A \cap B$, and assume that $a + b = 0$ implies $a = b = 0$ for $a \in A$, $b \in B$. Then $|A + B| \geq |A| + |B| - 1$. (The conditions guarantee $|A| + |B| - 1 \leq m$, so, in contrast with Theorem 12.3.1, there is no need for a minimum in the formulation of the inequality.)

   (b) The inequality in (a) is sharp.

   (c) The statement in (a) remains valid if $\mathbf{Z}_m$ is replaced by finite subsets of an abelian group.

* 4. Let p be a prime, $A \subseteq \mathbf{Z}_p$, $|A| = k$, and $A \hat{+} A = \{a + a' \mid a, a' \in A, a \neq a'\}$, so we consider now only the sums of different elements. Show $|A \hat{+} A| \geq \min(p, 2k - 3)$.

   Remarks: (1) This long-standing conjecture of Erdős and Heilbronn was first verified by Hamidoune and Da Silva. Later Alon, Ruzsa, and Nathanson found a much simpler proof.

   (2) The example $A = \{0, 1, \ldots, k - 1\}$ shows that this bound is best possible.

5. (a) Let F be a commutative field, $A, B \subseteq F$, $|A| = k$, $|B| = r$, and $G(x, y)$ a polynomial over F in two variables of degree $k + r - 2$ where the coefficient of $x^{k-1}y^{r-1}$ is not zero. Prove $G(a, b) \neq 0$ for some $a \in A$, $b \in B$.

(b) Generalize part (a) for n subsets instead of two and for a polynomial G in n variables.

S* 6. Let $p > 2$ be a prime, and let C and D be two subsets of the same size in $\mathbf{Z}_p$. Show that we can pair the elements of C and D so that the sums of the two elements in the pairs are all distinct.

7. Formulate and prove the generalization of Theorem 12.3.1 for more than two sets.

8. We consider Exercise 3.6.6 and its generalizations in the plane and in higher dimensions.

    * (a) Give a new proof to Exercise 3.6.6 based on Exercise 12.3.7.

    (b) Verify that we can always find, among any five points of the usual square lattice in the plane, two points such that their midpoint is a lattice point.

    (c) Let $f(n)$ be the smallest integer such that among any $f(n)$ lattice points in the plane, we can always find n whose center of gravity is a lattice point. Show $f(n) \geq 4n - 3$.

    Remark: The old conjecture $f(n) = 4n - 3$ was proved in 2004.

    (d) Let $f(n, d)$ be the smallest integer such that among any $f(n, d)$ points in the usual d dimensional lattice, we can always find n whose center of gravity is a lattice point. Prove

        (i) $2^d(n - 1) + 1 \leq f(n, d) \leq n^d(n - 1) + 1$

        (ii) $f(nm, d) \leq f(n, d) + n(f(m, d) - 1)$.

    Remark: The upper bound in (i) can be greatly improved to $c_d n$ where c_d is a constant depending only on d. The lower bound is sharp for $d = 1$ and $d = 2$ (see part (a) and the remark after part (c)). However, the lower bound can be improved for every $d > 2$ if $n \geq 3$ is odd (the lower bound gives the right value for every d if $n = 2^k$, see below). The exact value of $f(n, d)$ is known for $n > 2$ and $d > 2$ only in the cases

$$f(3, 3) = 19, \quad f(3, 4) = 41, \quad f(3, 5) = 91, \quad \text{and} \quad f(2^k, d) = (2^k - 1)2^d + 1.$$

9. Let p be a prime, $A \subseteq \mathbf{Z}_p$, and assume that the difference of two distinct elements of A is never a square in $\mathbf{Z}_p$ (so $a_i - a_j$, $i \neq j$, is always a quadratic non-residue mod p). Prove $|A| < \sqrt{p}$.

10. A set A of non-negative integers is called a *basis of order h* for the interval $[0, n-1]$ if every integer $0 \leq r \leq n - 1$ is the sum of h elements of A. Let $g(h, n)$ denote the minimal possible size of such a basis. Verify

$$\sqrt[h]{h! \, n} - h + 1 < g(h, n) < \sqrt[h]{h^h n} + h.$$

12.4. Schur's Theorem

This classical result of combinatorial number theory has its strange origin in the seemingly remote Fermat's Last Theorem, and its proof requires methods from graph theory. The topic has had intensive research ever since, but there still are many unsolved problems.

We deal first with the graph theoretical background. We start with the following well-known puzzle: Among any six people there are either three so that any two know each other, or there are three where no two know each other (the acquaintance is supposed to be mutual).

Rewording in terms of graph theory, we consider a complete graph (or clique) of six nodes corresponding to the six people, and an edge is colored red if its endpoints know each other, and is colored blue otherwise. Then the statement says that however we color the edges of a complete graph of six nodes, there is a monochromatic triangle.

To prove this, we pick a node A. Considering the five edges starting from A, (at least) three of them must be of the same color, say red. Let B, C, and D be the other endpoints of these edges. If there is a red edge between two of them, say edge BC is red, then ABC is a red triangle, otherwise BCD is a blue triangle.

We can generalize this puzzle: We color the edges of a complete graph of n nodes with t colors, and instead of a monochromatic triangle we want to find a complete graph of k nodes with edges of the same color (the original problem is a special case $t = 2$, $k = 3$). Ramsey's fundamental theorem asserts that we always have such a subgraph if n (depending on k and t) is large enough :

Theorem 12.4.1 (Ramsey's Theorem). *For any t and k there exists an integer $R(k, t)$ such that if $n \geq R(k, t)$ and we color the edges of a complete graph of n nodes with t colors, then there is a complete subgraph of k nodes with edges of the same color.* ♣

In the sequel $R(k, t)$ will denote the minimal integer with this property.

Solving the puzzle, we verified $R(3, 2) \leq 6$, and it is easy to check that we have here equality (see Exercise 12.4.1). We can read from the proof that $R(3, t) \leq 3t!$, moreover $R(3, t) \leq \lceil et! \rceil$, where $e = 2.71 \ldots$ is the base of the natural logarithm (see Exercise 12.4.2). We can improve the constant multiplier to $e - 1/24$ with more refined methods, but probably this is very far from the actual value of $R(3, t)$. We know the exact values of Ramsey numbers $R(k, t)$ only in very few cases, e.g. $R(3, 3) = 17$, and there is generally a large gap between the lower and upper estimates.

Proof. For a clearer exposition, we prove first the case $k = 3$ by induction on t, and turn to a general k afterwards. (The proof of Schur's Theorem will require only the case $k = 3$.)

I. We can start the induction either with $t = 1$ (clearly, $R(3, 1) = 3$), or with $t = 2$, as we verified $R(3, 2) \leq 6$ earlier. The idea used to prove the latter can serve as a general induction step.

Assume that $n = R(3, t - 1)$ exists, and color the edges of a complete graph of N nodes with t colors. If $N \geq 1 + t(n - 1) + 1$, then considering $t(n - 1) + 1$ edges starting from a node A, there will be at least n among them of the same color, say red, by the pigeonhole principle. If there is a red edge between two other endpoints of these edges, e.g. between B and C, then ABC is a red triangle. Otherwise, the n endpoints form a complete graph whose edges are colored with $t - 1$ colors, so it contains a monochromatic triangle by the induction hypothesis.

II. To prove the general case, it is worthwhile to formulate a more refined version of the problem. For a simpler wording, the *size* of a graph is its number of nodes, the *colors* are the integers 1, 2, ..., t, and a *graph of color j* is a complete graph where every edge has color j. Then the modified statement is:

For any t and k_1, ..., k_t, there exists an $n = R^*(k_1, k_2, ..., k_t)$ such that if we color the edges of a complete graph of n nodes arbitrarily with colors 1, 2, ..., t, then there results a complete subgraph of size k_j and of color j for some j. ($R^*(k_1, k_2, ..., k_t)$ is the smallest n with this property.)

The two problems can easily be deduced from each other: clearly, $R(k, t) = R^*(k, ..., k)$, and on the other hand, $R^*(k_1, ..., k_t) \leq R(k, t)$, where $k = \max(k_1, ..., k_t)$.

If every $k_i = 1$ or 2, then the modified statement is trivial. We claim that induction yields

$$(12.4.1) \qquad R^*(k_1, ..., k_t) \leq 1 + \sum_{j=1}^{t} [R^*(k_1, ..., k_j - 1, ..., k_t) - 1] + 1.$$

Let us color the edges of a complete graph of size N with t colors, where N is the value on the right-hand side of (12.4.1). Considering the edges starting from a node A, there will be at least $R^*(k_1, ..., k_j - 1, ..., k_t)$ among them of color j for some j by the pigeonhole principle. The other endpoints of these edges form a complete graph that contains a suitable monochromatic complete subgraph by the induction hypothesis. If the color of the subgraph is $i \neq j$, then we have a complete graph of size k_i and color i, so we are done. If its color is j, then we have a graph of size $k_j - 1$, and together with A it forms a complete graph of size k_j and color j. $\qquad \square$

Schur's Theorem refers to colorings of positive integers.

Theorem 12.4.2 (Schur's Theorem). *For any t there exists an $n = S(t)$ with the property that coloring the numbers 1, 2, ..., $n + 1$ with t colors arbitrarily, there will be some a and b of the same color such that $a + b$ has this color (we allow $a = b$).* ♣

In the sequel $S(t)$ will denote the smallest such n. That is, $S(t)$ is the biggest wrong integer: 1, 2, ..., $S(t)$ can still be colored with t colors so that the equation $x + y = z$ has no monochromatic solution. (In Ramsey's Theorem, $R(k, t)$ is the minimal good integer; we keep the traditional notation in both cases.)

Clearly, $S(1) = 1$, and we easily infer $S(2) = 4$. Besides these, the only values known exactly are $S(3) = 13$ and $S(4) = 44$. We discuss some lower and upper bounds for Schur numbers $S(t)$ in Exercise 12.4.3.

Proof. We show $S(t) < R(3, t)$, so the required property holds for an arbitrary coloring of 1, 2, ..., $R(3, t)$. Consider the complete graph having these numbers as nodes, and the graph-color of edge (i, j) is defined as the number-color of $|i - j|$. Then by Ramsey's Theorem, there results a monochromatic triangle in the graph, so the edges (i, j), (j, m), and (i, m) have the same graph-color for some $i < j < m$. This means that the integers $a = j - i$, $b = m - j$, and $a + b = m - i$ have the same number-color. $\qquad \square$

Now we turn to the connection between Schur's Theorem and Fermat's Last Theorem.

Consider the congruence $x^t + y^t \equiv z^t$ (mod p). If it has only trivial solutions where $xyz \equiv 0$ (mod p) for infinitely many primes p, then Fermat's Last Theorem follows for the exponent t. Indeed, if we have a counterexample of non-zero integers a, b, and c satisfying $a^t + b^t = c^t$, then they provide a non-trivial solution of the congruence for every prime $p > \max(|a|, |b|, |c|)$. But this contradicts that there are only trivial solutions for infinitely many primes.

It turns out, however, that this idea does not lead to a proof of Fermat's Last Theorem:

Theorem 12.4.3. *The congruence $x^t + y^t \equiv z^t$ (mod p) has a non-trivial (i.e. $xyz \not\equiv 0$ (mod p)) solution for every prime p large enough (depending on t).*   ♣

Proof. Let $p - 1 > S(t)$ and g be a primitive root mod p. We color the integers 1, 2, ..., $p - 1$ with colors 0, 1, ..., $t - 1$ as follows: An integer gets color r if it is congruent mod p to one of the numbers $g^r, g^{r+t}, g^{r+2t}, \dots$.

By Schur's Theorem, there is a monochromatic triple $a, b, a + b$, so

$$a \equiv g^{st+r}, \quad b \equiv g^{ut+r}, \quad a + b \equiv g^{vt+r} \ (\text{mod } p)$$

for some r, s, u, and v. Hence

$$g^{st+r} + g^{ut+r} \equiv g^{vt+r} \ (\text{mod } p).$$

Cancelling g^r (which is coprime to p), we obtain

$$(g^s)^t + (g^u)^t \equiv (g^v)^t \ (\text{mod } p),$$

so $x = g^s, y = g^u, z = g^v$ is a non-trivial solution of the congruence.   □

Schur raised also another problem concerning colorings of natural numbers, that was first solved by Van der Waerden. We state this result without proof:

Theorem 12.4.4 (Van der Waerden's Theorem). *Coloring the positive integers with two colors, there are arbitrarily long (finite) monochromatic arithmetic progressions.*   ♣

In fact, Van der Waerden proved the following finite variant involving more colors with a very tricky induction:

Theorem 12.4.4A (Van der Waerden's Theorem). *For any t and k there exists an $n = w(k, t)$ such that coloring the integers 1, 2, ..., n with t colors arbitrarily, there is a monochromatic arithmetic progression of k terms.*

Similar to the Ramsey numbers $R(k, t)$ and Schur numbers $S(t)$, there is a big gap between the lower and upper estimates for the (minimal) Van der Waerden numbers $w(k, t)$. The only exact values known are

$$w(3, 2) = 9 \quad w(4, 2) = 35 \quad w(5, 2) = 178 \quad w(6, 2) = 1132$$
$$w(3, 3) = 27 \quad w(4, 3) = 293 \quad w(3, 4) = 76$$

and trivially $w(k, 1) = k$ and $w(2, t) = t + 1$. For two colors, lower estimates of $w(k) = w(k, 2)$ are discussed in Exercise 12.4.11.

On the other hand, we can color the positive integers with two colors so that no infinite monochromatic arithmetic progression arises, moreover we can show that no infinite red and not even a three-term blue arithmetic progression occurs (see Exercise 12.4.7).

We conclude the section by mentioning a substantial generalization of Van der Waerden's Theorem. This famous conjecture of Erdős and Turán resisted all attempts for many decades, and was solved finally by Szemerédi. He thus deserved the biggest prize (1000 US dollars) offered and paid by Erdős for a solution of a mathematical problem. (Very recently, a $10000 problem of Erdős was solved, too, see the story after Theorem 5.5.4.) Szemerédi got an Abel Prize, one of the most prestigious honors in mathematics, in 2012 for his many fundamental contributions to number theory, combinatorics, and computer science.

Let us look at the conjecture of Erdős and Turán. Van der Waerden's Theorem states that coloring the natural numbers, or its sufficiently long initial segment, there will occur a long monochromatic arithmetic progression, but provides no information about its color. We feel, of course, that this should be the most frequent color, i.e. one having the largest density. Erdős and Turán had the idea that, independent of any coloring, if we take a sufficiently dense subsequence of the natural numbers, then it will contain a long arithmetic progression. The precise formulation of their conjecture is

Theorem 12.4.5 (Szemerédi's Theorem). *Consider a subset of* $\{1, 2, \ldots, n\}$ *of maximal size that does not contain a k-term arithmetic progression, and denote the number of its elements by* $r_k(n)$. *Then* $\lim_{n \to \infty} r_k(n)/n = 0$ *for any fixed k.* ♣

This implies Van der Waerden's Theorem: Coloring the integers $1, 2, \ldots, n$ with t colors, some color must occur at least n/t times. If n is large enough, then $1/t$ is bigger than $r_k(n)/n$ tending to 0, so $n/t > r_k(n)$, thus there must occur a k-term arithmetic progression of that color.

Another formulation of Szemerédi's Theorem is that any sequence of natural numbers having positive upper density must contain arbitrarily long (finite) arithmetic progressions. Erdős extended his conjecture for even less dense sequences, thinking that it is sufficient that the sum of reciprocals of elements be divergent. It was a great surprise in 2004 when the conjecture was verified for the sequence of primes (so there are arbitrarily long arithmetic progressions among the primes, see also Section 5.1), but the general conjecture is still open.

Exercises 12.4

1. Verify $R(3, 2) = 6$, $R(k, 1) = k$, $R(1, t) = 1$, and $R(2, t) = 2$.

2. Show (a) $R(3, t) \leq 3t!$ (b) $R(3, t) \leq \lceil et! \rceil$.

3. Prove the estimates for Schur numbers:

   (a) $S(t) < et!$

(b) $S(t+1) \geq 3S(t) + 1$

(c) $S(t) \geq (3^t - 1)/2$

* (d) $S(t+v) \geq 2S(t)S(v) + S(t) + S(v)$.

Remark: Part (b) is a special case of (d) for $v = 1$. Using (d) and $S(5) \geq 157$, we can slightly improve the lower bound in (c).

4. For a given n, find the largest $r = f(n)$ such that the integers $n, n+1, \ldots, r$ can be colored with two colors so that the equation $x + y = z$ has no monochromatic solution.

5. Prove that for any t there exists an n such that coloring the integers $1, 2, \ldots, n+1$ with t colors arbitrarily, there will be three (not necessarily distinct) integers of the same color so that their sum has this color.

6. Let j be fixed. Show that there exist two consecutive jth power residues modulo every sufficiently large prime p, i.e. both $x^j \equiv a - 1$ and $z^j \equiv a \pmod{p}$ are solvable for some $a \not\equiv 0, 1 \pmod{p}$.

7. Verify that the positive integers can be colored red and blue avoiding

   (a) infinitely long monochromatic arithmetic progressions

 * (b) both infinite red and three-term blue arithmetic progressions.

8. Show that coloring the natural numbers with finitely many colors arbitrarily, for every k there will arise infinitely many k-term arithmetic progressions all having the same color.

9. Demonstrate that coloring the natural numbers with finitely many colors arbitrarily, there arise arbitrarily long (finite) monochromatic geometric progressions.

10. Verify $w(3,2) = 9$.

** 11. Prove the lower estimates for $w(k) = w(k,2)$:

   (a) $w(k) \geq 2^{k/2}\sqrt{k-1}$

   S (b) $w(p+1) > p(2^p - 1)$, if p is a prime.

S** 12. Prove a lower bound for $r_3(n)$: For every sufficiently large n, there exist $n/e^{c\sqrt{\log n}}$ integers between 1 and n (where $c > 0$ is a suitable constant), that contain no three-term arithmetic progression.

12.5. Covering Congruences

We deal with another favorite problem of Erdős: We represent the set of non-negative integers as a union of finitely many arithmetic progressions with distinct differences (greater than 1):

$$\{0, 1, \ldots, n, \ldots\}$$

(12.5.1) $$= \{a_1, a_1 + m_1, a_1 + 2m_1, \ldots\} \cup \cdots \cup \{a_k, a_k + m_k, a_k + 2m_k, \ldots\},$$

$$\text{where } 1 < m_1 < \cdots < m_k.$$

An equivalent formulation is that we cover the integers with residue classes of distinct moduli (greater than 1): Every integer t is an element of at least one of the residue classes

(12.5.2) $a_1 \ (\mathrm{mod} \ m_1), \dots, a_k \ (\mathrm{mod} \ m_k), \quad 1 < m_1 < \cdots < m_k,$

so $t \equiv a_i \ (\mathrm{mod} \ m_i)$ for at least one i.

Such systems of arithmetic progressions or congruences are called *covering congruences*.

A simple example is

(12.5.3) $0 \ (\mathrm{mod} \ 2), \quad 0 \ (\mathrm{mod} \ 3), \quad 1 \ (\mathrm{mod} \ 4), \quad 1 \ (\mathrm{mod} \ 6), \quad 11 \ (\mathrm{mod} \ 12).$

This is the minimal number of moduli, and these are the only possible moduli for five congruences (see Exercise 12.5.4).

Erdős invented covering congruences to solve a seemingly remote problem, see Theorem 12.5.2. There arise many questions concerning covering congruences. The two oldest and at the same time most interesting ones are:

• Can all moduli be odd? This problem is still unsolved.

• Can all moduli be arbitrarily large, i.e. does there exist for any L covering congruences whose moduli are greater than L?

This was verified for values of L which reached $L = 40$ in 2008. In an extremely long and tricky construction by Nielsen, just explaining the notation took several pages.

It was a great surprise, however, when it turned out that the answer is negative, and there is an upper bound for the smallest modulus in covering congruences. Hough presented this result in 2013 at a conference in honor of the centennial of Erdős' birth.

It is a natural question to investigate *exact* or *disjoint* covering when the arithmetic progressions in (12.5.1), or equivalently, the residue classes in (12.5.2) are disjoint, i.e. every integer satisfies exactly one congruence in (12.5.2).

The next theorem shows that this is not possible:

Theorem 12.5.1. *The set of non-negative integers cannot be obtained as the disjoint union of finitely many arithmetic progressions with distinct differences greater than* 1.   ♣

We present two proofs. The first relies on elementary analysis with complex numbers. The second formulates an interesting equivalent statement about regular polygons and verifies it using geometric arguments.

First proof. We use a generating function, where z denotes a complex number of $|z| < 1$.

For a proof by contradiction, assume that (12.5.1) is a disjoint union. Then every $n \geq 0$ has a unique representation $n = a_i + r m_i$, where $1 \leq i \leq k$ and $r \geq 0$. Therefore

$$(z^{a_1} + z^{a_1 + m_1} + z^{a_1 + 2m_1} + \dots) + \dots + (z^{a_k} + z^{a_k + m_k} + z^{a_k + 2m_k} + \dots) =$$
$$= 1 + z + z^2 + \dots + z^n + \dots.$$

(We used the fact that the series can be rearranged arbitrarily because it is absolutely convergent for $|z| < 1$.)

Summing the infinite geometric series, we obtain

(12.5.4)
$$\sum_{i=1}^{k} z^{a_i} \frac{1}{1 - z^{m_i}} = \frac{1}{1 - z}.$$

If the complex variable z tends to an m_ith complex root of unity (on a path in the region $|z| < 1$), then the corresponding term $z^{a_i}/(1 - z^{m_i})$ on the left-hand side of (12.5.4) will be unbounded. Thus, if $z \to w = \cos(2\pi/m_k) + i\sin(2\pi/m_k)$, then the last term on the left-hand side is unbounded, whereas the other terms and the right-hand side are bounded, since w is not an m_ith root of unity for $i < k$ due to the maximality of m_k. This yields the desired contradiction.   □

Second proof. Assume again that (12.5.1) is a disjoint union. As the arithmetic progressions are periodic modulo the least common multiple $M = [m_1, \ldots, m_k]$ of their differences, our assumption is equivalent to the condition that each of the integers 1, 2, $\ldots$, M is an element of exactly one arithmetic progression.

We draw a regular M-gon, and label its vertices 1, 2, $\ldots$, M in that order. We choose distinct colors to the covering arithmetic progressions, and paint the vertices covered by a given arithmetic progression with its color. For example, if $M = 12$ and the color of the arithmetic progression 1 (mod 4) is red, then the vertices 1, 5, and 9 will be red.

The vertices covered by the arithmetic progression a_i (mod m_i) form a regular polygon of $n_i = M/m_i$ sides (allowing for when the polygon degenerates into a segment or a point for $n_i = 2$ and 1, resp.). Clearly, $n_1 > n_2 > \cdots > n_k$.

In this geometric formulation, we assumed the existence of a regular M-gon where the vertices can be colored with $k > 1$ colors so that the monochromatic vertices form regular (possibly degenerate) polygons of different numbers of sides.

We shall use a simple geometric fact, namely that the sum of vectors from the center of a regular n-gon to its vertices is zero for $n > 1$ (including the degenerate case $n = 2$). Indeed, rotating the sum vector $\mathbf{v}$ around the center by angle $2\pi/n$ does not change, since the polygon was mapped onto itself. On the other hand, $\mathbf{v}$ gets rotated by the given angle, so it can be only the zero vector.

For a clearer exposition, assume first $n_k = 1$. Let $\mathbf{s}$ and $\mathbf{s}_i$, $i = 1, \ldots, k$ be the sums of vectors leading from the center to the vertices of the M-gon and the n_i-gons formed from the vertices of color i. Then obviously $\mathbf{s} = \sum_{i=1}^{k} \mathbf{s}_i$, but by the previous remark, $\mathbf{s} = \mathbf{s}_1 = \cdots = \mathbf{s}_{k-1} = \mathbf{0}$, whereas $\mathbf{s}_k \neq \mathbf{0}$, which is a contradiction.

We can handle the general case with a refinement of the argument. Let t be fixed, and consider the transformation that maps vertex j of the regular M-gon into the vertex tj (mod M), $j = 1, \ldots, M$. We show that the images of the originally monochromatic vertices cover the vertices of a regular polygon with the same multiplicity. E.g. if $M = 12$ and $t = 2$, then the images of 1, 5, 9 corresponding to the arithmetic progression 1 (mod 4) will be 2, 10, 6 in this order, so we get the regular triangle 2, 6, 10; for 2 (mod 3), vertices 2, 5, 8, 11 are mapped into 4, 10, 4, 10, so the images cover the regular

2-gon 4, 10 twice; finally, for 4 (mod 6), vertices 4, 10 go to 8, 8, thus we have a 1-gon with double multiplicity.

This is the case also in general. For a_i (mod m_i), the vertices $a_i + jm_i$, $j = 0, 1, \ldots,$ $n_i - 1$, are mapped into $ta_i + jtm_i$ (mod M). Considering this arithmetic progression with difference tm_i modulo m, and arranging its elements into a suitable order, we see that starting from ta_i we get vertices of distance $(tm_i, M) = (t, n_i)m_i$ between the neighbors, and each vertex occurs (t, n_i) times. So the images cover the vertices of a regular polygon with the same multiplicity, and we get a 1-gon if and only if $n_i \mid t$.

Based on this, we choose $t = n_k$. Repeating our argument about the sums of vectors from the center to the vertices, we get that the sum vector is zero for the images of the n_i-gons for $i < k$ and of the original M-gon, but it is not zero for the images of the n_k-gon. Thus we arrived at the same contradiction as in the special case $n_k = 1$. $\square$

Now we turn to Romanoff's problem which was solved by Erdős using covering congruences.

Theorem 12.5.2. *There are infinitely many odd numbers that cannot be written as a sum of a power of two and an odd integer.* ♣

Proof. We shall verify a stronger statement. We construct an infinite arithmetic progression of odd integers none of which has such a representation.

We start with the following covering congruences a_i (mod m_i), $i = 1, 2, \ldots, 6$:

(12.5.5)   0 (mod 2), 0 (mod 3), 1 (mod 4), 3 (mod 8), 7 (mod 12), 23 (mod 24).

We use the fact that to every m_i there exists a prime p_i such that the order of 2 mod p_i is m_i, i.e. $o_{p_i}(2) = m_i$: the primes 3, 7, 5, 17, 13, and 241 have this property:
(12.5.6)
$$o_3(2) = 2, \quad o_7(2) = 3, \quad o_5(2) = 4, \quad o_{17}(2) = 8, \quad o_{13}(2) = 12, \quad o_{241}(2) = 24.$$

(To every $m \neq 6$ and 1, there exists a prime p such that the order of 2 mod p is just m. This means that we cannot use (12.5.3), but any covering congruences are suitable that do not contain 6 among the moduli. It is clear that for different values of m there always belong different primes p.)

Consider the above values a_i, m_i, p_i, and choose s to satisfy $2^{s-1} > \max_i p_i$ (for covering congruences (12.5.5)–(12.5.6) we can take $s = 9$).

We show that taking any solution $x = c$ of the simultaneous system of congruences

(12.5.7) $x \equiv 2^{a_i}$ (mod p_i), $i = 1, \ldots, k,$ $x \equiv 1$ (mod 2^s),

we cannot write c as a sum of a power of two and an odd integer. The moduli in (12.5.7) are pairwise coprime, thus the system is solvable and the solutions form an infinite arithmetic progression of odd numbers, which proves the theorem.

Assume that for some solution $c = 2^n + p$, where p is a prime. Since a_i (mod m_i) are covering congruences, $n \equiv a_i$ (mod m_i) for some i. We know that the order of 2 mod p_i is m_i, and c satisfies (12.5.7), thus

$$2^n \equiv 2^{a_i} \equiv c \pmod{p_i}.$$

This implies $p = c - 2^n \equiv 0$ (mod p_i), so only $p = p_i$ is possible.

To achieve a contradiction, we show that $c = 2^n + p_i$ does not satisfy the last congruence in (12.5.7), i.e. $2^n + p_i \not\equiv 1 \pmod{2^s}$. If $n \leq s - 1$, then this is guaranteed by $1 < 2^n + p_i < 2^{s-1} + 2^{s-1} = 2^s$. If $n \geq s$, then clearly $2^n + p_i \equiv p_i \not\equiv 1 \pmod{2^s}$. $\square$

Exercises 12.5

(We use the notation (12.5.1) and (12.5.2).)

1. Verify $\sum_{i=1}^{k} 1/m_i \geq 1$ for any covering congruences.

2. Show that replacing a modulus m_i by one of its divisors (different from 1 and the other moduli), the new congruences keep the covering property.

3. Consider minimal covering congruences, where deleting any congruence will destroy the covering property. Demonstrate that each m_i divides the least common multiple of the other moduli m_j.

4. Prove that two, three, or four residue classes cannot form covering congruences, and for five residue classes only the moduli in (12.5.3) are possible.

5. Construct covering congruences where the minimal modulus is 3.

6. We can infuse life into the notion of disjoint covering congruences (DCC) if we allow the repetition of moduli: $a_i \pmod{m_i}$, $i = 1, \ldots, k$, where $1 < m_1 \leq \ldots \leq m_k$, and every integer is an element of exactly one residue class. Verify the statements about DCC:

    (a) $\sum_{i=1}^{k} 1/m_i = 1$

    (b) $m_k = m_{k-1}$

    (c) to every k there exist DCC satisfying $m_1 < m_2 < \cdots < m_{k-1}$.

7. Prove that infinitely many even numbers cannot be written as a sum of a power of three and a prime. In general, to every odd number $a > 1$ and to every even number $b > 2$ there exist infinitely many even and odd numbers, resp., that cannot be written in the form $a^n + p$ and $b^n + p$, where p is a prime.

12.6. Additive Complements

Two infinite sets A and B of non-negative integers are *additive complements* of each other if every sufficiently large integer can be written as $a + b$, where $a \in A$, $b \in B$.

Consider, for example, the unique decimal representation of a positive integer $n = c_0 + 10c_1 + 10^2 c_2 + \cdots + c_k 10^k$ (where $0 \leq c_i \leq 9$), and let A consist of 0 and of positive integers with $0 = c_0 = c_2 = c_4 = \ldots$, and let B consist of 0 and of positive integers with $0 = c_1 = c_3 = c_5 = \ldots$ (e.g. $3010 \in A$, $70005 \in B$). Then clearly every non-negative integer has a unique representation $a + b$, so A and B are additive complements of each other. (We shall generally omit the phrases "additive" and "of each other" for brevity.)

We establish first a simple density condition necessary for A and B to be complements, and check how sharp this condition is. Then we investigate for two important special sets, the powers of two and the primes, how rare their complements can be.

Let $A(n)$ and $B(n)$ denote the number of elements not greater than n in the sets A and B. Let $f(n)$ denote how many integers $0 \leq t \leq n$ can be written as $t = a + b$. Then $f(n) \leq A(n)B(n)$, since $a \leq t \leq n$ and $b \leq t \leq n$ in such representations of t. (This estimate is crude from two points of view: some t may have more than one representation, and many sums $a+b$ will be larger than n.) If A and B are complements, then every $t > t_0$ can be written in the form $t = a + b$, so $f(n) \geq n - t_0$. Combining the lower and upper estimates for $f(n)$, we obtain $A(n)B(n) \geq n - t_0$ for every n. Dividing by n and letting $n \to \infty$, we obtain that additive complements satisfy

$$(12.6.1) \qquad \liminf_{n \to \infty} \frac{A(n)B(n)}{n} \geq 1.$$

As our estimates were very generous, we might feel that (12.6.1) cannot hold with equality and it would seem to be absurd that not only lim inf but even lim sup can reach 1 (which means that the limit is 1). Therefore it is quite surprising that all this occurs, and in fact there are lots of various constructions. We present the first such example found by Danzer.

Theorem 12.6.1. *There exist additive complements A and B satisfying*

$$(12.6.2) \qquad \lim_{n \to \infty} A(n)B(n)/n = 1. \qquad \clubsuit$$

Proof. We choose A as a rapidly growing sequence with certain divisibility properties:

$$(12.6.3) \qquad a_k = (k^2)! + k.$$

Clearly $a_k \equiv k \pmod{d}$ for $d \leq k^2$. It follows that

$$(12.6.4) \qquad a_k, a_{k-1}, \ldots, a_{k-d_k+1} \text{ is a complete residue system mod } d_k$$

if $d_k \leq (k - d_k + 1)^2$. For every k, we select a relatively large d_k with this property such that

$$(12.6.5) \qquad d_k \leq d_{k+1}$$

and

$$(12.6.6) \qquad \lim_{k \to \infty} d_k/k = 1.$$

We can take, for example, $d_k = \lfloor k - \sqrt{k} \rfloor$.

Let now n be arbitrary and define k by

$$(12.6.7) \qquad ka_k \leq n < (k+1)a_{k+1}.$$

Then by (12.6.4), there exists $0 \leq s < d_k$ satisfying $n \equiv a_{k-s} \pmod{d_k}$, so

$$(12.6.8) \qquad n = a_{k-s} + rd_k.$$

From (12.6.7) and $0 < a_{k-s} \leq a_k$, we obtain

$$(12.6.9) \qquad \frac{(k-1)a_k}{d_k} \leq r = \frac{n - a_{k-s}}{d_k} < \frac{(k+1)a_{k+1}}{d_k}.$$

Let B_k be the set of integers rd_k where r satisfies (12.6.9), and let

$$(12.6.10) \qquad B = \bigcup_{k=1}^{\infty} B_k = \bigcup_{k=1}^{\infty} \left\{ rd_k \mid \frac{(k-1)a_k}{d_k} \leq r < \frac{(k+1)a_{k+1}}{d_k} \right\}.$$

Then A and B are complements by (12.6.8).

To verify (12.6.2), we estimate $A(n)$ and $B(n)$ from above.

Since $n < (k+1)a_{k+1} < a_{k+2}$ by (12.6.7) and (12.6.3), thus

$$(12.6.11) \qquad\qquad\qquad A(n) \leq k+1$$

(in fact, $A(n) = k$ or $k+1$).

By (12.6.10), the smallest element in B_{k+2} is at least $(k+1)a_{k+2}$, which is greater than n by (12.6.7), therefore we do not have to consider B_{k+2} when checking $B(n)$. So

$$(12.6.12) \qquad B(n) \leq B_{k+1}(n) + B_k(n) + |B_{k-1}| + \left| \bigcup_{i=1}^{k-2} B_i \right|.$$

Let us examine the terms on the right-hand side of (12.6.12) one by one.

By (12.6.10), the smallest element in B_{k+1} is at least ka_{k+1}, so B_{k+1} has a role in $B(n)$ only if

$$(12.6.13) \qquad\qquad ka_{k+1} \leq n, \quad \text{or} \quad a_{k+1} \leq \frac{n}{k}.$$

Even in this case, $n < (k+1)a_{k+1}$ by (12.6.7), hence at most the multiples of d_{k+1} between ka_{k+1} and $(k+1)a_{k+1}$ count in $B_{k+1}(n)$, so

$$(12.6.14) \qquad B_{k+1}(n) \leq \frac{(k+1)a_{k+1} - ka_{k+1}}{d_{k+1}} + 1 = \frac{a_{k+1}}{d_{k+1}} + 1 \leq \frac{n}{kd_{k-1}} + 1$$

(we used (12.6.5) and (12.6.13) for the last inequality).

Similarly, $B_k(n)$ counts the multiples of d_k between $(k-1)a_k$ and n, so

$$(12.6.15) \qquad B_k(n) \leq \frac{n - (k-1)a_k}{d_k} + 1 \leq \frac{n - (k-1)a_k}{d_{k-1}} + 1.$$

Also,

$$(12.6.16) \qquad |B_{k-1}| \leq \frac{ka_k - (k-2)a_{k-1}}{d_{k-1}} + 1 \leq \frac{ka_k}{d_{k-1}} + 1.$$

Finally, for $i \leq k-2$, every element in B_i is less than $(k-1)a_{k-1}$, so

$$(12.6.17) \qquad\qquad \left| \bigcup_{i=1}^{k-2} B_i \right| \leq (k-1)a_{k-1} - 1.$$

From (12.6.12), (12.6.14), (12.6.15), (12.6.16), and (12.6.17), we get

$$(12.6.18) \qquad B(n) \leq \frac{\frac{n}{k} + n + a_k}{d_{k-1}} + (k-1)a_{k-1} + 2.$$

By (12.6.7), $a_k \leq n/k$, so (12.6.18) implies

$$(12.6.19) \qquad B(n) \leq n \left(\frac{1 + \frac{2}{k}}{d_{k-1}} + \frac{(k-1)a_{k-1} + 2}{n} \right).$$

Combining (12.6.11) and (12.6.19), we have

$$(12.6.20) \qquad \frac{A(n)B(n)}{n} \leq \frac{(k+1)(1 + \frac{2}{k})}{d_{k-1}} + \frac{(k+1)((k-1)a_{k-1} + 2)}{n}.$$

If n, and thus also k, tends to infinity, then the first fraction on the right-hand side of (12.6.20) tends to 1 by (12.6.6), and the second fraction tends to 0 by (12.6.7) and (12.6.3), so

$$(12.6.21) \qquad \limsup_{n \to \infty} \frac{A(n)B(n)}{n} \leq 1.$$

Since A and B are complements, (12.6.1) holds, thus, combining (12.6.1) and (12.6.21), we get the desired formula (12.6.2). $\qquad \square$

We say that B is a *completely economical complement* (CEC) of A if it is a complement of A and (12.6.2) holds. By Theorem 12.6.1, $A = \{(k^2)! + k \mid k = 1, 2, \dots\}$ has a CEC. Ruzsa proved that every $A = \{a_1 < a_2 < \cdots\}$ satisfying $\lim_{k \to \infty} a_{k+1}/(ka_k) = \infty$ has a CEC (thus much denser sets than A in Theorem 12.6.1 have this property even without any divisibility requirements).

Next we examine how rare complements are for powers of two and for the primes. Ruzsa proved that the powers of two have a CEC. We verify a slightly weaker result:

Theorem 12.6.2. *The powers of two $W = \{2, 4, 8, \dots\}$ have a complement M satisfying*

$$(12.6.22) \qquad M(n) < cn/\log_2 n,$$

where c is an effectively computable constant. ♣

As $W(n) = \lfloor \log_2 n \rfloor$, $W(n)M(n)/n < c$, which is just slightly worse than (12.6.2). The set $W_s = \{s, s^2, s^3, \dots\}$ consisting of the powers of any integer $s > 1$ has a CEC.

Proof. Since 2 is a primitive root mod 9, it is a primitive root mod 3^r for every r (see part Y2 in the proof of Theorem 3.3.5). This means that if $(3, n) = 1$, there exists k, $0 < k \leq \varphi(3^r) < 3^r$, satisfying $n \equiv 2^k \pmod{3^r}$. If $3 \mid n$, then we have $n - 1 \equiv 2^k \pmod{3^r}$. Thus for every n and r, there exist v and $0 < k < 3^r$ satisfying

$$(12.6.23) \qquad n = 2^k + 3^r v \quad \text{or} \quad n = 2^k + 3^r v + 1.$$

Accordingly, the complement M will consist of suitable integers of the form $3^r v$ and $3^r v + 1$.

For a given n, we first choose r and then check which values of v are needed.

Since $k < 3^r$ implies $2^k < 2^{3^r}$, v is positive in (12.6.23) if $2^{3^r} \leq n$. Therefore we choose r to satisfy

$$(12.6.24) \qquad 2^{3^r} \leq n < 2^{3^{r+1}}.$$

Then by (12.6.23) and (12.6.24),

$$v \le 3^r v < n < 2^{3^{r+1}},$$

so let

(12.6.25) $M = \bigcup_{r=1}^{\infty} M_r, \quad \text{where} \quad M_r = \left\{ 3^r v, 3^r v + 1 \mid 0 < v < 2^{3^{r+1}} \right\}.$

By the previous considerations, M is a complement of W.

   We show now that (12.6.22) holds.

   Let

(12.6.26) $K = \{ 3^r v \mid 0 < r, 0 < v < 2^{3^{r+1}} \}.$

Then

(12.6.27) $M(n) \le 2|K|.$

   We divide K into two parts K_1 and K_2 depending on $v \le T$ and $v > T$, resp., where we choose a suitable T later (as a function of n).

   In K_1, there are T possible values for v, and at most $\log_3 n$ values for r, so

(12.6.28) $|K_1| \le T \log_3 n.$

   By (12.6.26), $T < v < 2^{3^{r+1}}$ in K_2, so

(12.6.29) $3^{r+1} > \log_2 T.$

At most $\lfloor n/3^r \rfloor$ values of v belong to 3^r, hence

$$|K_2| < \sum_{r \ge r_0} \frac{n}{3^r} = \frac{3}{2} \cdot \frac{n}{3^{r_0}},$$

where r_0 is the smallest value of r satisfying (12.6.29). This implies

(12.6.30) $|K_2| < \frac{9}{2} \cdot \frac{n}{\log_2 T}.$

   By (12.6.27), (12.6.28), and (12.6.30), we have $M(n) < 2T \log_3 n + 9n/\log_2 T$. Choosing $T = \lfloor n/(\log_2 n)^2 \rfloor$ for example, we arrive at (12.6.22). $\square$

Now we find a rare complement to the primes. The best known result is due to Erdős:

Theorem 12.6.3. *The set P of the prime numbers has a complement R satisfying*

$$(12.6.31) \qquad R(n) < c \log^2 n$$

(*where c is an explicitly computable constant and* log *denotes the natural logarithm*). ♣

Since $P(n) = \pi(n) \sim n/\log n$, then $P(n)R(n)/n < c \log n$, which is significantly weaker than (12.6.2). Ruzsa verified that (12.6.2) is not attainable, so P has no CEC.

The main line of the proof. We construct a probability space that consists of certain sequences R of positive integers, show that any sequence R is a complement of P with probability 1, and $R(n) \sim c \log^2 n$ holds with probability 1. This implies that there exists an R meeting the requirements of the theorem. (This argument verifies only the existence of a suitable sequence without explicitly constructing one. Moreover it guarantees that nearly all sequences are suitable, which should be understood, of course, as a function of the probability in question.)

Let $0 \le \alpha_i \le 1$, $i = 1, 2, \ldots$ be real numbers. Then there exists a probability space consisting of certain sequences R of positive integers, where the probability of $n \in R$ is α_n for every positive integer n and the events $n \in R$ and $m \in R$ are independent for any $n \ne m$. We can imagine this as choosing the integers $1, 2, \ldots$ in the sequences independently with probabilities $\alpha_1, \alpha_2, \ldots$

Let

$$(12.6.32) \qquad \alpha_i = \min(1, d(\log i)/i),$$

where we will specify the constant $d > 0$ later.

We sketch first a proof that a sequence R is a complement of P with probability 1.

Let Q_n be the event that n cannot be written as $n = p + r$, where p is a prime and $r \in R$, and we denote the probability of Q_n by q_n. R will be a complement of P if and only if only finitely many events Q_n occur. By the Borel–Cantelli lemma this has probability 1 if the infinite series of probabilities q_n is convergent, or

$$(12.6.33) \qquad S = \sum_{n=1}^{\infty} q_n < \infty.$$

We compute q_n. For a prime p, $n \ne r + p$ is equivalent to $n - p \notin R$, which has probability $1 - \alpha_{n-p}$. The event Q_n means that n cannot be written as $n = r + p$ with any prime p, so

$$(12.6.34) \qquad q_n = \prod_{p < n}(1 - \alpha_{n-p}).$$

By (12.6.34) and $1 - x \le e^{-x}$, the sum S in (12.6.33) obeys

$$(12.6.35) \qquad S = \sum_{n=1}^{\infty} q_n = \sum_{n=1}^{\infty} \prod_{p<n}(1 - \alpha_{n-p}) \le \sum_{n=1}^{\infty} e^{-\sum_{p<n} \alpha_{n-p}}.$$

The exponent of e on the right-hand side of (12.6.35) is about

$$(12.6.36) \qquad\qquad -d \sum_{p<n} \frac{\log(n-p)}{n-p}$$

by (12.6.32). It can be proved that

$$(12.6.37) \qquad\qquad \sum_{p<n} \frac{\log(n-p)}{n-p} > h \log n$$

with a suitable constant h if n is large enough. Hence the quantity in (12.6.36) is less than $-dh \log n$, so by (12.6.35),

$$S < \sum_{n=1}^{\infty} e^{-dh \log n} = \sum_{n=1}^{\infty} n^{-dh},$$

which is convergent if d is chosen to satisfy $dh > 1$.

Seemingly, (12.6.37) is similar to the relation $\sum_{p<n}(\log p)/p \sim \log n$ in Theorem 5.6.3, as both contain terms of type $(\log k)/k$. However, due to the thinning of the primes, the latter sum is dominated by larger terms $(\log k)/k$ belonging to small values of k, whereas the situation is just the opposite in (12.6.37). To verify (12.6.37), we need that the primes are sufficiently dense even in later relatively small intervals.

Now we turn to sketching the proof that $R(n) \sim c \log^2 n$ holds with probability 1. We can write $R(n) = \sum_{i=1}^{n} \xi_i$ where the random variable ξ_i is 1 if $i \in R$ and is 0 if $i \notin R$. Then the expectation

$$\sum_{i=1}^{n} E(\xi_i) = \sum_{i=1}^{n} \alpha_i \sim \sum_{i=1}^{n} d \frac{\log i}{i} \sim d \int_1^n \frac{\log x}{x} \, dx = \frac{d \log^2 n}{2}.$$

Thus it suffices to show that $\sum_{i=1}^{n} \xi_i \sim \sum_{i=1}^{n} E(\xi_i)$ with probability 1. This is true in general if $E(\xi_i)$ and the standard deviations $D(\xi_i)$ satisfy certain conditions which hold in this case. $\qquad\square$

Finally, we state Lorentz's result about complements of general sets without proof:

Theorem 12.6.4. *For any A, there exists a complement B satisfying*

$$B(n) < 10 \sum_{i=a_1}^{n} \frac{\log A(i)}{A(i)}. \qquad\qquad \clubsuit$$

Exercises 12.6

1. Generalize the example at the beginning of this section to a number system with an arbitrary base $c > 1$ instead of 10 and for an arbitrary grouping of the places instead of the even-odd distribution. Verify that these sets A and B are always complements, and compute $\liminf_{n \to \infty} A(n)B(n)/n$.

2. Let W be the set of powers of 2 and $P_1 = \{ p, p + 1 \mid p \text{ is a prime } \}$, so we include the numbers $p + 1$ into P_1. Are W and P_1 complements?

3. Decide for each of the following conditions whether or not it is necessary or sufficient for the set $A = \{a_1 < a_2 < \ldots\}$ to have a finite complement, so every sufficiently large integer is the sum of an element in A and an element in B for some suitable finite set B.

   (a) $a_{i+1} - a_i$ is bounded.

   (b) A contains an infinite arithmetic progression.

   (c) $\liminf_{n \to \infty} A(n)/n > 0$.

   (d) $\lim_{n \to \infty} A(n)/n = 1$.

* 4. Let A consist of the numbers $a_k = 6^k + k$, and B consist of the multiples of d_k between $6^k(1 - 1/k)$ and 6^{k+1}, where d_k is an integer of the form $2^i 3^j$ satisfying $d_k < k - 5 \log_6 k$, but also $d_k \sim k$ and $d_{k+1} \geq d_k$. Verify that A and B are completely economical complements.

5. Show that Theorem 12.6.4 guarantees a complement S to the primes with $S(n) < c \log^3 n$ (which is thus weaker than Theorem 12.6.3).

6. Prove that any infinite set A has a complement B of density zero, i.e. $B(n)/n \to 0$, as $n \to \infty$.

Answers and Hints

A.1. Basic Notions

1.1.

1. The six digit number is 1001 times the three digit number, and 1001 is divisible by 91.

2. Show that in the product $a^2 - b^2 = (a-b)(a+b)$, both factors are even and exactly one of them is divisible by 4.

   Another option: $(2k + 1)^2 - (2m + 1)^2 = 4k(k + 1) - 4m(m + 1)$, and both terms are multiples of 8 on the right-hand side.

3. $\overline{bca} = 100b + 10c + a = 10 \cdot \overline{abc} - 999a$.

4. Multiply $5a + 9b$ by a suitable integer so that adding an appropriate multiple of 23 we obtain just $3a + 10b$.

5. True: (b), (d), (f).

6. (i) Apply the identity $a^n - b^n = (a - b)(a^{n-1} + a^{n-2}b + \cdots + b^{n-1})$.

   (ii)–(iii) Apply (i) replacing b by $-b$.

7. $c = \pm 3$.

8. $11^{n+2} + 12^{2n+1} = 12(144^n - 11^n) + 133 \cdot 11^n$.

   We can also use induction.

9. $n = 4k + 2$. (It can be shown that there are no other appropriate n.)

10. $(b-1)^2 \mid b^k - 1$ is equivalent to $b-1 \mid b^{k-1} + b^{k-2} + \cdots + 1$. Rewrite the right-hand side as

$$(b^{k-1} - 1) + (b^{k-2} - 1) + \cdots + (1 - 1) + k.$$

Here, the first k terms are divisible by $b - 1$.

11. If $a \geq b$, then $2^a + 1 = 2^{a-b}(2^b - 1) + 2^{a-b} + 1$. Continuing, we obtain $2^b - 1 \mid 2^d + 1$ for some $d < a$. Then $2^b - 1 \leq 2^d + 1 \leq 2^{b-1} + 1$ implying $b \leq 2$.

Another way: If b has an odd prime divisor $c > 1$, then $2^c - 1 \mid 2^{ac} - 1$, and $2^c - 1 \mid 2^b - 1 \mid 2^a + 1 \mid 2^{ac} + 1$, thus $2^c - 1 \mid 2$, a contradiction. If b is a multiple of 4, then $15 = 2^4 - 1 \mid 2^b - 1 \mid 2^a + 1$, but this is impossible since $3 \mid 2^a + 1 \iff a$ is odd, $5 \mid 2^a + 1 \iff a = 4k + 2$.

12. (a) If $a = bq$, then $|a| = |b| \cdot |q| \geq |b| \cdot 1$ for $q \neq 0$.

(b) Part (a) implies that a has $2 \cdot |a|$ divisors at most.

13. The largest and second largest proper divisors are less than or equal to the half and the one third of the number. Answers:

(a) the positive even numbers

(b) the positive multiples of 3 and/or 4 (only $3k = k + k + k$, $4k = 2k + k + k$, and $6k = 3k + 2k + k$ are possible).

14. Denoting the digits backwards by $a_0, \ldots, a_s$,

$$\overline{a_s a_{s-1} \ldots a_1 a_0} = a_s 10^s + a_{s-1} 10^{s-1} + \cdots + a_1 10 + a_0.$$

Observe:

(a) $10^k - 1$ is divisible by 9.

(b) 10^k is divisible by 4 and 25 for $k \geq 2$.

(c) 10^k is divisible by 8 and 125 for $k \geq 3$.

(d) $10^k + 1$ or $10^k - 1$ is divisible by 11 depending on whether k is odd or even.

15. No, check the divisibility by 3.

16. Yes, prove by induction that to any k there exists a k-digit number divisible by 2^k and consisting only of digits 1 and 2.

17. (b) $\binom{n}{k}$ is an integer.

18. The first player has a winning strategy for every $n > 1$.

19. Factor the numbers into the product of a power of two and an odd number and use the pigeonhole principle. We can also use induction.

20. In $0 = 0 \cdot q$, the number q is not unique.

21. (a) $n = 4k + 2$. (b) $n = \pm 4$.

22. (a) Divisible; the quotient is of the required form after eliminating the square root in the denominator.

(b) $1 + \sqrt{2} \mid 1$.

(c) The powers of $1 + \sqrt{2}$ are units.

(d) Infinitely many.

(e) If $\pm 1 = c^2 - 2d^2 = (c+d\sqrt{2})(c-d\sqrt{2})$, then $c+d\sqrt{2} \mid 1$. To prove the converse, show that if $c + d\sqrt{2} \mid r + s\sqrt{2}$, then $c^2 - 2d^2 \mid r^2 - 2s^2$.

(f) If there were another unit, then multiplying it by a suitable $\pm(1 + \sqrt{2})^k$ we obtain a unit $u + v\sqrt{2}$ with

$$1 < u + v\sqrt{2} < 1 + \sqrt{2}.$$

Using (e), we get a contradiction in all the four cases of the signs of u and v.

(g) Both occur infinitely often; this follows basically from (e).

23. (d) (ii) and (iv) are true in any integral domain, (i) and (iii) are true if and only if there is an identity element. (If there is no identity, $a \mid b, b \mid a \iff a = b = 0$.)

1.2.

1. Answer: 97. Hint: The three-digit number divides the difference of the two numbers.

2. There are only m possible remainders, so there must be infinitely many powers of two all giving the same remainder when divided by m.

3. Consider the remainders of $c_1, c_1 + c_2, \ldots, c_1 + c_2 + \cdots + c_n$ when divided by n.

4. Given m, consider the integers having as digits only 1s and having at most $m + 1$ digits: $1, 11, 111, \ldots$ There must occur two among them with the same remainder when divided by m, hence their difference is of the required form and is a multiple of m.

5. Let r_k be the remainder of φ_k on division by m. The pairs (r_k, r_{k+1}) can assume only m^2 distinct values, therefore $(r_t, r_{t+1}) = (r_s, r_{s+1})$ for some $t > s$. Show that $(r_k, r_{k+1}) = (r_{k+t-s}, r_{k+t-s+1})$ for every k, i.e. the sequence of the remainders r_n is periodic (with a period $t - s$). As $r_0 = 0$, also $r_{j(t-s)} = 0$ for every j, so $m \mid \varphi_{j(t-s)}$.

6. (a) Every integer is of the form $3k$ or $3k \pm 1$, so its square is of the form $3s$ or $3s+1$. This means that a square can have a remainder 0 or 1 on division by 3.

   (b) $0, 1$. (c) $0, \pm 1$. (d) $0, 1, 4$.

7. Examine the remainder of the sum on division by 3 or 4.

8. (a) No. Examine the remainders of divisions by 4 and 5.

   (b) Similar to (a), one can show that there is no such square with eight or more digits and a four- or six-digit number must terminate with 4. Finally, check the divisibility by 11 and $111/3 = 37$. Answer: 7744 is the only solution.

9. Verify that an odd power of a number gives the same remainder as the number itself when divided by 3.

10. Answer: 16 (so the product is always a multiple of 2^{16} but not of 2^{17} in some cases).

11. $\lfloor\sqrt{n}\rfloor = k$ holds exactly for $k^2 \le n < (k+1)^2$. Of these, k^2, $k^2 + k$, and $k^2 + 2k$ are divisible by k. Answer: $3(10^5 - 1) = 299997$.

12. $\lfloor a + b\rfloor - (\lfloor a\rfloor + \lfloor b\rfloor) = 0$ or 1.

13. No: e.g. $|r| \ge 4$ if $12 = 4q + r$.

14. Let t be the base of the number system. If $d \mid t - 1$, then the remainder on division by d equals the remainder of the sum of the digits. If $d \mid t^k$, then the remainder equals the remainder of the number composed from the last k digits. If $d \mid t + 1$, then the remainder equals the remainder of the alternating sum of the digits (the last digit has to be taken with a positive sign).

15. This is the special case $t = 100, d = 99$ of the previous exercise.

16. Consider the remainder on division by 9. Answer: 8.

17. We convert each digit in base 9 into a two-digit number in base 3 (with first digit 0 if necessary). We can apply a similar procedure if one base is a power of the other (with positive integer exponent).

18. Answer: $n = 8$. Hint: $t^3 \le n \le (t + 1)^2 - 1$ implies $t = 2$.

19. From $t \mid 735$, $t \ge 6$, and $t < 10$ we get $t = 7$.

20. (a) We can measure every integer gram up to $2^{10} - 1 = 1023$ with weights of $1, 2, 4,$ $\ldots, 2^9$ grams. This is the maximum. When measuring, there are two options for each weight: whether or not we put it onto the pan. Thus ten weights can measure at most $2^{10} - 1$ values (we subtract 1 for the case when we put no weight onto the pan).

    (b) We can measure every integer gram up to $(3^{10} - 1)/2$ with weights of $1, 3, 9, \ldots,$ 3^9 grams: in base three representation we have to convert the digits 2 to -1. There is no better stock of weights: when measuring, there are three options for each weight (left pan, right pan, no pan) but the result has to be divided by 2 due to the symmetry of the two pans.

21. The limit is $\log_2 10 = 3.3219\ldots$.

22. Apply a suitable modification of the proof of Theorem 1.2.2.

23. Though the numbers increase rapidly in the beginning, we will get 0 in finitely many steps. The reason is that we gradually "lose" all digits.

1.3.

1. $14 = 3794 \cdot (-44) + 2226 \cdot 75$.

2. (a) If $(3n + 5, 7n + 12) = d$, then $d \mid -7(3n + 5) + 3(7n + 12) = 1$, so $d = 1$.

    (b) If $(3n^2 + 1, 4n^2 + 3) = d$, then $d \mid -4(3n^2 + 1) + 3(4n^2 + 3) = 5$, but $5 \nmid 3n^2 + 1$, thus $d = 1$.

    (c) If $(n! - 1, (n + 1)! - 1) = d$, then $d \mid (n + 1)! - 1 - (n + 1)(n! - 1) = n$, hence $d \mid n! - (n! - 1) = 1$.

(d) If $(7^n - 2, 7^{n+1} - 5) = d$, then $d \mid (7^{n+1} - 5) - 7(7^n - 2) = 9$, but $7^n - 2 = (2 \cdot 3 + 1)^n - 2 = 3k + 1 - 2 = 3k - 1$, thus d is not divisible by 3.

3. 1 if n is odd, and 2 if n is even.

4. (a) 5 or 10.　(b) 5, 15, or 45.

5. 6, 10, 15, or 21, 66, 77, etc.

6. True: (a), (c).

7. Answer: (a, b). Hint: $b \mid ka \iff \frac{b}{(a,b)} \mid k$.

8. (a) True: Since $(a+n, b+n) \mid (a+n)-(b+n) = a-b$, n works if $a+n = k(a-b)+1$.
   (b) True.　(c) False, $a = 1$, $b = 4$ is a counterexample.

9. (a) Infinitely many; if u, v is appropriate, then the pair $u + tb$, $v - ta$ also works for any integer t.
   (b) 1.
   (c) (a, b).

10. (b) Use that δ and δ_1 divide each other.

11. Verify first $c(a, b) \mid (ca, cb)$. Then prove that q is a unit in the equality $c(a, b)q = (ca, cb)$.

12. (a) These are exactly the numbers coprime to 10 (i.e. which are divisible neither by 2 nor by 5). Hint: Apply the argument of Exercise 1.2.4 and then Theorem 1.3.9.
    (b) The smallest one is the repunit consisting of 3^{1000} digits. Hint: Show by induction on k that the smallest repunit multiple of 3^k has 3^k digits.

13. Use $r \mid s \Rightarrow c^r - 1 \mid c^s - 1$ several times and the representation $(n, k) = nu + kv$.

14. (a) Show that if n and k are powers of two and $k < n$, then $a^k + 1 \mid a^n - 1$.
    (b) $a^{(n,k)} + 1$ if both $n/(n, k)$ and $k/(n, k)$ are odd, and 1 or 2 otherwise depending on whether a is even or odd.

15. The second neighbors are coprime. The third neighbors with indices divisible by 3 have gcd 2, the others are coprime.

16. Use the identity $\varphi_{m+n} = \varphi_{m-1}\varphi_n + \varphi_m\varphi_{n+1}$. Based on this, we can prove $k \mid n \Longrightarrow \varphi_k \mid \varphi_n$ by induction on n/k. For the converse and the claim concerning the gcd, verify that $a = bq + r$ implies $(\varphi_a, \varphi_b) = (\varphi_b, \varphi_r)$. An alternative method: Show that for every m, the indices of the Fibonacci numbers divisible by m are just the multiples of the index of the smallest Fibonacci number with this property.

17. We denote the lengths of the two segments by a and b, and k and n are suitable positive integers.
    (a) If $a/b = k/n$, then $a/k = b/n$ is a common measure. Conversely, if c is a common measure, $a = kc$ and $b = nc$, so $a/b = k/n$.
    (b) Infinitely many; dividing a common measure by any integer n, we get a common measure again.

(c) The analog of the division algorithm: We measure the smaller segment on the larger one as many times as possible, so $a = bq+r$ where q is a positive integer, r is a real number, and $0 \le r < b$. If two segments are commensurable, so $a = kc$ and $b = nc$ (with a common measure c), then the Euclidean algorithm with a and b is essentially the same as the similar procedure with the integers k and n, therefore it terminates. Conversely, if the Euclidean algorithm for the segments terminates, then the last non-zero remainder is a common measure.

(d) The existence of such a special common measure follows from the Euclidean algorithm.

(e) We start the Euclidean algorithm by measuring the side of length b of the square $ABCD$ from A on the diagonal AC of length a. We obtain an endpoint E with $AE = b$ and $EC = r$. The perpendicular to the diagonal in E intersects side BC in F. Then $r = EC = EF = FB$. In the next step of the Euclidean algorithm we divide b by r in the following way. We first measure BF on BC and then perform the division algorithm for the hypotenuse CF and the leg CE of the isosceles right triangle EFC. But this leads to the original state on a smaller scale: we have to compare the diagonal and the side of a (smaller) square. This shows that the Euclidean algorithm goes on for ever.

1.4.

1. Answer: (a) and (b) 3.   (c) 5.   (d) 7. Hint: Check the remainder on division by 3, 5, and 7.

2. No; if the difference d is positive and $c > 1$ is an arbitrary element in the arithmetic progression, then e.g. $c + cd$ is composite.

3. Answer: 3 years old. Hint: Consider the remainders on division by 3.

   Remark: We have no information about the ages of the two older grandchildren: e.g. 3, 5, 7, or 3, 7, 11, or 3, 13, 17 are all triples satisfying the requirements. It is an unsolved problem whether or not there are infinitely many such triples. However, the smallest element of every such triple must be 3, as claimed in this exercise.

4. (a) $a - 1 \mid a^k - 1$ and if $k = rs$, then $a^r - 1 \mid a^k - 1$.

   (b) If $k = rs$ with s odd, then $a^r + 1 \mid a^k + 1$.

5. Answer: $t = 2$, $k = 1$. Hint: Check the divisibility by $t + 1$ or t.

6. Answer: (a), (d), (e) $n = 1$.   (b) $n = 2, 4$.   (c) There is no such n. Hint: Check the divisibility by 3 for (a), and factor the other four expressions.

7. (a) If $n = ab$ with $0 < a \le b$, then $a \le \sqrt{n}$, thus only $a = 1$ is possible.

   (b) If this smallest divisor d had a non-trivial positive divisor s, then $s \mid n$ and $1 < s < d$ yield a contradiction.

   (c) If $n = dk$ where d is the minimal divisor greater than 1, then d is a prime by (b) and k is a prime by (a).

8. Use the prime property of 17.

9. The irreducible elements are the numbers $4k + 2$, and there are no primes.

10. (a) Consider the divisibility $p \mid p^2$ (and use the argument seen at the solution of Exercise 1.1.23a).

    (b) Follow part I in the proof of Theorem 1.4.3.

1.5.

1. If $a = p_1 \dots p_r$, then $|a| \geq 2^r$ since $|p_i| \geq 2$.

2. (a) $2t, \pm 2^k$, and $2^k p$ where t is an odd number and p is any odd integer irreducible among the integers.

   (b) E.g. $2^2 \cdot 3^{1998}$.

3. First proof: An irreducible element is not necessarily prime (and there are no primes at all).

   Second proof: The implication $p_1 \mid q_1 - p_1 \Rightarrow p_1 \mid q_1$ occurring in the last step is false among the even numbers as here $p_1 \nmid p_1$.

4. $1000 = 20 \cdot 50 = 10 \cdot 10 \cdot 10$.

5. We use the fact that the non-zero elements of F have a unique decomposition as $2^k 5^m t$ where the exponents k and m are integers and t is coprime to 10.

   (a) Units: $\pm 2^k 5^m$. Irreducibles: $2^k 5^m p$ where $p \neq \pm 2, \pm 5$ is irreducible in the integers.

   (b) The factorization of $2^k 5^m t$ in F is essentially the same as the decomposition of t among the integers.

   (c) Let $f(2^k 5^m t) = |t|$ and $f(0) = 0$.

6. We obtain $p_1 \mid q_1 q_3 \dots q_s$ in the last step. To show its impossibility, use the induction hypothesis again for $a = q_1 q_3 \dots q_s$.

7. Let $a = \pm p_1^{k_1} \dots p_r^{k_r}$ where the p_i are pairwise distinct positive irreducibles, $k_i > 0$, and $k = k_1 + \dots + k_r$. Then the number of decompositions is $2^{k-1} k! / (k_1! \dots k_r!)$.

8. Assuming that p is an irreducible element dividing ab, establish the decomposition of ab into a product of irreducible factors from the decompositions of a and b. Observe that an associate of p must occur in the decomposition of ab.

9. The appropriate triples p_1, p_2, p_3 are $5, 2, 2; -5, -2, -2; 5, 2, -3; 5, -3, 2; -5, -2, 3; -5, 3, -2$. Hint: After ordering, we obtain $p_2 p_3 = (p_1 - p_2 - p_3)(p_2 + p_3)$. By the Fundamental Theorem, this can hold only if $p_2 + p_3 = \pm p_2, \pm p_3, \pm 1$, or $\pm p_2 p_3$.

10. Answer: 2 and 3. Hint: Writing $x^3 + y^3 = p^\alpha$, we can assume that x and y are coprime. Factoring the left-hand side, both factors must be powers of p. Express xy from these two equalities.

1.6.

1. An integer n is a kth power if and only if the exponents of all primes are multiples of k in the standard form of n.

2. (a) Let p be an arbitrary prime divisor of the factor a in the product ab. Since $(a, b) = 1$, $p \nmid b$, so p occurs with the same exponents in the standard forms of a and ab. Now apply Exercise 1.6.1.

   (b) The two factors will be associates of kth powers except if the product is zero.

   (c) We have to assume that the factors are pairwise coprime.

3. Rely on Exercise 1.6.2a.

4. Answer: 3 and 7. Hint: Factor the numerator and argue as in Exercise 1.6.2a.

5. (a) If $a_1 \mid a$ and $b_1 \mid b$, then $a_1 b_1 \mid ab$ follows from the elementary properties of divisibility. For the converse, use Theorem 1.6.2. Consider an arbitrary prime divisor p of ab and let the (possibly 0) exponents of p be α, β, and γ in the standard forms of a, b, and c. The condition $c \mid ab$ implies $\gamma \leq \alpha + \beta$. Thus, we have to show that $\gamma = \alpha' + \beta'$ for some $0 \leq \alpha' \leq \alpha$ and $0 \leq \beta' \leq \beta$.

   (b) Apply the argument of (a) knowing that either α, or β is 0. An alternative way: Assume $a_1 b_1 = a_2 b_2$ where $a_i \mid a$ and $b_i \mid b$. Then $a_1 \mid a_2 b_2$ and $(a_1, b_2) = 1$, thus $a_1 \mid a_2$. We obtain the converse divisibility similarly, therefore (using positivity) $a_1 = a_2$.

   (c) For example, any common divisor $c > 1$ of a and b can be represented as $c = 1 \cdot c = c \cdot 1$.

   (d) Use the arguments of (a) and (b).

   (e) $(a, b) \mid c \mid [a, b]$.

6. Use Theorem 1.6.2.

7. (a) 2^{30}.   (b) $2^{10} \cdot 3^2$.   (c) $2^3 \cdot 3 \cdot 5 \cdot 7 = 840$.

8. These are the squares. Hint: Use the formula for $d(n)$ and Exercise 1.6.1. Another way: Form pairs of divisors matching every $d \mid n$ to its complementary divisor n/d. This match is not perfect if a divisor is equal to its complementary divisor.

9. Answer: 20. Hint: Examine which guards touched a lock and apply the previous exercise.

10. (b) Equality holds if and only if the exponents of all primes are odd in the standard form of n.

11. (a)–(b) Check how many divisors of n can be larger than $n/2$ and $n/3$.

    (c) Form pairs of divisors whose product is n. The smaller (more precisely, not greater) element in each pair is at most $\sqrt{n}$. Another possibility: Apply the argument in (a) and (b) for a general n/k and choose the optimal value of k.

12. Answer: $n^{d(n)/2}$. Hint: Form pairs of divisors.

13. Answer: $n+1$. Hint: (i) $n+1$ such divisors are $2^i 5^{n-i}, i = 0, 1, \ldots, n$. (ii) Among $n+2$ divisors two must contain 5 with the same exponent by the pigeonhole principle, and so the larger divisor is a multiple of the smaller one.

14. (a) $a \mid b$.

    (b) 8.

    (c) 2^r where r is the number of distinct prime factors in b/a.

    (In (b) and (c), we considered the pairs x, y and y, x as different solutions for $x \neq y$.)

15. Use arguments similar to those in the proof of $(a, b)[a, b] = ab$ (Theorem 1.6.6/III).

16. True: (b), (d).

17. (a) $a \mid [a, b] \mid a + b \implies a \mid b$, and $b \mid a$ follows by symmetry.

    (b) and (d) Divide by (a, b) and apply Exercise 1.6.16b.

    (c) For example, $a = 10k$, $b = 15k$; or $a = u(u + v)$, $b = v(u + v)$.

18. Each equality holds if and only if every common prime divisor of a and b occurs with the same exponent in the standard forms of a and b.

19. Let α, β, and γ be (the possibly 0) exponents of a prime p in the standard forms of a, b, and c. To prove (a), we have to verify $\max(\alpha, \min(\beta, \gamma)) = \min(\max(\alpha, \beta), \max(\alpha, \gamma))$. We can check this in the three cases separately when α is the smallest, middle, or largest among the three exponents. We can prove also (b) along similar lines.

20. (a) Using the notation of the previous exercise, both conditions mean that two exponents of α, β, and γ are equal and the third exponent is not smaller.

    (b) Infinitely many.

    (c) The analog of (a) remains valid if we replace gcd everywhere by lcm. This means for the exponents that two of α, β, and γ are equal and the third one is not larger. The number of solutions is the product of the values δ belonging to the distinct prime divisors of abc where $\delta = 3\alpha + 1$ if $\alpha = \beta = \gamma$, and $\delta = 2\min(\alpha, \beta, \gamma) + 1$ otherwise. (There is a unique solution if and only if $(a, b, c) = 1$.)

21. Factor $p^4 - 1$ as long as you can, and verify the divisibilities by 16, 3, and 5 separately.

22. Factor $a^6 - b^6$ as long as you can, and verify the divisibilities by 7, 8, and 9 separately.

23. Factor the expression, and show the divisibility for each of the prime power factors of 360.

24. Verify the divisibility for each prime power factor in the standard form of the divisor separately. Apply various forms of $a - b \mid a^m - b^m$ and the binomial theorem for the divisibility by 101.

25. (a) 275. (b) The last digit is not zero.

26. (a) Every prime occurs in the standard form of $n!$ with an exponent less than n: if $p^s \le n < p^{s+1}$, then

$$\alpha_p = \sum_{k=1}^{\infty} \left\lfloor \frac{n}{p^k} \right\rfloor = \sum_{k=1}^{s} \left\lfloor \frac{n}{p^k} \right\rfloor \le \sum_{k=1}^{s} \frac{n}{p^k} = \frac{n(p^s - 1)}{p^s(p - 1)} < \frac{n}{p - 1} \le n.$$

(b) $c = 2,\ n = 2^j$.

27. (a) $\binom{n}{k} = (n/k)\binom{n-1}{k-1}$. Thus $k \mid n\binom{n-1}{k-1}$ and $(k, n) = 1$ implies $k \mid \binom{n-1}{k-1}$. This means that $\binom{n}{k}/n = \binom{n-1}{k-1}/k$ is an integer.

(b) False, $\binom{10}{4}$ is a counterexample.

(c) (c1) n is a prime.   (c2) $n = 2^j$.   (c3) $n = 2^j - 1$.

(d) No: $k\binom{n}{k} = n\binom{n-1}{k-1}$. Thus $n \mid k\binom{n}{k}$. If $(n, \binom{n}{k}) = 1$, then this implies $n \mid k$, a contradiction.

28. Exactly the powers of two are appropriate.

29. First solution: Choose a prime that occurs in the standard form of $n! + k$ with a higher exponent than in k.

Second solution: Every integer has a prime divisor greater than $n/2$ and it divides none of the other numbers.

30. 9.

31. The squarefree numbers (i.e. those that are not divisible by any square greater than one).

32. Prove by contradiction. Reduce the problem to the case when the two kth powers are coprime. Show that their difference divides the double of both kth powers. Thus, it also divides 2 which is impossible.

33. (a) $(a/b)^5 = 100 \Rightarrow a^5 = 100b^5$. Examine the exponent of 5 (or of 2) in the standard forms of the two sides.

(b) $6^{a/b} = 18 \Rightarrow 6^a = 18^b$. We may assume $a, b > 0$. Check the exponents of 2 and 3 in the standard forms of the two sides.

35. (a) Yes. (b) No.

A.2. Congruences

2.1.

1. Apply the method in Example E1.

2. Answer: 999. Hint: $999 \equiv -1 \pmod{1000}$.

3. The proof of divisibility by 11:

$$10 \equiv -1 \pmod{11} \implies 10^k \equiv (-1)^k \pmod{11},$$

hence

$$\overline{a_s a_{s-1} \ldots a_1 a_0} = a_s 10^s + a_{s-1} 10^{s-1} + \cdots + a_1 10 + a_0 \equiv$$
$$\equiv a_0 - a_1 + a_2 \pm \cdots + (-1)^s a_s \pmod{11}.$$

4. True: (a), (d), (e), (h).

5. Answer: 50. Hint: We obtain the last digits of the squares by squaring the 101 digits, i.e. all possible remainders modulo 101. To determine the pairwise incongruent values, examine the coincidences induced by squaring. Use Exercise 2.1.4h.

6. The theorem is false, e.g. $\binom{7}{4} \not\equiv \binom{8}{4}$. The proof violated the rule that you must not replace the numerator of a fraction by a congruent value even if both the original and the new fractions are integers.

7. Using $a \equiv b \pmod{m}$, demonstrate that $(a^m - b^m)/(a-b) = a^{m-1} + a^{m-2}b + \cdots + b^{m-1}$ is divisible by m.

8. Show $a \equiv b \pmod 3$ and use it to prove $3 \nmid (a^n - b^n)/(a-b)$.

9. (b) Prove by induction on k, using (a) and $\binom{n}{k} = \binom{n-1}{k} + \binom{n-1}{k-1}$. Another option: To avoid the difficulties with fractions, multiply by the denominator $k!$. Since $(k!, p) = 1$, we obtain an equivalent congruence with the same modulus. To prove this fraction-free version, take the product of the congruences $p - j \equiv -j \pmod p$.

   (c) Apply a suitable modification of either method indicated in part (b).

10. Answer: $p = 5$. Hint: Verify $\binom{3p}{p} \equiv 3 \pmod p$.

11. (a) Cancelling the left-hand side by p, the new denominator $(p-1)!$ is coprime to p. Hence, multiplying by $(p-1)!$, we get an equivalent congruence. This can be proved as in the previous two exercises.

    (b)–(c) Apply similar methods as in (a).

2.2.

1. (a) 3. (b) 5. (c) 2. Hint: The modulus is coprime to the given numbers and divides their difference.

2. (a) $6^2 \cdot 5^{m-2} \cdot m!$ (b) $6 \cdot 5^{\varphi(m)-1} \cdot \varphi(m)!$

   (We considered two residue systems as distinct even if they differed only in the order of the elements.)

3. Both properties depend only on the difference d of the arithmetic progression:
   (a) $d \mid m$. (b) $(d, m) = 1$.

4. (a) m is odd.

   (b) Every m is suitable.

   (c) $m = 2$.

   (d) $(m, 10) = 1$.

   (e) $m = 2$.

   (f) $m = 3^k$.

   (g) m is squarefree.

   (Parts (a) and (d) can be considered as special cases of Exercise 2.2.3b.)

5. (a) $(m, 15) = 1$.

   (b) Every m is suitable.

   (c) $m = 2$.

   (d) $(m, 20) \leq 2$.

   (e) Every m is suitable. This can be verified similar to the proof of 2.2.4g but by a considerably simpler argument.

6. True: (b).

7. (a) The remainder is 0 for m odd and $m/2$ for m even. Hint: Demonstrate that the result does not depend on which complete residue system we consider. Then examine e.g. the least non-negative remainders or the ones of least absolute value. Another way: Form suitable pairs from the elements of a complete residue system.

   (b) Use the result of (a). If m is odd, then we can always exhibit examples both for $a_i + b_i$ forming and not forming a complete residue system.

   (c) The remainder of the sum of elements in a reduced residue system is 0 for $m > 2$. For the sums $a_i + b_i$, we have the same results as seen at the complete residue systems.

8. (a) m is either odd or is a multiple of 4. (b) m is odd.

9. (a) $m = 2^k$. Hint: We get a complete residue system if and only if the given numbers are pairwise incongruent, i.e. $(i+1)+(i+2)+\cdots+j = (i+j+1)(j-i)/2$ is not divisible by m for $0 \leq i < j \leq m-1$. For $m = 2^k$, use the opposite parity of the two factors to show the impossibility of such a divisibility. If m is not a

power of two, i.e. $m = 2^k(2s + 1)$ with $s > 0$ (the exponent k may be 0), then $(2^k - s) + (2^k - s + 1) + \cdots + (2^k + s)$ is divisible by (in fact, is equal to) m. The largest term satisfies the condition $2^k + s < m$, but for the smallest term, $2^k - s \leq 0$ may occur. In this case, deleting all negative terms, their negatives, and 0, we obtain a forbidden sum within the given limits still divisible by m.

(b) m is even.

10. True: (a), (c), (e). Hint for (c) and (e): Show that both assertions follow from the claim:

If $(r, k) = 1$, then there exists an s satisfying $s \equiv r \pmod{k}$ and $(s, m) = 1$.

Proof of the claim: If every prime divisor of m divides k, then $(r, k) = 1 \Rightarrow (r, m) = 1$, thus we can choose $s = r$. Otherwise, let $q_1, \ldots, q_t$ be those prime divisors of m that do not divide k. Assume that $q_1, \ldots, q_j$ are the ones among these that divide r (also $j = 0$ or $j = t$ may occur). Then $s = r + q_{j+1} \ldots q_t k$ satisfies the requirements.

11. (b) Answer: $m/(a, m)$.

Hint: $ar_i + b \equiv ar_j + b \pmod{m} \iff r_i \equiv r_j \pmod{m/(a, m)}$.

12. (a) $(a, m) = 1$ or 2 for $m = 4k + 2$, and $(a, m) = 1$ otherwise.

(b) $p_1 \cdot \cdots \cdot p_s \mid b$ where $p_1, \ldots, p_s$ are the distinct prime divisors of m.

13. $(k, m) = 1$.

14. (c) Use (b).

2.3.

1. Form pairs of the elements of a (cleverly chosen) reduced residue system or use the formula for $\varphi(n)$.

2. (a) 3, 4, 6.

(b) 5, 8, 10, 12.

(c) There is no such n.

(d) 61, 77, 93, 99, 122, 124, 154, 186, 198.

3. (a) $1285 = 5 \cdot 257$. Hint: $\varphi(2^{11}) = 2^{10}$ shows that the minimal number is not greater than 2^{11}. A smaller suitable integer can only be the product of primes of the form $2^k + 1$.

(b) 3^{11}. Hint: Use the following: (i) $2 \cdot 3^{10} + 1$ is composite (17 divides it); (ii) If $3^j \mid p - 1$ for a prime $p(> 2)$, then $p \geq 2 \cdot 3^j + 1$.

4. 100, 80, 50, 40.

5. (a) Use the standard forms of k and n and the formula for φ. Be careful that only positive exponents should occur in each standard form.

(b) It follows from (a).

(c) For the least computation, verify the identity

$$\varphi((a,b))\varphi([a,b]) = (\varphi(a), \varphi(b))[\varphi(a), \varphi(b)].$$

6. Rewrite $\varphi(a)/\varphi(b) = a/b$ as

(A.2.1)
$$\prod_{\substack{p|a \\ p \text{ prime}}} \left(1 - \frac{1}{p}\right) = \prod_{\substack{q|b \\ q \text{ prime}}} \left(1 - \frac{1}{q}\right).$$

If a and b have the same prime divisors, then (A.2.1) clearly holds. To prove the converse, assume that (A.2.1) is true in some other case, too. Delete the common factors $1 - 1/p = 1 - 1/q$ and multiply by the common denominator (i.e. by the product of all remaining primes p and q). Then the largest prime will divide only one side thus yielding a contradiction.

7. True: (a).

8. Let the standard form of k be $k = \prod_{i=1}^{r} p_i^{\beta_i}$, $\beta_i > 0$. Then an appropriate n is

$$n = \prod_{i=1}^{r} p_i^{\alpha_i} \quad \text{where} \quad \alpha_i = \begin{cases} \beta_i, & \text{if } p_i \mid \prod_{j=1}^{r}(p_j - 1) \\ \beta_i + 1, & \text{otherwise.} \end{cases}$$

9. Use that both $r \mid n$ and $(r, n) = 1$ are true only for $r = 1$. Equality holds if and only if n is 1, 4, or a prime. Hint: In every other case there exists a number r, $1 < r < n$, neither coprime to n nor dividing n; e.g. $r = n - p$ where p is the smallest prime divisor of n.

10. (a) and (c) Use the formula for $\varphi(n)$.

    (b) The columns in table 2.3.1 are not complete residue systems mod b.

11. (a) The multiples of the least prime divisor of n are not coprime to n. Equality holds if and only if n is the square of a prime.

    (b) (b1) n is a prime. (b2) 10. (b3) 15, 49. (b4) There is no such n.

12. 1, 2, and 3. Hint: Verify $\varphi(n) \mid n \iff n = 2^{\alpha}3^{\beta}$ where either $\alpha \geq 0$ and $\beta = 0$, or $\alpha > 0$ and $\beta > 0$.

13. Prove by contradiction. Use the formula for φ. The largest prime divisor remaining after cancellation will divide only one side.

14. Write the fractions $1/n, 2/n, \ldots, n/n$ in reduced form, and count how many times a denominator occurs.

15. Using the formula for $\varphi(n)$, prove $\varphi(n) \geq \sqrt{n}/2$.

    Another option: All primes are coprime to n except its prime divisors and there are many primes up to n (see Section 5.4).

16. Denote by $2 = p_1 < p_2 < \ldots$ the sequence of (positive) primes and let p_j be the smallest prime not dividing k. Then $n = (p_j - 1)k$ works.

17. Let $2 = p_1 < p_2 < \cdots < p_{1000}$ be the first 1000 primes and P their product. Then $n_i = P(p_i - 1)/p_i$ satisfy the requirements.

18. Answer: $n \leq 3$. Hint: Compare the exponents of 2 in the standard forms of $\varphi(n!)$ and $k!$.

19. $m = 2^k$, p, or $2p$ where $p > 2$ is a prime.

2.4.

1. $\varphi(n) \leq n$ implies $\varphi(n) \mid n!$. We can solve the exercise without using the Euler–Fermat Theorem. Among $1, 2, 2^2, \ldots, 2^n$ there must be two numbers congruent modulo n by the pigeonhole principle: $2^i \equiv 2^j \pmod{n}$ with some $0 \leq i < j \leq n$. Since $(2, n) = 1$, we can cancel 2^i to obtain $2^{j-i} \equiv 1 \pmod{n}$ with $1 \leq j - i \leq n$. Finally, $j - i \mid n!$ implies the assertion of the exercise.

2. Answer: 49. Hint: $(1793, 10^2) = 1$ implies $1793^{k\varphi(100)} \equiv 1 \pmod{100}$. Compute $\varphi(100)$ and use $1793 \equiv -7 \pmod{100}$.

3. Apply Fermat's Little Theorem for $p = 13$ several times.

4. Prove that one of the numbers is divisible by 7.

5. Exhibit the standard form of the divisor and verify the divisibility for each prime power factor separately by the Euler–Fermat Theorem. Do not forget the cases where the prime power is not coprime to a.

6. Demonstrate that the remainder of a 30th power can be only 0 or 1 modulo 11 and modulo 9.

7. Show that the remainder of an 88th power is 0 or 1 modulo 23.

8. If neither of r_i and r_j is divisible by p, then, multiplying $r_i^{2p-3} \equiv r_j^{2p-3} \pmod{p}$ by $r_i r_j$, we infer $r_i \equiv r_j \pmod{p}$, i.e. $i = j$ by Fermat's Little Theorem.

9. (a) Examine the cases $p \nmid a$ and $p \mid a$ separately as in the proof of Theorem 2.4.1B.

   (b) Let k be the maximum of the exponents in the standard form of m. Then $i, j \geq k$, $i \equiv j \pmod{\varphi(m)} \implies a^i \equiv a^j \pmod{m}$. Hint: Verify $a^i \equiv a^j \pmod{p^\alpha}$ for every prime power factor p^α in the standard form of m. Also use $\varphi(p^\alpha) \mid \varphi(m)$ (see Exercise 2.3.5a).

10. True: (a), (c).

    (a) Use the Euler–Fermat Theorem ($a = 133$, $m = 1000$), or modify the method sketched at Exercise 2.4.1.

    (b) Check the divisibility by 4.

    (c) Start using $136^k \equiv 136 \pmod{1000} \iff 136^{k-1} \equiv 1 \pmod{125}$.

11. Hint: $a^k \equiv a \pmod{d} \iff a^{k-1} \equiv 1 \pmod{d/(a,d)}$.

12. The repunits are the numbers $(10^k - 1)/9$. Thus we have to determine the integers m satisfying $10^k \equiv 1 \pmod{9m}$ for some (positive) k.

13. It is sufficient to show that every odd positive prime divisor p of $n^2 + 1$ is of the form $4k + 1$. To do this, raise $n^2 \equiv -1 \pmod{p}$ to the power $(p-1)/2$ and use Fermat's Little Theorem.

    We can also solve the problem without Fermat's Little Theorem. Assume that some positive integer a of the form $4k-1$ divides n^2+1 for some n. Consider the smallest such a. We shall get a contradiction by finding a positive integer b less than a also of the form $4k - 1$ and dividing some integer $s^2 + 1$.

    As the divisibility $a \mid n^2 + 1$ depends only on the remainder of n on division by a, we may assume $0 \le n \le a - 1$ (or even $|n| \le a/2$).

    Let $n^2 + 1 = aq$. Then $aq = n^2 + 1 \le (a-1)^2 + 1 < a^2$, so $(0 <)q < a$.

    If n is even, then $n^2 + 1$ is of the form $4k + 1$, hence q is of the form $4k - 1$.

    If n is odd, then $n^2 + 1$ is of the form $8k + 2 = 2(4k + 1)$, hence $q/2$ is of the form $4k - 1$.

    We obtained that the positive number q or $q/2$ of the form $4k - 1$ and less than a divides $n^2 + 1$, contradicting the minimality of a.

14. By Fermat's Little Theorem, $n^{40} \equiv n^4 \pmod{19}$. Thus, the condition can be written as $a^4 \equiv -b^4 \pmod{19}$. Raise this congruence to the 9th power.

15. In the special case $m = p$, assertions (a) and (b) are just the second form of Fermat's Little Theorem. Assertion (c) shows that $a^m \equiv a \pmod{m}$ may hold for every a even with a composite m. (These composite integers are called *universal pseudoprimes* or *Carmichael numbers*. We discuss them more in detail in Section 5.7.) Hints:

    (a) In the case of a squarefree m, verify $a^{\varphi(m)+1} \equiv a \pmod{p}$ for every prime divisor p of m. If m is not squarefree, so the square of a prime p divides m, then the congruence does not hold e.g. for $a = p$.

    (b) Use the result of Exercise 2.4.9b.

    (c) Check $a^{1729} \equiv a \pmod{k}$ for every prime (power) divisor k of 1729.

16. 2.4.1B: It is sufficient to prove $a^p \equiv a \pmod{p}$ for the elements of a complete residue system, e.g. for $a = 1, 2, \ldots, p$. Using induction, assume that the congruence is true for some $a = k$. Expanding $(k + 1)^p$ by the binomial theorem, we obtain that the congruence holds for $a = k + 1$.

    2.4.1A: Let $(a, p) = 1$. We may divide the congruence $a^p \equiv a \pmod{p}$ (just proved) by a, i.e. also $a^{p-1} \equiv 1 \pmod{p}$ is valid.

2.5.

2. (a) $x \equiv 11, 28, 45 \pmod{51}$.

   (b) $x \equiv 9, 38, 67, 96 \pmod{116}$.

   (c) $x \equiv 1011 + 11111k \pmod{55555}, 0 \le k \le 4$.

(d) $x \equiv (2^{k+3} + 4)/3 \pmod{2^{k+2} + 1}$ if k is even and there is no solution if k is odd.

(e) $x \equiv 0, 11 \pmod{19}$. Hint: By Fermat's Little Theorem, we obtain the congruence $x(8x + 7) \equiv 0 \pmod{19}$. Use again that 19 is a prime.

(f) $x \equiv 79 \pmod{100}$. Hint: Since $(27, 100) = 1$, only solutions coprime to 100 are possible. Thus we can use the Euler–Fermat Theorem.

3. We get 25 and 74 from the congruence $13x \equiv 31 \pmod{49}$.

4. Answer: 67. Hint: The Euler–Fermat Theorem implies $3^{280} \equiv 1 \pmod{100}$, thus we have to solve $3x \equiv 1 \pmod{100}$.

5. Sufficient: (a), (c), (f).

6. True: (a), (b).

7. m.

2.6.

1. (a) 93.

   (b) The system $x \equiv 4 \pmod{12}$, $x \equiv 8 \pmod{15}$ has no solution.

2. (a) Every digit can occur. (b) 3 or 7.

3. Apply the method shown in Example E1: Transform each congruence into a system of congruences where the moduli are the prime powers in the standard form of the original modulus. Handling a congruence with prime power modulus, we generally have to distinguish two cases according to whether or not the solution is coprime to the modulus. Answers:

   (a) $x \equiv 20 \pmod{176}$
   (b) $x \equiv 60 \pmod{333}$ and $x \equiv 208 \pmod{333}$
   (c) $x \equiv 91 \pmod{105}$.

4. (a) 1.   (b) 2.

5. Instead of the resulting congruence modulo 1000, investigate the simultaneous system modulo 125 and modulo 8. Answer: 016.

6. 1166.

7. (a) 25, 76.   (b) 376, 625.

8. (a) Answer: 36. Hint: Instead of $x^2 \equiv x \pmod{10^{20}}$, consider the system of congruences with the corresponding prime power moduli. Show that the congruence $x(x - 1) \equiv 0$ has two solutions modulo a prime power.

   (b) Answer: 135. Hint: Find the number of solutions of $x^3 \equiv x \pmod{10^{20}}$ similar to (a).

9. There are $24 \cdot 60 = 1440$ minutes in a day, so we have $x \equiv 39^{38^{37}}$ (mod 1440). Using $1440 = 2^5 \cdot 3^2 \cdot 5$, consider the congruence for moduli 2^5, 3^2, and 5. Answer: 13 hours and 21 minutes.

10. Proceed as in the solution of Exercise 2.2.14b-c.

11. Let $p_1, \ldots, p_K$ be distinct primes and consider the system $x + i \equiv 0 \left(\text{mod } p_i^2\right), i = 1, 2, \ldots, K$.

12. (a) Solutions are $x = a + b + c$ and $x = ab + bc + ca$.

    (b) *Necessity*: Apply Theorem 2.6.1 for the subsystems consisting of two congruences. *Sufficiency*: Let $a = da_1$, $b = db_1$, and $c = dc_1$ where a_1, b_1, and c_1 are pairwise coprime and $x = dx_1$. Divide the congruences by d (including the moduli). The variable in the resulting system is x_1 and the moduli are a_1, b_1, and c_1. The moduli are pairwise coprime, therefore this system is solvable, and thus so is the original system.

13. *Necessity*: Apply Theorem 2.6.1 for the subsystems consisting of two congruences. *Sufficiency*: Prove by induction on k. The subsystem of the first $k - 1$ congruences is solvable by the induction hypothesis for $k - 1$. Let c be a solution. Thus we need to verify the solvability of

$$x \equiv c \ (\text{mod } [m_1, \ldots, m_{k-1}]), \qquad x \equiv c_k \ (\text{mod } m_k).$$

To check the criterion of Theorem 2.6.1, apply the generalization of Exercise 1.6.19b for more terms and use the conditions $(m_k, m_i) \mid c_k - c_i$ and $m_i \mid c_i - c$ for $1 \leq i \leq k - 1$.

14. No. Rewrite the congruence as a system of congruences with prime (power) moduli. The product of the numbers of solutions of these congruences cannot be 14.

15. (a) *Necessity*: The number of elements has to be $\varphi(k) = n$. We have $n \mid c$ for the number c representing 0 (mod n), and $(c, k) = 1$, so $(k, n) = 1$. *Sufficiency*: Let $r_1, \ldots, r_n$ be a complete residue system modulo n and $s_1, \ldots, s_n$ a reduced residue system modulo k (by assumption, $\varphi(k) = n$). Then the systems

$$x \equiv r_i \ (\text{mod } n), \quad x \equiv s_i \ (\text{mod } k), \quad i = 1, 2, \ldots, n$$

are solvable due to $(k, n) = 1$. Picking one solution for each, these integers satisfy the requirements.

    (b) Necessity is obvious. To verify sufficiency, we can apply the method in (a) directly if $(k, n) = 1$. In the general case, however, we have to pair the elements of the two reduced residue systems so that the resulting systems of congruences are solvable. This can be guaranteed by proving the claim: If $d \mid n$, then every reduced residue class modulo d contains the same number of elements as a reduced residue system modulo n.

16. (a) Verify first that $(a_i + n, a_j + n)$ must be a divisor of $S = (a_1 - a_2)(a_1 - a_3)(a_2 - a_3)$ for any n and $i \neq j$. Let p be a prime divisor of S and choose n modulo p so that at most one of $a_1 + n$, $a_2 + n$, and $a_3 + n$ is divisible by p (for $p > 3$ we can get that none of them is a multiple of p). The system composed of these

congruences for the various prime divisors of S is solvable since the moduli are pairwise coprime.

(b) For example, 1, 2, 3 ,4.

(c) Refine the method of (a) by choosing the odd prime divisors of the product $S = \prod_{1 \le i < j \le 4}(a_i - a_j)$ and 4 as moduli.

(d) Now we have to choose n so that for any prime divisor p of S, at most two of the numbers $a_i + n$ should be multiples of p.

(e) Both assertions are true for five numbers and both are false for six numbers.

2.7.

1. (a) Answer: 2 for $m = 4$ and 0 for $m > 4$. Hint: If m is the product of two distinct integers greater than 1, then both occur as factors in $(m-1)!$, so $m \mid (m-1)!$. The remaining case is $m = p^2$ where p is a prime. If $p > 2$, then both p and $2p$ are factors in $(m-1)!$.

   (b) Answer: 2 for $m = 4$, $p-1$ for $m = 2p$ where $p > 2$ is a prime, and 0 otherwise. Hint: Verify first $\varphi(m) \ge p^\alpha$ for $m = p^\alpha t$ where $p \nmid t$ and $t > 2$. This implies that the remainder is 0 unless $m = 2^\alpha$, p^α, or $2p^\alpha$ (where $p > 2$ is a prime). If $m = p^\alpha$ or $2p^\alpha$ with $\alpha \ge 2$, then both $p^{\alpha-1}$ and $2p^{\alpha-1}$ occur in the product $(\varphi(m))!$, so $m \mid (\varphi(m))!$. Similarly, if $m = 2^\alpha$ with $\alpha \ge 3$, then both $2^{\alpha-1}$ and 2 appear as factors in $(\varphi(m))!$. Finally, for $m = 2p$, investigate the remainders of $(\varphi(m))! = (p-1)!$ separately modulo p and modulo 2.

   (c) Answer: -1 for $m = 4$, p^α, and $2p^\alpha$ where $p > 2$ is a prime, and 1 in all other cases. Hint: Form pairs as in the proof of Wilson's Theorem. An element c in the reduced residue system causes a problem if it is the pair of itself, i.e. $c^2 \equiv 1$ (mod m). Let H denote the set of these wrong elements c. Then the remainder r we are looking for equals the remainder of the product of the elements in H. The main difficulty is that $c^2 \equiv 1$ (mod m) holds not only for $c \equiv \pm 1$ (mod m) for most composite m. The exceptions are $m = 4$, p^α, and $2p^\alpha$. Then H contains no other elements than $c \equiv \pm 1$ (mod m), so $r \equiv -1$ (mod m). For all other moduli, show by the Chinese Remainder Theorem that H has more than two elements. Let $d \not\equiv 1$ (mod m) be any element in H and pair the elements of H by the rule $c \mapsto cd$ (mod m). Show that this implies $r \equiv d$ or 1 (mod m). Forming the pairs within H by another element $d' \not\equiv 1$ (mod m), we infer that only $r \equiv 1$ (mod m) is possible.

2. Answer: 7 and 17. Hint: Use Wilson's Theorem for m prime. For composite m, observe that $(m-6)!$ is not coprime to m if $m - 6 \ge m/2$.

3. We have to show that the products $a_1 b_1, \ldots, a_m b_m$ do not form a complete residue system modulo m.

   (a) Let m be a prime, $m = p$. If $p = a_i = b_j$ holds with $i \ne j$, then $a_i b_i \equiv a_j b_j \equiv 0$ (mod p). If $p = a_i = b_i$, then the remaining elements a_j and b_j, form two reduced residue systems modulo p. By Wilson's Theorem,

$$\prod_{j \neq i} a_j b_j \equiv \prod_{j \neq i} a_j \prod_{j \neq i} b_j \equiv (-1)(-1) = 1 \not\equiv -1 \ (\text{mod } p).$$

Thus the products $a_j b_j$ ($j \neq i$) cannot form a reduced residue system modulo p.

(b) Verify first that for any k, $k \mid m$, the multiples of k among the elements a and b must be multiplied by each other. If m is not squarefree, i.e. $p^2 \mid m$ for some prime p, then each product $a_i b_i$ is either coprime to p, or is a multiple of p^2 by the previous observation. Thus the residue class $(p)_m$ is one that cannot be represented. If m is squarefree and p is an odd prime divisor of m, then show that the multiples of m/p among the elements a and b form two complete residue systems modulo p. This reduces the problem to (a).

4. Replace the factors $p - c > (p - 1)/2$ in $(p - 1)! \equiv -1 \ (\text{mod } p)$ by $-c$ and take a square root using the prime property of p.

5. Factoring p^{p-1} from $(p^2 - 1)!$, the remaining part is the product of $p + 1$ reduced residue systems modulo p (the $(p+1)$-st system comes from the coefficients of the numbers divisible by p).

6. $(p - 3)/2$.

7. Answer: 10000. Hint: Examine the remainders modulo 101 and modulo 100 separately and solve the resulting system of congruences.

8. Answer: 3, 4, 5, 9. Hint: First get rid of the factorial: subtracting an appropriate multiple of the first number from the second one, we see that the required gcd d divides $3n(n + 3)$. Using that the remainder of $(n + 2)!$ modulo $n + 3$ is 0 or -1, show $d = 3$ for $n \geq 4$.

9. The answer for both questions is $m \leq 3$. Hint: Clearly, it is sufficient to prove that there is no such reduced residue system for $m > 3$. If $m = p > 3$ is a prime, then only $1!, 2!, \ldots, (p - 1)!$ could work, but $(p - 2)! \equiv 1!$ by Wilson's Theorem. If m is composite and p is its smallest prime divisor, then $(k!, m) \neq 1$ for $p \leq k$. It is easy to see that $p \leq \varphi(m)$, thus there are less than $\varphi(m)$ factorials coprime to m.

10. The divisibility by 31 is not affected by multiplying the sum by $(a_1 a_2 a_3)^{27}$ coprime to 31. Now use Wilson's Theorem and Fermat's Little Theorem.

11. Answer: 0, ± 1. Hint: Verify that if no element in the arithmetic progression is a multiple of p, then the elements either form a reduced residue system modulo p, or all have the same remainder on division by p. Use Wilson's Theorem and Fermat's Little theorem, in the two cases.

12. Answer: $x = 1$, $z = 2$. Hint: Replace every factor $1 \leq i \leq x - 1$ in $x!$ by the congruent number $-(z - i)$. Then

$$x! \, (z - x)! \equiv (-1)^{x-1} x (z - 1)! \ (\text{mod } z).$$

Use Wilson's Theorem and Exercise 2.7.1a.

13. Answer: $p \leq 5$. Hint: For a proof by contradiction, assume $(p-1)! + 1 = p^k$ for some prime $p > 5$. After transformation, we obtain

(A.2.2) $$(p-2)! = \frac{p^k - 1}{p-1} = p^{k-1} + p^{k-2} + \cdots + 1.$$

Considering (A.2.2) modulo $(p-1)$, we obtain $0 \equiv k \pmod{p-1}$ using Exercise 2.7.1a and $p \equiv 1 \pmod{p-1}$. This yields $k \geq p-1$, implying $p^k \geq p^{p-1} > (p-1)! + 1$, a contradiction.

2.8.

1. For m even.

2. (a) We have to solve the congruence $13x \equiv 1 \pmod{100}$. Answer: (77).
   (b) $100 - \varphi(100) - 1 = 59$.
   (c) 19.
   (d) Yes.

3. Answers: (a) 2. (b) 4. (c) 8. (d) Let $m = 2^\alpha t$ with t odd and let t have k distinct prime divisors. Then the answer is 2^k for $\alpha \leq 1$, 2^{k+1} for $\alpha = 2$, and 2^{k+2} for $\alpha \geq 3$. Hint: We have to determine the number of solutions of $x^2 \equiv 1 \pmod{m}$. First examine the special cases where m is a power of a prime (treat the odd primes and 2 separately). In the general case, convert the problem into a system of congruences modulo the prime powers in the standard form of m.

4. (a)–(b) Apply the definition of zero divisor or Theorem 2.8.5.
   (c) Prime powers.
   (d) The sum is (0) for m odd and $(m/2)$ for m even. The product is (2) for $m = 4$ and (0) for $m > 4$.
   (e) The integers not squarefree, i.e. which are divisible by the square of at least one prime.

5. (a) We have to verify first of all that the operations are well defined, so the sum and product of two such residue classes are again residue classes of this type. The identities hold among all residue classes modulo 20, so they are valid automatically also in the subset H. The zero element is $(0)_{20}$, and the negative of $(4s)_{20}$ is a $(-4s)_{20} = (20-4s)_{20}$. The identity element is $(16)_{20}$, the inverses of $(16)_{20}$ and $(4)_{20}$ are themselves, whereas $(8)_{20}$ and $(12)_{20}$ are the inverses of each other.
   (b) $(a)_{40}(20)_{40} = (0)_{40}$ for every $(a) \in K$, so every (non-zero) element is a zero divisor. This implies that there is no identity element and thus K is not a field. (K is a commutative ring as can be verified similar to part (a).)
   (c) Let $1 < k < m$ and $k \mid m$.
      (i) The multiples of k among the residue classes modulo m form a commutative ring R under the addition and multiplication of residue classes.

(ii) If $(k, m/k) = 1$, then this ring R has an identity element.

(iii) If $(k, m/k) = 1$ and m/k is a prime, then R is a field.

(iv) If $(k, m/k) \neq 1$, then every non-zero element in R is a zero divisor, so there is no identity element.

6. Only raising to a third power is okay. Details:

   (a) Gcd: the residue class on the right-hand side generally depends on which elements were chosen to represent the residue classes $(a)_m$ and $(b)_m$.

   (b) Third power: the definition makes sense.

   (c) Cube root: the residue class on the right-hand side generally depends on which element was chosen to represent the residue class $(a)_m$ and that can depend on the choice of the representative whether or not $\sqrt[3]{a}$ is an integer.

   (d) Arithmetic mean: the situation is similar to (c). To make a more subtle analysis, we have to distinguish cases according to the parity of m. If m is odd and we represent the residue classes with elements providing an integer value for $(a + b)/2$, then this determines the residue class $\big((a + b)/2\big)_m$ uniquely. This makes it possible to define (slightly artificially) the arithmetic mean of any two residue classes. If m is even, then picking arbitrary representatives from the two residue classes, $(a + b)/2$ will be uniformly either always an integer, or never an integer. However, $\big((a+b)/2\big)_m$ will not be unique even in the first case. This means that there is no way to define the arithmetic mean of two residue classes if m is even.

   (e) Exponentiation: the residue class on the right-hand side generally depends on which element was chosen to represent the residue class $(b)_m$.

7. Modify suitably the argument in the proof of Theorem 2.4.1. Let $g_1, \ldots, g_k$ be all elements in G. Show that $ag_1, \ldots, ag_k$ enumerates all elements in G. This implies $(ag_1)(ag_2)\ldots(ag_k) = g_1 g_2 \ldots g_k$. Multiplying by the inverse of $g_1 g_2 \ldots g_k$, we obtain the statement of the exercise.

8. Following the proof of Wilson's Theorem, pair every element with its inverse. The assertion follows if there are at most two elements (including the identity) whose square is the identity. If there are more than two such elements, devise another pairing among them, similar to the end of the hint of Exercise 2.7.1c.

A.3. Congruences of Higher Degree

3.1.

1. (a) 2.   (b) 4.   (c) 0.   (d) 60.

2. To the ring $\mathbf{Z}_m$, apply the theorem that a polynomial over a (commutative) ring is divisible by $x - \alpha$ if and only if α is a root of the (corresponding) polynomial (function).

3. Only (c) is true.

4. (a) E.g. $f = x^2(x-1)\ldots(x-11)$.   (b) $37 \cdot 36 \cdot \binom{36}{11}$.

5. If i is a solution, then $f(i)^{p-1} \equiv 0 \pmod{p}$, whereas if i is not a solution, then $f(i)^{p-1} \equiv 1 \pmod{p}$ by Fermat's Little Theorem.

6. Rely on the proof of Wilson's Theorem in this section: the product is $(-1)^{j+1} a_{p-1-j}$ where a_{p-1-j} is the relevant coefficient of the polynomial f used in the proof.

7. Replace x^{p-1} in f by 1 as long as possible.

8. Treat the problem among polynomials over the field $\mathbf{Z}_p$. The polynomial function belonging to the polynomial f is described by the values assumed at the p elements of the field. The interpolation by Lagrange or Newton guarantees a unique polynomial g of degree at most $p-1$ (or g is the zero polynomial) that assumes the required values at the given elements of the field. This means that now g and f assume the same values at each place. There are several methods for the construction of the interpolation polynomial but we always need all values of f. This means that we have to know all the roots and thus also the number of solutions. Therefore, we cannot use the interpolation polynomial for determining the number of solutions.

9. Assume that both polynomials g_1 and g_2 meet the requirements and consider $h = g_1 - g_2$. The degree of h modulo p is at most $p-1$ by the conditions. However, every c is a solution of $h(x) \equiv 0 \pmod{p}$, so there are p solutions. The only possibility to avoid a contradiction is that h has no degree modulo p, i.e. every coefficient of h is a multiple of p.

10. Modify the first proof of Theorem 3.1.3 using Exercise 2.4.15b.

3.2.

1. (a) 1.   (b) 2.   (c) 12.   (d) 46. (We can exlude 23 as possible order even without any computations using $43 \equiv -2^2 \pmod{47}$ and Fermat's Little Theorem.)

2. There is an appropriate a only in (c).

3. 9, 21, and 63.

4. Use $(a^i)^t = a^{it}$ and assertion (i) in Theorem 3.2.2. The most difficult part (c) (containing (a) and (b) as special cases) can be verified as follows:

$$1 \equiv (a^i)^t = a^{it} \pmod{m} \iff k \mid it \iff \frac{k}{(i,k)} \,\Big|\, t\frac{i}{(i,k)} \iff \frac{k}{(i,k)} \,\Big|\, t.$$

5. (a) 10 and 30 (show examples that both really do occur).

   (b) 36.

7. 16.

8. (a) $p \mid a^3 - 1 = (a-1)(a^2 + a + 1)$ but $p \nmid a - 1$.

   (b) Answer: 6. Hint: $(1+a)^2 \equiv a \pmod{p}$ by (a).

9. 16.

10. (a) The modulus of the congruences is m:

$$a^n \equiv 1 \iff o_m(a) \mid n \atop a^k \equiv 1 \iff o_m(a) \mid k \quad \iff \quad o_m(a) \mid (n,k) \iff a^{(n,k)} \equiv 1.$$

   (b) By (a), the common divisors of $a^n - 1$ and $a^k - 1$ are the same as the divisors of $a^{(n,k)} - 1$.

11. For a proof by contradiction, assume that both $a^n \equiv 1$ and $a^k \equiv -1 \pmod{m}$ hold for some $m > 2$. Then $o_m(a) \mid n$ implies that $o_m(a)$ is odd. Further, $a^{2k} \equiv 1 \pmod{m}$ yields $o_m(a) \mid 2k$. Hence, $o_m(a) \mid k$, so $a^k \equiv 1 \pmod{m}$, a contradiction.

12. To prove $a^s \equiv -1 \pmod{p} \implies o_p(a)$ is even, follow the previous hint. This part is true for any modulus $m > 2$ instead of p. For the converse, let $o_p(a) = 2k$, then $a^k \equiv -1 \pmod{p}$. This is false in general for composite moduli, consider e.g. $m = 15$ and $a = 4$.

13. (b) Use that $a^k \equiv 1 \pmod{[m,n]}$ holds if and only if both congruences $a^k \equiv 1 \pmod{m}$ and $a^k \equiv 1 \pmod{n}$ are valid.

14. Answer: 7. Hint: We ask how many $x \not\equiv 1 \pmod{1000}$ satisfy $x^2 \equiv 1 \pmod{1000}$. Instead of mod 1000, consider the system $x^2 \equiv 1 \pmod{125}$, $x^2 \equiv 1 \pmod{8}$.

15. (a) $(ab)^{[u,v]} = a^{[u,v]}b^{[u,v]} \equiv 1 \cdot 1 = 1 \pmod{m}$, so $o(ab) \mid [u,v]$. Thus $o(ab) = uv$ can occur only for $(u,v) = 1$. To prove the converse, assume $(ab)^t \equiv 1 \pmod{m}$; we have to show $uv \mid t$. To eliminate a, raise the congruence to the uth power: $1 \equiv a^{tu}b^{tu} \equiv b^{tu} \pmod{m}$. This implies $o(b) = v \mid tu$. Since $(u,v) = 1$, we infer $v \mid t$. Similarly, $u \mid t$, thus $uv = [u,v] \mid t$.

   (b) We proved $o(ab) \mid [u,v]$ in (a). The other divisibility can be verified using the ideas in the second part of (a).

16. Let $d = (o(a), o(b))$ and raise the congruence to powers of exponents $o(a)/d$ and $o(b)/d$, resp.

17. Observe that the order of a modulo $a^n - 1$ is just n.

18. Show that $ab \equiv 1 \pmod{m}$ implies $o_m(a) = o_m(b)$, and so $o_m(a) + o_m(b)$ is even. We have to treat separately the case of $a \equiv b \pmod{m}$, i.e. $a^2 \equiv 1 \pmod{m}$. This means $o_m(a) = 2$ (which is even) or $a \equiv 1 \pmod{m}$ (of order 1).

19. (a) The remainder is 1 for $a \equiv 1 \pmod{p}$ and 0 otherwise.

   (b) The remainder is 1 if $o(a)$ is odd and -1 if $o(a)$ is even.

20. (a) Let $a/b = 0.c_1c_2c_3 \ldots$ be the decimal representation of the rational number a/b. We obtain the digits c_i from the following divisions:

$$10a = c_1b + r_1 \quad \text{where} \quad 0 \le r_1 < b$$

(A.3.1)

$$10r_1 = c_2b + r_2 \quad \text{where} \quad 0 \le r_2 < b$$

$$10r_2 = c_3b + r_3 \quad \text{where} \quad 0 \le r_3 < b$$

$$\vdots$$

If some $r_i = 0$, then the algorithm terminates and we obtain a finite decimal fraction. Otherwise we have $r_h = r_j$ for some $h < j$ since r_i can assume only the values 1, 2, ..., $b - 1$. Then (A.3.1) implies $c_{h+1} = c_{j+1}$, $r_{h+1} = r_{j+1}$, so $c_{h+2} = c_{j+2}$, $r_{h+2} = r_{j+2}$, etc. This means that the decimal fraction is periodic.

For the converse, assume that the decimal representation of the real number $0 < \alpha < 1$ is finite

(A.3.2a) $$\alpha = 0, u_1 u_2 \ldots u_k, \qquad u_k \neq 0,$$

or periodic

(A.3.2b) $$\alpha = 0, u_1 u_2 \ldots u_k v_1 \ldots v_n v_1 \ldots v_n \ldots$$

where $u_1 u_2 \ldots u_k$ is the non-periodic part (which is empty in the case of pure periodicity, i.e. $k = 0$) and $v_1 \ldots v_n$ is the (smallest) period. Then (A.3.2a) means $\alpha = \overline{u_1 \ldots u_k}/10^k$, and (A.3.2b) implies that $\alpha(10^{n+k} - 10^k)$ is an integer.

(b) We verified this in (a) essentially.

(c) A purely periodic decimal representation can be transformed into a fraction with denominator $10^n - 1$ by the procedure in (a). As b is obtained from this denominator after (eventual) cancellation, so $(b, 10) = 1$.

For the converse, consider the algorithm in (A.3.1). Introducing $r_0 = a$, we have

$$\begin{aligned} r_0 &\equiv a & \pmod{b} \\ r_1 &\equiv 10a & \pmod{b} \\ r_2 &\equiv 10r_1 \equiv 10^2 a & \pmod{b}, \end{aligned}$$

and similarly $r_i \equiv 10^i a \pmod{b}$ in general.

The equality $r_h = r_j$ $(h < j)$ means $10^h a \equiv 10^j a \pmod{b}$. By $(10a, b) = 1$, this is equivalent to $10^{j-h} \equiv 1 \pmod{b}$. Therefore, $r_i = r_0 = a$ for some $i > 0$, so the period starts right after the decimal point and its length is equal to the number of pairwise incongruent powers of 10 which is $o_b(10)$.

(d) The equivalence follows from the previous parts as the rational numbers not yet discussed must form the remaining mixed periodic case. The lengths of the period and of the non-periodic part can be shown as in (c).

3.3.

1. All elements of the reduced residue classes represented by: (a) 3, 5. (b) 3, 7. (c) 5, 11.

2. Take e.g. the solution of the system $x \equiv 2 \pmod{11}$, $x \equiv 3 \pmod{14}$, it is $x \equiv 101 \pmod{154}$.

3. (a) Follow the arguments of (Y1) and (Y2) in the proof of Theorem 3.3.5. Find first a primitive root modulo 5, e.g. 2 is suitable. Then test whether or not 2 is a primitive root modulo 25; it suffices to check $2^{5-1} \not\equiv 1 \pmod{25}$ which

holds. Since 2 is a primitive root modulo 5^2, therefore it is a primitive root for every power of 5, as well.

(b) We search for a number in the form $a = 2 + 5t$. It is not a primitive root modulo 25 if and only if $1 \equiv (2 + 5t)^{5-1} \equiv 2^4 + 4 \cdot 8 \cdot (5t) \pmod{25}$, so $t \equiv 1 \pmod 5$. This gives $a \equiv 7 \pmod{25}$. We have to show that if a is not a primitive root modulo 25, then it cannot be a primitive root modulo 625 either; this can be done using Theorem 3.3.2.

4. True: (b), (d), (e), (f).

5. (a) If g is a primitive root, then $g^{(p-1)/2} \equiv -1 \pmod p$. So
$$(g_1 g_2)^{(p-1)/2} \equiv (-1)(-1) = 1 \pmod p.$$

(b) Show that if g is a primitive root and $gh \equiv 1 \pmod p$, then h is a primitive root. Thus, such a pair g and h and a primitive root t meet the requirements as $ght \equiv t \pmod p$.

(c) These are the Fermat primes, those of the form $2^k + 1$ (the exponent k must be a power of two, see Exercise 1.4.4).

6. Let $p > 2$ be a prime and g a primitive root mod p. Then $1, g, \ldots, g^{p-2}$ form a reduced residue system mod p, so
$$(p-1)! \equiv 1 \cdot g \cdot \cdots \cdot g^{p-2} = \left(g^{(p-1)/2}\right)^{p-2} \equiv (-1)^{p-2} = -1 \pmod p.$$

7. The remainder of the sum is 0 if $p-1 \nmid k$ and $p-1$ if $p-1 \mid k$. Hint: The remainder is the same if we consider another reduced residue system instead of $1, 2, \ldots, p-1$. Thus, compute the sum $1^k + g^k + \cdots + g^{(p-2)k}$ where g is a primitive root mod p. If $g^k \equiv 1 \pmod p$, the the remainder of the sum is clearly $p - 1$. Otherwise apply the formula for the sum of a (finite) geometric series.

8. Answer: 1 for $p > 3$ and 2 for $p = 3$. Hint: Form pairs of the primitive roots so that the product of the elements in a pair is congruent to 1 mod p.

9. (a) Use Exercise 3.2.4c.

(b) Consider the values $j = t(p-1)/d$ in (a) satisfying $0 \le j < p-1$. By $0 \le t < d$ and $(t, d) = 1$, we have $\varphi(d)$ such numbers.

10. One direction follows easily from Exercise 3.2.4a. For the other, write a and b as powers of a primitive root g and use Exercise 3.2.4c. Another option: For $(c, p) = 1$, the number of elements with order $o_p(c)$ is the same as the number of powers of c with order $o_p(c)$.

11. Every proposition remains valid if we replace $p - 1$ by $\varphi(m)$.

12. (a) It is sufficient to verify
$$5^{2^{\alpha-3}} = 1 + t2^{\alpha-1} \qquad \text{where } t \text{ is odd (and } \alpha \ge 3).$$
We can prove this by induction on α.

(b) The congruence is false mod 4.

(c) The given $\varphi(m)$ numbers are coprime to m and we easily infer from (a) and (b) that they are pairwise incongruent modulo m.

13. Let g_i be primitive roots mod $p_i^{\alpha_i}$, $i = 1, 2, \ldots, r$. Then u_i can be chosen as the solution of the system $x \equiv g_i \pmod{p_i^{\alpha_i}}$, $x \equiv 1 \pmod{m/p_i^{\alpha_i}}$. For m even, use Exercise 3.3.12c. Let α be the exponent of 2 in the standard form of m. For $\alpha = 1$, there is no need for any change in the formula. For $\alpha = 2$, we have to insert a factor u^j into the product of powers where $0 \le j < 2 = \varphi(4)$. For $\alpha \ge 3$, we need an extra factor $u^j v^k$ where $0 \le j < 2$ and $0 \le k < 2^{\alpha-2}$. The values of u and v are the solutions of the systems $x \equiv -1 \pmod{2^\alpha}$, $x \equiv 1 \pmod{m/2^\alpha}$, and $x \equiv 5 \pmod{2^\alpha}$, $x \equiv 1 \pmod{m/2^\alpha}$.

14. (a) For a polynomial F with integer coefficients, let $\deg F$ denote the degree of F and $N(F)$ the number of solutions of $F(x) \equiv 0 \pmod p$. Theorem 3.1.2 implies $N(F) \le \deg F$. If $x^{p-1} - 1 = fh$, then every element of a reduced residue system satisfies (at least) one of the congruences $f(x) \equiv 0 \pmod p$ and $h(x) \equiv 0 \pmod p$ by Fermat's Little Theorem and the prime property of p. Hence

$$p - 1 \le N(f) + N(h) \le \deg f + \deg h = p - 1.$$

Thus we have equality everywhere, so $N(f) = \deg f$.

(b) Apply (a) for the polynomials f_i.

(c) $o_p(c) = q^\beta$ if and only if $f_1(c) \equiv 0 \pmod p$ but $f_2(c) \not\equiv 0 \pmod p$. The existence of such a c now follows from (b).

(d) Let $d = q_1^{\beta_1} \ldots q_r^{\beta_r}$ be the standard form of d. By (c), there exist c_i with $o_p(c_i) = q_i^{\beta_i}$ $(i = 1, 2, \ldots, r)$. Then $o_p(c_1 \ldots c_r) = d$ by Exercise 3.2.15a.

3.4.

1. The condition implies $p \mid 7^3 - 2 = 11 \cdot 31$. Thus, the only candidates are $p = 11$ and $p = 31$. Since 7 is a primitive root mod 11 but not mod 31, the only solution is $p = 11$.

2. (a) 0. (b) $(p-1)/2$. (c) $(p+1)/2$.

3. (a) We exhibit powers of g congruent to ab in two ways:
$$g^{\mathrm{ind}(ab)} \equiv ab \equiv g^{\mathrm{ind}\,a} \cdot g^{\mathrm{ind}\,b} = g^{\mathrm{ind}\,a + \mathrm{ind}\,b} \pmod p.$$

Hence, the exponents of g in the first and last terms are congruent mod $p - 1$.

(b) We argue as in (a):
$$g^{\mathrm{ind}(a^k)} \equiv a^k \equiv \left(g^{\mathrm{ind}\,a}\right)^k = g^{k \cdot \mathrm{ind}\,a} \pmod p.$$

4. Follow the method of the previous exercise.

5. $o_p(a)$.

6. This is just a reformulation of asssertion (i) in Theorem 3.3.4.

7. (a) By the previous exercise, both conditions are equivalent to a being a primitive root mod p.

(b) By Exercise 3.2.4c, $o_p(a) = (p-1)/(\text{ind } a, p-1)$ independent of the choice of the primitive root.

8. Use Exercise 3.4.6.

9. Start from the hint to Exercise 3.4.7b.

10. The upper row in each table lists the least positive representatives of the reduced residue classes mod p in increasing order and the lower row contains their indices with base g.

   (a) $p = 7, g = 3$:

1	2	3	4	5	6
0	2	1	4	5	3

   (b) $p = 11, g = 2$:

1	2	3	4	5	6	7	8	9	10
0	1	8	2	4	9	7	3	6	5

   (c) $p = 17, g = 3$:

1	2	3	4	5	6	7	8	9	10	11	12	13	14	15	16
0	14	1	12	5	15	11	10	2	3	7	13	4	9	6	8

11. For $p \mid a$, the multiples of p serve as k. For $(a, p) = 1$, let g be a primitive root mod p. Then the solutions of the system $x \equiv g \pmod{p}$, $x \equiv \text{ind}_g a \pmod{p-1}$ can be taken as k. The exercise can be solved also without primitive roots, relying just on Fermat's Little Theorem: Choose the solutions of the system $x \equiv a \pmod{p}$, $x \equiv 1 \pmod{p-1}$ as k.

3.5.

1. (a) No solution.

   (b) $x \equiv 51 \pmod{101}$. Hint: Use Fermat's Little Theorem.

   (c) $x \equiv \pm 2 \pmod{23}$. Hint: We get $x^2 \equiv 4 \pmod{23}$ after the usual reduction.

   (d) $x \equiv 0, \pm 6, \pm 7 \pmod{17}$.

   (e) $x \equiv 0, 2, 5, 6 \pmod{13}$.

   (f) $x \equiv \pm 5 \pmod{11}$. Hint: As $x \equiv 0 \pmod{11}$ is not a solution, we can replace x^{20} by 1 during the reduction.

2. (a) Answer: 12. Hint: Add the numbers of solutions of $x^{30} \equiv 1 \pmod{73}$ and $x^{45} \equiv 1 \pmod{73}$ and subtract the number of common solutions. The latter are the solutions of $x^{(30,45)} \equiv 1 \pmod{73}$.

   (b) Answer: $(k+1, 30)$ if $31 \mid k+1$, and $(k+1, 30) - 1$ otherwise. Hint: The left-hand side can be written as $(x^{k+1} - 1)/(x - 1)$. Thus the solutions are the same as the solutions of $x^{k+1} \equiv 1 \pmod{31}$ except perhaps for $x \equiv 1$. Therefore, we have to check separately for which k does $x \equiv 1 \pmod{31}$ satisfy the original congruence.

3. $a \equiv 0, \pm 1 \pmod{p}$.

4. The condition for solvability is $(k, p-1) \mid \text{ind}_g g = 1$. The number of solutions is $(k, p-1) = 1$.

5. $x \equiv cb_i \pmod{p}, i = 1, \ldots, r$.

6. (a) 1.   (b) ± 1.

7. $(k, p-1) = 1$.

8. For 3 and the primes of the form $3t - 1$.

9. Use any of the two criteria in Theorem 3.5.3 or Definition 3.5.2 (in the latter case we need Fermat's Little Theorem for (b)).

10. $(k, p-1) = 2$.

11. (a) Answer: 1 if $p-1 \mid k$, and 0 otherwise. Hint: Put $d = (k, p-1)$. The kth power residues can be written as g^{rd} where $0 \le r < (p-1)/d$. Apply the formula for the sum of a finite geometric series. Another way: In the sum of Exercise 3.3.7, every kth power residue occurs $(k, p-1)$ times. A third possibility: Observe that the kth power residues are just the roots (with multiplicity one) of the polynomial $x^{\frac{p-1}{(k,p-1)}} - 1$ over $\mathbf{Z}_p$. Apply the law connecting the roots and the coefficients (Viète's formulas) for this polynomial.

    (b) Answer: -1 or 1 according as $\dfrac{p-1}{(k, p-1)}$ is even or odd. Hint: Form pairs from the kth power residues so that the product of the elements in each pair is congruent to 1. Two other options: Write the kth power residues as in the first hint to (a), or apply the third hint to (a).

12. See the hint to Exercise 3.5.9. Generalization: a is both a kth and an nth power residue if and only if it is a $[k, n]$th power residue.

3.6.

1. A homogeneous system of linear equations always has a non-trivial solution if there are more variables than equations. (This holds not just modulo p but over any field.)

2. Apply Chevalley's Theorem.

3. (a) It suffices to solve the problem for a prime power modulus p^α by the Chinese Remainder Theorem. For $\alpha > 1$, take $x_1 = p^{\lfloor \alpha/2 \rfloor}$, $x_2 = x_3 = 0$. For $\alpha = 1$, the congruence $x_1^2 + x_2^2 + x_3^2 \equiv 0 \pmod{p}$ has a non-trivial solution (e.g. by Chevalley's Theorem). We may assume $|x_i| \le p/2$, so $0 < x_1^2 + x_2^2 + x_3^2 < p^2$. Thus, $x_1^2 + x_2^2 + x_3^2$ is a multiple of p but not of p^2.

    (b) We have to refine the procedure in (a) only in one case: if $\alpha > 1$ is odd, then let $x_i = p^{(\alpha-1)/2} y_i$ and apply to y_i the previous argument used for $\alpha = 1$.

4. The case $p = 2$ is obvious. For $p > 2$, by Chevalley's Theorem, there exist integers u_i, $1 \leq i \leq 5$, yielding a non-trivial solution of $\sum_{i=1}^{5} x_i^4 \equiv 0 \pmod{p}$. If e.g. $u_1 \not\equiv 0 \pmod{p}$, then $v_i = u_1^{p-2} u_i$ is another solution with $v_1 \equiv 1 \pmod{p}$. We may assume $|v_i| \leq (p-1)/2$ also for the other v_i. Thus $\sum_{i=1}^{5} v_i^4$ is a multiple of p and

$$0 < \sum_{i=1}^{5} v_i^4 \leq 1 + 4\left(\frac{p-1}{2}\right)^4 < \frac{p^4}{4}.$$

5. (a) Let γ_{ij} be the exponent of the prime q_i in c_j ($1 \leq i \leq k, 1 \leq j \leq t$). Apply Chevalley's Theorem for the polynomials $f_i(x_1, \ldots, x_t) = \sum_{j=1}^{t} \gamma_{ij} x_j^2$ and $p = 3$.

   (b) The corresponding condition is $t \geq (m-1)k + 1$.

6. Verify first that if the proposition holds for $n = r$ and $n = s$, then it is true for $n = rs$. Taking any $2r - 1$ from the $2rs - 1$ numbers, we can select r of them whose sum is a multiple of r. We repeat this for $2r - 1$ from the remaining $2rs - 1 - r$ numbers, etc. Show that we obtain $2s - 1$ groups of size r where the sums $U_1, U_2, \ldots, U_{2s-1}$ of the elements in the groups are all multiples of r. Apply now the proposition about s for $U_1/r, U_2/r, \ldots, U_{2s-1}/r$.

   We thus reduced the problem to the case $n = p =$ prime. Let $f_1 = \sum_{j=1}^{2p-1} c_j x_j^{p-1}$, $f_2 = \sum_{j=1}^{2p-1} x_j^{p-1}$ and apply Chevalley's Theorem.

7. (a) For a proof by contradiction, assume that $x_j = a_j$, $j = 1, 2, \ldots, t$, is the only solution. We have to modify the polynomial G in the proof of Chevalley's Theorem to

$$G(x_1, \ldots, x_t) = \prod_{j=1}^{t} \left(1 - (x_j - a_j)^{p-1}\right).$$

   (b) Assume that there are s solutions $\underline{a}_1, \ldots, \underline{a}_s$. Form the polynomials G_v for every solution $\underline{a}_v$ ($v = 1, \ldots, s$) as described in (a). Let $G = \sum_{v=1}^{s} G_v$. Following the proof of Chevalley's Theorem, we obtain $F^* = G$. Comparing the degrees yields that the coefficient of the term $(x_1 \ldots x_t)^{p-1}$ in G must be 0 modulo p, i.e. $s(-1)^t \equiv 0 \pmod{p}$.

8. (a) The determinant of the matrix A is

$$\begin{vmatrix} -b & a & 0 & \cdots & 0 \\ 0 & -b & a & \cdots & 0 \\ 0 & 0 & -b & \cdots & 0 \\ \vdots & \vdots & \vdots & \ddots & \vdots \\ a & 0 & 0 & \cdots & -b \end{vmatrix} = (-b)^{p-1} + (-1)^{p-2} a^{p-1} \equiv 0 \pmod{p},$$

   so $r(A) \leq p - 2$. On the other hand, the minor belonging to the upper left corner is $(-b)^{p-2} \not\equiv 0 \pmod{p}$, so $r(A) \geq p - 2$. This implies $r(A) = p - 2$ and the number of solutions is $p - 1 - (p - 2) = 1$. (Of course, the result is well known from Theorem 2.5.5.)

(b) Every element of the matrix is 1, hence its rank is 1, implying that there are $p - 2$ solutions (cf. Exercise 3.5.3).

(c) Similar to (a), we get rank $p-2$, so there is one solution (this follows also from Exercise 3.5.7). The solution is $x \equiv a^{p-2} \pmod{p}$.

9. We have to prove that the determinant of the matrix is 0 (mod p).

(a) The sum of each row is 0.

(b) Denoting the ith row by r_i, we get $r_1 + r_2 - r_3 - r_4 + r_5 + \cdots = 0$.

10. Let A_f, A_g, and A_h be the matrices corresponding to the three polynomials. We get A_g by putting the last row of A_f above the other rows without changing the order of the others. We obtain A_h by reflecting A_f through the main diagonal and then make a first row from the last one. These transformations do not affect the rank of the matrix, so the numbers of solutions are the same for the three congruences. We can easily solve the problem without the Kőnig–Rados Theorem, too. Since $f(j) \equiv jg(j) \pmod{p}$ for $(j, p) = 1$, the first two congruences have the same solutions. Similarly,

$$f(a) \equiv 0 \pmod{p} \iff h(a^{-1}) \equiv 0 \pmod{p}$$

where a^{-1} is the multiplicative inverse of a, i.e. $aa^{-1} \equiv 1 \pmod{p}$.

11. We can eliminate the terms of degree higher than $p-1$ by the reduction described in Theorem 3.1.3. Since it is easy to see whether or not $x \equiv 0 \pmod{p}$ is a solution, we can concentrate on finding the solutions coprime to p. Thus we can replace x^{p-1} by 1 by Fermat's Little Theorem. If every coefficient d_j in the resulting polynomial $h = d_0 + d_1 x + \cdots + d_{p-2} x^{p-2}$ is a multiple of p, then $h(x) \equiv 0 \pmod{p}$ is true for every x. Finally, if $d_0 \equiv \cdots \equiv d_{i-1} \equiv 0 \pmod{p}$ but $d_i \not\equiv 0 \pmod{p}$, then we can apply the Kőnig–Rados Theorem to the polynomial $h_1 = h/x^i$. The congruences $h(x) \equiv 0 \pmod{p}$ and $h_1(x) \equiv 0 \pmod{p}$ will have the same same reduced residue classes as solutions.

3.7.

1. (a) 1.   (b) 0.   (c) 12.   (d) 73.   (e) 15.

2. Use Theorem 3.7.1.

3. (a) The condition of solvability is $a \equiv 1 \pmod{11}$ and there are ten solutions. Hint: Use Fermat's Little Theorem and Theorem 3.7.1.

   (b) It is solvable if and only if $a \equiv 1 \pmod{8}$ and there are four solutions.

4. The proposition follows from the proof of Theorem 3.7.1.

5. (a) $x \equiv 32 \pmod{7^3}$.   (b) No solution.   (c) $x \equiv 2 + 49j \pmod{7^3}$.

A.4. Legendre and Jacobi Symbols

4.1.

1. First proof: The congruence $x^2 \equiv c^2 \pmod{p}$ is solvable as $x \equiv c \pmod{p}$ is a solution.

   Second proof: $(c^2)^{(p-1)/2} = c^{p-1} \equiv 1 \pmod{p}$.

   Third proof: $\left(\frac{c^2}{p}\right) = \left(\frac{c}{p}\right)^2 = 1.$

2. (a) 1.   (b) -1.   (c) -1.

3. The sum is 0. The product is 1 for $p \equiv 1 \pmod 4$ and -1 for $p \equiv -1 \pmod 4$.

4. The solution of $x^2 \equiv a \pmod p$ must be congruent to an element j from the reduced residue system $\pm 1, \pm 2, \ldots, \pm\left(\frac{p-1}{2}\right)$. So, $a \equiv |j|^2 \pmod p$. Further, there are $(p-1)/2$ quadratic residues, therefore no two of the given $(p-1)/2$ numbers can be congruent. This can be verified directly, too. Assuming $u^2 \equiv v^2 \pmod p$ for some $1 \le u < v \le (p-1)/2$, we have $p \mid (v-u)(v+u)$. However, $1 \le v - u < v + u \le (p-1)$, thus none of the factors is a multiple of p, which contradicts the prime property of p.

5. We show $a \equiv b \equiv 0 \pmod{77}$, which implies $5929 = 77^2 \mid a^2 + b^2$. For a proof by contradiction, if e.g. a is not a multiple of (say) 7, then $7 \mid a^2 + b^2$ and 7 being a prime implies that b is not divisible by 7 either. Using $a^2 \equiv -b^2 \pmod 7$ and $\left(\frac{-1}{7}\right) = -1$, we get a contradiction:

$$1 = \left(\frac{a}{7}\right)^2 = \left(\frac{a^2}{7}\right) = \left(\frac{-b^2}{7}\right) = \left(\frac{-1}{7}\right)\left(\frac{b}{7}\right)^2 = -1.$$

6. Apply Wilson's Theorem.

7. $(\pm a^{(p+1)/4})^2 = a^{(p+1)/2} = a \cdot a^{(p-1)/2} \equiv a \cdot 1 = a \pmod p$.

8. (a) If $o_p(a) = 2t - 1$, then $(a^t)^2 \equiv a \pmod p$.   (b) $p = 4k + 3$.

9. (a) If $o_p(g) = p - 1$, then $g^{(p-1)/2} \not\equiv 1 \pmod p$.

   (b) $p = 2^k + 1$ (i.e. the Fermat primes, see Exercise 1.4.4 and Section 5.2).

10. 32.

11. (a) Since $p + 1 = 4k$, $1 = \left(\frac{1}{p}\right) = \left(\frac{p+1}{p}\right) = \left(\frac{4k}{p}\right) = \left(\frac{2}{p}\right)^2\left(\frac{k}{p}\right) = \left(\frac{k}{p}\right).$

    (b) Argue as in (a).

12. For $p \le 11$, at least one of the congruences is of the type $x^2 \equiv 0 \pmod p$. Otherwise, observe that the product of the five numbers is a square. Hence, the product of the five corresponding Legendre symbols is 1.

13. (a) $x \equiv 1$ and $6 \pmod{13}$. Hint: Eliminate the linear term by completing the square.

(b) $x \equiv -3 \pmod{17}$.

(c) $x \equiv 0, \pm 8 \pmod{23}$. Hint: x^{25} can be replaced by x^3 according to Fermat's Little Theorem. Factoring out x, we get a congruence of degree four that can be reduced to a quadratic congruence by introducing a new variable.

(d) No solution. Hint: $x \equiv 0 \pmod{19}$ is not a solution, so multiplying by x and replacing x^{18} by 1 is an equivalent transformation.

14. (a) Apply the multiplicative property of the Legendre symbol.

    (b) Let $n(p) = n$ and r the smallest integer satisfying $rn > p$. Then $0 < rn - p < n$, thus $1 = \left(\frac{rn-p}{p}\right) = \left(\frac{rn}{p}\right) = -\left(\frac{r}{p}\right)$ yielding $r \geq n$. The assertion now follows from $(r-1)n < p$.

15. (a) $\left(\frac{i^2}{p}\right) = 1$ for $(i, p) = 1$ and 0 for $p \mid i$.

    (b) Verify that replacing i by ai in $S(a, p)$ the sum remains the same, but is equal to $S(1, p)$ after factoring out $\left(\frac{a^2}{p}\right) = 1$.

    (c) For a fixed i, the values $i + a$ form a complete residue system mod p, so the sum of the corresponding Legendre symbols is 0 by Exercise 4.1.3.

    (d) This follows from the previous three parts.

16. (a) Observe that $\left(\frac{c}{p}\right) + 1$ is 2, 0, or 1 according to c being a quadratic residue, a non-residue, or a multiple of p.

    (b) It follows from (a) using Exercises 4.1.3 and 4.1.15d and the formula for $\left(\frac{-1}{p}\right)$.

4.2.

1. Solvable: (c), (e), (f). Use Wilson's Theorem for (c). A congruence with a composite modulus is solvable if and only if there is a solution for every prime power divisor of the modulus.

2. (a) $p = 8k + 1$ or $8k + 3$. Hint: $\left(\frac{-2}{p}\right) = \left(\frac{-1}{p}\right)\left(\frac{2}{p}\right)$.

    (b) $p = 12k \pm 1$ or $p = 3$. Hint: To apply reciprocity, we need the remainder of $p > 3$ mod 4, and afterwards we need the remainder of p mod 3. Therefore, it is best to distinguish cases according to the remainder of p mod 12.

    (c) $p = 6k + 1$ or $p = 3$.

    (d) $p = 5k \pm 1$ or $p = 5$.

    (e) $p = 8k \pm 1$ or $8k + 3$. Hint: Factor $x^4 - 4$.

    (f) $p = 4k + 1$. Hint: Apply Theorem 3.5.1. Distinguish cases according to the remainder of p mod 8 and use the formulas of $\left(\frac{-1}{p}\right)$ and $\left(\frac{2}{p}\right)$.

    (g) Every p. Hint: Use (e) and (f) or apply Theorem 3.5.1.

    (h) Every prime except the ones of the form $24k + 17$.

3. Follow the hint to Exercise 4.1.5, and apply that 1999 is a prime and $\left(\frac{-2}{1999}\right) = -1$.

4. The condition is equivalent to $(2c)^8 \equiv -2^7 \pmod{43^{100}}$. The solvability of $x^8 \equiv -2^7 \pmod{43}$ follows from Theorem 3.5.1 and $\left(\frac{-2}{43}\right) = 1$. The conversion to modulus 43^{100} is based on Exercise 3.7.2 (or Theorem 3.7.1). Finally, a residue class obtained as a solution must contain even elements as the modulus is odd.

5. (a) If $8c^2 \equiv 1 \pmod{p}$, then

$$1 = \left(\frac{1}{p}\right) = \left(\frac{8c^2}{p}\right) = \left(\frac{2}{p}\right)^3 \left(\frac{c}{p}\right)^2 = \left(\frac{2}{p}\right).$$

We prove the second assertion by contradiction: If every prime factor of $8c^2 - 1$ were of the form $8k + 1$, then also their product (with the corresponding multiplicity), i.e. $8c^2 - 1$ itself would be of the form $8k + 1$.

   (b) Argue as in (a) using $\left(\frac{3}{p}\right) = 1$.

   (c) Working with $\left(\frac{-1}{p}\right)$, we obtain that $p \equiv 1 \pmod{4}$ for every odd prime divisor p of $c^2 + 4$. This implies $p \equiv 1$ or $5 \pmod{8}$ and $\pmod{12}$. Since $c^2 + 4 \equiv 5 \pmod{8}$ and $\pmod{12}$, we cannot have $p \equiv 1 \pmod{8}$ and $\pmod{12}$, for every prime divisor.

6. (a) By reciprocity,

$$\prod_{i=1}^{5} \left(\frac{a_i}{p_i}\right) = (-1)^{\binom{r}{2}},$$

where r is the number of p_i of the form $4k - 1$. Further, $\binom{r}{2}$ is odd if and only if $r = 2$ or 3.

   (b) The condition implies $\left(\frac{p_i}{p_j}\right) = 1$ for every $i \neq j$. Hence, at most one of the primes p_i can be of the form $4k - 1$.

7. (a) Denoting the middle number by c, the sum is

$$S = (c - 9)^2 + (c - 8)^2 + \cdots + (c + 9)^2 = 19(c^2 + 30).$$

As $\left(\frac{-30}{19}\right) = -1$, only the first power of 19 divides S. So S cannot be a power.

   (b) As in (a), it suffices to show that $a = (1 - p^2)/12$ is a quadratic non-residue mod p. Observe $\left(\frac{a}{p}\right) = \left(\frac{36a}{p}\right) = \left(\frac{3}{p}\right)$.

8. For example, $f = (x^2 + 1)(x^2 - 17)(x^2 + 17)$ is suitable.

4.3.

1. (a) 1.   (b) -1.   (c) -1.   (d) 1.

2. (a) Let $m = p_1 \ldots p_r$. If $x^2 \equiv a \pmod{m}$ is solvable, then $x^2 \equiv a \pmod{p_i}$ is solvable for every i. Thus $\left(\frac{a}{p_i}\right) = 1$ for every i implying $\left(\frac{a}{m}\right) = \left(\frac{a}{p_1}\right) \ldots \left(\frac{a}{p_r}\right) = 1$.

   (b) For example, $m = 9, a = 2$; or $m = 15, a = 8$, etc.

   (c) $m = p^{2k+1}$ (where p is a prime, $k \geq 0$).

3. The case $p = 2$ is obvious. Otherwise $p \equiv 1 \pmod 4$, implying $\left(\frac{-1}{p}\right) = 1$. So we can reduce the problem to $a, b > 0$. Let (say) a be odd, then $\left(\frac{a}{p}\right) = \left(\frac{p}{a}\right) = \left(\frac{a^2+b^2}{a}\right) = \left(\frac{b^2}{a}\right) = 1$ (for $a > 1$).

4. Both sums equal -1.

   Hint to (b): Verify $\left(\frac{k}{2k+1}\right) = \left(\frac{-2}{2k+1}\right)$.

5. (a) If $a \equiv 1 \pmod 4$, then $\left(\frac{a}{m}\right) = \left(\frac{m}{a}\right) = \left(\frac{n}{a}\right) = \left(\frac{a}{n}\right)$. If $a = 2^k t$ with $k \geq 2$ and t odd, then $m \equiv n \pmod 4$ guarantees that the pairs t, m and t, n behave alike concerning reciprocity. Also, $\left(\frac{2}{m}\right) = \left(\frac{2}{n}\right)$ if $m \equiv n \pmod 8$, i.e. $k \geq 3$. If $k = 2$ (or any even number), then $\left(\frac{2}{m}\right)$ and $\left(\frac{2}{n}\right)$ play no role.

   (b) Any odd $m > 1$ (coprime to a) and $n = m + 2a$ are suitable in both cases.

6. (a) 0 or $\varphi(m)$. Hint: If every $\left(\frac{r}{m}\right) = 1$, then the sum S is clearly $\varphi(m)$. Otherwise take any c with $\left(\frac{c}{m}\right) = -1$ and replace every r by cr. Verify that the resulting sum equals both S and $-S$.

   (b) -1 if m is an odd power of a prime of the form $4k + 3$, and 1 in every other case.

7. (a) m is a square. Hint: The squares clearly meet the requirement. If m is not a square, then there is a prime p occuring at an odd exponent in the standard form of m, i.e. $m = p^k t$ with $(t, p) = 1$ and k odd. Let c be a quadratic non-residue mod p. Then $\left(\frac{a}{m}\right) = -1$ for a solution a of the system $x \equiv c \pmod p$, $x \equiv 1 \pmod t$.

   (b) a is a square. Hint: Argue as in (a) using reciprocity. Be careful to handle the negative and/or even numbers a, as well.

A.5. Prime Numbers

5.1.

1. If (say) $r_1, \ldots, r_m$ is a complete residue system modulo $m > 1$, then $n + r_1, \ldots, n + r_m$ also form a complete residue system modulo m for any n, hence m divides one of these elements. If $m \mid n + r_i$ and $n + r_i > m$, then $n + r_i$ cannot be a prime.

2. (a) Let $n \geq 7$ be odd. Then $n - 3 \geq 4$ is even, so $n - 3 = p_1 + p_2$, i.e. $n = 3 + p_1 + p_2$.

   (b) If an even number is the sum of three primes, then one of the primes must be 2 and $n - 2 = p_1 + p_2 \iff n = 2 + p_1 + p_2$.

3. Every $n \geq 8$. — Every $n \geq 40$ and $n = 18, 24, 30, 34$, and 36.

4. Only the pair 5 and 2 has this property.

5. (c) Use that for $(p, d) = 1$, the first p terms of the arithmetic progression form a complete residue system modulo p.

6. *Mersenne and Fermat*: We saw in Exercise 1.4.4 that $2^k - 1$ is composite if k is composite, and $2^k + 1$ is composite if k is not a power of two.

$n^2 + 1$: If $n > 1$ is odd, then $n^2 + 1 > 2$ is even, and in general, if $k < n \equiv k$ $(\bmod\ k^2 + 1)$, then $n^2 + 1 \equiv k^2 + 1 \equiv 0\ (\bmod\ k^2 + 1)$, hence $n^2 + 1$ is composite.

Repunit: If k is composite, then the repunit of k digits is composite.

$333\ldots31$: These numbers are of the form $(10^k - 7)/3$ and are primes for $2 \le k \le 8$. If, however, $k = 2 + 30r$, then, by Fermat's Little Theorem, $10^k - 7 \equiv 10^2 - 7$ $(\bmod\ 31)$, hence $(3, 31) = 1$ implies

$$\frac{10^k - 7}{3} \equiv \frac{10^2 - 7}{3} = 31 \equiv 0\ (\bmod\ 31),$$

thus we obtain a multiple of 31. Similarly, infinitely many of them are divisible by 17: 10 is a primitive root modulo 17, thus $10^s \equiv 7\ (\bmod\ 17)$ for some s, and then $17 \mid (10^k - 7)/3$ for $k = s + 16r$.

Fibonacci: Every third element is even. For any m, there are infinitely many Fibonacci numbers divisible by m (see Exercise 1.2.5).

7. Apply the theorem about interpolation polynomials: Prescribing the values at k places, there exists exactly one suitable polynomial of degree at most $k - 1$ (with coefficients from the given field).

8. (a) If $a \equiv b\ (\bmod\ f(b))$, then $f(a) \equiv 0\ (\bmod\ f(b))$.

    (b) (i) Equivalently, for a polynomial g with integer coefficients, $g(n)$ cannot be a constant times a prime for every n. This can be shown as in (a).
    (ii) If a polynomial with complex coefficients assumes rational values at more places than its degree, then it must have rational coefficients. This can be proved using the interpolation polynomials.
    (iii) Fix integer values for all but one variable, thus reducing the problem to the case of a single variable.

9. (a) It follows by induction on n from the argument in the proof of Theorem 5.1.1.

    (b) The last digits in the integer $\lfloor 10^{2^{2^j}} c \rfloor$ are p_j.

    (c) c can be computed (probably) only if we know in advance the prime numbers.

10. E.g. $K = (10^4)!$ is suitable.

5.2.

1. (a) Verify $F_{n+1} = F_n(F_n - 2) + 2$, then use induction.

    (b) Use part (a).

    (c) Every Fermat number has a prime factor that does not divide any other Fermat number.

    (d) The nth prime cannot be larger than F_{n-1}.

2. For a prime $F_n(\neq 5)$, show that both $\left(\frac{5}{F_n}\right)$ and $\left(\frac{10}{F_n}\right)$ are -1. The converse can be proved exactly the same way as in Theorem 5.2.2.

3. The only if part follows exactly the same way as in Theorem 5.2.2. The converse can be proved by contradiction: we can assume then that K_n has a prime divisor $q \leq \sqrt{K_n}$. Show that $o_q(3) = 2^n$ or $5 \cdot 2^n$. This implies $2^n \mid q - 1$ which combined with $q \leq \sqrt{K_n}$ yields the desired contradiction.

4. Apply the formula for $\varphi(N)$.

5. Answer: 5. Hint: Show first that k must be a power of two. Then apply Exercise 5.2.1a and the fact that F_5 is divisible by 641.

6. By Theorem 5.2.3, the smallest possible primes are 47, 233, 223, and 431, and these divide the given Mersenne numbers, as can be checked quickly by repeated squarings.

8. If $2^{2^n} \equiv -1 \pmod{q^2}$, then we obtain $o_{q^2}(2) = 2^{n+1} \mid \varphi(q^2) = q(q - 1)$ as in the proof of Theorem 5.2.1. This implies $o_{q^2}(2) \mid q - 1$, so $2^{q-1} \equiv 1 \pmod{q^2}$. The statement for the Mersenne numbers can be proved similarly.

9. Besides $(8, 9)$ only those pairs work where one element is a Fermat or Mersenne prime and the other element is a suitable power of two.

10. If $k \mid n$ holds in H, then $n/k = a + b\sqrt{3}$ for suitable integers a and b, and n/k is also rational. Using the irrationality of $\sqrt{3}$, it follows that $b = 0$ and a is integer. The converse is straightforward.

11. It is sufficient to show that if F_n is a prime, then $F_n \mid H_k$ for a suitable k. Observe that $o_{F_n}(6) \mid F_n - 1$, so $o_{F_n}(6) = 2^j$ for some j. Then $F_n \mid H_{j-1}$.

5.3.

1. Answer: 6003. (There are infinitely many primes in the reduced residue classes and each residue class represented by a prime divisor of 9999 contains a positive prime.)

2. The integer $A = 4p_1 \ldots p_r + 1$ does not necessarily have a prime divisor of the form $4k + 1$, since it can be the product of an even number of primes of the form $4k + 3$.

3. (a) Adapt the proof of Theorem 5.3.2.

   (b)–(h) Argue as in the proof of Theorem 5.3.3. Examine the possible forms of prime divisors of the following numbers (rely on Exercise 4.2.5 in parts (c), (d), (f), and (h)):
   (b) $n^2 + 2$; (c) $n^2 + 4$; (d) $n^2 - 2$ or $8n^2 - 1$; (e) $5n^2 - 1$; (f) $n^2 + 4$;
   (g) $(2n)^2 + 3$; (h) $12n^2 - 1$.

4. Infinitely many; the question refers to the arithmetic progression $10000k + 4321$.

5. Prove by contradiction: assume that the decimal fraction is periodic with a period of length k starting after an initial aperiodic part of m digits. We know that there are infinitely many primes having 1s as their last $2k$ digits, and the same holds with 3s as the last $2k$ digits. Therefore the period must consist purely of 1s on the one hand, and purely of 3s on the other hand, which is impossible.

6. The condition is $(a, b, c) = 1$. Necessity is obvious. Hint for sufficiency: Put $(a, b) = s$, then $(s, c) = 1$. By Dirichlet's theorem, $a + bk = sp$ for some k, where p is a prime greater than c. Apply again Dirichlet's Theorem to the arithmetic progression $sp + cn, n = 0, 1, \ldots$.

7. (a) $\left(\frac{c}{p}\right) = 1$ e.g. for primes of the form $p = 8 \cdot |c| \cdot k + 1$. We can verify this by using the standard form of $|c|$ and the properties of the Legendre (or Jacobi) symbol. (We also have to consider the cases when c is negative or even.)

   (b) Answer: c is not a square. Hint: Use Exercise 4.3.7b (or proceed along the lines of the solution seen there).

8. For distinct primes $p_1, \ldots, p_{n-1}$, $f = x\left(1 + k(x - p_1) \ldots (x - p_{n-1})\right)$ meets the requirement for some integer k: $v_1 = p_1, \ldots, v_{n-1} = p_{n-1}, v_n = 1$.

9. Let a and d be fixed positive coprime integers. Then $a_1 = a + rd$ is composite for some $r \geq 0$. Since $(a_1, d^s) = 1$ for every s, the assumption implies that $p_s = a_1 + k_s d^s$ is a prime for some k_s. These primes p_s are also of the form $a + kd$, and there are infinitely many distinct numbers among them since $k_s \neq 0$.

5.4.

1. Write a and b as $a = \lfloor a \rfloor + \{a\}$ and $b = \lfloor b \rfloor + \{b\}$, where $0 \leq \{a\}, \{b\} < 1$. Then $a + b = \lfloor a \rfloor + \lfloor b \rfloor + \{a\} + \{b\}$. If the sum of the last two terms is less than 1, then $\lfloor a + b \rfloor = \lfloor a \rfloor + \lfloor b \rfloor$, whereas if it falls between 1 and 2, then $\lfloor a + b \rfloor = \lfloor a \rfloor + \lfloor b \rfloor + 1$.

2. Show first that we can restrict ourselves to integer values of x and then observe that there are only finitely many positive integers less than the x_0 guaranteed by Theorem 5.4.3.

3. Proceed as in the proof of Theorem 5.4.2. Combine $\pi(p_n) = n$ with the upper bound for $\pi(x)$ to obtain $p_n > (1/c_2) \cdot n \cdot \log n$, if n is large enough. The other estimate is slightly more complicated. You need to verify $\log p_n < (1 + \varepsilon) \log n$. This leads to $p_n < (1/c_1 + \varepsilon) \cdot n \cdot \log n$ for any $\varepsilon > 0$ if n is large enough (depending on ε).

4. Part (a) is the logarithmic version of (b), so we need to prove only (a). We can use the inequalities

$$\log n \cdot \pi(n) \geq \sum_{p \leq n} \log p \geq \log f(n) \cdot \left(\pi(n) - \pi(f(n))\right)$$

and choosing e.g. $f(n) = n/(\log n)^2$ leads to the desired result.

5. (iii) is the logarithmic form of (iv). The implications (i)⇒(ii) and (i)⇒(iii) can be verified as in Theorem 5.4.2 and Exercise 5.4.4. The converses can be proven by similar arguments.

6. (a) The upper bound follows immediately from $S(n) \leq n \cdot \pi(n)$ and from the upper bound for $\pi(x)$. For the lower bound, start from $S(n) \geq (\pi(n) - \pi(cn)) \cdot (cn)$ where $0 < c < 1$, and show, with the help of the Prime Number Theorem, that $\pi(n) - \pi(cn) > c' \cdot n/\log n$ for some $c' > 0$. (We can use Theorem 5.4.3 instead. Then c must be chosen sufficiently small to guarantee the existence of a suitable c'.)

   (b) Using $p_k \sim k \log k$, show

   $$S(n) \sim \sum_{k=2}^{\pi(n)} k \log k \sim \int_2^{\pi(n)} t \log t \, dt.$$

   To evaluate the integral, apply

   $$\int t \log t \, dt = \frac{2t^2 \log t - t^2}{4} \quad \text{and} \quad \pi(n) \sim \frac{n}{\log n}.$$

7. The argument is based on the fact that there are many primes up to N, so they give rise to many sums and differences, but these sums and differences can assume only a few even values, hence, by the pigeonhole principle, some even integer must have many representations as such a sum or difference.

   Let us see this in detail for the sums; the differences can be handled similarly. The sum of any two odd primes not exceeding N is an even integer not greater than $2N$. The number of such sums is

   $$\binom{\pi(N) - 1 + 1}{2} \sim \frac{N^2}{2(\log N)^2},$$

   the number of even integers up to $2N$ is N. Therefore, (for N large enough) there exists an even integer that has at least

   $$\frac{N}{3(\log N)^2} > K$$

   representations as the sum of two primes.

8. The formula is based on Wilson's Theorem and its converse: For $j > 1$, we have $j \mid (j-1)! + 1 \iff j$ is a prime. This cannot be used to determine $\pi(n)$ in practice, since no quick methods are known to compute factorials or their remainders.

5.5.

1. Apply Chebyshev's Theorem.

2. Write the larger number in the form $n = p + (n - p)$ where p is the largest prime not exceeding n, then repeat the process for $n - p$ instead of n, etc. till you get 0 or 1. The primes thus representing n or $n - 1$ will be distinct by Chebyshev's Theorem.

3. (a) The integers n with $k+1$ digits and first digit 1 satisfy $10^k \le n < 2 \cdot 10^k$, hence there is a prime among them for every k by Chebyshev's Theorem.

    (b) Use part (A) in Theorem 5.5.5 instead of Chebyshev's Theorem.

4. (a) Let p be a prime satisfying $n/2 < p \le n$. Writing the fractions with a common denominator, the denominator and all but one numerators are divisible by p. Therefore, the sum cannot be an integer (it will be a fraction with a denominator divisible by p). The statement can be proved without Chebyshev's Theorem by examining the exponent of 2 in the least common denominator $\mathrm{lcm}(1, 2, \dots, n)$ and in the numerator (of the sum).

    (b) For $n \ge 2k - 1$, any proof of (a) works. For $n < 2k - 1$, the sum is less than 1.

5. As $\binom{2n}{k} = \binom{2n}{2n-k}$, we may assume $k < n$. Then

    $$\binom{2n}{n} = \binom{2n}{k} \cdot \frac{(2n - k) \dots (n + 1)}{(k + 1) \dots n}.$$

    Both the numerator and the denominator of the last fraction are products of $n - k$ factors, and every factor in the numerator is bigger than any factor in the denominator. Hence this fraction is larger than 1.

6. The moduli are pairwise coprime, hence this system of congruences is solvable. The solutions form a reduced residue class modulo $m = p_1 \dots p_K q_1 \dots q_K$ that contains (infinitely many) primes $p > m$, by Dirichlet's Theorem. By the construction of the congruences, $p - j$ is divisible by p_j, and $p + j$ is divisible by q_j. Further, $p - j > p_j$, $p + j > q_j$, hence each $p \pm j$ is composite.

7. (a) The numerator of $\binom{2n}{n}$ contains p as a factor, whereas the denominator and the other factors of the numerator are not divisible by p.

    (b) Both the numerator and the denominator are divisible by exactly the second power of p (the factors $3p$ and $4p$ in the numerator and the factors p and $2p$ in the denominator contain p). Generalization: If $2n/(2k + 1) < p \le n/k$ and $p > 2k$, then $\binom{2n}{n}$ is not divisible by p.

8. Let L be the number of primes between n and $2n$. By Exercise 5.5.7a, the product of these primes is the quantity C defined in the proof of Theorem 5.5.3 after (5.5.1), hence $C < (2n)^L$. On the other hand, (5.5.6) in the same proof implies $C > 4^{n/4}$ for n large enough, since the second term on the right-hand side of (5.5.7) can be neglected compared to the first term. The two inequalities thus obtained for C imply $4^{n/4} < (2n)^L$, and, taking logarithms, we get the statement for n large enough. This can be extended to every $n \ge 2$ by the argument seen in the hint to Exercise 5.4.2.

9. (a) Use the fact that the interval $(n, n+n^{2/3})$ contains a prime if n is large enough.

    (b) The condition $q_n = \lfloor \alpha^{3^n} \rfloor$ is equivalent to

    (A.5.1) $$\sqrt[3^n]{q_n} \le \alpha < \sqrt[3^n]{q_n + 1}.$$

    Using (A.5.1), choose the primes q_n so that that α should be a common element of a nested sequence of intervals. This can be done since nestedness is equivalent to $q_n^3 \le q_{n+1} < (q_n + 1)^3 - 1$.

(c) The formula in (b) gives no exact value for α, so we could prove only the existence of such an α.

10. (a) Following the proof of part (B) in Theorem 5.5.5, we get that, choosing a suitable $c > 0$, the intervals $(n, n + c \log n)$ contain no primes for infinitely many values of n.

(b) By the proof of Theorem 5.5.1, the interval $(n, n + K)$ is primefree for $n = (K + 1)! + 1$. We express K in terms of n, using the following estimates for $m!$

$$\left(\frac{m}{e}\right)^m < m! \le m^m.$$

(The upper bound is obvious, and the lower bound can be easily verified by induction.) Taking logarithms of the inequalities (or of Stirling's formula), we get $\log m! \sim m \log m$. In our case this means $\log n \sim K \log K$, yielding $K \sim \log n / \log \log n$.

Thus we proved that for any $\varepsilon > 0$ there exist infinitely many positive integers n such that the interval $(n, n + (1 - \varepsilon) \log n / \log \log n)$ contains no primes.

(c) By the Remark, the interval $(n, n + K)$ is primefree if $n - 1$ is the product of primes not greater than $K + 1$. By Lemma 5.4.5, $n \le 4^{K+1}$, and by Exercise 5.4.4b, even $n < e^{(1+\varepsilon)(K+1)}$ holds (the latter inequality requires the Prime Number Theorem). This gives $K > c \log n$, which is the result in (a), or using the sharper inequality, we get part (B) of Theorem 5.5.5.

11. Apply similar arguments as in the proof of part (B) of Theorem 5.5.5 (we need now inequalities in the opposite direction, of course). The only essential difference is that the inequality corresponding to (5.5.15) would need $\log p_j > \log N$ which is false since $p_j < N$. We can overcome this difficulty as follows: If $N > p_j > N/(\log N)^2$, then $\log p_j > (1 - \varepsilon') \log N$ for sufficiently large N. Therefore it is worthwhile to write and add the inequalities corresponding to (5.5.13) for these primes.

5.6.

1. Divergent: (a), (c), (e).

Denote the sequences by $A, B, \ldots, F$, and the number of elements not greater than n in them by $A(n), B(n), \ldots, F(n)$. Then

$$A(n) \sim c_1 n; \qquad B(n) \sim \sqrt{n}; \qquad E(n) \sim c_2 n; \qquad F(n) \sim c_3 \sqrt{n},$$

where the c_is are suitable positive constants (depending on L except for c_3). For the sequence D we have

$$D(n) \sim c_4 (\log n)^k$$

where k is the number of primes less than L. Here it is much simpler to prove the weaker result

$$c_5 (\log n)^k < D(n) < c_6 (\log n)^k.$$

2. Only (c) is divergent. The corresponding integrals are

(a) $\int \dfrac{dx}{x^{1.01}} = \dfrac{-100}{x^{0.01}}$

(b) $\int \dfrac{dx}{x(\log x)^2} = \dfrac{-1}{\log x}$

(c) $\int \dfrac{dx}{x \cdot \log x \cdot \log \log x} = \log \log \log x.$

3. Divergent: (b). Use arguments similar to the first proof of Theorem 5.6.1.

4. (a) Convergent: Rearrange the numbers according to their smallest prime divisors (these are distinct from the assumption). Then $a_n \geq p_n^2$, so

$$\sum_{n=1}^{\infty} \frac{1}{a_n} \leq \sum_{n=1}^{\infty} \frac{1}{p_n^2} < \infty.$$

(b) Convergent: By assumption, $a_n \geq 2^{2 \log n} = n^{2 \log 2}$. Since $\alpha = 2 \log 2 > 1$,

$$\sum_{n=1}^{\infty} \frac{1}{a_n} \leq \sum_{n=1}^{\infty} \frac{1}{n^{\alpha}} < \infty.$$

(c) Divergent: $a_n < cn$ with $c = 10^{1001}$, if n is large enough.

(d) Both convergence and divergence are possible.

(e) Convergent: Rearrange the elements according to the number of their divisors, so $d(a_n) \geq n$ by assumption. By Exercise 1.6.11c, this implies $n \leq 2\sqrt{a_n}$, i.e. $a_n \geq n^2/4$. Apply the fact that the sum of reciprocals of the squares is convergent.

5. It makes no sense; the value of the sum is strongly influenced by the first few terms. For example, adjoining 2 and 3 to the cubes, the sum of reciprocals will exceed the sum of reciprocals of the squares, whereas the cubes grow much faster than the squares. Thus the sequence of cubes (plus the numbers 2 and 3) is less dense than the sequence of squares.

6. Argue as in the third proof of Theorem 5.6.1.

7. If the sequence a_j does not tend to 0, then it is easy to see that the infinite series diverges and the infinite product is 0. Hence, we may assume that the sequence a_j tends to 0. Taking the logarithm of the infinite product, we have

$$\prod_{j=1}^{\infty}(1 - a_j) = 0 \iff \sum_{j=1}^{\infty} -\log(1 - a_j) = \infty.$$

Use the fact that $0 < a_j < 1/2$ implies $a_j < -\log(1 - a_j) < 2a_j$.

8. It is more convenient to prove the corresponding inequality for the logarithms of the two sides. Use Theorem 5.6.2 and the fact that $-\log(1 - a)$ can be well approximated by a for $0 < a \leq 1/2$.

9. (a) Divergent: For even numbers $n = 2k$ we have $np(n) = 4k$, and $\sum_k 1/(4k)$ is divergent.

(b) Convergent. Let q be a fixed prime and S_q the sum of reciprocals of the integers n satisfying $P(n) = q$. Verify

$$S_q = \frac{1}{q} \prod_{p \leq q} \frac{1}{1 - \frac{1}{p}}.$$

This implies $S_q < c(\log q)/q$, by Exercise 5.6.8. Hence,

$$\sum_{n=2}^{\infty} \frac{1}{nP(n)} = \sum_q \frac{S_q}{q} < c \sum_q \frac{\log q}{q^2} < \infty.$$

10. To prove the observation, let s be the period of the rational number and consider only those $i > i_0$ for which a_i falls into the periodic part. Show that there can be at most s such a_i with exactly t digits (for any t). Hence, $\sum_{i>i_0}^{\infty} 1/a_i < s \sum_{t=1}^{\infty} 1/10^{t-1} < \infty$.

5.7.

1. (a) Consider the step $r_k = r_{k+1}q_{k+2} + r_{k+2}$ in the algorithm. If we decrease the product after the equality sign using $r_{k+1} > r_{k+2}$ and $q_{k+2} \geq 1$, we obtain the desired inequality $r_k > 2r_{k+2}$.

   (b) $2\log_2 b$.

   (c) We get the smallest b, if $(a, b) = r_{s-1} = 1$ and the quotients q_i are minimal, i.e. $q_s = 2$ and $q_i = 1$ for $i < s$. Then, starting from the end, the algorithm gives

   $$r_{s-1} = 1, \quad r_{s-2} = 2, \quad r_{s-3} = r_{s-2} + r_{s-1}, \quad \ldots, \quad b = r_1 + r_2,$$

   thus $r_{s-j} = \varphi_{j+1}$ and $b = \varphi_{s+1}$, by the recursion for the Fibonacci numbers.

2. The gcd of the numerator and the denominator does not change during the process (even when halving the numerator, since the denominator is odd, hence so is the gcd). As the procedure is a variant of the Euclidean algorithm, we reach finally $(a, b) = d$. This d appears in the numerator, since the new numbers occur there after each step. Then the denominator v satisfies $(d, v) = (a, b) = d$, thus $d \mid v$.

3. $341 = 11 \cdot 31$. Note that

   $$\varphi(11) \mid 340 \Rightarrow 2^{340} \equiv 1 \pmod{11} \quad \text{and}$$

   $$2^5 \equiv 1 \pmod{31} \Rightarrow 2^{340} \equiv 1 \pmod{31}.$$

   This implies $2^{340} \equiv 1 \pmod{11 \cdot 31}$, so 341 is a pseudoprime of base 2. But

   $$3^{340} \equiv 3^{10} \not\equiv 1 \pmod{31} \Rightarrow 3^{340} \not\equiv 1 \pmod{341},$$

   so 341 is not a pseudoprime of base 3.

5. As p is odd, we have

   $$n = \frac{a^p - 1}{a - 1} \cdot \frac{a^p + 1}{a + 1} =$$
   $$= (a^{p-1} + a^{p-2} + \cdots + 1)(a^{p-1} - a^{p-2} + \cdots + 1),$$

implying that n is odd and composite. The validity of $a^{n-1} \equiv 1 \pmod{n}$ follows from $a^{2p} \equiv 1 \pmod{n}$ and $n \equiv 1 \pmod{2p}$; the latter can be verified by considering $n(a^2 - 1) = a^{2p} - 1$ modulo p and using that n is odd.

6. $561 = 3 \cdot 11 \cdot 17$. To prove $(a, 561) = 1 \implies a^{560} \equiv 1 \pmod{561}$, it is sufficient to verify this for the moduli 3, 11, and 17, which follow from Fermat's Little Theorem.

7. (a) $\Rightarrow$ (b): If n is not squarefree, then we get a contradiction following the relevant parts in the proof of Theorem 5.7.4 (but disregarding (5.7.2) there, of course). If $p \mid n$, then consider a primitive root $g \bmod p$ coprime to n (this can be guaranteed by a suitable system of congruences as seen in the proof of Theorem 5.7.4). Then

$$(g, n) = 1 \implies g^{n-1} \equiv 1 \pmod{n}$$
$$\implies g^{n-1} \equiv 1 \pmod{p}$$
$$\implies o_p(g) = p - 1 \mid n - 1.$$

(b) $\Rightarrow$ (c): Since n is squarefree, it is sufficient to verify $a^n \equiv a \pmod{p}$ for every prime divisor p of n. This is obvious for $p \mid a$. If $(p, a) = 1$, then Fermat's Little Theorem and $p - 1 \mid n - 1$ imply $a^{n-1} \equiv 1 \pmod{p}$, and multiplying by a we get the desired congruence.

(c) $\Rightarrow$ (a): If $(a, n) = 1$, then we can divide $a^n \equiv a \pmod{n}$ by a to get $a^{n-1} \equiv 1 \pmod{n}$.

8. Use condition (b) of Exercise 5.7.7.

9. (a) If luckily we get $1 < (a, n) < n$, then we verified not only the compositeness of n, but also found a non-trivial divisor. (This has, however, a very small probability, see part (b).)

   (b) Roughly 10^{-100}.

10. $(a - 1, n)$ (or $(a + 1, n)$) is a non-trivial divisor.

11. First we check whether or not n is a prime. Then we can clearly restrict ourselves to the case when n is odd and composite.

    We see, using a quick algorithm, if n is a perfect power: we check whether $\sqrt[k]{n}$ is an integer for some $2 \le k \le \log_2 n$. If $n = m^k$, then it suffices to factor m. The initial condition also holds for m, since $\varphi(m) \mid \varphi(n)$, hence the given multiple of $\varphi(n)$ is a multiple of $\varphi(m)$. Thus we may assume that n is not a perfect power.

    We choose (say) 1000 random values $n \nmid a$ and compute (a, n). If $(a, n) > 1$, then we can decompose n into the product of two non-trivial factors, by Exercise 5.7.9.

    If $(a, n) = 1$, then adapting the basic idea in Theorem 5.7.5 to our case, consider the sequence

$$a^e, a^{\frac{e}{2}}, a^{\frac{e}{4}}, \dots,$$

    where we know that e is a multiple of $\varphi(n)$. The remainder modulo n of the first element is 1 from the Euler–Fermat Theorem. In the squarefree part of the proof of Theorem 5.7.5, we only used that n is not a prime power, and we can show the same way that at least half of the elements in a reduced residue system modulo n generate a sequence of remainders where the 1s are followed by a remainder different from ± 1, and so we can factor n, by Exercise 5.7.10.

If $n = n_1 n_2$ with $n_i > 1$, then we repeat the entire process for n_1 and n_2 ($\varphi(n_i) \mid \varphi(n)$) implies that we can use the same exponent e), and proceed similarly till we get the prime factorization of n. Since the number of prime factors is at most $\log_2 n$ and each factorization requires at most $c \log_2 n$ steps, we get the complete factorization in not more than $c(\log_2 n)^2$ steps with a suitable constant c.

12. This idea does not work in practice since no quick methods are known for computing factorials or their remainders.

13. (a) Argue as in the part in the proof of Theorem 5.7.4 where we showed that there are at least as many witnesses as accomplices, provided there are witnesses at all.

   (b) Let $n > 1$ be odd. Choose (say) 1000 random values $a \not\equiv 0 \pmod{n}$ and check the validity of $a^{n-1} \equiv 1 \pmod{n}$. If it is false in at least one case, then n must be composite. If it is true in all the 1000 cases, then the probability of n not being a prime or a universal pseudoprime is less than 2^{-1000}.

14. We check R integers a. If n is a prime, then we always obtain remainders ± 1 and the probability of pure 1s is 2^{-R}. (Thus we can make an error also in the opposite direction at this test by declaring a prime falsely to be a composite integer.) If n is composite, then we can proceed as in the proofs of Theorems 5.7.4 and 5.7.5.

15. Apply a suitable modification of the argument in the hint to Exercise 5.2.3.

16. (a) Verify $o_n(a) = n - 1$.

   (b) Let the standard form of $n - 1$ be

   $$n - 1 = p_1^{\beta_1} \dots p_r^{\beta_r}.$$

   By assumption, $p_i^{\beta_i} \mid o_n(a_i)$. Then (e.g. by Exercise 3.2.4c) there are integers b_i satisfying $o_n(b_i) = p_i^{\beta_i}$, which implies $o_n(b_1 \dots b_r) = n-1$, by Exercise 3.2.15a.

   (c) For a proof of contradiction, assume that n is composite, hence it has a prime divisor $q \le \sqrt{n}$. Repeat the argument of part (b) for the modulus q instead of n. We obtain $o_q(b) = c > \sqrt{n} \ge q$ for some b, which is a contradiction.

17. We have to show that if a generates a good sequence, then

(A.5.2) $$a^{(n-1)/2} \equiv \left(\frac{a}{n}\right) \pmod{n}$$

holds.

If $a^r \equiv 1 \pmod{n}$, deduce that both sides of (A.5.2) are 1.

Turning to the case $a^{2^j r} \equiv -1 \pmod{n}$, compute the remainder of $a^{(n-1)/2}$. Then show for any prime divisor q of n, that $o_q(a)$ is an odd multiple of 2^{j+1}, thus $2^{j+1} \mid q-1$. Based on that, prove that the value of $\left(\frac{a}{q}\right)$ depends on the parity of $(q-1)/2^{j+1}$, and write $\left(\frac{a}{n}\right)$ using the standard form of n. Replace the primes q in the standard form of n by the expressions obtained from $2^{j+1} \mid q-1$, perform the multiplications and examine the divisibility by a suitable power of two to obtain that $\left(\frac{a}{n}\right)$ assumes the value in (A.5.2).

5.8.

1. This would be an unsigned, anonymous letter that could have been falsified by a third party in the name of A.

2. The invertibility of E means that the congruence $r^e \equiv s \pmod{N}$ has exactly one solution in r for any s. This congruence is equivalent to the system

(A.5.3) $\qquad r^e \equiv s \pmod{p}, \qquad r^e \equiv s \pmod{q}.$

    By Exercise 3.5.7, each of the two congruences in (A.5.3) has exactly one solution for every s if and only if $(e, p-1) = (e, q-1) = 1$, i.e. $(e, \varphi(N)) = 1$.

3. (a) It suffices to show that the congruence is valid both mod p and mod q. Let us see this mod p: For $p \mid r$, both sides are congruent to 0, and for $(p, r) = 1$, we have
$$r^{1+k\varphi(N)} \equiv r\left(r^{p-1}\right)^{k(q-1)} \equiv r \cdot 1 = r \pmod{p}.$$
   (b) $v \equiv 1 \pmod{[p-1, q-1]}$.

4. This causes no problem, since we use only the property that $r^p \equiv r \pmod{p}$ holds for every r (see Exercise 5.8.3a). (In this case, however, the product $(p-1)(q-1)$ is not the same as $\varphi(N)$, of course.)

5. Let $s \equiv r^e \pmod{N}$, where s and e are known, and we want to find the value of r. We raise s to the eth power, and then raise the result to the eth power, etc., till we get a number congruent to s:

(A.5.4) $\qquad s^{e^k} \equiv s \pmod{N}.$

    Since $(e, \varphi(N)) = 1$, we can take eth roots in (A.5.4), by Exercise 5.8.2, so
$$s^{e^{k-1}} \equiv r \pmod{N}.$$
    This means that if (A.5.4) occurs for a small k, then we can determine r. If $e^k \equiv 1 \pmod{\varphi(N)}$, then (A.5.4) holds by Exercise 5.8.3a, therefore the order of e modulo $\varphi(N)$ must not be small.

6. A and B can compute the value using the identities
$$g^{k_A k_B} = (g^{k_B})^{k_A} = (g^{k_A})^{k_B}.$$
    Others cannot do this (hopefully) because they do not know k_A or k_B.

7. (a) For a proof by contradiction, assume that two subset sums are equal. Cancelling the common terms, we get that all terms are distinct in the two sums. By (5.8.6), the largest term is itself larger, than the complete other sum, yielding a contradiction.
   (b) For a proof by contradiction, assume $\sum d_i = \sum d_j$ for some d_i and d_j. Then $\sum r c_i \equiv \sum r c_j \pmod{m}$, by (5.8.7). We can divide by r because $(r, m) = 1$, i.e. $\sum c_i \equiv \sum c_j \pmod{m}$. Since $m > \sum_{i=0}^{k-1} c_i$, we can replace congruence by equality which contradicts that C is sum injective.
   (c) It follows directly from the definition of sum injectivity.

(d) To get u, we need the values δ_i, i.e. which terms of the sum injective sequence sum to the given v. For (5.8.6), we can obtain them by the greedy algorithm, where we always take the largest possible c_i. For (5.8.7), we get the values c_i and the corresponding v' as the smallest positive solutions of the congruences $rx \equiv d_i$ and $rx \equiv v$ (mod m). Then we apply the previous procedure.

A.6. Arithmetic Functions

6.1.

1. To verify multiplicativity, apply the formula for $d(n)$ (Theorem 1.6.3) or use Exercise 1.6.5a-b. To disprove complete multiplicativity, find a pair of integers a, b satisfying $d(ab) \neq d(a)d(b)$ (and $(a, b) \neq 1$).

2. (a), (c) $f(n)$ and $h(n)$ are neither additive nor multiplicative.

   (b) $g(n)$ is completely multiplicative.

   (d) $k(n)$ is additive but not completely.

3. There is no such multiplicative h: By the conditions,

$$0 = h(6) = h(2)h(3) \Rightarrow h(10)h(15) = h(2)h(5)h(3)h(5) = 0 \neq 3.$$

There are, however, infinitely many additive, in fact completely additive functions h. Solving the system of equations

$$0 = h(2) + h(3), \qquad 1 = h(2) + h(5), \qquad 3 = h(3) + h(5),$$

we obtain $h(2) = -1$, $h(3) = 1$, and $h(5) = 2$. Let $h(7)$ be a parameter c, and let $h(p) = 0$ for all other primes; there is exactly one completely additive function h satisfying these conditions: If

$$n = 2^{\alpha_1} 3^{\alpha_2} 5^{\alpha_3} 7^{\alpha_4} t, \quad \text{where} \quad (t, 210) = 1 \quad \text{where} \quad \alpha_i \geq 0, \quad i = 1, 2, 3, 4,$$

then

$$h(n) = -\alpha_1 + \alpha_2 + 2\alpha_3 + c\alpha_4.$$

4. If there exists such a multiplicative function $f \neq 0$, then $f(1) = 1$ by Theorem 6.1.6, and if $q_j, \ldots, q_w$ are the prime powers in the standard form of n, then only $f(n) = c_j \ldots c_w$ is possible by Theorem 6.1.7. Verify that the function defined this way is multiplicative. We can proceed similarly for additive functions and in part (b).

5. True: (a), (d).

6. (a) A necessary and sufficient condition is $f(k) = 0$.

   (b) A necessary and sufficient condition is $f(k) = 0$ in this case, too. To prove sufficiency, consider the standard forms of a, b, and k, and compute $g(a) = f(ka)$, $g(b) = f(kb)$, and $g(ab) = f(kab)$ by Theorem 6.1.7. Since $(a, b) = 1$, a prime divisor of k cannot divide both a and b.

(c) For completely multiplicative functions, a necessary and sufficient condition is $f(k) = 1$ or 0. In the multiplicative case, this condition is necessary but not sufficient: consider

$$f(n) = \begin{cases} 0, & \text{if } n \equiv 4 \ (\text{mod } 8); \\ 1, & \text{otherwise,} \end{cases} \quad \text{for } k = 4,$$

then $f(k) = 0$, but g is not multiplicative, as $g(3)g(2) = 0 \neq 1 = g(6)$. A necessary and sufficient condition is $f(k) = 1$ or $f(kn) = 0$ for every n (we get $g = 0$ in the latter case).

7. (a) Apply the relation $ab = (a, b)[a, b]$.

    (b) Use the standard forms of the numbers.

    (c) Answer: f is the sum of an additive and a constant function.

    (d) The constant multiples of a multiplicative function always satisfy the equality. If $f(1) \neq 0$, then there are no other solutions. In the general case, all solutions are given by the functions

$$f(n) = \begin{cases} 0, & \text{if } K \nmid n \\ cg(\frac{n}{K}), & \text{if } K \mid n \end{cases}$$

where $g(n)$ is multiplicative, c is a constant, and K is a fixed positive integer.

8. This follows directly from the definitions of multiplicativity and additivity.

9. (a), (e) These are direct consequences of the definitions.

    (b)–(d) Demonstrate first that the product fg is completely additive or additive if and only if $f(a)g(b) + f(b)g(a) = 0$ for every pair a, b or for every coprime pair a, b, respectively. Answer to (d): If $f \neq 0$ and $g \neq 0$, then f and g assume 0 at every prime power apart from the powers of one or two primes, and in the last case, strict rules apply for the values assumed on the powers of the two primes.

    (f) It follows from Theorem 6.1.6.

10. (a) It follows directly from the definitions.

    (b) We can transform the condition into $(f(a) - g(a))(f(b) - g(b)) = 0$. In the multiplicative case, the two functions may have different values on the powers of a single prime p, but must be equal on all other prime powers.

11. If $f = 0$, then the condition implies $g = 0$, so the assertion is true trivially. Otherwise, looking at the values assumed at 1, we infer that only the constant 1 is possible as the sum of the two functions. Writing the definition of multiplicativity for $f = 1 - g$ and using additivity of g, we obtain $g(a)g(b) = 0$ for every $(a, b) = 1$. Hence g assumes 0 and f assumes 1 at every prime power except perhaps the powers of a single prime p. Therefore $(f^{1000} + g^{1000})(n) = 1$ and $(f^{1000}g^{1000})(n) = 0$ if $p \nmid n$. This makes it possible to check easily the desired multiplicativity and additivity.

12. We can argue as in the solution of Exercise 6.1.9d. We can start from the equalities:

    (a) If $h = f - g$ where f and g are multiplicative, then

    $$(f(a) - 1)(f(b) - 1) = (g(a) - 1)(g(b) - 1)$$

    for any (a, b) with $(a, b) = 1$.

    (b) If $h = fg$ where f is multiplicative and g is additive, then

    $$f(a)g(a)(f(b) - 1) + f(b)g(b)(f(a) - 1) = 0$$

    for any (a, b) with $(a, b) = 1$.

13. (a) Show that the function has value 0 at infinitely many pairwise coprime integers.

    (b) Let $f(1) = f(2) = 1$ and $f(n) = 0$ for $n > 2$.

    (c) If $f(p^{\nu_p}) \neq 0$ for infinitely many primes p with suitable exponents $\nu_p > 0$, then we infer as in part (a) that the function assumes every value of the range infinitely often. Hence, there can be only finitely many such primes p and we can take K as their maximum.

14. (a) False. A counterexample is $f(n) = 3$ if $2 \mid n$ but $4 \nmid n$, and $f(n) = 0$ otherwise. This f is additive, and $f(4) + f(8) = f(32)$, but f is not completely additive as $f(2) + f(6) \neq f(12)$.

    (b) True. If $(c, ab) = 1$, then $(ca, b) \geq (a, b) > 1$, and

    $$f((ca)b) = f(c(ab)) = f(c) + f(ab) = f(c) + f(a) + f(b) = f(ca) + f(b).$$

    (c) False. This is another formulation of the statement in part (a).

    (d) True. We can use similar arguments as in part (b).

    (e) False. A counterexample is $f(1) = f(2) = 1$ and $f(n) = 0$, for $n > 2$.

    It is worthwhile to analyze why we get different answers for parts (d) and (e): Adding $f(c)$ to the inequality $f(ab) \neq f(a) + f(b)$ in (d), the inequality remains valid, but multiplying the inequality $f(ab) \neq f(a)f(b)$ by $f(c) = 0$ in (e), we obtain equality.

15. Answer: $\varphi_2(n) = n \prod_{p \mid n} (1 - 2/p)$ (where p denotes a prime). Hint: Using simultaneous systems of congruences, prove that $\varphi_2(n)$ is multiplicative. Then it is sufficient to compute the values of the function at prime powers.

16. Verify that the functions on both sides are multiplicative; for the sum on the left-hand side, we can argue similarly as in the previous exercise. By multiplicativity, it is enough to check equality for prime powers.

6.2.

1. Let $a_1, \ldots, a_r$ be all positive divisors of a, and $b_1, \ldots, b_s$ be all positive divisors of b. If $(a, b) = 1$, then $a_i b_j$ are all positive divisors of ab, each occurring once, by Exercise 1.6.5a-b. Thus

$$\sigma(ab) = \sum_{i=1}^{r} \sum_{j=1}^{s} a_i b_j = \left(\sum_{i=1}^{r} a_i \right)\left(\sum_{j=1}^{s} b_j \right) = \sigma(a)\sigma(b).$$

Then, by multiplicativity, it suffices to compute the values of σ for prime powers.

2. Use the formulas for the functions, or rely on Exercise 1.6.5a-c.

3. Since $3 \nmid n\varphi(n)$, every prime divisor p of n is of the form $3k - 1$. Let α be the exponent of such a prime p in the standard form of n. If α is odd, then $1 + p$ is a factor of $\sigma(p^\alpha)$, so 3 divides $\sigma(n)$ which contradicts the assumption. Therefore every p occurs with an even exponent, thus n is a square.

4. Let $n = p_1^{\alpha_1} \ldots p_r^{\alpha_r}$ be the standard form of n. A sufficient condition for k is

$$p_i^{\alpha_i + 1} - 1 \mid p_i^{k\alpha_i + 1} - 1, \quad i = 1, \ldots, r.$$

This is definitely satisfied if

$$\alpha_i + 1 \mid k\alpha_i + 1, \quad \text{or} \quad \alpha_i + 1 \mid (k - 1)\alpha_i$$

for every i. Thus we can choose $k - 1$ as any common multiple of the integers $\alpha_i + 1$.

5. Answer: n. Hint: Write the sum of reciprocals using a common denominator, and observe that if d ranges over all divisors of n, then the same applies also for n/d.

6. (a) Answer: The squares and the doubles of squares. Hint: Apply the fraction-free form of the formula for $\sigma(n)$. Another option: Write n as $n = 2^k t$ where t is odd. Then only the number of odd divisors of n, i.e. the number of divisors of t is relevant to the problem. By Exercise 1.6.8, $d(t)$ is odd if and only if t is a square.

   (b) Answer: The products of distinct Mersenne primes. Hint for necessity: Let $n = p_1^{\alpha_1} \ldots p_r^{\alpha_r}$ be the standard form of n. Then

$$2^k = \sigma(n) = \prod_{i=1}^{r} (1 + p_i + p_i^2 + \cdots + p_i^{\alpha_i}).$$

Here every factor is a power of two, so it is even. Therefore $p_i > 2$ and α_i is odd for every i. Thus we can refine the factorization into

$$2^k = \prod_{i=1}^{r} (1 + p_i)(1 + p_i^2 + p_i^4 + \cdots + p_i^{\alpha_i - 1}).$$

Every factor on the right-hand side is a power of two, hence p_i is a Mersenne prime. We have to show $\alpha_i = 1$. For a proof by contradiction assume $\alpha_i > 1$. Then $1 + p_i^2 + p_i^4 + \cdots + p_i^{\alpha_i - 1} > 1$ is a power of two, so it is even, hence it has a factor $1 + p_i^2$ which is a power of two. But this is impossible since $1 + p_i^2$ is not even divisible by 4.

7. *First solution*: $\sigma(n) \neq 2p$ where p is an odd prime of the form $3k - 1$.

Second solution: $\sigma(n) \neq 3^s$ for $s > 1$.

Third solution: Use the fact that σ assumes odd values rarely.

Fourth solution: If $\sigma(x) \leq N$, then $x \leq N$. But $\sigma(x) > N$ also for many integers $x \leq N$, therefore at least that many values $y \leq N$ are missing from the range of σ.

Fifth solution: Find many pairs $x_i \neq x_j$ for which $\sigma(x_i) = \sigma(x_j)$, and apply similar considerations as in the fourth solution.

8. Only $n = 1$ has this property. Hint: Verify $n! < \sigma(n!) < (n + 1)!$ for every $n \geq 2$.

9. Observe that $n = ab$ implies that a or b is greater than or equal to $\sqrt{n}$. Equality holds if and only if n is a prime square.

10. (a) (a1) n is a prime.   (a2) No solution.   (a3) $n = 10, 49$.   (a4) $n = 21$.

    (b) Only for $c = 1$.

    (c) If $c = 2k + 1 > 7$ and $2k = p + q$ where p and q are distinct primes, then $n = pq$ is a solution.

11. (a) (a1) n is a prime.   (a2) No solution.   (a3) $n = 4$.   (a4) $n = 6$.

    (b) Only for $c = 2$.

    (c) $c = 4k$ where $k > 3$.

12. (a) Infinitely many. If we find a suitable pair a_0, b_0 and p is a common prime divisor of a_0 and b_0, then $a_k = a_0 p^k, b_k = b_0 p^k$ meet the requirement for any k. We can start with $a_0 = 6, b_0 = 8$ or $a_0 = 12, b_0 = 14$, etc.

    (b) Infinitely many. If n can be represented as $n = p_1 + p_2 = p_3 + p_4$ with distinct primes p_i, then $a = p_1 p_2$, $b = p_3 p_4$ satisfy the equation. The existence of infinitely many such n (those that have at least two representations as the sum of two distinct primes) can be proved as in Exercise 5.4.7. We note that the same idea also works for part (a) but there we could use a simpler argument.

13. All non-trivial divisors d of n satisfy $2 \leq d \leq n/2$.

Equality: (a) n is a prime or $n = 1$, (b) and (c) n is a prime or $n = 4$.

14. Answer: $n = 6$. Hint: We have to refine the method of the previous exercise.

15. (a) For (a1), use the formulas for the functions. For (a2), observe that $\varphi(n)$ is the signed sum of certain divisors of n. Equality holds in each case if and only if n is a prime.

    (b) If $n = p_1^{\alpha_1} \ldots p_r^{\alpha_r}$ is the standard form of n, then

$$\frac{\sigma(n)\varphi(n)}{n^2} = \prod_{i=1}^{r}\left(1 - \frac{1}{p_i^{\alpha_i + 1}}\right) \geq \prod_{i=1}^{r}\left(1 - \frac{1}{p_i^2}\right).$$

This implies (b1) using $p_i \geq i + 1$.

To prove (b2), show

$$\inf \frac{\sigma(n)\varphi(n)}{n^2} = \lim_{N \to \infty} \prod_{p \leq N}\left(1 - \frac{1}{p^2}\right)$$

and apply Exercise 5.6.6.

16. Let $n = 2^\alpha p_1^{\alpha_1} \ldots p_r^{\alpha_r}$ be the standard form of n where $p_i > 2$ ($\alpha = 0$ and/or $r = 0$ is allowed). The condition implies $\alpha_i = 1$, $\alpha \leq 2$, and $r \leq 1$, so $n = 1, 2, 4, p$, $2p$, or $4p$ where p is an odd prime. Checking these integers, we see that only the specified four values of n satisfy the condition.

17. Both functions assume only the values 0 and ± 1.

18. (a) 3. Hint: There is a multiple of 4 among any four consecutive integers.

    (b) Arbitrarily many. Hint: See Exercise 2.6.11.

19. Let $S(n)$ be the sum of the nth primitive complex roots of unity. It is sufficient to show that $S(n)$ is multiplicative and $S(p^\alpha) = \mu(p^\alpha)$ for every prime power p^α. The multiplicativity is a corollary of the observation: if $(k, m) = 1$, then the product of a kth and an mth primitive root of unity is a kmth primitive root of unity, and every kmth primitive root of unity has a unique decomposition into such a product. We can solve the exercise also using summation and inversion functions, see Exercise 6.5.9a.

20. 0.

21. (a) Use the formulas for the functions, or the fact that the divisors of n correspond to certain subsets of prime divisors counted with multiplicity. If n is squarefree, then equality holds, otherwise we have strict inequalities.

    (b) $k^{\omega(n)} \leq d_k(n) \leq k^{\Omega(n)}$.

22. True: (a).

23. We can proceed as for σ. See the proofs of Theorem 6.2.2 and the part of Theorem 6.2.8 concerning σ, or Exercise 6.2.1. Answer: If $n = p_1^{\alpha_1} \ldots p_r^{\alpha_r}$ is the standard form of n and $\nu \neq 0$, then

$$\sigma_\nu(n) = \prod_{i=1}^{r}(1 + p_i^\nu + p_i^{2\nu} + \cdots + p_i^{\nu\alpha_i}) = \prod_{i=1}^{r} \frac{p_i^{\nu(\alpha_i+1)} - 1}{p_i^\nu - 1}.$$

6.3.

1. Use Theorem 6.3.2.

2. (a) If $n = \prod_{i=1}^{r} p_i^{\alpha_i}$ is the standard form of n, then $2n = \sigma(n)$ is equivalent to

(A.6.1)
$$2\prod_{i=1}^{r} p_i^{\alpha_i} = \prod_{i=1}^{r}\left(1 + p_i + \cdots + p_i^{\alpha_i}\right)$$

The left-hand side of (A.6.1) is divisible by exactly the first power of 2, so exactly one factor on the right-hand side is even but is not a multiple of 4. Therefore only one exponent α_i is odd and the prime p_i belonging to it is necessarily of the form $4k + 1$, whereas the other exponents α_j are even.

(b) By part (a), $n = s^2 p$ where p is a prime of the form $4k + 1$. This implies immediately $n \equiv 1 \pmod 4$. If $3 \mid s$, then $9 \mid n$, thus $n \equiv 9 \pmod{36}$. If $3 \nmid s$, then $3 \nmid n$ as $p \neq 3$. The exponent of p is odd in the standard form of n, so $1 + p \mid \sigma(n)$, therefore $3 \nmid p + 1$. Hence only $p \equiv 1 \pmod 3$ is possible, so $n = s^2 p \equiv 1 \pmod 3$. Combining this with $n \equiv 1 \pmod 4$, we obtain $n \equiv 1 \pmod{12}$.

3. (a) We have to prove $2p^\alpha > \sigma(p^\alpha)$ which is equivalent to $p^\alpha(p - 2) > -1$.

   (b)
$$\frac{\sigma(p^\alpha q^\beta)}{p^\alpha q^\beta} = \left(1 + \frac{1}{p} + \cdots + \frac{1}{p^\alpha}\right)\left(1 + \frac{1}{q} + \cdots + \frac{1}{q^\beta}\right) <$$
$$< \frac{p}{p-1} \cdot \frac{q}{q-1} \le \frac{3}{2} \cdot \frac{5}{4} < 2.$$

   (c) Examples of abundant numbers: integers with standard form $\prod_{i=1}^k p_i^{\alpha_i}$ where p_i is the ith odd prime ($p_1 = 3$, $p_2 = 5$, etc.), $\alpha_1 \ge 3$, and the other exponents α_i are positive integers.
   Examples for deficient numbers: products of k distinct primes q_i satisfying
$$1 + \frac{1}{q_i} < \sqrt[k]{2}.$$

   (d) If $d_1, \ldots, d_t$ are all (positive) divisors of n, then the integers ad_i are distinct divisors of an, so

(A.6.2)
$$\frac{\sigma(an)}{an} > \frac{\sigma(n)}{n}$$

   for any $a > 1$. Other options to prove A.6.2 are to use the formula for σ or the fact that $\sigma(m)/m$ is the sum of reciprocals of the divisors of m.

   (e) The product of a deficient and an abundant number is abundant, whereas multiplying a deficient number by a sufficiently large prime, gives a deficient number.

4. By Exercise 6.2.6a, such a number must be of the form $n = 2^\alpha t^2$ with t odd. We have to show $\alpha = 0$. We can transform $\sigma(n) = 2n + 1$ into

(A.6.3)
$$(2^{\alpha+1} - 1)(\sigma(t^2) - t^2) = t^2 + 1.$$

If $\alpha \ge 1$, then the first factor on the left-hand side of (A.6.3) is of the form $4k - 1$, hence it has a prime factor of this type. But $t^2 + 1$ cannot have such a prime divisor.

5. (a) We can apply an argument similar to the proof of Theorem 6.3.2.

   (b) We have to show that $\sigma(n)$ is odd. Write $\sigma(n) = 2^v w$ where w is odd and get a contradiction for $v \ge 1$.

   (c) Assume that p^α is superperfect, and express $\sigma(\sigma(p^\alpha))$ using the standard form of $\sigma(p^\alpha)$.

6. (a) The harmonic mean of the divisors of n is
$$\frac{d(n)}{\sum_{d\mid n} \frac{1}{d}} = \frac{nd(n)}{\sigma(n)}.$$

(b) By part (a), it is enough to check that $d(n)$ is even for a perfect number n, so n cannot be a square. This holds as $\sigma(n)$ is odd if n is a square, so $\sigma(n) \neq 2n$.

(c) For a proof by contradiction, assume $1 + p + \cdots + p^\alpha \mid p^\alpha(\alpha + 1)$. As $(1 + p + \cdots + p^\alpha, p^\alpha) = 1$, we have $1 + p + \cdots + p^\alpha \mid \alpha + 1$. But this is impossible, since $1 + p + \cdots + p^\alpha > \alpha + 1$.

(d) Let $n = p_1 p_2 \ldots p_r$ where $p_1 < p_2 < \cdots < p_r$ are primes. If every p_i is odd, then

$$\frac{p_1 + 1}{2} \cdots \frac{p_r + 1}{2} \mid p_1 \cdots p_r$$

cannot hold since $(p_1 + 1)/2$ is coprime to every factor of the right-hand side. If $p_1 = 2$, then necessarily $p_2 = 3$. We see that $n = 6$ is harmonic, but if n has further prime factors, we get a contradiction as in the previous argument.

7. (a) If $a < b$ and $\sigma(a) = \sigma(b) = a+b$, then $\sigma(b) = a+b < 2b$ and $\sigma(a) = a+b > 2a$.

(b) Assume that $a = 2^k$ and b are amicable. Then

$$\sigma(2^k) = 2^{k+1} - 1 = \sigma(b) = 2^k + b,$$

thus $b = 2^k - 1$, and because both b and $\sigma(b)$ are odd, $b = u^2$. This yields $2^k - 1 = u^2$, which is already false modulo 4 for $k \geq 2$.

6.4.

1. We start with the canyon theorems. The proof for $\Omega(n)$ is the same as for $d(n)$ (Theorem 6.4.1), for $d_k(n)$ we have to modify the moduli of the system of congruences to 2^{K+k} and 3^{K+k}, and for $\omega(n)$ we choose two coprime moduli where each is a product of $K + 2$ distinct primes. For $\sigma(n)$ we can take n as a sufficiently large prime since then $\sigma(n) = n + 1$. Because $n + 1$ are $n - 1$ even,

$$\sigma(n - 1) > (n - 1) + \frac{n - 1}{2} \quad \text{and} \quad \sigma(n + 1) > (n + 1) + \frac{n + 1}{2}.$$

This also gives a proof for the peak theorem for $\varphi(n)$.

Turning to the other peak theorems and to the canyon theorem for $\varphi(n)$, we choose n as the product of the first r primes as we did for $d(n)$ (Theorem 6.4.2).

The peak theorem for $d_k(n)$ and $\sigma(n)$, and the canyon theorem for $\varphi(n)$ can be verified by a suitable modification of the proof of Theorem 6.4.2. Keeping the notation there, for $\Omega(n)$ and $\omega(n)$ we have to show $r - s > K$. This follows combining

$$n \leq p_1 \cdots p_{K+1} p_r^{r-K-1} < p_r^{r-K}$$

(for r large enough), and

$$n + 1 = q_1 \ldots q_s > p_r^s.$$

2. Follow the proof of Theorem 6.4.5.

3. (a) We can take n as a large power of the product of the first 101 primes.

(b) Let n be the product of the first r primes. Then, using the results of Section 5.4, we get

$$\log n \sim p_r \sim r \log r, \quad \text{and so} \quad r \sim \frac{\log n}{\log \log n}.$$

By $d(n) = 2^r$, we get the estimate stated in the exercise.

4. Let $\Omega(n) = s$, so $n = q_1 \ldots q_s$ where $q_i = q_j$ may occur. Because $q_i \geq 2$ we get $n \geq 2^s$. Equality holds if and only if n is a power of two.

5. Show that for a fixed r, the product of the first r primes is the smallest n for which $\omega(n) = r$. This means that $\omega(n)$ attains its maximal order of magnitude as a function of n exactly on the products of that type. The desired estimates now follow as in Exercise 6.4.3b.

6. (a) Apply Theorem 6.4.6 for $n^{0.99}/\varphi(n)$, or use $d(n)\varphi(n) \geq n$ and Theorem 6.4.5.

   (b) $\varphi(n) \geq \pi(n) - \omega(n)$.

   (c) Let n be any integer with $\omega(n) = r$, and let n_r be the product of the first r primes. Show

   $$\frac{\varphi(n)}{n} \geq \frac{\varphi(n_r)}{n_r} \quad \text{and} \quad \log \log n \geq \log \log n_r.$$

   Hence, it suffices to prove the statement for the numbers n_r. Using results on the distribution of primes, we obtain

   $$\log\left(\frac{\varphi(n_r)}{n_r}\right) = \log \prod_{i=1}^{r}\left(1 - \frac{1}{p_i}\right) = \sum_{i=1}^{r} \log\left(1 - \frac{1}{p_i}\right) \geq$$

   $$\geq -\sum_{i=1}^{r} \frac{1}{p_i} - \sum_{i=1}^{r} \frac{1}{p_i^2} > -\sum_{p \leq p_r} \frac{1}{p} - 2 > -\log \log p_r - c - 2,$$

   so

   $$\frac{\varphi(n_r)}{n_r} > \frac{1}{c' \log p_r}.$$

   Finally, apply $\log \log n_r \sim \log p_r$ (obtained by taking the logarithm of $\log n_r \sim p_r$ which is legal as both sides tend to infinity).

   (d) Apply Theorem 6.4.6 for $\sigma(n)/n^{1.01}$, or use $\sigma(n) \leq nd(n)$ and Theorem 6.4.5.

   (e) $\sigma(n)/n$ is the sum of reciprocals of divisors of n, thus

   $$\frac{\sigma(n)}{n} \leq \sum_{j=1}^{n} \frac{1}{j} \leq 1 + \log n.$$

   (f) Use arguments similar to those in part (c).

   Remark: By Exercise 6.2.15a, the statements on $\sigma(n)$ follow directly from the relevant statements on $\varphi(n)$ (and by Exercise 6.2.15b, this is almost true vice versa).

7. Use

   (a) $\displaystyle \lim_{n \to \infty} \prod_{p \leq n}\left(1 - \frac{1}{p}\right) = 0$

(b) $\displaystyle\lim_{n\to\infty}\prod_{p\le n}\left(1+\frac{1}{p}\right)=\infty.$

8. (a) Let $v_1, v_2, \ldots$ be all primes satisfying $k \mid v_i - 1$, and let $B_r = v_1 \ldots v_r$.
   If $(n, B_r) > 1$, then some v_i divides n, so

   $$k \mid v_i - 1 \mid \varphi(n).$$

   Therefore $k \nmid \varphi(n)$ can occur only if n is coprime to B_r. The number of such integers $n \le N$ is about

   $$\frac{\varphi(B_r)}{B_r} N$$

   for large N. Therefore it is enough to show that for any $\varepsilon > 0$ there is an r such that

   $$\frac{\varphi(B_r)}{B_r} = \prod_{i=1}^{r}\left(1 - \frac{1}{v_i}\right) < \varepsilon.$$

   This follows from the divergence of $\sum_{i=1}^{\infty} 1/v_i$ by Exercise 5.6.7.

   (b) If $\omega(n)$ is large, then a large power of two divides $\varphi(n)$, so there can be only a few such values $\varphi(n)$. If $\omega(n)$ is small, then $\varphi(n) > cn$ with some (small positive) constant c, thus $\varphi(n) \le N$ implies $n < N/c$. By part (a), it is also true for these integers n that $\varphi(n)$ is nearly always a multiple of a fixed large k (which can be e.g. the power of two used already), so again there can arise only few values $\varphi(n)$.

9. (a) We can follow the ideas used in Exercise 6.4.8a. Let $w_1, w_2, \ldots$ be all primes satisfying $k \mid w_i + 1$, and $C_r = w_1 \ldots w_r$.
   If n is divisible by exactly the first power of some w_i, then

   $$k \mid w_i + 1 \mid \sigma(n).$$

   Thus $k \nmid \sigma(n)$ can occur if either n is coprime to C_r or the square of some prime factor in C_r divides n. These integers n fall into certain residue classes mod C_r^2. The ratio of the number of these residue classes to the number of all residue classes modulo C_r^2 is

   $$\prod_{i=1}^{r}\left(1 - \frac{w_i - 1}{w_i^2}\right).$$

   Using the ideas seen in Exercise 6.4.8a, prove that this ratio can be arbitrarily small if r is large enough.

   (b) The situation is simpler than it was for $\varphi(n)$ since $\sigma(n) \le N$ implies $n \le N$. Therefore we need just the last step of the proof seen for $\varphi(n)$: as $\sigma(n)$ is nearly always the multiple of a large fixed k, an integer can occur only rarely among the values $\sigma(n)$.

6.5.

1. It follows from the definition of $d_j(n)$.

2. (a) Assume that f is multiplicative, and $(a, b) = 1$. To prove the multiplicativity of f^+, we use that, by Exercise 1.6.5.a-b, the divisors of ab have a unique representation as the product of a divisor of a and a divisor of b (which are coprime). Hence

$$f^+(ab) = \sum_{d|ab} f(d) = \sum_{a_1|a, b_1|b} f(a_1 b_1) =$$

$$= \sum_{a_1|a, b_1|b} f(a_1)f(b_1) = \left(\sum_{a_1|a} f(a_1)\right)\left(\sum_{b_1|b} f(b_1)\right) = f^+(a)f^+(b).$$

We can verify the converse similarly by induction on $n = ab$ or with the Möbius Inversion Formula.

(b) Replacing f by f^+, we just get the statement in (a).

3. (a) Answer: $f = 0$ and $e(n)$ defined among the examples after Definition 6.5.1. Hint: Verify that $f^+(p^2) = \left(f^+(p)\right)^2$ can hold for a prime p only if $f(p) = 0$.

(b) Answer: $f = 0$. Hint: Examine the values of f^+ assumed at $p_1 p_2$, $p_1 p_3$, $p_2 p_3$, $p_1^2 p_2$, $p_1^3 p_2$, etc. where p_1, p_2, and p_3 are distinct primes, and deduce $f(p^k) = 0$ for every prime power p^k.

4. (a) Use Exercise 6.5.2.

(b) It follows from part (a) and the complete multiplicativity of f. For $f(n) = n$, we obtain the formulas for $\sigma(n)$ and $\varphi(n)$.

5. (a) $\tilde{f}(n) = \begin{cases} c, & \text{if } n = 1 \\ 0, & \text{if } n > 1. \end{cases}$

(b) $\tilde{g}(n) = \begin{cases} 1, & \text{if } n = 2 \\ 0, & \text{if } n \neq 2. \end{cases}$

(c) $\tilde{\Omega}(n) = \begin{cases} 1, & \text{if } n \text{ is a prime power} \\ 0, & \text{otherwise.} \end{cases}$

(d) $\tilde{\omega}(n) = \begin{cases} 1, & \text{if } n \text{ is a prime} \\ 0, & \text{otherwise.} \end{cases}$

6. Let $n = \prod_{i=1}^{r} p_i^{\alpha_i}$ be the standard form of n. Then

$$f(n) = \sum_{i=1}^{r} f(p_i^{\alpha_i}) = \sum_{i=1}^{r} \sum_{\beta_i=0}^{\alpha_i} \tilde{f}(p_i^{\beta_i}) = \sum_{p^\beta | n} \tilde{f}(p^\beta).$$

The uniqueness of $\tilde{f}$ implies $\tilde{f}(k) = 0$ if k is not a prime power.

7. $f(n) = n$, by the Möbius Inversion Formula.

8. By the Möbius Inversion Formula,

$$\varphi(n) = \sum_{d|n} \mu(d)\frac{n}{d}.$$

9. (a) Let $T(n)$ be the sum of all nth roots of unity, and $S(n)$ the sum of the primitive nth roots of unity. Then $S^+(n) = T(n) = e(n)$ implies $S(n) = \mu(n)$. (We sketched another proof in Exercise 6.2.18.)

   (b) Let $T_k(n)$ be the sum of kth powers of all nth roots of unity, and $S_k(n)$ the similar sum for primitive nth roots of unity. Then $S_k(n) = \tilde{T}_k(n)$ and

   $$T_k(n) = \begin{cases} n, & \text{if } n \mid k \\ 0, & \text{if } n \nmid k. \end{cases}$$

   Exhibit $S_k(n)$ by the Möbius Inversion Formula, and also using Exercise 6.5.8. Another option: Since $S_k(n)$ and the function given in the exercise are multiplicative, it is sufficient to verify their equality at prime power places.

   (c) We convert the problem into the modulo p field. Let $V(k)$ be the sum of the solutions of the congruence $x^k \equiv 1 \pmod{p}$, and $U(k)$ the sum of the elements of order k. Prove $U^+(n) = V(n)$ and $V(d) = e(d)$ for $d \mid p - 1$. Deduce $U(d) = \mu(d)$ for $d \mid p - 1$, thus $U(p-1) = \mu(p-1)$.

10. (a) $\varphi(1)\varphi(2)\dots\varphi(n)$.   (b) $n!$.   (c) 1.   (d) 0.

11. The proof of Theorem 6.5.4 applies for this general case.

6.6.

1. $d_k(n)$.

2. It is well known that addition satisfies all requirements. For convolution playing the role of multiplication, the associative and commutative laws and the existence of an identity element follow from Theorem 6.6.2. The distributive law

   $$(f + g) * h = (f * h) + (g * h)$$

   can be verified easily (it is enough to check one of the two distributive laws since multiplication is commutative).

   No zero divisors: Show that if k and m are the smallest positive integers satisfying $f(k) \neq 0$ and $g(m) \neq 0$, then $(f * g)(km) \neq 0$.

3. Answer: k. Hint: List the equalities $(g * g * \cdots * g)(n) = f(n)$ for every n. If $n = 1$, then we get

   $$g(1) = \sqrt[k]{f(1)}.$$

   Considering $n = 2, 3, \dots$ one after the other, we get unique values for $g(2), g(3), \dots$

4. (a) Apply an argument similar to that in the hint for Exercise 6.5.2.

(b) The statement is true if $f = 0$ or $g = 0$, hence we may assume $f(1) = g(1) = 1$. If $f * g$ is completely multiplicative, then

$$(f * g)(p^2) = ((f * g)(p))^2$$

implies $f(p)g(p) = 0$ for every prime p, and $f(n)g(n) = 0$ for every $n > 1$ follows from the complete multiplicativity of f and g. To prove the converse, show that $f(p)g(p) = 0$ implies

$$(f * g)(p^k) = f(p^k) + g(p^k) = ((f * g)(p))^k.$$

5. Since the functions on both sides of the equality are multiplicative, it is enough to prove equality for prime powers. But it is more elegant to rely on the properties of convolution: Put $g(n) = n$, then $\sigma * \varphi = (g * 1) * (\mu * g) = g * g$, thus

$$\sum_{d|n} \sigma(d)\varphi(\tfrac{n}{d}) = (\sigma * \varphi)(n) = (g * g)(n) = \sum_{k|n} k \cdot \frac{n}{k} = nd(n).$$

6.

$$\sum_{n=1}^{\infty} \left|\frac{f(n)}{n^s}\right| = \sum_{n=1}^{\infty} \left|\frac{f(n)}{n^{s_0}}\right| \cdot \frac{1}{n^{s-s_0}} < c \sum_{n=1}^{\infty} \frac{1}{n^{s-s_0}} < \infty.$$

7. Apply Theorem 6.6.4.

8. Use Exercise 6.6.1 and Theorem 6.6.4.

9. Write the functions σ and μ as summation and inversion functions, and apply Exercise 6.6.7.

10. (a) By definition, the right-hand side is the limit of

(A.6.4)
$$\prod_{p \leq N} \left(\sum_{k=0}^{\infty} \frac{f(p^k)}{p^{ks}}\right)$$

when $N \to \infty$. Performing the multiplication of the finitely many absolutely convergent series in (A.6.4), by unique prime factorization and multiplicativity of f we obtain the infinite series $F_N(s)$ consisting of terms $f(n)/n^s$ where every prime divisor of n is less than or equal to N. Since the series

$$F(s) = \sum_{n=1}^{\infty} \frac{f(n)}{n^s}$$

is absolutely convergent,

$$\lim_{N \to \infty} F_N(s) = F(s).$$

(b) Since f is completely multiplicative,

$$\frac{f(p^k)}{p^{ks}} = \left(\frac{f(p)}{p^s}\right)^k,$$

hence we have infinite geometric series on the right-hand side of the formula in part (a).

11. Take the product form of the ζ function and note that the Dirichlet series $M(s)$ of the function μ is the reciprocal of ζ. Another possibility: Apply Exercise 6.6.10a for the function $f = \mu$.

12. (a) Answer: $\pi^4/36$. Hint: Apply Exercise 6.6.8a.

    (b) Answer: $5\pi^4/72$. Hint: Transform the Dirichlet series $T(s)$ belonging to the function $d^2(n)$ into an infinite product using Exercise 6.6.10a, then compute the infinite series occuring in the factors of the product to establish

$$T(s) = \frac{\zeta^4(s)}{\zeta(2s)}.$$

13. Answer: $15/\pi^2$. Hint: Apply Exercise 6.6.10a for $f = |\mu|$, and show that the infinite product equals $\zeta(s)/\zeta(2s)$.

14. (a) $\displaystyle\sum_{n=1}^{\infty} \frac{f(n)x^n}{1-x^n} = \sum_{n=1}^{\infty} f(n)\left(\sum_{j=1}^{\infty} x^{jn}\right) = \sum_{k=1}^{\infty} x^k\left(\sum_{d|k} f(d)\right) = \sum_{k=1}^{\infty} f^+(k)x^k.$

    (b) Apply part (a) for functions μ and φ, taking $x = 1/2$. Answer: (b1) $1/2$; (b2) 2.

6.7.

1. Answer: 1. Hint: As in the second proof of Theorem 6.7.5, this sum is an application of the Inclusion and Exclusion Principle for the number of integers among 1, 2, ..., n that have no prime divisors at all. As 1 is the only integer with this property, the sum equals 1. (The lesson of this story is that even a complicated argument can be useful sometimes: we computed the obvious number of prime-free integers with a complicated formula, and this made it possible to find a simple form for the intricate sum.) Another option: After checking a few small values of n, we guess the answer, and then prove it by induction.

2. Answer: $6/\pi^2$. Hint: Let $K(n)$ be the number of squarefree integers among 1, 2, ..., n. We have to determine

$$\lim_{n\to\infty} \frac{K(n)}{n}.$$

As in the second proof of Theorem 6.7.5, use the Inclusion and Exclusion Principle to establish

$$K(n) = \sum_{j\leq\sqrt{n}} \mu(j)\left\lfloor\frac{n}{j^2}\right\rfloor.$$

Omitting the floors causes an error term not greater than $\sqrt{n}$ that can be neglected compared to the main term

$$n\sum_{j\leq\sqrt{n}} \frac{\mu(j)}{j^2}.$$

3. (a) Applying Theorem 6.7.2 to the convolution $d_3 = d_2 * 1$, we obtain

$$D_3(n) = \sum_{j=1}^{n} d(j)\left\lfloor\frac{n}{j}\right\rfloor.$$

After dividing by n and deleting the floors, we have to estimate

(A.6.5)
$$\sum_{j=1}^{n} \frac{d(j)}{j}$$

apart from an error term. We can do this using Theorem 6.4.3 about the mean value of $d(n)$. Using the notation there, $d(j) = D(j) - D(j-1)$. We reorder the sum in (A.6.5) accordingly and apply Theorem 6.4.3 for $D(j)$. Thus we obtain

$$\sum_{k=2}^{n} \frac{\log j}{j} \sim \int_{2}^{n} \frac{\log t}{t}\, dt \sim \frac{\log^2 n}{2}$$

apart from error terms. We must also show that the error terms are negligible compared to the main term.

(b) We follow the proof of Theorem 6.7.3. Let $f_\nu(n) = n^\nu$ and apply Theorem 6.7.2 to the convolution $\sigma_\nu = 1 * f_\nu$, then we obtain

$$\Sigma_\nu(n) = \sum_{j=1}^{n} \sum_{k=1}^{\lfloor \frac{n}{j} \rfloor} k^\nu.$$

We can estimate the inner sum for k on the right-hand side with the integral criterion as usual (see the first proof of Theorem 5.6.1 or Exercise 5.6.2).

4. Since the mean value of σ is relatively small, there are many integers i among 1, 2, ..., n for which (say) $\sigma(i) \le 2n$. There are few such values $\sigma(i)$ by Exercise 6.4.9, so there must be one that is assumed many times by the function.

5. (a) The lower bound is obvious as $\Omega(i) \ge \omega(i)$. To establish the upper bound, represent $\Omega(i)$ and $\omega(i)$ with the help of their inversion functions (see Exercise 6.5.5c-d). After the usual rearrangement and omitting the floors, we obtain

$$\sum_{i=1}^{n} (\Omega(i) - \omega(i)) < n \sum_{r \le n}{}' \frac{1}{r},$$

where $\sum'$ indicates that the sum is taken only for the prime power values r with exponent greater than one. This sum is less than 1, see the solution of Exercise 5.6.1b.

(b) This follows from part (a) and the theorems in question.

6. Use Exercise 6.2.20a and apply the Hardy–Ramanujan Theorem for ω and (relying on Exercise 6.7.5b) for Ω.

7. The (surprising) answer is 0. Hint: We use that the Hardy–Ramanujan Theorem is valid also for Ω (see Exercise 6.7.5b). Assume $i = ab$ where a and b are less than $\sqrt{n}$. Then in most cases both $\Omega(a)$ and $\Omega(b)$ are about

$$\log \log \sqrt{n} \sim \log \log n,$$

thus $\Omega(i) \sim 2 \log \log n$. But there are only few such integers i (using Exercise 6.7.5b again).

8. For a precise formulation, replace ω by f (meeting the requirements) and modify $\log \log i$ to

$$\sum_{p \leq i} \frac{f(p)}{p}$$

in Theorem 6.7.7. The proof is the same as for Theorems 6.7.7 and 6.7.7A.

6.8.

1. If for a fixed m, the sequence $f(m^k) = kf(m)$, $k = 1, 2, \ldots$, is bounded, then $f(m) = 0$.

2. Let m be fixed, and consider the positive integers k coprime to m. Then $f(m) = f(km) - f(k)$. By Cauchy's criterion for convergence, for any $\varepsilon > 0$ and k large enough, we have $|f(km) - f(k)| < \varepsilon$. Thus only $f(m) = 0$ is possible.

3. Answer:

$$f = 0; \qquad g_c(n) = n^c; \qquad h_r(n) = \begin{cases} 1, & \text{if } n = 1 \\ r, & \text{if } n = 2 \\ 0 & \text{if } n > 2, \end{cases}$$

where c is any real number and $0 \leq r \leq 1$. Hint: If the function is positive everywhere, then we can apply Theorem 6.8.1 to its logarithm, and arrive at g_c. If the function has the value 0 somewhere than it must be 0 at every larger integer; show that the first appearance of 0 cannot occur later than at 3. This yields $f = 0$ and h_r. It is easy to see that the function cannot assume negative values.

4. Then $-f$ satisfies condition (6.8.1) in the proof of Theorem 6.8.1.

5. We can apply Theorem 6.8.1 to the real and imaginary parts of f separately.

6. (a) We can take the sequence

$$k_1, 2k_1, 2k_2, 3k_2, 3k_3, 4k_3, \ldots, jk_j, (j+1)k_j, \ldots,$$

where $(k_j, j(j+1)) = 1$ and the numbers k_j are large enough (compared to b_n). If f is monotone (say) increasing on the elements of the sequence, then

$$f(j) + f(k_j) = f(jk_j) \leq f((j+1)k_j) = f(j+1) + f(k_j)$$

by additivity. Subtracting $f(k_j)$, we get $f(j) \leq f(j+1)$, so f is monotone. Theorem 6.8.1 guarantees $f(n) = c \log n$.

(b) Consider the sequence

$$c_1, c_1 d_1, c_2, c_2 d_2, c_3, c_3 d_3, \ldots, c_j, c_j d_j, \ldots,$$

where every integer greater than 1 occurs infinitely often in the sequence d_1, $d_2, \ldots$, $(c_j, d_j) = 1$, and c_j is sufficiently large (compared to b_n). Let $m > 1$ be fixed and take $\varepsilon > 0$. Then using the construction of the sequence, the additivity of f, and the condition of the exercise, we can find a (large) j such that $m = d_j$ and

$$|f(m)| = |f(c_j d_j) - f(c_j)| < \varepsilon.$$

Hence, only $f(m) = 0$ is possible.

A.7. Diophantine Equations

7.1.

1. 3. $(10000 = 201 \cdot 47 + 7 \cdot 79 = 122 \cdot 47 + 54 \cdot 79 = 43 \cdot 47 + 101 \cdot 79.)$

2. 14.

3. 7. Hint: Eliminating x from the system of equations $7x+13y+15z = 500$, $x+y+z = 50$, we get $6y + 8z = 150$. Dividing by 2, we need solutions of the Diophantine equation $3y + 4z = 75$ satisfying $y \geq 0$, $z \geq 0$, and $y + z \leq 50$ (since $x \geq 0$).

4. 9.

5. A pair of integers x, y satisfies the Diophantine equation $ax + by = c$ with $b \neq 0$ if and only if x is a solution of the linear congruence $ax \equiv c \pmod{b}$ (and then y is determined uniquely from the equation). Observe that formula (2.5.5) in the proof of Theorem 2.5.4 is the same as the description of x' in formula (7.1.1) of Theorem 7.1.1 (after converting the notation). (We do not have to use the proof. We need only the statements of Theorem 2.5.4, though then the argument is slightly clumsier.)

6. (a) 0 or ∞. (b) 0 or 1.

7. $x = -3 - 5u - 10v$, $y = 3u + 3v + 1$, $z = 2v + 1$. Hint: As in the case of two variables, we solve for one of the variables, separate the integer part from the fraction, and introduce a suitable new variable, reducing the absolute values of the coefficients till we arrive at a fraction with denominator 1. Then, proceeding backwards, we express the original variables with the help of the two integer parameters obtained before.

8. A possible approach is to generalize the algorithm sketched after Theorem 7.1.1 (applied in the previous exercise), which also establishes the statement about solvability. Another option is to use induction on k. We can reduce an equation $a_1 x_1 + \cdots + a_k x_k = c$ with k variables to an equation with $k - 1$ variables: Let $d = (a_{k-1}, a_k)$, then the integers of the form $a_{k-1} x_{k-1} + a_k x_k$ are exactly the multiples of d, i.e. the numbers dy. Thus we can reduce the original equation to $a_1 x_1 + \cdots + a_{k-2} x_{k-2} + dy = c$ with $k - 1$ variables to which we can apply the induction hypothesis.

9. If the equation is solvable, then obviously any solution also satisfies the congruence for an arbitrary modulus m. If the equation has no solutions, then $(a_1, \ldots, a_k) \nmid c$, so the congruence modulo $m = (a_1, \ldots, a_k)$ is not solvable either.

10. This is true if and only if the integers a_i are coprime and at least one of them is positive. Hint: Necessity is obvious. To prove sufficiency, assume (e.g.) $a_1 > 0$. If we can find a solution in positive integers for some c_0, then increasing x_1 and keeping the other variables unaltered, we get a positive solution for every $c > c_0$ in the residue class of c_0 modulo a_1. Thus it is enough to show that every residue class modulo a_1 contains an element c for which there is a positive solution. We rely on the equivalence (in the precisely defined meaning discussed in Section 2.5) of the Diophantine equation

(A.7.1) $$a_1 x_1 + \cdots + a_k x_k = c$$

and the congruence

(A.7.2) $$a_2 x_2 + \cdots + a_k x_k \equiv c \pmod{a_1}.$$

We solve (A.7.2) for $c = 1, 2, \ldots, a_1$. (It is solvable, since $(a_1, \ldots, a_k) = 1$ guarantees its solvability for any c.) In congruences we can replace any integer by one congruent to it, so we can assume that the values $x_2, \ldots, x_k$ obtained in the a_1 congruences are all positive.

11. (a) Let $a > b$ and apply the key idea of the previous exercise: If c is assemblable, then also $c + tb$ is assemblable for any positive t. Thus we have to find the smallest assemblable element in every residue class modulo b. Since $(a, b) = 1$, the numbers $0a, 1a, 2a, \ldots, (b-1)a$ form a complete residue system modulo b, so the smallest assemblable elements in the residue classes are b, a, $2a, \ldots, (b-1)a$. The residue class of $(b-1)a$ enters last, so the largest number that is not assemblable is $(b-1)a - b = ab - a - b$.

    (b) Answer: $(a-1)(b-1)/2$. (This is an integer, since $(a, b) = 1$ implies that at least one of a and b is odd.) Hint: Show that if the sum of two positive integers is $ab - a - b$, then exactly one of them is assemblable.

12. It is more convenient to view the problem from an inverse perspective as cutting to pieces instead of assembling. Thus we claim that a large cube can be cut into exactly n cubes if (a) n is large enough; (b) $n \geq 48$, and we ask for the complete answer in (c) for the analog for squares.

    (a) We can easily cut a cube into 8 or 27 small (congruent) cubes, so with the repeated application of these steps, we can cut a cube into $1 + 7x + 26y$ cubes, where x and y are arbitrary non-negative integers. Since 7 and 26 are coprime, every sufficiently large n can be represented in this form.

    (b) Cutting a cube into 8 cubes, we can always increase the number of small cubes by 7. Hence it suffices to verify the statement for $48 \leq n \leq 54$.
    48: $48 = 27 + 3 \cdot 7$. We cut the cube into 27 cubes, and then we cut each of three small cubes into eight parts.
    49: For brevity, let us write Cuk for a cube if the length of its edge is k. We cut the lower half of Cu6 into four Cu3, the top row into thirty-six Cu1, and the remaining two rows into nine Cu2.
    50: $50 = 7 \cdot 7 + 1$.
    51: In Cu6, we form five Cu3 from the lower half plus one quarter in the upper half, select five Cu2 from the remaining part, and there are forty-one Cu1 left.

52: We cut a Cu3 from Cu4, and partition two of the remaining thirty-seven Cu1 into eight parts.

53: Using $53 = 1 + 2 \cdot 19 + 2 \cdot 7$, it is enough to show a procedure that increases the number of cubes by nineteen; we cut Cu3 into a Cu2 and nineteen Cu1.

54: We cut Cu8 into six Cu4, two Cu3, four Cu2, and forty-two Cu1.

(c) $n \neq 2, 3, 5$.

7.2.

1. Show that if $x^2 + y^2 = z^2$, then (at least) one of x, y, or z must be divisible by 3, 4, and 5. We consider divisibility by 5. The remainder of a square mod 5 is 0, 1, or -1. For a proof by contradiction, assume that none of x, y, and z is a multiple of 5, so the left-hand side of $x^2 + y^2 = z^2$ is congruent to 0 or ± 2, the right-hand side is congruent to ± 1 modulo 5, which is a contradiction. We can verify the divisibility by 3 and 4 similarly. Alternatively, we can use the characterization in Theorem 7.2.1 and apply similar arguments.

2. Answer: 8, 15, 17. Hint: The area is $xy/2$, so $xy = 120$. Checking all possible factorizations of 120, $x^2 + y^2$ is a square only for $8 \cdot 15$. Another option: By Theorem 7.2.1, we have to solve the equation $60 = d^2 mn(m-n)(m+n)$ with respect to conditions (7.2.4). Thus we get $d = 1, m = 4, n = 1$.

3. Answer: 6, 8, 10 and 5, 12, 13. Hint: By Theorem 7.2.1, the area is

$$d^2 mn(m-n)(m+n),$$

and the perimeter is

$$d\big(2mn + (m^2 - n^2) + (m^2 + n^2)\big) = 2md(m+n).$$

Equating them gives $dn(m-n) = 2$ after cancellation. The solutions satisfying also conditions (7.2.4) in Theorem 7.2.1 are $d = m = 2, n = 1$, and $d = 1, n = 2$, $m = 3$. Another option: We have to solve the system of Diophantine equations

$$\frac{xy}{2} = x + y + z, \qquad x^2 + y^2 = z^2.$$

Squaring the form $(xy/2) - z = x + y$ of the first equation, combining the result with the second equation, and dividing by xy, we obtain $z = (xy/4) - 2$. Substituting into the first equation, reordering, and factoring gives $(x-4)(y-4) = 8$. Since x and y are positive, we have only the decompositions $1 \cdot 8$ and $2 \cdot 4$ (apart from the order of factors).

4. Answer: Every $k \geq 3$. Hint: Use Theorem 7.2.1. Verify that 1 and 2 can be represented in none of the forms given there for x, y, and z. For integers greater than 2, it is sufficient to find a representation for 4 and the odd numbers, due to the multiplier d in the formula: $4 = 2 \cdot 2 \cdot 1$, and $2r + 1 = (r + 1)^2 - r^2$.

5. If x, y, z is a primitive Pythagorean triple, then $(y - x)^2$, z^2, and $(x + y)^2$ are coprime and form an arithmetic progression.

   Remark: The solutions $0 < u < w < v$ of the Diophantine equation $u^2 + v^2 = 2w^2$ and the solutions $0 < x < y < z$ of the Pythagorean equation $x^2 + y^2 = z^2$ can be deduced from each other by the substitutions $u = y - x$, $v = x + y$, $w = z$, and $x = (v - u)/2$, $y = (u + v)/2$, $z = w$. (x and y are integers as u and v must be of the same parity). Therefore, we can characterize all solutions of $u^2 + v^2 = 2w^2$ with three integer parameters.

7.3.

1. As the signs of x and y are irrelevant now, we can group the solutions in integers by four to obtain the essentially distinct solutions, except for the case $y = 0$ that occurs if and only if n is a square and then these two solutions form a group. Thus there are $\left\lceil \dfrac{f(n)}{4} \right\rceil$ essentially distinct solutions, where $f(n)$ is the number of solutions given in Theorem 7.3.1.

2. There are two solutions: we have to make 5 and 7, or 4 and 11 cuts parallel to the walls of the tin (we get $6 \cdot 8 = 48$ and $5 \cdot 12 = 60$ pieces). Hint : If we make $x - 1$ and $y - 1$ cuts parallel to the tin's walls, then there are $xy/2$ crispy and $(x - 2)(y - 2)$ soft pieces. Equating the two numbers, we obtain $(x - 4)(y - 4) = 8$ after ordering. Another option: In the first row running around the inside of the tin's walls, there are by eight more pieces than in the second such row. This means that apart from these two rows, there are altogether eight pieces inside that constitute a 2×4 or 1×8 rectangle.

3. The equation $2/p = 1/x + 1/y$ is equivalent to $(2x - p)(2y - p) = p^2$. Géza Ottlik's approach was different: Multiplying the original equation by xy, we get that one of the variables is divisible by p, say $x = kp$. Substituting it into the equation and solving for y, we find $p = 2k - 1$. This determines also x and y uniquely.

4. Answer: The denominator has a positive divisor not of the form $4k + 1$.

5. Using

$$\frac{1}{u} = \frac{1}{2u} + \frac{1}{2u}$$

it is enough to represent $4/n$ as a sum of two or three natural numbers for the given values of n.

$$n = 2s: \qquad \frac{4}{n} = \frac{1}{s} + \frac{1}{s}$$

$$n = 4s - 1: \qquad \frac{4}{n} = \frac{1}{s} + \frac{1}{s(4s - 1)}$$

$$n = 8s - 3: \qquad \frac{4}{n} = \frac{1}{2s} + \frac{1}{s(8s - 3)} + \frac{1}{2s(8s - 3)}$$

$$n = 24s - 15: \qquad \frac{4}{n} = \frac{1}{8s - 5} + \frac{1}{24s - 15}$$

$$n = 24s - 7: \qquad \frac{4}{n} = \frac{1}{6s} + \frac{1}{s(24s - 7)} + \frac{1}{6s(24s - 7)}.$$

6. Start with the wrong representation $a/b = 1/b + 1/b + \cdots + 1/b$, and apply the identity

$$\frac{1}{n} = \frac{1}{n+1} + \frac{1}{n(n+1)}$$

sufficiently many times.

7. Answer: No. Hint: Factoring the left-hand side of the Diophantine equation $x^4 - 4 = y^5$, the two factors are coprime for x odd, hence each is a fifth power. However, their difference is 4, which is impossible. If x is even, then the right-hand side of the equation is a multiple of 8, which is false for the left-hand side.

8. The only solution is $x = y = s = t = 0$. Hint: Assuming a non-trivial rational solution, we can convert it into an integer solution, one with $(x, y, s, t) = 1$. Examining parity, we get a contradiction. Another approach: A non-trivial integer solution leads to an equilateral triangle where all three vertices are lattice points. Show by area considerations that no such triangle exists.

9. The sum is divisible by 3, but not by 9.

10. $\pm 4, \pm 6$.

11. An ugly solution: Let the six numbers be n, $n + 1$, ..., $n + 5$, and partition them into two groups in all possible ways. We have to show that none of the resulting equations have integer solutions. Since we can easily find the integer (or even rational) roots of a polynomial with integer coefficients, the proof requires just some patient (and tedious) computation. We do not have to do this for all groupings, of course, for example, we immediately see by comparing the size of the factors that $n(n+1)(n+4)$ is smaller than $(n+2)(n+3)(n+5)$ for every $n \geq 0$, and also further similar considerations can speed up the work.

The following argument is much more elegant; Three of the six numbers are even, one more is a multiple of 3, and at most one more can be divisible by 5. Hence, one of the numbers must have a prime divisor greater than 5 for $n > 1$. This prime cannot divide any of the other five numbers, hence it divides only one of the two products.

A third option: If one of the numbers is divisible by 7, then we are done, as seen previously. Otherwise the six numbers form a reduced residue system mod 7. If there exist two equal products, then the product of all the six numbers is a square. However, the product of the six numbers is congruent to -1 mod 7, which is impossible for a square.

The third solution works also for 106 instead of 6, using Wilson's Theorem and the Legendre symbol $\left(\frac{-1}{p}\right)$ (as 107 is a prime of the form $4k - 1$). Also the first solution works for 106 (or any other number) in principle (or even in practice with a well-designed computer program). We can generalize the second solution, too: A classical theorem by Sylvester and Schur states that among k consecutive integers greater than k there always exists one having a prime divisor greater than k, hence this prime can divide only one of the two products. In the remaining cases, Chebyshev's Theorem guarantees such a prime that divides only one of the products.

Finally we note that the validity of the statement for any k consecutive integers instead of six follows from the hard theorem that the product of consecutive integers is never a power (see the Remark after Exercise 1.6.3).

12. There is a solution only for even m: $n = m + 1$ and $x = y = 2^{m/2}$. Hint: Rewrite the equation with the help of (x, y) and show $x = y$. Then the equation is of the form

(A.7.3) $$2^m = x^{2n-2m}.$$

Clearly, $x = 2^s$. Substituting it into (A.7.3), prove $m = 2s$ and $n = m + 1$.

13. (a) From the form $(x + 5)(y + 3) = 22$, we can obtain $2d(22) = 8$ solutions.

(b) No solution; consider the equation modulo 11.

(c)–(e) We have only the trivial solution $x = y = z = 0$. The good moduli are 3 or 8 for (c); 5, 7, 8, or 23 for (d); 11 for (e).

(f) $x = \pm 1, z = -2$. Hint: The two factors on the left-hand side are coprime for any integer x, thus both factors are cubes.

(g) $x = \pm 1, y = 0$. Hint: After simple transformations, we obtain that the product of two consecutive integers is almost a fourth power. Continuing by congruence considerations, we need one more factorization.

(h) Besides $y = x$, the only solutions are $x = 2$, $y = 4$, and $x = 4$, $y = 2$. Hint: Rewrite the equation with the help of (x, y), or take the logarithm and examine the behavior of the (real) function $f(z) = z/\log z$.

(i) $x = 5, y = 1$. Hint: Consider the equation modulo 31, and apply the facts about power residues.

14. (a) There is no such number system. Hint: $1 + x + x^2$ is always between two consecutive squares for $x > 1$.

(b) Base 3 is the only solution. Hint: $4(1 + x + x^2 + x^3 + x^4)$ is between two consecutive squares for $x > 3$.

(c) There is no such number system. Hint: The expression can be decomposed into two coprime factors where one of them cannot be a square.

1. $1 + i \mid a + bi \iff a \equiv b \pmod 2$.

2. (a) $\alpha = \gamma\varrho \iff \overline{\alpha} = \overline{\gamma}\,\overline{\varrho}$,

   (b) It follows from part (a).

   (c) Apply the definition either of a Gaussian irreducible, or of a Gaussian prime, or use Theorem 7.4.15.

3. By Exercise 7.4.2a, $\alpha \mid \overline{\alpha} \iff \overline{\alpha} \mid \alpha$, so $\alpha = \varepsilon\overline{\alpha}$ with a unit ε. Check that the absolute values of the two sides are always equal, and comparing the angles we get $\arg(\alpha) = k \cdot 45°$. This means that α is on one of the coordinate axes or the lines $y = \pm x$. (We can get the same result by substituting $\varepsilon = \pm 1, \pm i$ into $\alpha = \varepsilon\overline{\alpha}$ and solving the four equations.)

4. (a) Observe that a rational number a/b is a Gaussian integer if and only if it is an (ordinary) integer.

   (b) If $(a, b) = d$ in the integers, then we have to show that $a_1 = a/d$ and $b_1 = b/d$ are also coprime in the Gaussian integers. If a Gaussian integer γ is a common divisor of a_1 and b_1, then $N(\gamma)$ is a common divisor in the integers of $N(a_1) = a_1^2$ and $N(b_1) = b_1^2$, which implies $N(\gamma) = 1$, so γ is a unit. (Another option is to establish $1 = a_1 u + b_1 v$ with suitable integers u and v, so $\gamma \mid a_1$ and $\gamma \mid b_1$ imply $\gamma \mid 1$.)

5. True: (a), (c).

6. (Of course, any associate of the results below is correct.)

   (a) $2 - i$. Hint: Apply the Euclidean algorithm.

   (b) 2. Hint: Observe that $1 - i$ and $2 + i$ are Gaussian primes, $2 = \varepsilon(1 - i)^2$, and $2 + i \nmid 39$.

   (c) $1 + i$. Hint: The gcd δ also divides the sum and difference of the two numbers, and since $(4 + i, 2 + i) = 1$, we obtain $\delta \mid 2$. Hence $\delta = 1$, or 2, or $1 + i$. Show that the first two cases are not possible.

7. (a) True: (a1).

   (b) $(\alpha, \overline{\alpha}) = (a, b)$ or $(\alpha, \overline{\alpha}) = (1 + i)(a, b)$.

8. Verify that β is a friend of α if and only if $\beta = \varepsilon\overline{\alpha}$ and $(\alpha, \overline{\alpha}) = 1$. Thus α has no or four friends, and we can easily deduce the condition in (a).

9. $3^2(2 + i)^3(2 - i)(1 + i)^3(-1 - 4i)$. Hint: Decompose $(270, 2610) = 90$ into a product of Gaussian primes by Theorem 7.4.15. To find the factorization of the remaining part, $3 + 29i = \pi_1 \dots \pi_r$, consider the norms: $850 = N(\pi_1) \dots N(\pi_r)$. From the standard form of 850 (in the integers), we obtain $r = 4$, and the norms of the Gaussian primes π_i are 2, 5, 5, and 17. So $\pi_1 = 1 + i$, $\pi_2 = \pi_3 = 2 + i$ or $2 - i$ depending on whether or not $(3 + 29i)/(2 + i)$ is a Gaussian integer ($\pi_3 = \overline{\pi}_2$ is impossible, why?), etc.

10. True: (b), (c), (e).

11. We can use induction on $N(\alpha)$. The key step is: If α has two distinct decompositions into the product of Gaussian primes

$$\alpha = \pi_1 \dots \pi_r = \varrho_1 \dots \varrho_s, \qquad \text{where} \qquad \pi_i \neq \varepsilon \varrho_j,$$

and (say) $N(\pi_1) \le N(\varrho_1)$, then there is a unit ε such that

$$\alpha_1 = \varepsilon \alpha - \pi_1 \varrho_2 \dots \varrho_s$$

satisfies $N(\alpha_1) < N(\alpha)$, and also α_1 has two distinct factorizations into the product of Gaussian primes.

7.5.

1. $\left\lceil \dfrac{r(n)}{8} \right\rceil$, where $r(n)$ is the number of solutions given in Theorem 7.5.1 ($r(n) = 0$ if there are no solutions). Hint: Interchanging x and y, or modifying signs do not yield essentially different solutions. These give eight possibilities except when x or y is 0, or $|x| = |y|$ (these occur in the cases $n = k^2$ and $n = 2k^2$).

2. 16.

3. Answer: 7. Hint: The integers of the form $8k + 6$ cannot be represented as the sum or difference of two squares, thus $r \le 7$. We have to show that all the seven numbers between two consecutive integers of the form $8k + 6$ can be represented as desired in infinitely many cases.

4. By Theorem 7.5.1, the exponents of primes 7 and 11 in the standard form of $a^2 + b^2$ are even, thus they must be at least 2. Another option: 7 and 11 are Gaussian primes, hence

$$7 \mid a^2 + b^2 = (a + bi)(a - bi) \Longrightarrow 7 \mid a + bi \text{ or } 7 \mid a - bi$$
$$\Longrightarrow \frac{a + bi}{7} \text{ or } \frac{a - bi}{7} \text{ is a Gaussian integer}$$
$$\Longrightarrow 7 \mid a \text{ and } 7 \mid b,$$

and the same holds also for 11.

5. It is solvable if and only if the exponents of the primes of the form $4k - 1$ are even and the exponent of 2 is not one in the standard form of n. The number of solutions is the same as in Theorem 7.5.1 if n is a multiple of 4, and is half of that if n is odd.

6. An integer has such a representation if and only if it is not a multiple of 4 and has no prime divisors of the form $4k - 1$. Then the number of representations is 2^{r+2}, where r is the number of its odd prime divisors (all are of the form $4k + 1$).

7. (a) Depending on whether k is the length of the hypothenuse or of a leg, we need the number of essentially different solutions in positive integers x and y of

the Diophantine equations $x^2 + y^2 = k^2$ and $x^2 - y^2 = k^2$. We infer from Theorem 7.5.1 (and Exercise 7.5.1), that for the first equation, this number is

$$\frac{(2\beta_1 + 1)\dots(2\beta_r + 1) - 1}{2},$$

where $\beta_1, \dots, \beta_r$ are the exponents of the primes of the form $4t + 1$ in the standard form of k. For the second equation, we use Theorem 7.3.1 (and Exercise 7.3.1) to obtain the answer

$$\frac{1}{2}(d(k^2) - 1) \quad \text{if } k \text{ is odd;}$$

$$\frac{1}{2}\left(d\left(\frac{k^2}{4}\right) - 1\right) \quad \text{if } k \text{ is even.}$$

(b) By Exercise 7.5.6, the length of a hypothenuse can be k if and only if every prime divisor of $k > 1$ is of the form $4t + 1$, and then the number of triangles is $2^{\omega(k)-1}$. We get by similar arguments that the length of a leg can be k if and only if $k > 1$ is either odd, or $4 \mid k$, and there are $2^{\omega(k)-1}$ suitable triangles in both cases.

Instead of the above considerations, we could apply Theorem 7.2.1 characterizing the primitive Pythagorean triples.

8. Assume first that there is a prime q of the form $4k - 1$ occurring with an odd exponent $2w - 1$ in the standard form of n. Then the equation has no solutions, so we have to prove that n has as many (positive) odd divisors of the form $4k + 1$ as of the form $4k - 1$. Any odd divisor of n has a (unique) decomposition as tq^u, where $(t, 2q) = 1$ and $0 \le u \le 2w - 1$. Then one of the divisors tq^{2j} and tq^{2j+1} is of the form $4k + 1$, and the other is of the form $4k - 1$ (for any $0 \le j \le w - 1$).

Now we turn to the case when every prime q_ν of the form $4k - 1$ occurs with an even exponent $2w_\nu$ in the standard form of n. By Theorem 7.5.1, we have to verify

(A.7.4)

$$d'(n) - d''(n) = \prod_{\mu=1}^{r}(\beta_\mu + 1),$$

where β_μ are the exponents of primes of the form $4k + 1$ in the standard form of n. If we perform the previous pairing of divisors by q_1, then only those (odd positive) divisors are left where the exponent of q_1 is $2w_1$. Now we repeat the procedure by q_2 for these divisors, etc. Thus finally only those (positive) odd divisors remain unmatched where the exponent of every q_ν is $2w_\nu$. The number of such divisors is clearly the product on the right-hand side of (A.7.4), on the one hand, and as all these divisors are of the form $4k + 1$, their number is just $d'(n) - d''(n)$, on the other hand.

We can prove the statement of the exercise also in a single step by writing the difference $D = d'(n) - d''(n)$ as

$$D = \sum_{\substack{0 \le \beta'_\mu \le \beta_\mu \\ 0 \le \gamma'_\nu \le \gamma_\nu}} (-1)^{\gamma'_1 + \cdots + \gamma'_s} = \prod_{\mu=1}^{r}(\beta_\mu + 1)\prod_{\nu=1}^{s}(1 - 1 + \cdots + (-1)^{\gamma_\nu}).$$

9. Answer: π. Hint: Observe that $1 + \sum_{i=1}^{n} r(i)$ is just the number of lattice points inside or on the border of a circle around the origin of radius $\sqrt{n}$. Show that the number of these lattice points is asymptotically equal to the area of the circle as $n \to \infty$.

10. All solutions are $x = \pm 2, y = 2$ and $x = \pm 11, y = 5$. Hint: Factor the left-hand side of the equation in the Gaussian integers and find the possible values of the greatest common divisor of the two factors. It turns out that each factor must be the cube of a Gaussian integer. Finally, cube and compare the imaginary parts.

11. $\alpha = a + bi$ is not of this form if and only if b is odd or $a \equiv b \equiv 2 \pmod{4}$. Hint: Apply the argument in the proof of Theorem 7.3.1.

12. Each Gaussian prime in the standard form can be replaced by any of its associates (which can be compensated by modifying the extra unit factor).

13. True: (a), (c).

14. Answer: 5/6. Hint: Let $F(N)$ be the number of integers among $1, 2, \ldots, N$ that cannot be written as the sum of three squares. Prove

$$F(N) = \left\lfloor \frac{N+1}{8} \right\rfloor + \left\lfloor \frac{N+4}{8 \cdot 4} \right\rfloor + \left\lfloor \frac{N + 4^2}{8 \cdot 4^2} \right\rfloor + \ldots,$$

hence

$$\lim_{N \to \infty} \frac{F(N)}{N} = \frac{1}{8} \sum_{k=0}^{\infty} \frac{1}{4^k}.$$

15. Answer: 10. Hint: Verify, using the Three Squares Theorem that at most ten odd squares suffice, and relying on the Two Squares Theorem, show that infinitely many integers of the form $8k + 2$ cannot be represented as the sum of less than ten squares.

16. If $n = 4^k(8m + 7)$, then $n - (2^k)^2$ is the sum of three squares.

17. Exactly the positive integers $n = 4^k(16m + 14)$ have no such representation. Hint: Show that n can be written in the required form if and only if $2n$ is the sum of three squares.

18. Yes, it is solvable. Hint: We have to show that the number can be written as the sum of four squares with at least one of them divisible by 3.

19. It follows from Chevalley's theorem (or from Exercise 3.6.2) that the congruence $X^2 + Y^2 + Z^2 \equiv 0 \pmod{p}$ has a non-trivial solution X, Y, Z. If $Z \not\equiv 0 \pmod{p}$, then multiplying the congruence by Z^{p-3} (for $p > 2$), we obtain

$$1 + c^2 + d^2 \equiv 0 \pmod{p}, \quad \text{where} \quad c = XZ^{(p-3)/2} \quad \text{and} \quad d = YZ^{(p-3)/2}.$$

20. We can use the solvability of $x^2 + 1 \equiv 0 \pmod{p}$ instead of Lemma 7.5.5, and use the identity

(A.7.5) $\qquad (a_1^2 + a_2^2)(b_1^2 + b_2^2) = (a_1 b_1 + a_2 b_2)^2 + (a_1 b_2 - a_2 b_1)^2$

instead of Lemma 7.5.4. We note that there is no need to prove that m is odd (though the argument is valid), and (A.7.5) is just an expanded form of the identity $N(\alpha)N(\beta) = N(\beta\overline{\alpha})$ for the norms of Gaussian integers.

21. (a) Consider those vectors $\mathbf{d} = C\mathbf{s} - \mathbf{t}$, where the components of $\mathbf{s}$ and $\mathbf{t}$ satisfy

$$0 \le s_i < u_i, \quad 0 \le t_i < v_i, \quad i = 1, 2, \ldots, k.$$

By the pigeonhole principle, there must be two of the $\mathbf{d}$ that are congruent modulo p. Then the difference of the vectors $\mathbf{s}$ belonging to them can be taken as $\mathbf{x}$, and the difference of the relevant vectors $\mathbf{t}$ plays the role of $\mathbf{z}$.

(b) Apply part (a) for the case

$$k = 2, \qquad u_1 = u_2 = v_1 = v_2 = \lceil \sqrt{p} \rceil, \qquad C = \begin{pmatrix} c & d \\ -d & c \end{pmatrix}$$

where $1 + c^2 + d^2 \equiv 0 \pmod{p}$. We obtain

$$0 < x_1^2 + x_2^2 + z_1^2 + z_2^2 < 4p \quad \text{and} \quad p \mid x_1^2 + x_2^2 + z_1^2 + z_2^2.$$

(c) If $2p$ is nice, then we can proceed exactly as when we showed that m is odd in the proof of Theorem 7.5.3.
If $3p = a_1^2 + a_2^2 + a_3^2 + a_4^2$, then let b_i be the residue of least absolute value mod 3 of a_i, and apply (7.5.10) in Lemma 7.5.4. Then $9p$ is the sum of four squares where each is a multiple of 3, thus cancelling by 9 we get that also p is nice. (In this step, we basically repeated the proof of Theorem 7.5.3 in the special case $m = 3$.)

7.6.

1. If n is the sum of s terms of 600th powers, then n is the sum of the same number of 200th powers as

$$n = x_1^{600} + \cdots + x_s^{600} = (x_1^3)^{200} + \cdots + (x_s^3)^{200}.$$

2. As in the proof of Theorem 7.6.5, the keys are congruences with suitable moduli.

   (a) (a1) Prove by induction on j that $31 \cdot 16^j$ cannot be written as a sum of less than 16 fourth powers.

   (a2) The integers $64t + 32$ are not the sums of 31 eighth powers.

   (a3) $G(24) \ge G(8)$ follows as in Exercise 7.6.1.

   (a4) The integers $625t + 125$ require at least 125 hundredth powers.

   (a5) Check the numbers $625t + 312$.

   (b) We can generalize parts (a1)–(a3) to the cases $k = 2^r$ and $k = 3 \cdot 2^r$ with $r \ge 2$. Prove that the remainder of a^k modulo 2^{r+2} can only be 0 or 1 as there is no primitive root for this modulus.
   Part (a4) can be generalized to $k = \varphi(p^\alpha)$, where $p > 2$ is a prime and $\alpha \ge 2$. Apply the Euler–Fermat Theorem (as in the proof of Theorem 7.6.5).
   The generalization of part (a5) works for $k = \frac{1}{2}\varphi(p^\alpha)$, where $p > 2$ is a prime and $\alpha \ge 2$. Verify

$$a^{\varphi(p^\alpha)/2} \equiv 0 \text{ or } \pm 1 \pmod{p^\alpha}$$

   for any a.

We get the following lower bounds for $G(k)$ in the cases $p > 2$ is a prime, $\alpha \geq 2$, and $r \geq 2$:

$$G(3 \cdot 2^r) \geq G(2^r) \geq 2^{r+2}$$
$$G(p^\alpha - p^{\alpha-1}) \geq p^\alpha$$
$$G\left(\frac{p^\alpha - p^{\alpha-1}}{2}\right) \geq \frac{p^\alpha - 1}{2}.$$

We note that these are the only known lower bounds for $G(k)$ besides those in Theorem 7.6.4.

3. Let R be a large number and form the sums

$$x_1^k + \cdots + x_{k+1}^k, \quad x_i \text{ are integers}, \quad 0 \leq x_i \leq R, \quad i = 1, 2, \ldots, k+1.$$

Demonstrate that there are many more sums than values they can have. Thus there must be an n that has many such representations.

4. (a) Performing the operations on the left-hand side, there remain only terms of the type a_i^4 and $a_i^2 a_j^2$ ($i < j$) with coefficients 6 and 12. We obtain the same result after squaring on the right-hand side.

   (b) Let $n = 6q + r$ where $0 \leq r \leq 5$. By Theorem 7.5.3, $q = x_1^2 + x_2^2 + x_3^2 + x_4^2$. Write each x_i as a sum of four squares. Applying the identity in part (a), we can represent $6q$ as a sum of 48 fourth powers, and r is the sum of at most five terms 1^4.

5. The integers $8t + 6$ cannot be written as $x^2 \pm y^2$, thus two squares are not sufficient. To verify the second part of the statement, transform the Diophantine equation $x^2 + y^2 - z^2 = n$ into $z^2 - y^2 = x^2 - n$, and select the value of x arbitrarily with the restriction that $x^2 - n$ should not be of the form $4t + 2$ (for any n, all even or all odd integers can be taken as x, and occasionally both the odd and even numbers are suitable). Apply Theorem 7.3.1 (and the fact that if an integer is the difference of two squares, then so is its negative). We can proceed similarly for the other equation.

7.7.

1. (a) If $m = qk$, then

$$x^m + y^m = z^m \implies (x^q)^k + (y^q)^k = (z^q)^k.$$

   (b) It follows from part (a), as any $k > 2$ has an odd prime divisor or is a multiple of 4.

2. (a) No solution, which follows from the case $k = 4$ of Fermat's Last Theorem.

   (b) There are infinitely many solutions. Looking for solutions in the form $x = 2^\alpha$, $y = 2^\beta$, and $z = 2^\gamma$, we get

$$2^{3\alpha} + 2^{4\beta} = 2^{5\gamma}.$$

   Choosing $\alpha = 4\nu$ and $\beta = 3\nu$, the condition $12\nu + 1 = 5\gamma$ has to be satisfied.

Remark: The arguments can be generalized to Diophantine equations of the type $x^k + y^m = z^n$:

(i) If $(k, m, n) \geq 3$, then there are no solutions in positive integers.

(ii) If $(km, n) = 1$, then there are infinitely many solutions in positive integers. Also, if two of the exponents are given arbitrarily, then we can find infinitely many third exponents for which the equation is solvable in positive integers. For example, let us fix k and m, let a and b be positive integers, and define c as $c = a^k + b^m$. Multiplying this equality by c^s where s is any common multiple of k and m, we see that $x = ac^{s/k}, y = bc^{s/m}, z = c$ is a positive integer solution of $x^k + y^m = z^{s+1}$.

3. All solutions are $k = 2, x = y = z - 1$.

4. (a) No solution; the equation obtained after multiplying by a common denominator contradicts the case $k = 4$ of Fermat's Last Theorem.

(b) All solutions are $x = vwd, y = uwd, z = uvd$, where u, v, w is a (primitive) Pythagorean triple (with w as hypotenuse) and d is an arbitrary positive integer. Hint: We can check by a simple substitution that these are solutions. Conversely, assume that x, y, z is a solution. We can restrict ourselves to the case $(x, y, z) = 1$. Let $(x, y) = w, (x, z) = v$, and $(y, z) = u$. Verify that u, v, and w are pairwise coprime, hence $x = vwx_1, y = uwy_1$, and $z = uvz_1$, where x_1, y_1, and z_1 are pairwise coprime. Substituting these values into the equation, prove $x_1 = y_1 = z_1 = 1$ and $u^2 + v^2 = w^2$.

(c) All solutions are $x = a^2 d, y = b^2 d, z = (a + b)^2 d$, where a, b, d are positive integers and we can assume $(a, b) = 1$ (this guarantees the uniqueness of the parametric representation). Hint: This can also be reduced to the case $(x, y, z) = 1$. After two squarings, we obtain

$$xy = \left(\frac{z - x - y}{2}\right)^2,$$

where $(x, y) = 1$, so x and y are (coprime) squares.

(d) All solutions are $x = a^3 d, y = b^3 d, z = (a + b)^3 d$, where a, b, d are positive integers and $(a, b) = 1$. Hint: The first cubing yields

$$x + y + 3\sqrt[3]{x}\sqrt[3]{y}(\sqrt[3]{x} + \sqrt[3]{y}) = z.$$

Replace $\sqrt[3]{x} + \sqrt[3]{y}$ by $\sqrt[3]{z}$, and cube again to get

$$xyz = \left(\frac{z - x - y}{3}\right)^3.$$

After dividing by $(x, y, z)^3$, the three factors on the left-hand side are pairwise coprime, so each is a cube.

5. Use the characterization of the primitive Pythagorean triples.

(a) For the equation $x^4 + y^2 = z^2$ we need $x^2 = 2mn$ or $x^2 = m^2 - n^2$, whereas the equation $x^2 + y^2 = z^4$ requires $z^2 = m^2 + n^2$. There are infinitely many ways to choose m and n so that $2mn$, $m^2 - n^2$, and $m^2 + n^2$, should resp. be squares (and $m > n > 0, (m, n) = 1$, and $m \not\equiv n \pmod 2$ are valid).

(b) Prove by infinite descent.

6. All solutions are $x = \pm 1, y = \pm 1$. Hint: Transform the equation into $x^4 + (y^2 - 1)^2 = y^4$.

7. Only base 7 works. Hint: After several factorizations and applying the characterization of primitive Pythagorean triples, we can reduce the equation to Exercises 7.3.13g and 7.7.6.

8. We can proceed similarly as in Exercise 7.4.3. Answer: 0 and

$$c, \quad c\omega, \quad c(1 + 2\omega), \quad c(1 + \omega), \quad c(1 - \omega), \quad c(2 + \omega),$$

where c is an integer. This formula characterizes Eulerian integers as complex numbers with angles $k\pi/6$ where k is an integer.

9. If $\alpha = a + b\omega, \beta = c + d\omega$, then the identity in the exercise is just an expanded form of

$$|\alpha|^2 \cdot |\beta|^2 = |\alpha\beta|^2.$$

(Of course, we can verify the identity by performing multiplications and comparing the terms resulting on the two sides, but this would be an ugly solution and would not reveal the background of the exercise.)

10. (a) The simplest demonstration of the relation between the two equations uses norms of Eulerian integers.

(b) The equations are solvable if and only if every prime of the form $3t - 1$ appears with an even exponent in the standard form (in $\mathbf{Z}$) of n.

Counting the number of solutions, we distinguish those that differ only in signs or the order of terms. Assume that the equations are solvable and put

$$L = \prod_{\mu=1}^{r} (\beta_\mu + 1),$$

where $\beta_1, \ldots, \beta_r$ are the exponents of the primes of the form $3t + 1$ in the standard form (in $\mathbf{Z}$) of n ($L = 1$ if n has no such prime divisor).

Then $x^2 - xy + y^2 = n$ has $6L$ solutions, and the equation $x^2 + 3y^2 = n$ has $6L$ or $2L$ solution according as n is divisible by 4 or is odd. ($n = 4s + 2$ is not possible since 2 (as a prime of the form $3t - 1$) must occur with an even exponent in the standard form of n.)

We can verify the statement about the equation $x^2 - xy + y^2 = n$ as in the proof of the Two Squares Theorem 7.5.1.

The second equation can be reduced to the first one by establishing a bijection (as in part (a)) between the solutions of $x^2 + 3y^2 = n$ and those solutions of $x^2 - xy + y^2 = n$ where x is even. For $4 \mid n$, we have to show that x is even in all solutions of $x^2 - xy + y^2 = n$. If n is odd, observe that the corresponding x is even for exactly two of the six associates of an Eulerian integer with norm n.

11. All solutions are $x = \pm 10, y = 7$. Hint: Follow the ideas used in Exercise 7.5.11. Start with factoring the left-hand side in the Eulerian integers, then both factors are cubes apart from units and the Eulerian primes in their gcd.

12. (a) Let k_μ denote the number of elements in a complete residue system modulo μ, and let R be the rhombus lattice of the Eulerian integers. Multiplying R by μ, we obtain the rhombus lattice R_μ consisting of the multiples of μ. Thus the vectors defining the sides of the fundamental rhombus in R_μ are μ and $\omega\mu$. The Eulerian integers in every such fundamental rhombus of R_μ form a complete residue system modulo μ. Therefore k_μ is roughly the ratio of the areas of the fundamental rhombuses in R_μ and R, which ratio is $|\mu|^2 = N(\mu)$. We can get rid of the word "roughly" by considering the number of points of the two lattices in a large circle or square H. Let H have area A, the number of points of lattices R and R_μ in H be n and n_μ, and the areas of the fundamental rhombuses be a and a_μ. Since there are k_μ Eulerian integers in every fundamental rhombus of R_μ,

(A.7.6)
$$k_\mu \sim \frac{n}{n_\mu}$$

if $A \to \infty$. On the other hand,

$$n \sim \frac{A}{a}, \qquad n_\mu \sim \frac{A}{a_\mu} = \frac{A}{aN(\mu)},$$

so

(A.7.7)
$$\frac{n}{n_\mu} \sim \frac{\frac{A}{a}}{\frac{A}{a_\mu}} = \frac{a_\mu}{a} = N(\mu).$$

By (A.7.6) and (A.7.7), the constants k_μ and $N(\mu)$ are asymptotically equal, so they must be equal.

(b) By part (a), the number of elements is all right. To show that the elements are pairwise incongruent modulo μ, use

$$\mu \mid j \Longrightarrow p = N(\mu) \mid j^2 \Longrightarrow p \mid j.$$

(c) Apply the argument in the proof of the Euler–Fermat Theorem 2.4.1.

13. No solution. Hint: Multiplying the equation by uvw, we obtain $u^2w + v^2u = w^2v$. Introducing $u^2w = c$ and $v^2u = d$, we get $cd(c + d) = (uvw)^3$. We can see in the usual way that the factors on the left-hand side be pairwise coprime, thus c, d, and $c + d$ are (non-zero) cubes, which contradicts Fermat's Last Theorem for $k = 3$.

14. By the formula for the Pythagorean triples, the area of the triangle is

$$d^2mn(m + n)(m - n),$$

where $m > n > 0$, $(m, n) = 1$, and $m \not\equiv n \pmod 2$.

(a) The area is (measured by) a square number if and only if $mn(m+n)(m-n)$ is a square. The conditions imply that the four factors are (positive and) pairwise coprime, thus each is a square. This, however, contradicts Lemma 7.7.3.

(b) $d = 1$ by the assumption, thus we obtain as in the previous argument that m, n, $m + n$ are cubes, which contradicts Theorem 7.7.10.

(c) There are infinitely many such triangles. For every Pythagorean triangle we can find a similar triangle with this property: if m and n are given, choose $d = mn(m + n)(m - n)$. We can express this also without the parametric characterization: If the area of a triangle is A, then enlarging its size by A, the new triangle has area A^3.

(d) For k even, part (a) implies that the area cannot be a kth power. For k odd, we get by the arguments in parts (b) and (c) that in the case of coprime side lengths the area cannot be a kth power, but for every Pythagorean triangle we can find a similar one such that its area is a kth power.

7.8.

1. If $m = 0$, then $x = \pm 1$ and y is arbitrary; if $m = -1$, then $x = \pm 1$, $y = 0$, or $x = 0$, $y = \pm 1$; if $m \leq -2$ or $m = k^2 > 0$, then $x = \pm 1$, $y = 0$. Hint: For $m = k^2$, the factorization $(x - ky)(x + ky) = 1$ is in the integers, so either both factors are 1 or both are -1.

2. This is Pell's equation $10y^2 + 1 = x^2$, so there are infinitely many such squares.

3. If we multiply a solution of $x^2 - my^2 = r$ by a solution of Pell's equation $x^2 - my^2 = 1$ (as seen in the proof of Theorem 7.8.2), then we again get a solution of $x^2 - my^2 = r$.

4. Infinitely many solutions: (a1), (a2), (b1). No solution: (b2) (this follows from considering $x^2 - 3y^2 = -1$ modulo 3 or modulo 4).

5. Infinitely many. Hint: Multiplying $n(n - 1) = 2y^2$ by 4, we get Pell's equation $z^2 - 8y^2 = 1$ (the condition $z = 2n - 1$ imposes no restriction since z is odd in all solutions of $z^2 - 8y^2 = 1$). Another option: One of n and $n - 1$ is a square, the other is the double of a square, and both resulting equations $u^2 - 2v^2 = \pm 1$ have infinitely many solutions.

6. Infinitely many. Hint: Multiplying $x^2 + (x + 1)^2 = z^2$ by 2, we get $(2x + 1)^2 - 2z^2 = -1$. Another option: The parametric characterization of the primitive Pythagorean triples leads to the equations $u^2 - 2v^2 = \pm 1$.

7. No solution: (a), (b), (d), (e). Infinitely many solutions: (c), (f). Hint: The insolvability can be shown by congruences with suitable moduli. Modulus 8 works in all the four cases, but 3, 7, 9, and 3, can also be applied in the order of the list. In (c), we easily see the solution $x = 4$, $y = 1$, thus there are infinitely many solutions by Exercise 7.8.3. In (f), after multiplying by 3, we get $z^2 - 6y^2 = 3$. As $z = 3$, $y = 1$ is a solution, there are infinitely many solutions. It is clear that $3 \mid z$ in every solution, so also $x = z/3$ is an integer.

8. The equation is solvable if and only if $p \equiv 1 \pmod 4$ or $p = 2$. Hint: A congruence modulo 4 implies necessity immediately. To prove sufficiency, consider the

solution $x > 0$, $y > 0$ of $x^2 - py^2 = 1$ where x is minimal. Show that x must be odd, and transform the equation into

(A.7.8) $$\frac{x+1}{2} \cdot \frac{x-1}{2} = p\left(\frac{y}{2}\right)^2.$$

One of the factors on the left-hand side of (A.7.8) is a square, and the other factor is a square multiplied by p. Hence $u^2 - pv^2 = \pm 1$, but the $+$ sign is impossible, as x was a minimal solution.

9. The statement about signs is obvious. To establish the congruences, consider a non-trivial solution satisfying $(x, y, z) = 1$, and derive $(z, a) = (y, a) = 1$. Multiplying the equation by $bz^{2(\varphi(|a|)-1)}$, we obtain

$$(byz^{\varphi(|a|)-1})^2 \equiv -bc \pmod{|a|}.$$

The other two congruences can be obtained the same way, or we can refer to symmetry.

10. Infinitely many. Hint: A necessary condition is that $28k^2 + 1$ be a square and this happens infinitely often. Show that then $2 + 2\sqrt{28k^2 + 1}$ is a square. For the proof, divide $r^2 - 1 = 28k^2$ by 4, and factor the left-hand side. The factors are consecutive integers, one of them a square, and the other one 7 times a square. Finally, verify that the factor $(r + 1)/2$ has to be a square, thus the number $2r + 2$ in the exercise is a square.

11. If $x = u^2$, then we can rewrite the equation as $(u^2 - 1)(u^2 + 1) = 2y^2$. Verify that $u^2 - 1$ is a square, which is impossible as $u \neq \pm 1$. (This equation also arose in the solution of Exercise 7.7.7.) The case $y = v^2$ is the same as Exercise 7.3.13g.

7.9.

1. If a term 1 occurs in a partition of $n + 1$, then delete one such term, otherwise decrease the least term by one. Thus we get every partition of n at most twice. The second step produces only partitions with 1 occurring in them at most once, thus we cannot get all partitions of n if $n > 1$. Hence, equality holds only for $n = 1$.

2. (a) ∞.   (b) $-\infty$.

3. These integers are $(3k^2 \pm k)/2$ by Theorem 7.9.5.

4. 2^{n-1}. Hint: Consider a representation $n = x_1 + x_2 + \cdots + x_r$ (where r and $x_1, \ldots, x_r$ are positive integers), and, starting from the origin, draw segments of lengths x_1, $\ldots, x_r$ in this order one after the other onto the interval $[0, n]$. Then the endpoints of the segments different from 0 and n form a subset of $H = \{1, 2, \ldots, n-1\}$ in the interval (the extreme cases are possible when all or none of these points are marked). Thus we established a bijection between the representations of n and the subsets of H. So the number of representations is equal to the number of subsets in a set of $n - 1$ elements.

5. Subtracting 1 from each term in a partition of n with r terms, we get a partition of $n - r$ with at most r terms (there will be fewer than r terms if and only if 1 occurred in the original partition). Verify that this is a bijection between the two types of partitions.

6. $\dfrac{x^r}{\prod_{i=1}^{r}(1 - x^i)}$.

7. We can proceed either by establishing a suitable bijection, or by applying generating functions.

   (a) *Bijection*: Consider a partition $n = x_1 + \cdots + x_r$ with distinct integers x_i. Every x_i has a unique decomposition as $2^\alpha t$, where $\alpha \geq 0$ and t is odd. Factoring out the common values of t, we obtain $n = 1u_1 + 3u_2 + 5u_3 + \ldots$, where every u_j is a non-negative integer. This can be considered as a partition of n containing u_1 terms of 1, u_2 terms of 3, etc.

   To illustrate the procedure, consider the partition $23 = 10 + 6 + 4 + 3$. Then

   $$23 = 2^1 \cdot 5 + 2^1 \cdot 3 + 2^2 \cdot 1 + 2^0 \cdot 3$$
   $$= 2^1 \cdot 5 + (2^1 + 2^0) \cdot 3 + 2^2 \cdot 1 = 1 \cdot 4 + 3 \cdot 3 + 2 \cdot 5$$

   leads to the partition $23 = 5 + 5 + 3 + 3 + 3 + 1 + 1 + 1 + 1$.

   Verify that the above map is a bijection between the two types of partitions of n.

   Generating functions: The appropriate generating functions are

   $$U(x) = \prod_{i=1}^{\infty}(1 + x^i) \quad \text{and} \quad W(x) = \prod_{j=1}^{\infty}\frac{1}{(1 - x^{2j-1})}.$$

   Rewriting $U(x)$ using the identity

   $$(1 + x^i) = \frac{1 - x^{2i}}{1 - x^i},$$

   we get $W(x)$ after cancellation. For a precise proof, one has to work either with formal power series and formal infinite products, or has to manage properly the limit process in the infinite products.

   (b) To establish a bijection, write the terms as $k^\alpha t$ with $k \nmid t$.

   The generating functions are

   $$U_k(x) = \prod_{i=1}^{\infty}(1 + x^i + \cdots + (x^i)^{k-1}) \quad \text{and} \quad W_k(x) = \prod_{\substack{t=1 \\ k \nmid t}}^{\infty}\frac{1}{(1 - x^t)}.$$

8. *First proof*: By Exercise 7.9.6, the coefficient of x^n in the power series expansion of

   $$\frac{x^r}{(1 - x)(1 - x^2)\ldots(1 - x^r)}$$

   is the number of partitions of n where the biggest term is r. Thus the sum of these coefficients for all r is just $p(n)$.

Second proof: The coefficient of x^n is influenced only by the first n terms on the right-hand side. Giving them a common denominator and adding, we obtain

$$-1 + \frac{1}{(1-x)(1-x^2)\dots(1-x^n)}.$$

By Theorem 7.9.2, the coefficient of x^n equals the number of partitions of n from summands $1, 2, \dots, n$, which is just $p(n)$.

9. The derivative of the logarithm of $V(x) = \prod_{i=1}^{\infty}(1-x^i)$ is

(A.7.9)
$$\frac{V'(x)}{V(x)} = \sum_{i=1}^{\infty} \frac{-ix^{i-1}}{1-x^i}.$$

(Taking the logarithm and differentiating term by term are legal for $|x| < 1/2$.) Multiply (A.7.9) by $-xV(x)$, and apply

$$\sum_{i=1}^{\infty} \frac{ix^i}{1-x^i} = \sum_{i=1}^{\infty} i(x^i + x^{2i} + \dots) = \sum_{j=1}^{\infty} \sigma(j)x^j.$$

Then

(A.7.10)
$$-xV'(x) = V(x) \sum_{j=1}^{\infty} \sigma(j)x^j.$$

Finally, substitute the formulas

$$V(x) = 1 - x - x^2 + x^5 + x^7 - x^{12} - x^{15} + \dots$$
$$xV'(x) = -x - 2x^2 + 5x^5 + 7x^7 - 12x^{12} - 15x^{15} + \dots$$

into (A.7.10) and perform the multiplication of the two power series on the right-hand side of (A.7.10).

A.8. Diophantine Approximation

8.1.

1. (a) With a common denominator, the numerator $as - br \neq 0$, thus $|as - br| \geq 1$.
   (b) $|as - br| = 1$ holds infinitely often, since the Diophantine equations $as - br = \pm 1$ have infinitely many solutions.

2. Since $d = \alpha - r/s \neq 0$, so $|d| > 1/(ks)^2$ if k is large enough.

3. (a) For any $s > 1$, at most one suitable fraction with denominator s can fit.
   (b) It follows from part (a).

4. (a) For any k, a fraction with denominator either 2^k or 2^{k+1} meets the requirement.
   (b) $\alpha = 1/3$.
   (c) There is a fraction with denominator 3^k for any k.
   (d) $\alpha = 1/2$.

(e) The squares of fractions r/s approximating $\sqrt{\alpha}$ well have this property.

(f) $\alpha = (1 + \sqrt{5})^2/4$.

5. Use that the fractions r/s approximating α well satisfy $r^2 \sim \alpha s^2$.

6. We can argue similarly as in the proof of Theorem 8.1.6. To get rid of the square root, multiply the difference by $\sqrt{2} + r/s$.

7. If r/s approximates α well, then

   (a) $a(r/s) + b$ approximates $a\alpha + b$ well

   (b) r^2/s^2 approximates α^2 well.

8. (a) 0 and 1. (b) and (c) The complete interval $(-1, 1)$.

9. (a) Draw an interval of length $\varepsilon/2^i$ around the ith element.

   (b) Cardinality: There is a bijection between these ternary fractions and all real numbers in $[0, 1)$ written as binary fractions (replace digit 2 by 1). Measure zero: We obtain the Cantor set by deleting the middle third of the interval $[0, 1)$, then deleting the middle thirds of both remaining intervals, then deleting the middle thirds of the four remaining intervals, etc. The total length of the remaining intervals after m steps is

$$1 - \frac{1}{3} - \frac{2}{9} - \cdots - \frac{2^{m-1}}{3^m} = 1 - \frac{1}{3} \cdot \frac{1 - (\frac{2}{3})^m}{1 - \frac{2}{3}} \to 0, \quad \text{if } m \to \infty.$$

10. (a) This is a direct consequence of the definition.

   (b) Cover each of the k sets with intervals of total length ε/k.

   (c) Cover the ith set with intervals of total length $\varepsilon/2^i$ $(i = 1, 2, \ldots)$.

   (d) Every set is the union of its one-element subsets, which are of measure zero. This union has measure zero e.g. for the Cantor set, but not for the interval $[0, 1]$.

8.2.

1. (a) Both proofs of Theorem 8.2.1 can be adapted to the space; for the second proof, we have to apply a three-dimensional variant of Lemma 8.2.2.

   (b) In the n-dimensional case, we have to assume that the volume of H is at least $2^n \Delta$. (Here Δ is the absolute value of the determinant formed from the coordinate vectors of the n sides of the fundamental parallelepiped.)

2. Both proofs of Theorem 8.2.1 can be modified to verify this statement. For the second proof, we need a generalization of Lemma 8.2.2 (we keep the notation used there): If the intersection of any $r + 1$ sets K_P is empty, then $t \leq r\Delta$. Following the arguments of the first or second proof, we obtain r non-trivial lattice points, no two of which are symmetric about the center O. Their mirror images with respect to O yield another r lattice points.

3. Argue as in the proof of Theorem 8.2.4. For a prime $p = 3k + 1$, we have $c^2 \equiv -3$ (mod p) with a suitable c. Then $p \mid x^2 + 3y^2$ for the lattice points (8.2.6) in the proof of Theorem 8.2.4. Applying Minkowski's theorem for a suitable ellipse, we obtain a non-trivial lattice point satisfying $x^2 + 3y^2 < 3p$. Since $x^2 + 3y^2 = 2p$ is impossible modulo 3, $x^2 + 3y^2 = p$.

4. Using the notation in Theorem 8.2.1, now L is the usual square lattice, $\Delta = 1$, and H is the parallelogram bounded by the lines

$$a_{11}x_1 + a_{12}x_2 = \pm b_1, \qquad a_{21}x_1 + a_{22}x_2 = \pm b_2,$$

its area is $4b_1 b_2/|D|$. Thus the statement follows from Minkowski's theorem.

5. Based on the three-dimensional variant of Minkowski's theorem (see Exercise 8.2.1a), we can proceed as in the proof of Theorem 8.2.3. Consider the lattice

$$x = s\alpha_1 - r_1, \qquad y = s\alpha_2 - r_2, \qquad z = s.$$

The volume of the fundamental parallelepiped is $\Delta = 1$. The approximation requirement can be written as $|zx^2| < c^2$, $|zy^2| < c^2$, where $c = 2/3$. This set in the space is not convex (and is unbounded), so we should consider octahedrons

$$\frac{1}{a^2}|z| + 2a|x| \le \sqrt[3]{12}, \qquad \frac{1}{a^2}|z| + 2a|y| \le \sqrt[3]{12}$$

instead, with suitable values $a > 1$, also using the inequality between arithmetic and geometric means.

8.3.

1. (a) $4, 1, 4, 2$

   (b) $1, 1, 2, 1, 2, 1, 2, \ldots$

   (c) $2, 4, 4, 4, \ldots$

   (d) $1, 1, 1, 1, \ldots$

2. (a) $43/30$. (b) $(1 + \sqrt{3})/2$.

3. Use the good approximation of the fractions r_n/s_n in Theorem 8.3.3, and observe that $(s_{n-1}, s_n) = 1$ follows from (8.3.11).

4. By Exercise 8.3.1d, every digit in the continued fraction expansion of $(1 + \sqrt{5})/2$ is 1. Hence, the fractions r_n/s_n in Theorem 8.3.3 satisfy $r_n = \varphi_{n+2}$ and $s_n = \varphi_{n+1}$ by recursion (8.3.8a)–(8.3.8b).

5. Use (8.3.8a), (8.3.8b), and (8.3.10) in Lemma 8.3.4.

6. Denoting the original number by α and the one formed from the purely periodic part by β, we obtain the finite continued fractions

$$\alpha = C(c_0, c_1, \ldots, c_{M-k}, \beta) \quad \text{and} \quad \beta = C(c_{M-k+1}, \ldots, c_M, \beta).$$

We obtain the statement by simplifying the multiple-decked fractions and performing some further rearrangements.

8.4.

1. Dense: (b), (d), (f), (g).

2. We draw countably many subintervals in $[0, 1)$ of lengths tending to zero around every rational point $0 < r < 1$, and arrange them in a single sequence of intervals J_1, $J_2, \ldots$. To verify that the fractional parts of a sequence of numbers u_i is everywhere dense, we have to show that every interval J_s contains at least one $\{u_i\}$.

   We shall obtain α as a common point of nested closed intervals. We start (say) with $H_0 = [2, 3]$. If the interval H_{k-1} is given, then we define H_k as a subinterval of H_{k-1}: We choose an exponent n_k such that the n_kth powers of the elements x in H_{k-1} cover a complete interval T between two consecutive integers, and H_k consists of numbers x satisfying

   $$x \in H_{k-1}, \qquad x^{n_k} \in T, \quad \text{and} \quad \{x^{n_k}\} \in J_k.$$

   Then α can be chosen as a common point of the intervals H_k.

3. If a point P_n is very close to a point $Q = (v_1, \ldots, v_k)$ in the k-dimensional unit cube, then $\{n\alpha_j\} - v_j$ has small absolute value for every $1 \leq j \leq k$, so the absolute value of their linear combination is also small. If $1, \alpha_1, \ldots, \alpha_k$ are linearly dependent, then we get a condition for a suitable linear combination of the coordinates v_i that cannot hold for an arbitrary Q.

4. (a) We build the new sequence by always taking the first element of the old one that was not yet used and that falls into the following intervals in this order:

   $$\left[0, \frac{1}{2}\right), \left[\frac{1}{2}, 1\right), \left[0, \frac{1}{3}\right), \left[\frac{1}{3}, \frac{2}{3}\right), \left[\frac{2}{3}, 1\right), \left[0, \frac{1}{4}\right), \ldots.$$

   (b) Every second element in the new sequence should have e.g. a very small fractional part.

5. True: (b), (c).

6. (a) Let k be an arbitrary positive integer. If $10^k \leq m < 10^{k+1}$, then

   $$\{\log_{10} m\} > \frac{1}{2} \iff m > 10^k \sqrt{10}.$$

   This means that many more than half of the fractional parts fall into the interval $(1/2, 1)$, if $n = 10^{k+1}$ (and many less than half, if $n = \lfloor 10^k \sqrt{10} \rfloor$).

   (b) The ratio $1/(2\pi)$ is irrational, thus the angles n (measured in radians) are uniformly distributed on the unit circle by Theorem 8.4.5. Therefore, the uniform distribution of the values $\{\sin n\}$ would mean that the set of real numbers x for which $\{\sin x\}$ falls into a fixed subinterval I in $[0, 1]$ of length d would occupy a part on the unit circle of size d times the perimeter. However, we can easily check that this is false e.g. for the interval $I = [1/2, \sqrt{3}/2]$.

7. We have to show that $54321 \cdot 10^v \leq t^n < 54322 \cdot 10^v$ holds for suitable natural numbers n and v. Take the base-10 logarithm and apply Theorem 8.4.1 for $\alpha = \log_{10} t$.

A.9. Algebraic and Transcendental Numbers

9.1.

1. (a) A suitable polynomial is $x^{20} - 7$.

   (b) Square $\alpha - 3 = \sqrt{2}$.

   (c) Squaring $\alpha - \sqrt{3} = \sqrt{2}$, rearranging the result, and squaring again, we get a suitable polynomial with integer coefficients.

   (d) Cube $\alpha - \sqrt{2} = \sqrt[3]{4}$, rearrange the result and square it.

   (e) Cube $\alpha = \sqrt[3]{2} + \sqrt[3]{4}$, and rewrite the part including cube roots on the right-hand side as
   $$3\sqrt[3]{2}\sqrt[3]{4}(\sqrt[3]{2} + \sqrt[3]{4}) = 3 \cdot 2\alpha.$$

   (f) We can eliminate the square roots by successive squarings.

2. Assume that α is a root of a polynomial $f(x) = a_0 + a_1 x + \cdots + a_n x^n$ with rational coefficients and $a_n \neq 0$. Then the given numbers are roots of the polynomials

   (a) $f(-x) = a_0 - a_1 x + a_2 x^2 + \cdots + (-1)^n x^n$

   (b) $f(x)$ (if α is a root of a polynomial with real coefficients, then also $\overline{\alpha}$ is a root of the same polynomial)

   (c) $x f(1/x) = a_n + a_{n-1} x + \cdots + a_0 x^n$ (we can assume $a_0 \neq 0$, $\alpha \neq 0$)

   (d) $f(x - r) = a_0 + a_1(x - r) + \cdots + a_n(x - r)^n$

   (e) $f(x/r) = a_0 + a_1(x/r) + \cdots + a_n(x/r)^n$ (we may clearly assume $r \neq 0$)

   (f) $f(x^k) = a_0 + a_1 x^k + \cdots + a_n x^{kn}$.

3. If $\zeta(2) = \pi^2/6$ were algebraic, then π would be algebraic by parts (e) and (f) of the previous exercise.

4. Assume that $f(\alpha) = a_0 + a_1 \alpha + \cdots + a_n \alpha^n$ is algebraic, so
   $$b_0 + b_1(a_0 + a_1 \alpha + \cdots + a_n \alpha^n) + \cdots + b_s(a_0 + a_1 \alpha + \cdots + a_n \alpha^n)^s = 0$$

   for some rational numbers $b_0, b_1, \ldots, b_s$ not all zero. Performing the operations, we see that α is a root of a non-zero polynomial with rational coefficients, which is a contradiction.

5. If there is such an h, then all its roots, including all roots of g, are algebraic. Conversely, if the roots of g are the algebraic numbers $\alpha_1, \ldots, \alpha_r$ (counted with multiplicity), and α_j is a root of a non-zero polynomial f_j with integer coefficients ($j = 1, \ldots, r$), then $h = f_1 \ldots f_r$ meets the requirements.

6. The statement follows immediately from the definitions of algebraic numbers and linear dependence.

7. A complex number α is a root of the polynomials (a) $x - \alpha$;   (b) $(x - \alpha)(x - \overline{\alpha})$.

9.2.

1. (a)–(e) The degree equals $\deg \alpha$, except if $r = 0$ in (e). For a proof, choose the polynomial f in the hint to Exercise 9.1.2 as m_α, and verify that the polynomials $f(-x)$, etc. given in the hint are irreducible over **Q**.

   (f) $\deg \sqrt[k]{\alpha} \leq k \deg \alpha$.

2. Find first a non-zero polynomial f with rational coefficients such that $f(\alpha) = 0$, and check the irreducibility of f over **Q**. If f is irreducible, then $f = m_\alpha$, thus $\deg \alpha = \deg f$. If f is reducible, then decompose it into the product of irreducible factors, and determine which factor has α among its roots. We can often verify irreducibility using the Schönemann–Eisenstein criterion, and for polynomials of degree two or three it is sufficient to check whether or not the polynomial has a rational root.

   (a) 7.

   (b) 3. Express $1/2 = \cos 60°$ using $\cos 20°$.

   (c) 3. See hint to Exercise 9.1.1e.

   (d) 2. There is a perfect square under the big square root sign.

   (e) 4.

   (f) 4. Add 1 and apply the summation formula for this geometric series of four terms.

3. If $\alpha = r + \sqrt{s}$, then α is a root of the polynomial $(x - r)^2 - s$ irreducible over **Q**. For the converse, use the quadratic formula.

4. (a) Apply that if α is algebraic and r is rational, then $\deg(\alpha + r) = \deg \alpha$ (see Exercise 9.2.1d).

   (b) If α is a non-real complex number, then the numbers $s(\alpha + r)$ are everywhere dense in the complex plane when r and s assume all rational numbers.

5. (a) $\deg \alpha_i \leq \deg f$ for every i.

   (b) Equality holds if and only if f is irreducible over **Q**.

   (c) Write f as a product of irreducible polynomials (over **Q**): $f = f_1 \ldots f_k$, where $k \geq 2$ since f is reducible. Let $\deg f_j = n_j$. Then $n_1 + \cdots + n_k = n$ and

   (A.9.1) $$\sum_{i=1}^{n} \deg \alpha_i = \sum_{j=1}^{k} n_j^2.$$

   Show that the sum on the right-hand side of (A.9.1) is maximal if and only if $k = 2$, one of n_1 and n_2 is 1, and the other is $n - 1$.

6. $m_\alpha = x^6 + 5x^5 + 10x^2 + 5x - 10$. Hint: The conditions imply $f = g m_\alpha$, where $\deg g = 1$. Hence, f has a rational root. Determine it by the rational root test, and divide f by the suitable root factor (the best way to do this is to apply Horner's scheme).

7. We have $[m_\alpha, m_\beta] \mid f$, and if $m_\alpha \neq m_\beta$, then $(m_\alpha, m_\beta) = 1$ due to the irreducibility of minimal polynomials.

8. If f were irreducible, then the conditions would imply $f = m_\alpha \mid g$.

9.3.

1. (a) Let α be algebraic and β transcendental. If $\alpha + \beta$ were algebraic, then $\beta = (\alpha + \beta) - \alpha$ would be algebraic, which is a contradiction.

   (b) For example, $\pi + (1 - \pi)$ is algebraic, $\pi + (1 + \pi)$ is transcendental.

   (c) The only difference compared to addition is that the product of an algebraic and a transcendental number can be algebraic but only in the exceptional case where the algebraic factor is 0.

2. (a) Both α and β are algebraic.

   (b) Both α and β are transcendental.

   (c) At least one of α and β is transcendental (find examples for each of the possible cases).

   (d) α and β are algebraic, or $\alpha = 0$ and β is transcendental.

   (e) Both α and β are transcendental.

   (f) Both α and β are transcendental, or one of them is 0 and the other is transcendental.

   (g) At least one of α and β is transcendental.

   (h) Both α and β are algebraic. Hint: Solving the system of equations $\alpha + \beta = c$, $\alpha\beta = d$, the quadratic formula (or Theorem 9.3.6) yields that α and β are algebraic.

   In the rational/irrational variant, there are changes only at (d) and (h):

   (d) $\alpha = 0$ and $\beta(\neq 0)$, or $\alpha = \sqrt{r}$ and $\beta = s\sqrt{r}(\neq 0)$

   (h) $\alpha = s + \sqrt{r}$ and $\beta = s - \sqrt{r}$ where $r > 0$ and s are arbitrary rational numbers.

3. Algebraic: (b).

4. (a) At most one of them can be algebraic; use Exercise 9.3.2a,d,h.

   (b) Apply Theorem 9.3.3 for (b1) and Theorem 9.3.5 for (b2) observing $e^{i\pi} = -1$.

5. Algebraic: (a), (b).

7. Show that $\log_{10} n$ is irrational, and apply Theorem 9.3.5.

8. Assume that for some positive integers k and m, $k \neq m$, $\alpha^k + \beta^k = c$ and $\alpha^m + \beta^m = d$ are algebraic and not both are 0. Then

$$(c - \beta^k)^m = (d - \beta^m)^k.$$

So β is a root of a non-zero polynomial with algebraic coefficients, hence β itself is algebraic by Theorem 9.3.6. We get similarly that α is algebraic, and therefore $\alpha^n + \beta^n$ is algebraic for every n.

9. *Algebraic*: Show that there exist infinitely many positive integers that are not rational powers of α. These must be powers of α with transcendental exponents by Theorem 9.3.5.

Transcendental: The number of powers of α with transcendental exponents has the cardinality of the continuum but only countably many of them can be algebraic numbers.

9.4.

1. (b) The number α defined in Theorem 9.4.2 is a Liouville number, so part (a) implies that there are infinitely many Liouville numbers. Continuum: We obtain as in the proof of Theorem 9.4.2 that the infinite series formed of any infinite subsequence of the sequence $10^{-k!}$ is a Liouville number.

2. (a) Let $f = f_1 \ldots f_k$ be the decomposition of f into the product of irreducible polynomials over $\mathbf{Q}$. Then we can reduce the Diophantine equation (9.4.12) to a system of equations

$$g_j(y,z) = y^{n_j} f_j\left(\frac{z}{y}\right) = b_j, \qquad j = 1, 2, \ldots, k,$$

where $\prod_{j=1}^{k} b_j = b$. If $b \neq 0$, then there are only finitely many possibilities for (say) b_1, and for each b_1 the first equation can have only finitely many solutions by (the original) Theorem 9.4.5. If $b = 0$, then at least one $b_j = 0$, and the jth equation (with $b_j = 0$) can have only finitely many solutions for every possible j.

   (b) We used only these properties in the proof of Theorem 9.4.5.

3. Follow the proof of Theorem 9.4.5. If z_i/y_i has no bounded subsequence, then interchange the roles of z_i and y_i, and consider $f(y_i/z_i)$ instead of $f(z_i/y_i)$. It suffices to apply Theorem 9.4.4 in the special case (say) $\kappa = 0.99$.

4. Use that if α is a multiple root of a polynomial f, then α is a root of the derivative of f, too.

9.5.

1. Use ideas similar to those in the proof of Theorem 9.5.1.

2. (a) Using the power series of $\sin x$ and $\cos x$, expand $\sin 1$ and $\cos 1$, and argue as in the proof of Theorem 9.5.1.

   (b) Modify the proof of Theorem 9.5.2: replace $\sin(\pi x)$ by $\sin(rx)$ in the integral I, and let a be a common denominator of the rational numbers $1/r$, $\cos r$, and $\sin r$.

(c) Express $\sin(2x)$ and $\cos(2x)$ using $\tan x$. This implies that if $\tan r$ is rational, then both $\sin(2r)$ and $\cos(2r)$ are rational, which contradicts part (b).

3. In the proof of Theorem 9.5.2, the integral-free expression is 0 after every second integration by parts since $\sin \pi = \sin 0 = 0$. Thus considering two consecutive integrations by parts as a single step, there will arise always just one new integral-free expression, and its denominator is π^2 times the previous one. Hence, the assumption $\pi^2 = a/b$ will lead to a contradiction by computing the integral

$$\pi a^{n+1} \int_0^1 \sin(\pi x) f(x) \, dx$$

following the ideas seen at Theorem 9.5.2.

9.6.

1. The numbers $\bar{\alpha}$ and α share the same minimal polynomial. The other three numbers can be obtained from α and $\bar{\alpha}$ by addition; subtraction and multiplication by i; multiplication and taking a square root.

2. Only (c) is an algebraic integer. Hint for (d): Assume that $\cos 1°$ is an algebraic integer, and show that then so is $\sin 1°$. The addition formulas show that $\cos k°$ and $\sin k°$ are algebraic integers for every integer k. This is, however, false e.g. for $k = 30$.

3. True: (a), (c), (e), (f), (h).

4. Yes, it is solvable, e.g. $x = y = 1$, $z = \sqrt[n]{2}$ is a non-trivial solution.

5. True: (a), (c), (d).

6. As α is algebraic, it satisfies

$$a_0 + a_1 \alpha + \cdots + a_n \alpha^n = 0$$

with suitable integers a_i, where $a_n \neq 0$. Multiplying by a_n^{n-1} and arranging the result by the powers of $a_n \alpha$, we obtain that $a_n \alpha$ is an algebraic integer, i.e. α is a quotient of an algebraic integer and the integer a_n. Applying the procedure for $1/\alpha$ instead of α, we get that α is a quotient of an integer and an algebraic integer (and if $\alpha = 0$, then this holds trivially).

7. The constant term is ± 1 in the minimal polynomial of α (with integer coefficients and leading coefficient 1).

8. (a) For example, $\beta_n = (\sqrt{2} - 1)^n$.

   (b) If both $1/\alpha$ and α/β are algebraic integers, then so is their product $1/\beta$. For the converse, let $\beta_n = \sqrt[n]{\alpha}$.

   (c) Let $a_0 + a_1 x + \cdots + a_{n-1} x^{n-1} + x^n$ be the minimal polynomial of an algebraic integer α (where every a_i is an integer). Then the minimal polynomial of α/b is $a_0 + a_1 b x + \cdots + a_{n-1} b^{n-1} x^{n-1} + b^n x^n$. Rewriting it with a leading coefficient 1, the constant term can be an integer only if $b^n \mid a_0$, thus there exist only finitely many such integers b (since $\alpha \neq 0$ implies $a_0 \neq 0$).

9. The answer is yes for both questions. Take e.g. $\cos\varphi + i\sin\varphi$, where (a) $\cos\varphi = 1/3$; (b) $\cos\varphi = \sqrt{2} - 1$.

10. (a) The numbers $a + b\sqrt[q]{2}$, where a and b are integers, are everywhere dense on the real line by Theorem 8.4.1.

    (b) We obtain from the quadratic formula that the real part of a non-real algebraic integer can only be a fraction with denominator 2. Hence the algebraic integers of degree 2 are not dense in the complex plane. The ones of degree 4 are, however, dense: the numbers $(a + b\sqrt{2}) + i(c + d\sqrt{2})$ where a, b, c, and d are integers have mostly degree 4, and are dense in the plane.

11. (a) If r is rational, then $\alpha = \cos r° + i\sin r°$ is a complex root of unity, and so it is an algebraic integer. Therefore $2\operatorname{Re}\alpha = 2\cos r°$ is an algebraic integer. If $2\cos r°$ is rational then it must be an integer. Hence $\cos r°$ is 0, $\pm 1/2$, or ± 1. We can solve the problem without referring to algebraic integers. If r is rational, then nr is an integer multiple of 360 for some positive integer n, i.e. $\cos(nr°) = 1$. Using
    $$\cos(n\alpha) = 2\cos((n-1)\alpha)\cos\alpha - \cos((n-2)\alpha),$$
    verify by induction that $2\cos(n\alpha)$ is a polynomial in $2\cos\alpha$ with integer coefficients and leading coefficient 1. Thus if $\cos(nr°) = 1$, then $2\cos r°$ is a root of a polynomial with integer coefficients and leading coefficient 1. All rational roots of such a polynomial can only be integers, so $2\cos r°$ must be an integer.

    (b) At least one of r and $\sin r°$ is irrational, except if r is an odd multiple of 30 or is divisible by 180.
    Assume that $\tan r°$ is defined, so r is not an odd multiple of 90. Then at least one of r and $\tan r°$ is irrational, except if r is an odd multiple of 45 or is divisible by 180.
    The result for the sine follows immediately from part (a) because $\sin r° = \cos(90 - r)°$. This implies the statement for the tangent by the hint to Exercise 9.5.2c.

A.10. Algebraic Number Fields

10.1.

1. In the chain of extensions $L \subseteq F \subseteq M$, one of the two links must have degree 1 by the Tower Theorem 10.1.3.

2. (a) 2; (b) ∞; (c) ∞.

3. (a) One of the directions is obvious, and the other follows from Theorem 9.3.6.
    (b) (b1) 1; (b2) 2; (b3) 2; (b4) 3.

4. (a) True: (a1).

   (b) $m_{\vartheta,M} \mid m_{\vartheta,L}$ and $\deg_M \vartheta \le \deg_L \vartheta$.

10.2.

1. To establish $\mathbf{Q}(\alpha) = \mathbf{Q}(\beta)$, it suffices to verify $\alpha \in \mathbf{Q}(\beta)$ and $\beta \in \mathbf{Q}(\alpha)$ by Theorem 10.2.2.

2. (a) It follows from Theorem 10.2.2.

   (b) $\mathbf{Q}(\alpha)$ is a subspace in the finite dimensional vector space $\mathbf{Q}(\vartheta)$ over $\mathbf{Q}$. In a finite dimensional vector space V, a subspace U satisfies $U = V$ if and only if $\dim U = \dim V$.

   (c) The numbers ϑ and α can be mutually expressed in the form prescribed by Definition 10.2.1 if and only if the given condition holds.

3. True: (b), (d). (When verifying these, do not forget that ϑ can be transcendental.)

4. (a) $12 + 2\sqrt[3]{2} + 9\sqrt[3]{4}$; (b) $\dfrac{1}{2}\sqrt[3]{4}$; (c) $\dfrac{9}{17} - \dfrac{1}{17}\sqrt[3]{2} + \dfrac{2}{17}\sqrt[3]{4}$.

5. (a) 4; (b) 10; (c) 7; (d) 4. Hint: It is worthwhile to operate with the two observations: (i) If α is an element of a finite extension, then $\deg \alpha$ divides the degree of the extension and (ii) If α is algebraic and a number of degree k is an element of $\mathbf{Q}(\alpha)$, then $k \mid \deg \alpha$.

6. (a) $\emptyset$; (b) $\mathbf{Q}(\sqrt[3]{7})$; (c) $\mathbf{Q}(\sqrt{5})$. Hint to (b): Use that $\mathbf{Q}(\sqrt[3]{7})$ is a subset of the intersection and the degree of the intersection (over $\mathbf{Q}$) divides the degrees of both extensions. An alternative approach: Write an element of the intersection both as an element of $\mathbf{Q}(\sqrt[6]{7})$ and as an element of $\mathbf{Q}(\sqrt[9]{7})$ in the form given in Theorem 10.2.3. The two representations can be considered as writing the same element of $\mathbf{Q}(\sqrt[18]{7})$ according to Theorem 10.2.3, and the answer follows from the uniqueness of the representation.

7. (a) $\mathbf{Q}$; (b) $\mathbf{Q}(\sqrt[3]{3})$; (c) $\mathbf{Q}(\sqrt{2})$.

8. Observe that $|\vartheta| = 1$ implies

$$\operatorname{Re} \vartheta = \frac{\vartheta + \overline{\vartheta}}{2} = \frac{1}{2}\left(\vartheta + \frac{1}{\vartheta}\right).$$

(Be aware during the proof that ϑ can be transcendental.)

9. Comparing the extensions and their degrees, we obtain $\mathbf{Q}(\alpha) = \mathbf{Q}(\sqrt[7]{5})$. (Use Theorems 10.2.5 and 10.2.3 and Exercise 10.2.2.)

10. Answer: k and $k/2$ (the latter can occur only for even k). Hint: Apply the Tower Theorem for the chain $\mathbf{Q} \subseteq \mathbf{Q}(\beta^2) \subseteq \mathbf{Q}(\beta)$. (If k is even, exhibit examples to demonstrate that both values can occur.)

11. Answer: ± 1. Hint: As in Exercise 10.2.8, consider the chain $\mathbf{Q} \subseteq \mathbf{Q}(\operatorname{Re} \vartheta) \subseteq \mathbf{Q}(\vartheta)$. Another option: Show that ϑ and $1/\vartheta$ share the same minimal polynomial and 1 or -1 is a root of this minimal polynomial.

12. Parts (a) and (b) follow from the proofs of Theorem 10.2.6 (or 9.3.1) and Theorem 10.2.7 (or 9.3.6).

13. (a) As ϑ is transcendental, $(g_1 h_2 - g_2 h_1)(\vartheta) = 0$ holds if and only if $g_1 h_2 - g_2 h_1 = 0$.

  (b) By part (a),
$$\frac{g(x)}{h(x)} \mapsto \frac{g(\vartheta)}{h(\vartheta)}$$
  is a bijection between the algebraic fractions over $\mathbf{Q}$ and $\mathbf{Q}(\vartheta)$ that preserves the operations.

10.3.

1. (a) Verify as in the proof for $I(\sqrt{2})$ in Theorem 10.3.5 that there is a division algorithm with respect to the absolute value of the norm in $I(\sqrt{3})$.

  (b1) The irreducible factors on the two sides are associates:
$$5 + 3\sqrt{3} = (2 + \sqrt{3})(1 + \sqrt{3}) \quad \text{and} \quad -4 + 3\sqrt{3} = (2 - \sqrt{3})(1 + 2\sqrt{3}).$$

  (b2) Each decomposition contains a reducible factor.

  (c) Apply Theorem 10.3.8. We obtain all (non-associate) primes from the decompositions of the positive prime numbers:
  (c1) $3 = (\sqrt{3})^2$; $2 = \varepsilon(1 + \sqrt{3})^2$, where $\varepsilon = 2 - \sqrt{3}$ is a unit.
  (c2) If $p \equiv \pm 5 \pmod{12}$, then p is a prime.
  (c3) If $p \equiv \pm 1 \pmod{12}$, then p is a product of two non-associate primes.

  (d) If solvable, there are infinitely many solutions, see Exercise 7.8.3. The equation is solvable if and only if every prime number of the form $12k \pm 5$ occurs with an even exponent in the standard form of n and the sum of the exponents of 2, 3, and the primes of the form $12k - 1$ is even. Hint: Use the result of part (c) and follow the proof of the Two Squares Theorem. Every unit has norm $+1$. The reason why we have to examine the exponents of 2, 3, and the prime numbers of the form $12k - 1$ is that the primes in $I(\sqrt{3})$ occurring in their decompositions have negative norms, therefore, if the sum of the exponents is odd, then not n but $-n$ can be written in the form $x^2 - 3y^2$.

2. (a) We can verify as in the Gaussian integers that there is a division algorithm with respect to the norm.

  (b) It follows from Theorem 10.3.8 that all (non-associate) primes are obtained from the decompositions of the positive prime numbers:
  (b1) $2 = -(\sqrt{-2})^2$.
  (b2) If $p \equiv 5$ or $7 \pmod 8$, then p is a prime.
  (b3) If $p \equiv 1$ or $3 \pmod 8$, then p is a product of two non-associate primes.

(c) Answer: $x = \pm 5$, $y = 3$. Hint: The two factors on the left-hand side of $(x + \sqrt{-2})(x - \sqrt{-2}) = y^3$ can share only $\sqrt{-2}$ as a common prime factor, which implies that x is even, but this is impossible by checking the original equation modulo 4. Thus the two factors are coprime, and since the only units are ± 1, which are cubes themselves, each factor must be a cube itself. So we can get the answer by comparing the coefficients of $\sqrt{-2}$ in $x + \sqrt{-2} = (a + b\sqrt{-2})^3$.

3. Consider e.g. the decompositions

    (a) $(1 + \sqrt{15})(1 - \sqrt{15}) = (-2) \cdot 7$

    (b) $(1 + \sqrt{26})(1 - \sqrt{26}) = (-5) \cdot 5$

    (c) $(2 + \sqrt{-6})(2 - \sqrt{-6}) = 2 \cdot 5$

    (d) $(2 + \sqrt{-10})(2 - \sqrt{-10}) = 2 \cdot 7$.

4. We follow the pattern seen for the Gaussian and Eulerian integers (Theorem 7.4.8). If $t \not\equiv 1 \pmod 4$, then the elements of $I(\sqrt{t})$ form a rectangular lattice in the complex plane, where the lengths of the horizontal and vertical sides of the fundamental rectangle are 1 and $\sqrt{|t|}$. The division algorithm requires that every element of $\mathbf{Q}(\sqrt{t})$ falls inside a unit circle around some lattice point. This is satisfied if the circles cover the entire plane, i.e. $\sqrt{|t|} < \sqrt{3}$, which means $t = -1$ or -2. Further, it is definitely not satisfied if a segment on the vertical side bisector of a fundamental rectangle remains uncovered, i.e. $\sqrt{|t|} > \sqrt{3}$, which means $t < -3$ (since $-3 \equiv 1 \pmod 4$, so $t = -3$ does not come up now). We can argue similarly also in the case $t \equiv 1 \pmod 4$. Then we have a parallelogram lattice where the length of the horizontal side of the fundamental parallelogram is 1, the corresponding altitude is $\frac{1}{2}\sqrt{|t|}$, and its foot is the midpoint of the horizontal base.

5. Use Theorem 10.3.8(vii).

6. Show $n^2 + n + k = N(\alpha_n)$ for every $0 \le n \le k - 2$, where α_n is irreducible in $I(\sqrt{-4k + 1})$. Deduce that if $N(\alpha_n)$ were a composite integer for some n, then $N(\alpha_n)$ would have two essentially distinct decompositions into the product of irreducible elements in $I(\sqrt{-4k + 1})$.

7. Irreducibility follows immediately from the properties of the norm. For (b), deduce from the condition that to every $\beta \in I(\sqrt{t})$ there exists an integer b satisfying $\alpha \mid \beta - b$. Then the prime property of α follows from

$$\alpha \mid \beta\gamma \Longrightarrow \alpha \mid bc$$
$$\Longrightarrow \pm p = N(\alpha) \mid b^2 c^2$$
$$\Longrightarrow p \mid b \text{ or } p \mid c$$
$$\Longrightarrow \alpha \mid b \text{ or } \alpha \mid c$$
$$\Longrightarrow \alpha \mid \beta \text{ or } \alpha \mid \gamma.$$

8. Since β^2/α^2 is an algebraic integer, so is its square root β/α. As $\beta/\alpha \in \mathbf{Q}(\sqrt{t})$, $\beta/\alpha \in I(\sqrt{t})$.

9. (a) $5 = -(\sqrt{-5})^2$.

   (b) We saw in the proof of Theorem 10.3.5 that 2 is irreducible and

   $$2 \mid 6 = (1 + \sqrt{-5})(1 - \sqrt{-5}), \quad \text{but} \quad 2 \nmid 1 \pm \sqrt{-5},$$

   thus 2 is not a prime.

   (c) By Theorem 10.3.7, we have to check that $\left(\frac{-5}{p}\right) = -1$ holds exactly for the primes p of the given forms.

   (d) We see from (c) that these are not primes. They are irreducible since such a p (and in general, any integer of the form $10s \pm 3$) cannot be the norm of an element in $I(\sqrt{-5})$.

   (e) Verify that these prime numbers p are the norms of some elements in $I(\sqrt{-5})$, $p = a^2 + 5b^2$ where a and b are integers, since then $p = (a + b\sqrt{-5})(a - b\sqrt{-5})$. Using the solvability of the congruence $x^2 \equiv -5 \pmod{p}$ and following the proof of Theorem 8.2.4 or applying Thue's lemma in Exercise 7.5.21a, show that a small multiple of p can be written in the form $a^2 + 5b^2$, and deduce that this holds also for p itself.

10.4.

1. (a) $\pm\sqrt{2} \pm \sqrt{3}$

   (b) $\sqrt{2}(\pm 1 \pm i)$

   (c) $\cos 20°$, $\cos 140°$, $\cos 260°$

   (d) $\cos k° + i \sin k°$, where $1 \le k \le 360$, k is an integer and $(k, 360) = 1$.

2. (a) Applying Viète's formula for the sum of the roots of the minimal polynomial of ϑ, we obtain that $\vartheta_{(1)} + \vartheta_{(2)}$ is rational, so $\vartheta_{(2)} \in \mathbf{Q}(\vartheta_{(1)})$.

   (b) There exists a real $\vartheta_{(j)}$, thus $\mathbf{Q}(\vartheta_{(j)}) \subseteq \mathbf{R}$, so $\mathbf{Q}(\vartheta_{(j)}) \ne \mathbf{Q}(\vartheta)$.

   (c) In the chain $\mathbf{Q} \subseteq \mathbf{Q}(\vartheta_{(j)}) \cap \mathbf{Q}(\vartheta_{(k)}) \subseteq \mathbf{Q}(\vartheta_{(j)})$, the product of the degrees of the two links is 3. Therefore it suffices to prove that any two extensions $\mathbf{Q}(\vartheta_{(j)})$ are distinct. Show that if two of the three extensions $\mathbf{Q}(\vartheta_{(j)})$ coincide, then the third must be equal to them. This, however, contradicts part (b).

3. (We abbreviate the relative conjugates by R.C.)

   (a) R.C.: $1 \pm \sqrt[4]{2}$, $1 \pm i\sqrt[4]{2}$, $N(\alpha) = -1$.

   (b) R.C.: $1 \pm \sqrt{2}$ with double multiplicity, $N(\beta) = 1$.

   (c) R.C.: $(1 \pm \sqrt[4]{2})(1 + \sqrt{2})$, $(1 \pm i\sqrt[4]{2})(1 - \sqrt{2})$, $N(\gamma) = -1$.

4. Adapt the proof of Theorem 10.3.4. (Be careful: the relative conjugates of ε are generally outside $\mathbf{Q}(\vartheta)$, but their product divided by ε is inside.)

5. (a) For example, $(3 + 4i)/5$ is suitable.

(b) Let the quadratic field be of the form $\mathbf{Q}(\sqrt{t})$, where $t \neq 1$ is a squarefree integer, and let $p > 2$ be a prime satisfying $\left(\frac{t}{p}\right) = 1$. Then the congruence $x^2 - t \equiv 0 \pmod{p}$ is solvable, and so the same is true for $x^2 - t \equiv 0 \pmod{p^2}$. Let c be a solution. Then $(c + \sqrt{t})/p$ is not an algebraic integer, but its norm is an integer.

10.5.

1. (a) -4; (b) -3; (c) -108; (d) $2^{n-1} n^n (-1)^{(n-1)(n-2)/2}$.

   Hint for (d): The discriminant is the square of a Vandermonde determinant that should be computed by the usual row-column multiplication of matrices.

2. (a) Then

   $$\begin{pmatrix} \beta_1 \\ \beta_2 \\ \vdots \\ \beta_n \end{pmatrix} = C \begin{pmatrix} \omega_1 \\ \omega_2 \\ \vdots \\ \omega_n \end{pmatrix}$$

   where the elements of matrix C are integers, thus

   $$\Delta(\beta_1, \ldots, \beta_n) = \Delta(\omega_1, \ldots, \omega_n)(\det C)^2$$

   by Theorem 10.5.3(iii).

   (b) By part (a), we obtain each discriminant by multiplying the other one by a positive integer.

3. The discriminant of $\mathbf{Q}(\sqrt{t})$ (with a squarefree integer $t \neq 1$) is $4t$ if $t \not\equiv 1 \pmod 4$, and t if $t \equiv 1 \pmod 4$.

4. Argue as in Exercise 10.5.2.

5. (a) The condition is $a, b, c, d \in \mathbf{Z}$ and

   $$\left| \begin{smallmatrix} a & b \\ c & d \end{smallmatrix} \right| = \pm 1$$

   (b) Let $\omega = \cos(2\pi/3) + i \sin(2\pi/3)$. The Eulerian rationals $a + b\omega$ and $c + d\omega$ form an integral basis if and only if a, b, c, d satisfy the same condition as in part (a).

6. These are the fields $\mathbf{Q}(\sqrt{t})$, where $(t \neq 1$ is a squarefree integer and) $t \equiv 1 \pmod 4$.

7. It follows directly from the definition of discriminant. Another option: Apply Theorem 10.5.3(iii) for $\alpha_i = \vartheta^{i-1}$, and note that $\Delta(1, \vartheta, \ldots, \vartheta^{n-1})$ is the square of a Vandermonde determinant with real generators.

8. (a) Take, for example, $1/2$ and $2i$ in $\mathbf{Q}(i)$.

   (b) $\Delta(r_1 \alpha_1, \ldots, r_n \alpha_n) = \Delta(\alpha_1, \ldots, \alpha_n)$ if the product of the rational numbers $r_1, \ldots, r_n$ is 1.

A.11. Ideals

11.1.

1. (a) (2)

    (b) $(1 + i)$

    (c) Not an ideal

    (d) $(1 + i)$

    (e) Not an ideal

    (f) (7).

2. (a) $(2x - 1)$

    (b) $([x^2 - 2][x^2 - 3])$

    (c) Not an ideal

    (d) $(x - 3, 2)$

    (e) Not an ideal.

3. Let R be a field and $I \neq 0$ an ideal in R. We have to show $I = R$. If $a \neq 0$ is an element of I and b is an element of R, then $c = b/a \in R$, so $ca = b$, $b \in I$, thus $I = R$. For the converse, pick an element $a \neq 0$ in R. Then, by the condition, $(a) = R$, thus $b \in (a)$ for every $b \in R$. This means $ca = b$ for some $c \in R$, so division works and R is a field.

4. Let $I = (\xi_1 2^{1/k_1}, \dots, \xi_n 2^{1/k_n})$, where $\xi_1, \dots, \xi_n \in U$. Then every element in I is of the form $\xi 2^{1/m}$, where $\xi \in U$ and $m = \text{lcm}[k_1, \dots, k_n]$. Since $2^{1/(m+1)}$ is not of this form, $I \neq K$, and K cannot be generated by finitely many elements.

5. Show that the generators of one of the ideals can be expressed with the help of generators of the other ideal, and vice versa.

6. (a) (a1): 4 (a2): 9 (a3): 5.
        Field: (a2), (a3).

    (b) If $\alpha \neq 0$, then $G/(\alpha)$ has $N(\alpha)$ elements, and $G/(\alpha)$ is a field if and only if α is a Gaussian prime. Hint: To determine the number of residue classes modulo α, see the hint for Exercise 7.7.12. To characterize the fields, argue as we proved that $\mathbf{Z}/(m)$ is a field if and only if m is a prime number (Theorem 2.8.4; we have to check, of course, that all necessary preliminary theorems can be adapted from integers to Gaussian integers).

7. (a) Proceed as in the proof that $(2, x)$ is not a principal ideal in $\mathbf{Z}[x]$ (see the paragraph about E3 before Definition 11.1.4).

    (b) (b1): 2, it is a field.   (b2): 6, it is not a field.   (b3): 121, it is a field.

8. (a) Field: (a2).

    (b) $F[x]/(g)$ is a field if and only if g is irreducible over F.

(c) The factor ring has four elements (the residue classes can be represented by the remainders $a_0 + a_1 x$, where $a_i = 0$ or 1), and we can easily check that the three non-zero elements have inverses. Another option: The factor ring is isomorphic to $S = \mathbf{Z}_2[x]/(x^2 + x + 1)$, where $\mathbf{Z}_2$ is the field of residue classes modulo 2, and S is a field by part (b).

9. (a) Follow the proof of the special case $\vartheta = \sqrt{2}$ seen in the Example after Theorem 11.1.6. The key observation is that each residue class of $\mathbf{Q}[x]$ modulo the principal ideal (m_ϑ) can be uniquely characterized by the common remainder of the polynomials in the class on division by m_ϑ, and the only computational rule for the remainders is that the multiples of m_ϑ do not count. This corresponds perfectly to the usual representation of the elements in $\mathbf{Q}(\vartheta)$ and to the computation method there that uses only $m_\vartheta(\vartheta) = 0$. An alternative approach: The map $f \mapsto f(\vartheta)$ from $\mathbf{Q}[x]$ onto $\mathbf{Q}(\vartheta)$ is a ring homomorphism with image $\mathbf{Q}(\vartheta)$ and kernel (m_ϑ). Thus the statement follows from the homomorphism theorem for rings.

(b) By part (a), let $M = L[x]/(f)$. The irreducibility of f implies that M is a field, the set of residue classes constant$+(f)$ corresponds to L^*, and the residue class $x + (f)$ plays the role of ϑ.

10. (a) It is sufficient to verify the statement for principal ideals since $\alpha \in I$ implies $(\alpha) \subseteq I$, and so the number of residue classes modulo I is less than or equal to the number of residue classes modulo (α). Let $\alpha \neq 0$, and we show that there are only finitely many remainders modulo α. Let $\omega_1, \ldots, \omega_n$ be an integral basis in $I(\vartheta)$. Then every $\xi \in I(\vartheta)$ can be written as $\xi = k_1 \omega_1 + \cdots + k_n \omega_n$, where $k_i \in \mathbf{Z}$, $i = 1, \ldots, n$. Since $\alpha \mid N(\alpha)$, every residue class modulo α has a representative ξ satisfying $0 \le k_i < |N(\alpha)|$, $i = 1, \ldots, n$.

(b) The numbers of elements of the factor rings R/A_j form a strictly decreasing sequence. This is impossible, however, as $A_2 \neq 0$ and so R/A_2 has only finitely many elements.

(c) If an ideal $I \neq 0$ were not finitely generated, then it would contain a strictly increasing chain of ideals

$$(a_1) \subset (a_1, a_2) \subset (a_1, a_2, a_3) \subset \ldots.$$

11.2.

1. (a) (5)

   (b) (60).

2. (a) 4

   (b) 16.

3. Rephrase the statement with divisibility according to Theorem 11.2.1.

4. (a) Both 2 and $1 + \sqrt{-5}$ are common divisors, but there is no common multiple of them among the common divisors.

   (b) $(2), (1 + \sqrt{-5}), (1)$.

   (c) For example, $\alpha = 2, \beta = 1 + \sqrt{-5}$.

11.3.

1. (a) They have minimal polynomials with integer coefficients, leading coefficient 1, and constant term 1 or -1 (see Exercise 9.6.7).

   (b) $\alpha = \sqrt{\alpha}\sqrt{\alpha}$.

2. (a) $a \neq 0$ and b is arbitrary or $a = b = 0$.

   (b) Every $a \neq 0$ is a unit, thus there are no irreducible or prime elements.

   (c) Fundamental Theorem: It is an empty statement, as it refers to elements different from 0 and units. Principal ideal domain: A field contains only the trivial ideals (0) and (1) (see Exercise 11.1.3), and these are principal ideals. Euclidean ring: As division can be performed, we can always achieve a zero remainder (and so any function can be chosen as f).

3. (a) Only 2 is irreducible.

   (b) The procedure yields a unit that has no irreducible divisors.

   (c) We can construct a suitable f as in the hint to Exercise 1.5.5c.

   (d) $(0), (1), (2), (2^2), (2^3), \ldots$

4. We have to check the requirements of Definition 11.1.1.

5. Hint for necessity: If $R[x]$ is a principal ideal domain, then (a, x) is a principal ideal for every (non-zero) constant polynomial a.

6. Show that if $f(c)$ is minimal for an element $c \neq 0$, then c is a unit. So $c \mid c$, or $ec = c$ for some e. Applying the lack of zero divisors, show that e is an identity element.

7. True: (a).

8. The division algorithm with remainders of least absolute value satisfies the condition. If ($b \neq 0$ and) $a = bq + r$, where $|r| \leq |b|/2$, then $f(r) < f(b)$.

9. (a) Prove first that every ideal in R is finitely generated, and then show $(a, b) = (d)$, where $d = \gcd\{a, b\}$.

   (b) It follows from part (a) by Exercise 11.1.10a.

10. For sufficiency, see Exercise 10.3.4 and the hint for it. For necessity, assume that $I(\sqrt{t})$ is a Euclidean ring for some $t < -3$ and take an element $\beta \neq 0, \pm 1$ for which $f(\beta)$ is minimal. Verify $N(\beta) \leq 3$. This implies $t = -7$ or $t = -11$ (for $t < -3$).

11.4.

1. (a) H is not an ideal if A and B are the ideals in Examples E1 or E2 before Definition 11.4.3: e.g. $2 \cdot 3 + [x+3][x-2] = x^2 + x$ and $3 \cdot 3 - [1+\sqrt{-5}][1-\sqrt{-5}] = 3$, are not of the form ab.

   (b) If $A = (\alpha)$, then
   $$\sum_{i=1}^{n} a_i b_i = \sum_{i=1}^{n} [r_i \alpha] b_i = \alpha \sum_{i=1}^{n} r_i b_i = \alpha b.$$

2. (a), (b) The proof is the same as for integers.

   (c) It follows from (iv) of Theorem 11.4.2.

   (d) Apply part (c).

   (e) To prove necessity, let $A = (1)$, and use part (c).

3. (a) If $\alpha = \beta \gamma$, then the proof of (iii) in Theorem 11.4.2 yields $(\alpha) = (\beta)(\gamma)$. Conversely, if $(\beta)C = (\alpha)$, then by the hint to Exercise 11.4.1b, every element of $(\beta)C$, thus also α is divisible by β.

   (b) $\gamma = \alpha/\beta$ obtained in part (a) meets the requirements.

4. (a) We have to verify that D is an ideal, D contains A and B, and if some ideal C contains A and B, then $D \subseteq C$.

   (b) Negative: $A \subseteq A + B = (0) \implies A = (0)$.

   (c) The elements of the ideal $A(B,C)$ are of the form $\sum_{i=1}^{n} a_i[b_i + c_i]$, so removing the brackets, we get elements of (AB, AC). For the other inclusion, the elements of (AB, AC) can be written as
   $$\sum_{i=1}^{n} a_i b_i + \sum_{j=1}^{k} a'_j c_j = \sum_{i=1}^{n} a_i[b_i + 0] + \sum_{j=1}^{k} a'_j[0 + c_j],$$
   so they are in $A(B,C)$, too.

5. The least common multiple of A and B is a common multiple that divides all common multiples. We can rephrase this by (11.4.5) to inclusion: M is the largest ideal that is a subset of both A and B, i.e. $M = A \cap B$.

6. For example, $A = (x)$ and $B = (2, x)$ in $\mathbf{Z}[x]$.

7. Prove (11.4.12)$\Rightarrow$(11.4.11) by contradiction. For the converse, apply (11.4.11) for $A = (a)$ and $B = (b)$.

8. (a) (a1): It has only the two trivial divisors.

      (a2): The only non-trivial divisor is the ideal in (a1).

      (a3): It has two non-trivial divisors: $(2, 1 + \sqrt{-5})$ and $(3, 1 + \sqrt{-5})$.

   (b) (b1): $(2, 1 + \sqrt{-5})$

      (b2): (1).

   (c) Irreducible: (c1), (c3).

9. True: (b), (c), (d).

10. (a) $A = (2, x)$, $B = (4, x^2)$, $C = (4, 2x, x^2)$.

   (b) By Exercise 11.4.1b, $(\alpha)B = \{\alpha b \mid b \in B\}$, $(\alpha)C = \{\alpha c \mid c \in C\}$, and since $\alpha \neq 0$ and R has no zero divisors, we have $\alpha b = \alpha c \Rightarrow b = c$.

11. (a) Check the requirements in the definition of ideals.

   (b) Verify that I is a maximal ideal, i.e it satisfies (11.4.9). It follows that I cannot have other decompositions than the three listed in the exercise. To show $I \cdot I = I$, use $x^\alpha = x^{\alpha/2} x^{\alpha/2}$.

12. (a) If $P = AB$, then $P \subseteq A$ and $P \subseteq B$. Further (11.4.11), equivalent to (11.4.12), implies $A \subseteq P$ or $B \subseteq P$. Therefore $P = A$ or $P = B$.

   (b) For example, $(4, x)$ in $\mathbf{Z}[x]$.

   (c) Assume that M is a maximal ideal, and there exist $a \notin M$ and $b \notin M$ satisfying $ab \in M$. Let (a, M) and (b, M) be the smallest ideals containing a and b, besides M. Since M is maximal, we have $(a, M) = (b, M) = R$. Then

   $$R = RR = (a, M)(b, M) \subseteq (ab, M) = M,$$

   a contradiction.

   (d) For example, (x) in $\mathbf{Z}[x]$.

   (e) Maximal ideals: Establish a bijection between the ideals of R containing I and the ideals of the factor ring R/I, and apply Exercise 11.1.3. Prime ideals: Rephrase condition (11.4.12) in the terms of the factor ring R/I.

11.5.

1. Both conditions are equivalent to $A \mid (\alpha)$.

2. (a) The quotient $N(\alpha)/\alpha$ is both an algebraic integer and an element in $\mathbf{Q}(\vartheta)$.

   (b) By part (a) (or by Theorem 11.5.5) A contains a non-zero integer, and so its integer multiples are in A. All integers in A are obtained as (integer) multiples of the least such positive integer a, therefore $A = (a)$.

3. It follows from Theorem 11.5.8.

4. (a) There is an integer $c > 1$ in every prime ideal P by Exercise 11.5.2. Factor c into the product of prime numbers. Since P is a prime ideal, it must contain at least one of them. If P contained two distinct positive prime numbers, then also their combination by suitable integers giving 1 would be in P which is impossible.

   (b) It follows from part (a).

   (c) Yes.

   (d) No, by Exercise 11.5.3.

5. Use the properties of greatest common divisors and least common multiples of ideals.

6. Express the greatest common divisor of ideals $(\alpha)^2$ and $(\beta)^2$ in two different ways.

7. (a) $(21) = (3, 4 + \sqrt{-5})(3, 4 - \sqrt{-5})(7, 4 + \sqrt{-5})(7, 4 - \sqrt{-5})$. (Of course, we can describe these prime ideals with other generators, as well, e.g. $(3, 4 + \sqrt{-5}) = (3, 1 + \sqrt{-5}) = (3, 1 - 2\sqrt{-5}) = (2 - \sqrt{-5}, 1 + \sqrt{-5})$, etc.)

   (b) $p = 2$ and 3.

   (c) $p = 2, 5$, and primes of the form $20k + 1$, $20k + 3$, $20k + 7$, and $20k + 9$.

8. Both properties are equivalent to the fact that every ideal in $I(\vartheta)$ is a principal ideal (see Exercise 11.3.9b, and Theorems 11.4.2(iii) and 11.5.8).

11.6.

1. $(2, \sqrt{-6})(3 + \sqrt{-6}) = (3, \sqrt{-6})(2 - \sqrt{-6})$.

2. Both conditions are equivalent to $I(\vartheta)$ being a principal ideal domain (cf. Exercise 11.3.9b).

3. Use the fact that $ku = 1 + hv$ with suitable integers u and v.

4. (a) No solution.

   (b) $x = 0, y = 1$.

   (c) $x = \pm 985, y = 99$.

   (d) $x = \pm 36, y = 11$. Note that $-35 \equiv 1 \pmod 4$, so a and b are not necessarily integers in $a + b\sqrt{-35} \in I(\sqrt{-35})$.

A.12. Combinatorial Number Theory

12.1.

1. (a) $\lfloor n/2 \rfloor + 1$, i.e. $h + 1$, if $n = 2h$ or $2h + 1$. Hint: The integers in the interval $[n/2, n]$ form a suitable set. To prove that no bigger set exists, observe that if $u + v = a_k$ where $0 < u < v$, then at most one of u and v can occur among the integers a_i.

   (b) Given r, let A_r consist of the numbers $a_1 + a_2 + \cdots + a_r + a_s$ where $s = r + 1$, $r + 2, \ldots$ (hence A_0 is the original sequence), and let $A_r(n)$ denote the number of elements in A_r not exceeding n. By the assumption, the sequences A_r are pairwise disjoint, hence

$$n \geq \sum_{i=0}^{t} A_i(n) \geq (t + 1)A_t(n)$$

for any n and t. On the other hand,

$$A_t(n) = A\left(n - \sum_{i=1}^{t} a_i\right) - t \geq A(n) - \sum_{i=1}^{t} a_i - t.$$

From the two inequalities we obtain

$$A(n) \leq \frac{n}{t+1} + \sum_{i=1}^{t} a_i + t.$$

Dividing by n, the right-hand side tends to $1/(t+1)$ as $n \to \infty$, and since t was arbitrary, this proves $A(n)/n \to 0$.

2. The integers not divisible by 3 form such a set of the desired density. To prove that no larger density is possible, observe that any interval $[r, 4r]$ can contain at most $2r + 1$ numbers a_i, because the sums $a_j + a_{j+1}$ where $a_j \leq 2r$ fall into the interval $[2r+1, 4r]$, except perhaps the last one, and by assumption they differ from the numbers a_i in this interval. Thus we obtain the desired result by dividing the interval $[1, n]$ into subintervals of the type $[r, 4r]$.

3. To construct a set of maximal size, take $\lceil k/2 \rceil$ numbers and let t be their sum. If we delete the first two terms and insert the sum of the deleted terms as a new term, then the sum does not change. Repeat the process as long as possible. To prove that no larger number of representations can occur, apply that the last term and the number of terms are different in every representation of t.

4. By the condition, an integer $1 \leq j \leq n$ is divisible by at most one a_i, hence $\sum_{i=1}^{k} \lfloor n/a_i \rfloor \leq n$, and so $n \sum_{i=1}^{k} 1/a_i < n + k$.

5. $\dfrac{1}{[a_i, a_{i+1}]} = \dfrac{(a_i, a_{i+1})}{a_i a_{i+1}} \leq \dfrac{a_{i+1} - a_i}{a_i a_{i+1}} = \dfrac{1}{a_i} - \dfrac{1}{a_{i+1}}.$

6. (a) The integers of the form $3j + 1$ satisfy the condition and this establishes the lower bound. To verify the upper bound, observe that if t^2 is the largest square not exceeding n and $u + v = t^2$, then at most one of u and v can occur among the numbers a_i.

   (b) Find eleven residues modulo 32 such that the sum of no two of them is congruent to a square mod 32.

7. Consider the numbers whose last and every second digit in base 5 are 0 or 2 (the other digits are arbitrary), and prove that they satisfy the condition. Then k is approximately n^c where $c = (1 + \log_5 2)/2 = 0.71\dots$.

8. The primes form a suitable set, hence $s(n) \geq \pi(n)$. To show $s(n) < \pi(n) + 2n^{2/3}$, let C be the set of integers between 1 and $n^{2/3}$ and let D be the union of C and the primes not exceeding n. First verify that every number up to n can be represented as $n = cd$ where $c \in C$, $d \in D$ (the representation is generally not unique). Then fix such a representation $a_i = c_i d_i$ for every a_i, and construct a bipartite graph with $|C| + |D| \leq \pi(n) + 2n^{2/3}$ vertices where the two groups of vertices are C and D, and the number a_i is represented by the edge between vertices c_i and d_i. If the number of edges is not less than the number of vertices, then the graph contains a

circuit. Since the graph is bipartite, the circuit has an even number of edges, and by the construction, the product of numbers a_i corresponding to every second edge is equal to the product of numbers a_i corresponding to the other edges in the circuit (as both products are equal to the product of all numbers appearing in the vertices of the circuit).

9. $\pi(n)$. Hint: The primes clearly satisfy the condition, hence the maximum is at least $\pi(n)$. Assume that there are $\pi(n) + 1$ such numbers a_i. Then for every a_i we can find a prime that occurs in the standard form of a_i with a larger exponent than in the standard form of all other numbers a_j. By the pigeonhole principle, there must be a prime that plays this role for two different numbers a_i which is a contradiction.

10. $2n/3$. Hint: The $2n/3$ numbers not relatively prime to 6 (that is, those that are divisible by at least one of 2 and 3) satisfy the condition. If we pick more than $2n/3$ elements, then by the pigeonhole principle there must be an s for which at least five a_i occur among the numbers $6s + 1, \ldots, 6s + 6$. Show that there must be three of them that are pairwise relatively prime.

    Remark: We can generalize the exercise, replacing three by r: Determine the maximum of k if among any r numbers a_i there must be two that are not coprime. For example, the numbers divisible by at least one of the first $r - 1$ primes form such a set. (Why?) Erdős conjectured that this set yields the maximum (for every n large enough compared to r). This long-standing unsolved problem was finally solved by Ahlswede and Khachatrian in 1994.

11. Dividing by the gcd of the integers a_i, we can assume that they are relatively prime. If some of them are divisible by k, e.g. $k \mid a_i$ and $k \nmid a_j$, then $k \mid \dfrac{a_i}{(a_i, a_j)}$ (since k is a prime) and $\dfrac{a_i}{(a_i, a_j)} \geq k$. If no a_i is divisible by k, then there are two of them, say a_i and a_j, that are congruent mod k. Hence $\dfrac{a_i}{(a_i, a_j)} \equiv \dfrac{a_j}{(a_i, a_j)} \pmod{k}$, thus the larger of the two quotients must be greater than k.

12. Let $a_1, \ldots, a_k$ be a suitable set for $n = 2^j$. Then the set $1, 2, \ldots, 2^{t-1}, 2^t a_1, \ldots, 2^t a_k$ will work for $2^{j+t} \leq n < 2^{j+t+1}$.

13. The optimal choice is $c = \sqrt{3}$ in Chebyshev's inequality. Then we can replace 8/3 in (12.1.9) and (12.1.10) by $3\sqrt{3}/2$. A further improvement is possible if we replace (12.1.6) by a better estimate: (12.1.10) implies $k \leq (1 + \varepsilon) \log_2 n$ with an arbitrarily small $\varepsilon > 0$ for n large enough, hence the term 1 on the right-hand side of (12.1.7) can be nearly omitted. In total, this means that 2 at the end of (12.1.2) can be replaced by any constant larger than $\log_2(3\sqrt{3}/2) = 1.377\ldots$ for n large enough.

12.2.

1. We apply the greedy algorithm, and always pick the first element which does not ruin the Sidon property. Assume that we have already chosen $a_1 < a_2 < \cdots < a_s < n$. We cannot choose d as a_{s+1} if $d + a_i = a_j + a_k$, or $d = a_j + a_k - a_i$ for some i, j, $k \le s$. (The case $d + d = a_j + a_k$ cannot occur, since then $d < a_k$ and so we would have chosen d in the sequence earlier, instead of a_k.) This excludes at most s^3 (in fact, less than $s^3/2$) elements. This means that if $s < n^{1/3}$, then we can still find a new element $a_{s+1} \le n$.

2. To verify the Sidon property, assume $a_i + a_j = a_k + a_l$, so

$$2p(i + j - k - l) + (\langle i^2 \bmod p \rangle + \langle j^2 \bmod p \rangle - \langle k^2 \bmod p \rangle - \langle l^2 \bmod p \rangle) = 0.$$

The second term is divisible by $2p$ and has absolute value less than $2p$, so it must be 0. Then the first term is also 0. This means $i - k = l - j$ and $i^2 - k^2 \equiv l^2 - j^2$ (mod p). A calculation shows that either $i = k$ and $j = l$, or $i = l$ and $j = k$.

3. Apply a simplified version of the proof of Theorem 12.2.2 for the field of p^2 elements and its subfield of p elements.

4. Let g be a primitive root modulo p, and let a_i be the solution of the system of congruences $x \equiv i$ (mod $p - 1$), $x \equiv g^i$ (mod p) modulo $p(p - 1)$, $i = 1, 2, \ldots, p - 1$.

5. We can take a Sidon set S_1 between 1 and n_1 having about $\sqrt{n_1}$ elements by Theorem 12.2.1. Let n_2 be much larger than n_1. We leave the interval $(n_1, n_1 + n_2]$ empty, choose a Sidon set in the interval $(n_1 + n_2, n_1 + 2n_2]$ of about $\sqrt{n_2}$ elements, delete (at least one member of) those pairs whose difference is less than $< n_1$, and denote the remaining set by S_2. By the Sidon property, we deleted fewer than $2n_1$ elements. Therefore we selected about $\sqrt{n_2} + \sqrt{n_1} - 2n_1 \approx \sqrt{n_2}$ elements up to $n_1 + 2n_2$. Verify that $S_1 \cup S_2$ is a Sidon set. Choose an n_3 much bigger than $n_1 + 2n_2$, place a Sidon set of size about $\sqrt{n_3}$ between $n_1 + 2n_2 + n_3$ and $n_1 + 2n_2 + 2n_3$, delete the elements with differences less than $n_1 + 2n_2$, etc. Continuing the procedure we obtain an infinite Sidon set meeting the requirements.

6. (a) Generalize the method of Exercise 12.2.3 to the field of p^h elements.

   (b) The h-fold sums are all distinct and fall between 1 and nh.

7. It is sufficient to prove that every positive integer has a unique representation as $a_i - a_j$ with $i > j$. We always define two new elements of the sequence. They should be big enough to avoid that their differences with previously constructed elements should coincide with differences of two previously constructed elements, and the difference of these two elements should be the smallest positive integer that has not yet appeared as a difference of two elements.

8. Let A and B consist of the numbers which have 0 digits at every odd or even place in their binary representation counted backwards.

12.3.

1. (a) Let $A = \{a_1 < a_2 < \cdots < a_k\}$. Then $a_1 + a_1 < a_1 + a_2 < a_2 + a_2 < a_2 + a_3 < \cdots < a_k + a_k$ are $2k - 1$ distinct sums. If $|A + A| = 2k - 1$, then every $a_i + a_j$, thus $a_i + a_{i+2}$ is among the above sums, and comparing magnitudes yields that it can only equal $a_{i+1} + a_{i+1}$, so $a_{i+1} = (a_i + a_{i+2})/2$.

   (b) If $A = \{a_1 < a_2 < \cdots < a_k\}$, $B = \{b_1 < b_2 < \cdots < b_r\}$, and $k \geq r$, then $a_1 + b_1 < a_1 + b_2 < a_2 + b_2 < a_2 + b_3 < \cdots < a_r + b_r < a_{r+1} + b_r < \cdots < a_k + b_r$ are $k + r - 1$ distinct sums. In the case of equality, every other $a_i + b_j$ coincides with one of the above sums. By estimating magnitudes, we can easily identify the sums $a_2 + b_1, a_1 + b_3, a_3 + b_2$, etc. We obtain that B and the first r elements of A form arithmetic progressions with the same difference. We can extend this to any consecutive r elements of A by modifying the initial sequence of $k + r - 1$ sums suitably.

   (c) Prove by induction on t.

2. Delete from B all non-zero elements that are not coprime to m, and follow the first proof of Theorem 12.3.1. To show that the estimate is sharp, consider e.g. $m = p^2$, $A = \{0, p, 2p, \ldots, (p-1)p\}$, and $B = \{0, p, 2p, \ldots, 1, p+1, 2p+1, \ldots, (p-1)p+1\}$.

3. (a) Follow the first proof of Theorem 12.3.1. We have to use the condition when showing the impossibility of $A + b = A$ for $b \neq 0$.

   (b) We have equality e.g. for $A = \{0, 1, \ldots, k-1\}$ and $B = \{0, 1, \ldots, r-1\}$ (where $k + r \leq m + 1$).

   (c) The same proof applies also for the general case.

4. We can argue as in the second proof of Theorem 12.3.1: let $A = B$, $C = A \hat{+} A$, and

$$f_1(x, y) = (x + y)^m (x - y)^2 \prod_{c \in C}(x + y - c),$$

   where $m + |C| = 2k - 4$.

5. (a) As in the second proof of Theorem 12.3.1, reduce the terms $x^i y^j$ where $i \geq k$ or $j \geq r$, and apply Lemma 12.3.2.

   (b) Let $|A_i| = k_i$, $i = 1, \ldots, n$, and let $G(x_1, \ldots, x_n)$ be a polynomial over F in n variables and of degree $\sum_{i=1}^{n}(k_i - 1)$. Assume that the coefficient of the term $\prod_{i=1}^{n} x_i^{k_i - 1}$ is not zero. Then $G(a_1, \ldots, a_n) \neq 0$ for some $a_i \in A_i$, $i = 1, \ldots, n$.

6. If $C = D = \mathbf{Z}_p$, then $c = d$ works since $2u = 2v$ in $\mathbf{Z}_p$ implies $u = v$ as p is odd. If $|C| = |D| = n < p$, then apply Exercise 12.3.5b for $A_1 = \cdots = A_n = D$ and

$$G(x_1, \ldots, x_n) = \prod_{1 \leq j < i \leq n}(x_i - x_j)(x_i + c_i - x_j - c_j).$$

7. Let p be a prime and $A_i \subseteq \mathbf{Z}_p$, $i = 1, \ldots, n$. Then

$$|A_1 + \cdots + A_n| \geq \min(p, |A_1| + \cdots + |A_n| + 1 - n).$$

This follows from Theorem 12.3.1 by induction on n.

8. (a) We have to show that among any $2n - 1$ integers, there exist n such that their sum is a multiple of n. As seen in Exercise 3.6.6, it is sufficient to prove this when n is a prime p. We can assume $0 \leq a_1 \leq a_2 \leq \ldots \leq a_{2p-1} \leq p - 1$. If there are p equal numbers a_i, then their sum is divisible by p. Otherwise, switching to $\mathbf{Z}_p$, let $A_i = \{a_i, a_{i+p-1}\}$, $i = 1, \ldots, p - 1$, then $|A_i| = 2$. By Exercise 12.3.7, $|A_1 + \cdots + A_{p-1}| = p$, so every element in $\mathbf{Z}_p$, thus a_{2p-1}, can be written as $a^{(1)} + \cdots + a^{(p-1)}$, where $a^{(i)} \in A_i$, so $a^{(1)} + \cdots + a^{(p-1)} + a_{2p-1}$ is a multiple of p.

   (b) The midpoint of lattice points P and Q is a lattice point if and only if both the first and second coordinates of P and Q have the same parity. By the pigeonhole principle, among any five lattice points there must be two with this property.

   (c) Take $n-1$ lattice points of each type where the coordinates modulo n are $(0, 0)$, $(0, 1)$, $(1, 0)$, and $(1, 1)$. We cannot select n out of these $4n - 4$ lattice points so that the averages of both the first and second coordinates are integers.

   (d) (i) The lower bound can be verified by generalizing the construction in (c). The upper bound follows from the pigeonhole principle since among that many lattice points there are always n such that considering any coordinate they are congruent modulo n. (ii) Argue similarly as we showed in Exercise 3.6.6 that if the statement there is valid for two integers, then it is true also for their product.

9. Let $|A| = k$, c a quadratic non-residue mod p, and consider the k^2 sums $a_i + ca_j$. If $k^2 > p$, then two sums must be equal, which yields $a_i - a_r = c(a_s - a_j)$. Then (exactly) one of $a_i - a_r$ and $a_s - a_j$ is a quadratic residue mod p.

10. Generalize the observations before Theorem 12.3.3.

12.4.

1. The last three equalities are obvious, and we have proved $R(3, 2) \leq 6$. Thus we have to show that we can color the edges of a complete graph of five vertices with two colors so that no monochromatic triangle arises. Coloring the sides and diagonals of a pentagon red and blue, resp., meets this requirement.

2. In part I of the proof of Theorem 12.4.1 we verified

(A.12.1) $$R(3, t) \leq t(R(3, t - 1) - 1) + 2.$$

   This implies $R(3, t) \leq tR(3, t - 1)$, and we get (a) by induction. We can prove also the sharper statement (b) by induction if we use (A.12.1) and

   $$\lceil et! \rceil = t! \left(1 + \frac{1}{1!} + \frac{1}{2!} + \cdots + \frac{1}{t!}\right) + 1$$

   obtained from the infinite series expansion of e.

3. (a) Combine $S(t) < R(3, t)$ (see the proof of Theorem 12.4.2) and part (b) in the previous exercise.

(b) Take a bad coloring of the integers $1, 2, \ldots, n = S(t)$ with t colors, one where the equation $x + y = z$ has no monochromatic solution, color each of the numbers $n + 1, \ldots, 2n + 1$ with the $(t + 1)$st color and repeat the coloring of the first n numbers for $2n + 2, \ldots, 3n + 1$ (i.e. $2n + 1 + i$ has the same color as i). Show that this is a bad coloring of the integers $1, 2, \ldots, 3n + 1$ with $t + 1$ colors.

(c) Prove by induction using part (b).

(d) We generalize the construction in (b). Let ν be a bad coloring of $1, \ldots, n = S(t)$ with t colors, and ϱ a bad coloring of $1, \ldots, r = S(v)$ with v other colors. Then we can obtain a bad coloring of $1 \leq m \leq 2nr + n + r$ with $t + v$ colors: Write m as $m = i(2n + 1) + j$, where $1 \leq j \leq 2n + 1$, and let the color of m be $\nu(j)$ or $\varrho(i)$ according as $1 \leq j \leq n$ or $n + 1 \leq j \leq 2n + 1$ (i.e. we repeat the coloring of $1, 2, \ldots, n$ defined by ν in the first halves of the intervals of length $2n + 1$, and the elements in the second halves of the intervals uniformly get the color of the serial number of the interval in the coloring of $1, 2, \ldots, r$ defined by ϱ).

4. $5n - 1$.

5. Apply the proof of Theorem 12.4.2 with $R(4, t)$ instead of $R(3, t)$.

6. If $B^t + C^t \equiv D^t \pmod{p}$ for some $BCD \not\equiv 0 \pmod{p}$ and $CF \equiv 1 \pmod{p}$, then $(BF)^t + 1 \equiv (DF)^t \pmod{p}$.

7. (a) Use longer and longer red and blue intervals.

(b) We order all arithmetic progressions into one sequence and color an element blue in each progression one after the other so that the next blue number is at least the double of the previous one. A more concrete construction: Just the integers $n! + n$ are blue. Then every arithmetic progression $a + md, m = 1, 2, \ldots$ contains a blue number since for $n = a + d$ we have $(a + d)! + a + d \equiv a \pmod{d}$. Therefore there result no infinite red arithmetic progressions, and as the blue numbers grow very quickly, they cannot even form a three-term arithmetic progression.

8. If $m = w(k, t) + 1$, then we get a k-term monochromatic arithmetic progression (k-MCAP) less than m. Consider the integers $m, 2m, \ldots, (m - 1)m$ and apply Theorem 12.4.4A again (in fact, we color the multipliers of m). Then we get a new k-MCAP of multiples of m not exceeding $(m - 1)m$, etc. Among these infinitely many k-MCAP there are infinitely many of the same color since the number of colors is finite.

9. Apply Van der Waerden's Theorem for the exponents of powers of two.

10. RRBBRRBB shows that eight numbers do not suffice. To prove the sufficiency of nine numbers, we must distinguish a few cases. It is worthwhile to rely on symmetry (of numbers and colors): we may assume that 5 is red, 1 is blue, and 9 is either red, or blue, then we consider the colors of 3 and 7, etc.

11. (a) There are 2^n colorings of $1, 2, \ldots, n$ with two colors. We estimate the number of colorings containing a k-term monochromatic arithmetic progression (k-MCAP). Counting by the first terms and differences, there are at most

$n^2/2(k-1)$ such k-term arithmetic progressions, each can have two colors, and we can color the other numbers in 2^{n-k} ways. Therefore, altogether at most $n^2 2^{n-k}/(k-1)$ colorings may contain a k-MCAP (we counted some bad colorings several times, of course). Thus, if $n^2 2^{n-k}/(k-1) < 2^n$, so $n < 2^{k/2}\sqrt{k-1}$, then there must be a coloring without a k-MCAP.

(b) Consider a finite field F with 2^p elements, let Δ be a generator of its multiplicative group, and W a $(p-1)$-dimensional subspace in F (considered as a vector space over $\mathbf{Z}_2$). We color k red if and only if $\Delta^k \in W$. In this coloring of $1, 2, \ldots, p(2^p - 1)$, there is no $p+1$-MCAP.

12. We use the number system with base d where we shall specify d later. Consider those positive integers up to n where every digit is less than $d/2$ and the sum of the squares of digits is a given q. Show that such a set contains no three-term arithmetic progression, and we can choose q and d so that the set should be as large as required in the exercise.

12.5.

1. A residue class $a_i \pmod{m_i}$ contains M/m_i numbers from $1, 2, \ldots, M = [m_1, \ldots, m_k]$, and as every integer is contained in at least one residue class, we have $\sum_{i=1}^{k} M/m_i \geq M$.

2. The old residue class is a subset of the new one.

3. Let m_i be arbitrary and let L denote the least common multiple of the other moduli m_j. By the condition, there is an integer c contained in none of the residue classes with $j \neq i$. Then this holds for $c + L$, too. Therefore, both integers must be in the residue class $a_i \pmod{m_i}$, so $c \equiv c + L \pmod{m_i}$ and $m_i \mid L$.

4. Rely on the previous three exercises.

5. For example, choose the divisors of 120 greater than 2 as moduli.

6. (a) Proceed similarly as in Exercise 12.5.1.

   (b) It follows from either proof of Theorem 12.5.1.

   (c) Let $m_i = 2^i$ for $1 \leq i \leq k-1$ (and $m_k = m_{k-1}$).

7. For example, the even multiples of 3 have no such representation except for the numbers of the form $3^n + 3$. If we also consider 1 as a power of 3, then we have to exclude the numbers of the form $1 + p$, where p is a prime (of the form $6k - 1$), but there still remain infinitely many non-representable numbers (as the primes occur rarely). We can proceed similarly in the general case replacing 3 by a and $b/2$. (This shows that $b = 2$ was the only difficult case, see Theorem 12.5.2.)

12.6.

1. The precise formulation: Decompose the set of non-negative integers into the disjoint union of two arbitrary infinite subsets I and J, and write an integer $n > 0$ in the number system with base c: $n = \sum_{v=0}^{V} \gamma_v c^v$, $0 \le \gamma_v < c$. Let $A = \{n \mid \gamma_i = 0 \text{ for } i \in I\}$ and $B = \{n \mid \gamma_j = 0 \text{ for } j \in J\}$. These are complements as every positive integer has a representation in the number system with base c. Every such construction satisfies $\liminf_{n\to\infty} A(n)B(n)/n = 1$ (but we can easily check $\limsup_{n\to\infty} A(n)B(n)/n > 1$).

2. No, this follows from Theorem 12.5.2.

3. (a) Necessary and sufficient. (b) Sufficient, but not necessary, see e.g. the red set in Exercise 12.4.7b. (c) Necessary, but not sufficient. (d) Neither necessary, nor sufficient.

4. We can proceed as in the proof of Theorem 12.6.1. Since $a_t \equiv t \pmod{2^i 3^j}$ for $i, j \le t$, the numbers a_{k-s} for $\log_6 k + 1 \le s \le \log_6 k + d_k$ form a complete residue system mod $d_k = 2^i 3^j$ if $d_k < k - 5 \log_6 k$. We can guarantee the conditions $d_k \sim k$ and $d_k \le d_{k+1}$ (needed to estimate the number of elements), since if we order the integer $2^i 3^j$ into an increasing sequence, then the quotient of consecutive elements tends to 1 because the fractional parts of the values $\log_2(2^i 3^j)$ are dense in $[0, 1]$ by Theorem 8.4.1.

   If $a_k \le n < a_{k+1}$, then $n = a_{k-s} + r d_k$, where $6^k(1 - 1/k) < r d_k < 6^{k+1}$. Thus, choosing these values $r d_k$ into B, we get a complement of A. Here $A(n) = k$. Concerning $B(n)$, we have to find a good estimate for the number of integers $r d_j$ satisfying $k \ge j \ge v = \lfloor k - 2 \log_6 k \rfloor$, and use the common denominator d_v. There are at most 6^v terms belonging to $j < v$.

5. Now $A(n) = \pi(n) \sim n / \log n$, so
$$S(n) <\sim 10 \sum_{i=2}^{n} \frac{\log^2 n}{n} \sim 10 \int_2^n \frac{\log^2 x}{x} \, dx \sim \frac{10(\log n)^3}{3}.$$

6. Apply Theorem 12.6.4. Since $(\log A(i))/A(i) \to 0$, for any $\varepsilon > 0$ there is an i_0 such that $(\log A(i))/A(i) < \varepsilon/20$ for $i \ge i_0$. Then
$$B(n) < 10 \sum_{i=a_1}^{n} \frac{\log A(i)}{A(i)} < C + 10 \sum_{i=i_0}^{n} \frac{\log A(i)}{A(i)} < C + \frac{10n\varepsilon}{20} < \varepsilon n.$$

Historical Notes

Continuing the historical comments in the text, we give the birth and death dates, nationalities, and some results in number theory for those mathematicians from the past whose names occurred in the book. This short summary is very subjective for two reasons. First, it contains only mathematicians who played an important role in the branches of number theory discussed in this book. Many great practitioners of number theory are missing. Second, what we mention or praise are not necessarily the most important results of the mathematicians listed and we say nothing about their activities in other branches of mathematics. Thus, the summary below is by no means a valuation of the mathematicians appearing in it, it is just a small supplement adding some historical background to the number theory material discussed in the book.

Chebyshev, Pafnuti Lvovich, 1821–1894, Russian. He was the first to prove that there is always a prime between $(2 \leq)n$ and $2n$, and he determined the order of magnitude of the number of primes up to x. His famous inequality plays an important role in probability theory and is connected to Turán's proof of the Hardy–Ramanujan Theorem, which became a starting point of probabilistic number theory.

Chevalley, Claude, 1909–1984, French. He achieved important results in algebraic number theory.

Dedekind, Richard, 1831–1916, German. He developed the notion of ideals introduced by Kummer as a basic tool of investigating rings both from algebraic and number theoretic aspects.

Diophantus of Alexandria, lived around 250 CE, Greek. His name is preserved in algebraic equations with (generally) integer coefficients when also the solutions are required to be integers (or occasionally, rational numbers), and also in Diophantine approximation, which plays an important role in the theory of Diophantine equations.

Dirichlet, Peter Lejeune, 1805–1859, German. He applied analytic methods in number theory effectively. He proved that if the first term and the difference of an arithmetic progression are coprime, then the progression contains infinitely many primes. Dirichlet series are important tools in the theory of arithmetic functions.

Eratosthenes, 276?–194? BCE, Greek. His name is preserved in a sieve method for finding primes.

Erdős, Paul, 1913–1996, Hungarian. One of the most influential mathematicians of the twentieth century, "traveling ambassador of mathematics, great master of problem solving, and uncrowned monarch of problem posing", as characterized by Ernst Straus who was a close coworker of both Erdős and Einstein(!). He became internationally known at the age of 18 with his simple proof of Chebyshev's theorem. He initiated, among other things, the probabilistic constructions, the characterization of arithmetic functions, and several topics in combinatorial number theory.

Euclid, lived around 300 BCE, Greek. Mathematicians were educated using his monumental work *Elements* for more than two thousand years. It contains thirteen books, three of which deal with number theory and contain the formula for even perfect numbers and the proof of the infinitude of primes. We still use the Euclidean algorithm to find the greatest common divisor of large integers.

Euler, Leonhard, 1707–1783, Swiss. Encyclopedist of great format, champion of analytic methods. In number theory, he introduced the function φ, discovered the Euler–Fermat Theorem as a generalization of Fermat's Little Theorem, elaborated the theory of quadratic congruences, solved the case for cubes of Fermat's Last Theorem, proved the divergence of the sum of reciprocals of primes, and achieved important results for partitions.

Fermat, Pierre, 1601–1665, French. Founder of modern number theory (though his official profession was in law). His famous Last Theorem remained a conjecture for more than 350 years, during which the attempts to prove it enriched mathematics with many effective, new methods. Andrew Wiles proved Fermat's Last Theorem in 1994. Fermat's Little Theorem and its generalization by Euler are fundamental in the theory of congruences. Fermat primes are related to the Euclidean constructibility of regular polygons. Fermat discovered which numbers can be represented as sums of two squares and showed that Pell's equation has infinitely many solutions.

Gauss, Carl Friedrich, 1777–1855, German. Perhaps the greatest and most versatile mathematician of all times. He was just 15 when he conjectured (but could not prove) the Prime Number Theorem. He published his book *Disquisitiones Arithmeticae* in 1801 containing among other things the detailed theory of quadratic congruences. Gauss introduced the standard notation for congruences and the Gaussian integers, which served later as a base to the theory of algebraic number fields. He proved the Three Squares Theorem and the criterion for constructibility of regular polygons.

Gelfond, Alexander Osipovich, 1906–1968, Russian. He and Schneider verified (at the same time, but independently) Hilbert's conjecture stating that an algebraic number (different from 0 and 1) raised to an irrational algebraic exponent is always transcendental.

Goldbach, Christian, 1690–1764, German. The famous Goldbach conjecture appears in one of his letters to Euler.

Hadamard, Jacques, 1865–1963, French. He and de la Vallée Poussin proved first (at the same time, but independently) the Prime Number Theorem.

Hardy, Geoffrey, 1877–1947, English. He achieved significant results in the theory of primes and in additive number theory. He discovered and helped Ramanujan.

Hermite, Charles, 1822–1901, French. He was the first to prove the transcendence of e in 1873.

Hilbert, David, 1862–1943, German. In his famous talk at the mathematical congress in Paris in 1900, he sketched 23 problems of fundamental importance which exerted a great influence on twentieth century mathematics. Several Hilbert problems are related to number theory. Hilbert was the first to prove the existence of $g(k)$ in Waring's problem.

Jacobi, Carl, 1804–1851, German. The Jacobi symbol obtained as a generalization of Legendre symbol bears his name.

Kalmár, László, 1905–1976, Hungarian. His main area of research was mathematical logic. In number theory, he and Erdős gave a simple proof for the upper bound on the number of primes up to x.

Kőnig, Gyula, 1849–1913, Hungarian. His main area was set theory. In number theory, he was the coauthor of the Kőnig–Rados theorem about the solvability and number of solutions of congruences of higher degree modulo a prime.

Kronecker, Leopold, 1823–1891, German. He achieved important results about ideals of algebraic number fields.

Kummer, Ernst, 1810–1893, German. By introducing ideals, he made significant progress on Fermat's Last Theorem.

Lagrange, Joseph Louis, 1736–1813, French. His proof of the Four Squares Theorem was a nice contribution to number theory.

Lamé, Gabriel, 1795–1870, French. A good mathematician, but remembered mostly for his erroneous proof to Fermat's Last Theorem.

Legendre, Adrien-Marie, 1752–1833, French. We find his name in the Legendre symbol and in Legendre's formula for the standard form of $n!$.

Lindemann, Ferdinand, 1852–1939, German. He proved in 1882 the transcendence of π settling thus the 2000-year-old problem of (the impossibility of) squaring the circle (with Euclidean constructions).

Liouville, Joseph, 1809–1882, French. He was the first to construct a transcendental number. He rendered a great service to mathematics by analyzing the mathematical legacy of Galois who sketchily wrote his ideas in the last night before he was killed in a duel at age 21. Liouville recognized and disseminated Galois' revolutionary discoveries.

Lucas, Edouard, 1842–1891, French. He elaborated an efficient procedure to test Mersenne numbers. Its improved version by Lehmer is used today in computer searches for large Mersenne primes.

Mersenne, Marin, 1588–1648, French. An excellent organizer who corresponded intensively with Fermat, Descartes, and several other leading mathematicians of the era. He was interested in the primes bearing his name mainly because of their connection to perfect numbers. His list of them contains surprisingly few errors (we had to wait more than 200 years for the mathematical and technical tools necessary to check it).

Minkowski, Hermann, 1864–1909, German. Founder of the geometry of numbers with his famous theorem about lattice points.

Möbius, Ferdinand, 1790–1868, German. The function μ introduced by him plays an important role for arithmetic functions and primes (and the Möbius strip occurs to most people hearing his name).

Poussin, Charles de la Vallée, 1866–1962, Belgian. He and Hadamard verified first (at the same time, but independently) the Prime Number Theorem.

Rados, Gusztáv, 1862–1942, Hungarian. In number theory, he was the coauthor of the Kőnig–Rados theorem about the solvability and number of solutions of congruences of higher degree modulo a prime.

Ramanujan, Srinivasa, 1887–1920, Indian. An uneducated mathematical genius who did not explain his results with the usual steps of mathematical reasoning. Hardy helped him to develop his intuitive mental gift at Cambridge University in England. His diaries are still sources for new research.

Ramsey, Frank Plumpton, 1903–1930, English. During his short life, he was equally excellent as economist, philosopher, and mathematician. He discovered his famous theorem in graph theory while investigating mathematical logic.

Rényi, Alfréd, 1921–1970, Hungarian. A leading mathematician in probability theory, founder and first director of the Mathematical Research Institute of the Hungarian Academy of Sciences that now bears his name. In number theory, he found important new results related to Goldbach's conjecture.

Riemann, Bernhard, 1826–1866, German. He sketched the principles leading to the Prime Number Theorem, which was proved using his ideas by Hadamard and de la

Vallée Poussin independently in 1896. Improving Euler's ideas, Riemann pointed out the central significance of the zeta function (that bears his name) in examining the distribution of primes. The celebrated Riemann Hypothesis about this function is still unsolved.

Schneider, Theodor, 1911–1988, German. He and Gelfond solved (at the same time, but independently) Hilbert's problem about the powers of algebraic numbers with an irrational algebraic exponent.

Schur, Issai, 1875–1941, German (forced to emigrate by the Nazis being a Jew). His famous theorem states that coloring a sufficiently large initial segment of the natural numbers using finitely many colors, the equation $x + y = z$ has a monochromatic solution.

Schnirelmann, Lev Demidovich, 1905–1938, Russian. Introducing a special notion for density, he achieved significant results about Goldbach's conjecture.

Thue, Axel, 1863–1922, Norwegian. He has important achievements in Diophantine approximation and in the theory of Diophantine equations.

Turán, Paul, 1910–1976, Hungarian. He gave a simple proof of the Hardy–Ramanujan theorem which argument became a starting point for applications of probability theory to number theory. He achieved outstanding results in analytic number theory and for partitions.

Vinogradov, Ivan Matveyevich, 1891–1975, Russian. He proved a slightly weaker version of the odd Goldbach conjecture that every sufficiently large odd integer is the sum of three primes. He improved significantly the previous upper bounds on $G(k)$ in Waring's problem.

Waerden, Bartel Leendert van der, 1903–1996, Dutch. He proved that coloring the natural numbers using finitely many colors there always arise arbitrarily long (finite) monochromatic arithmetic progressions.

Waring, Edward, 1736–1798, English. He initiated the investigation of representing integers as sums of kth powers. This area is called today Waring's problem.

Wilson, John, 1741–1793, English. His name appears in the theorem about the residue modulo p of $(p - 1)!$.

Tables

Primes 2–1733

2	127	283	467	661	877	1087	1297	1523
3	131	293	479	673	881	1091	1301	1531
5	137	307	487	677	883	1093	1303	1543
7	139	311	491	683	887	1097	1307	1549
11	149	313	499	691	907	1103	1319	1553
13	151	317	503	701	911	1109	1321	1559
17	157	331	509	709	919	1117	1327	1567
19	163	337	521	719	929	1123	1361	1571
23	167	347	523	727	937	1129	1367	1579
29	173	349	541	733	941	1151	1373	1583
31	179	353	547	739	947	1153	1381	1597
37	181	359	557	743	953	1163	1399	1601
41	191	367	563	751	967	1171	1409	1607
43	193	373	569	757	971	1181	1423	1609
47	197	379	571	761	977	1187	1427	1613
53	199	383	577	769	983	1193	1429	1619
59	211	389	587	773	991	1201	1433	1621
61	223	397	593	787	997	1213	1439	1627
67	227	401	599	797	1009	1217	1447	1637
71	229	409	601	809	1013	1223	1451	1657
73	233	419	607	811	1019	1229	1453	1663
79	239	421	613	821	1021	1231	1459	1667
83	241	431	617	823	1031	1237	1471	1669
89	251	433	619	827	1033	1249	1481	1693
97	257	439	631	829	1039	1259	1483	1697
101	263	443	641	839	1049	1277	1487	1699
103	269	449	643	853	1051	1279	1489	1709
107	271	457	647	857	1061	1283	1493	1721
109	277	461	653	859	1063	1289	1499	1723
113	281	463	659	863	1069	1291	1511	1733

Primes 1741–3907

1741	1993	2221	2437	2689	2909	3187	3433	3659
1747	1997	2237	2441	2693	2917	3191	3449	3671
1753	1999	2239	2447	2699	2927	3203	3457	3673
1759	2003	2243	2459	2707	2939	3209	3461	3677
1777	2011	2251	2467	2711	2953	3217	3463	3691
1783	2017	2267	2473	2713	2957	3221	3467	3697
1787	2027	2269	2477	2719	2963	3229	3469	3701
1789	2029	2273	2503	2729	2969	3251	3491	3709
1801	2039	2281	2521	2731	2971	3253	3499	3719
1811	2053	2287	2531	2741	2999	3257	3511	3727
1823	2063	2293	2539	2749	3001	3259	3517	3733
1831	2069	2297	2543	2753	3011	3271	3527	3739
1847	2081	2309	2549	2767	3019	3299	3529	3761
1861	2083	2311	2551	2777	3023	3301	3533	3767
1867	2087	2333	2557	2789	3037	3307	3539	3769
1871	2089	2339	2579	2791	3041	3313	3541	3779
1873	2099	2341	2591	2797	3049	3319	3547	3793
1877	2111	2347	2593	2801	3061	3323	3557	3797
1879	2113	2351	2609	2803	3067	3329	3559	3803
1889	2129	2357	2617	2819	3079	3331	3571	3821
1901	2131	2371	2621	2833	3083	3343	3581	3823
1907	2137	2377	2633	2837	3089	3347	3583	3833
1913	2141	2381	2647	2843	3109	3359	3593	3847
1931	2143	2383	2657	2851	3119	3361	3607	3851
1933	2153	2389	2659	2857	3121	3371	3613	3853
1949	2161	2393	2663	2861	3137	3373	3617	3863
1951	2179	2399	2671	2879	3163	3389	3623	3877
1973	2203	2411	2677	2887	3167	3391	3631	3881
1979	2207	2417	2683	2897	3169	3407	3637	3889
1987	2213	2423	2687	2903	3181	3413	3643	3907

Prime Factorization

The table below contains the prime factorization of integers less than 1100 and not divisible by 2, or 3, or 5.

$49 = 7^2$	$377 = 13 \cdot 29$	$637 = 7^2 \cdot 13$	$871 = 13 \cdot 67$
$77 = 7 \cdot 11$	$391 = 17 \cdot 23$	$649 = 11 \cdot 59$	$889 = 7 \cdot 127$
$91 = 7 \cdot 13$	$403 = 13 \cdot 31$	$667 = 23 \cdot 29$	$893 = 19 \cdot 47$
$119 = 7 \cdot 17$	$407 = 11 \cdot 37$	$671 = 11 \cdot 61$	$899 = 29 \cdot 31$
$121 = 11^2$	$413 = 7 \cdot 59$	$679 = 7 \cdot 97$	$901 = 17 \cdot 53$
$133 = 7 \cdot 19$	$427 = 7 \cdot 61$	$689 = 13 \cdot 53$	$913 = 11 \cdot 83$
$143 = 11 \cdot 13$	$437 = 19 \cdot 23$	$697 = 17 \cdot 41$	$917 = 7 \cdot 131$
$161 = 7 \cdot 23$	$451 = 11 \cdot 41$	$703 = 19 \cdot 37$	$923 = 13 \cdot 71$
$169 = 13^2$	$469 = 7 \cdot 67$	$707 = 7 \cdot 101$	$931 = 7^2 \cdot 19$
$187 = 11 \cdot 17$	$473 = 11 \cdot 43$	$713 = 23 \cdot 31$	$943 = 23 \cdot 41$
$203 = 7 \cdot 29$	$481 = 13 \cdot 37$	$721 = 7 \cdot 103$	$949 = 13 \cdot 73$
$209 = 11 \cdot 19$	$493 = 17 \cdot 29$	$731 = 17 \cdot 43$	$959 = 7 \cdot 137$
$217 = 7 \cdot 31$	$497 = 7 \cdot 71$	$737 = 11 \cdot 67$	$961 = 31^2$
$221 = 13 \cdot 17$	$511 = 7 \cdot 73$	$749 = 7 \cdot 107$	$973 = 7 \cdot 139$
$247 = 13 \cdot 19$	$517 = 11 \cdot 47$	$763 = 7 \cdot 109$	$979 = 11 \cdot 89$
$253 = 11 \cdot 23$	$527 = 17 \cdot 31$	$767 = 13 \cdot 59$	$989 = 23 \cdot 43$
$259 = 7 \cdot 37$	$529 = 23^2$	$779 = 19 \cdot 41$	$1001 = 7 \cdot 11 \cdot 13$
$287 = 7 \cdot 41$	$533 = 13 \cdot 41$	$781 = 11 \cdot 71$	$1003 = 17 \cdot 59$
$289 = 17^2$	$539 = 7^2 \cdot 11$	$791 = 7 \cdot 113$	$1007 = 19 \cdot 53$
$299 = 13 \cdot 23$	$551 = 19 \cdot 29$	$793 = 13 \cdot 61$	$1027 = 13 \cdot 79$
$301 = 7 \cdot 43$	$553 = 7 \cdot 79$	$799 = 17 \cdot 47$	$1037 = 17 \cdot 61$
$319 = 11 \cdot 29$	$559 = 13 \cdot 43$	$803 = 11 \cdot 73$	$1043 = 7 \cdot 149$
$323 = 17 \cdot 19$	$581 = 7 \cdot 83$	$817 = 19 \cdot 43$	$1057 = 7 \cdot 151$
$329 = 7 \cdot 47$	$583 = 11 \cdot 53$	$833 = 7^2 \cdot 17$	$1067 = 11 \cdot 97$
$341 = 11 \cdot 31$	$589 = 19 \cdot 31$	$841 = 29^2$	$1073 = 29 \cdot 37$
$343 = 7^3$	$611 = 13 \cdot 47$	$847 = 7 \cdot 11^2$	$1079 = 13 \cdot 83$
$361 = 19^2$	$623 = 7 \cdot 89$	$851 = 23 \cdot 37$	$1081 = 23 \cdot 47$
$371 = 7 \cdot 53$	$629 = 17 \cdot 37$	$869 = 11 \cdot 79$	$1099 = 7 \cdot 157$

Mersenne Numbers

Mersenne numbers are the integers $M_p = 2^p - 1$ where $p > 0$ is a prime. We discuss them in detail in Section 5.2 where we list the 51 primes of this form known in 2019.

The table contains the prime factorization of Mersenne numbers with exponents between 10 and 100.

$$2^{11} - 1 = 23 \cdot 89$$

$$2^{13} - 1 = 8191$$

$$2^{17} - 1 = 131071$$

$$2^{19} - 1 = 524287$$

$$2^{23} - 1 = 47 \cdot 178481$$

$$2^{29} - 1 = 233 \cdot 1103 \cdot 2089$$

$$2^{31} - 1 = 2147483647$$

$$2^{37} - 1 = 223 \cdot 616318177$$

$$2^{41} - 1 = 13367 \cdot 164511353$$

$$2^{43} - 1 = 431 \cdot 9719 \cdot 2099863$$

$$2^{47} - 1 = 2351 \cdot 4513 \cdot 13264529$$

$$2^{53} - 1 = 6361 \cdot 69431 \cdot 20394401$$

$$2^{59} - 1 = 179951 \cdot 3203431780337$$

$$2^{61} - 1 = 2305843009213693951$$

$$2^{67} - 1 = 193707721 \cdot 761838257287$$

$$2^{71} - 1 = 228479 \cdot 48544121 \cdot 212885833$$

$$2^{73} - 1 = 439 \cdot 2298041 \cdot 9361973132609$$

$$2^{79} - 1 = 2687 \cdot 202029703 \cdot 1113491139767$$

$$2^{83} - 1 = 167 \cdot 57912614113275649087721$$

$$2^{89} - 1 = 618970019642690137449562111$$

$$2^{97} - 1 = 11447 \cdot 13842607235828485645766393$$

Fermat Numbers

Fermat numbers are the integers $F_n = 2^{2^n} + 1$, where $n \geq 0$ is an integer. We discuss them in detail in Section 5.2.

F_n is a prime for $0 \leq n \leq 4$:

$$F_0 = 3, \qquad F_1 = 5, \qquad F_2 = 17, \qquad F_3 = 257, \qquad F_4 = 65537.$$

No primes are known among the Fermat numbers for $n \geq 5$.

The prime factorizations of F_5, F_6, and F_7 are

$$F_5 = 641 \cdot 6700417$$

$$F_6 = 274177 \cdot 67280421310721$$

$$F_7 = 59649589127497217 \cdot 5704689200685129054721.$$

The complete prime factorization of F_n is known also for $8 \leq n \leq 11$, but for no greater n.

F_n is known to be composite for $12 \leq n \leq 32$ and for some greater values of n.

No non-trivial divisor of F_{20} has been determined so far.

We do not know whether F_{33} is prime or composite.

Index

We generally indicate the first occurrence only. The data include the typical notation (if it exists), the serial number of the definition, theorem, etc. explaining the notion or denomination, and finally the page number in parentheses.

D3.2.1 means Definition 3.2.1, and letters T, L, E instead of D refer to the theorem, lemma, and exercise with the given number. P1.3.3 stands for the proof of Theorem 1.3.3, 9.6.E3 denotes Example 3 in Section 9.6, and 5.8 means Section 5.8. This latter can mean the entire section or a part of it. In some cases, there is only a page number pointing directly to the occurence of the expression in question, e.g. "Diffie–Hellman principle (160)".

We add a sign "−" or "+" to the number of definition theorem, etc., if the notion is introduced not in the given definition, theorem, etc., but just before or after it, resp., without a new serial number. E.g. D1.4.1− indicates at "trivial divisor" that this phrase is explained *before* Definition 1.4.1. Similarly, T6.7.3+ shows that we find the meaning of "average order of magnitude" for the function $\sigma(n)$ *after stating* the theorem (still *before the proof*), whereas P9.3.6+ indicates that we can look up "algebraically closed field" *after the proof* of Theorem 9.3.6.

We often include also important theorems besides the definition, e.g. for "$\sigma(n)$" we refer both to Definition 6.2.1 explaining this function and Theorem 6.2.2 establishing a formula for it. In some other cases, we list the related theorems in separate lines, e.g. at "mean value" we enumerate the mean value theorems for several arithmetic functions.

If an important notion appears in various topics, we generally list all of them, see e.g. at "unit" and "norm". (If the notation is the same, we indicate it only once.)

For information about notation used in the book, please consult part "Technical details" in the Introduction. We add that as mentioned in another part of the Introduction, exercises marked with one or two asterisks are considered hard or extra hard, resp., by our judgement, and a letter S indicates that a detailed solution can be found online at `www.ams.org/bookpages/amstext-48`.

Selected Published Titles in This Series

For a complete list of titles in this series, visit the
AMS Bookstore at **www.ams.org/bookstore/amstextseries/**.